Statistics and Experimental Design for Toxicologists and Pharmacologists

Fourth Edition

STATISTICS AND EXPERIMENTAL DESIGN FOR TOXICOLOGISTS AND PHARMACOLOGISTS

Fourth Edition

Shayne C. Gad

Gad Consulting Services
Cary, North Carolina

CRC Press
Taylor & Francis Group
Boca Raton London New York

CRC Press is an imprint of the
Taylor & Francis Group, an **informa** business
A TAYLOR & FRANCIS BOOK

CRC Press
Taylor & Francis Group
6000 Broken Sound Parkway NW, Suite 300
Boca Raton, FL 33487-2742

First issued in paperback 2019

ISBN-13: 978-0-8493-2214-3 (hbk)
ISBN-13: 978-0-367-39248-2 (pbk)
Library of Congress Card Number 2005041955

Library of Congress Cataloging-in-Publication Data

Gad, Shayne C., 1948-
    Statistics and experimental design for toxicologists and pharmacologists / by Shayne C. Gad.-- 4th
ed.
        p. cm.
    Includes bibliographical references and index.
    ISBN 0-8493-2214-6 (alk. paper)
    1. Toxicology--Statistical methods. 2. Toxicology, Experimental--Statistical methods. I. Title.

RA1199.4.S73G33 2005
615.9'0072'7--dc22
    2005041955

Visit the Taylor & Francis Web site at
http://www.taylorandfrancis.com

and the CRC Press Web site at
http://www.crcpress.com

Dedication

This book is dedicated to my beloved children,

Samantha, Katina, and Jake

Preface

Statistics and Experimental Design for Toxicologists and Pharmacologists, Fourth Edition was designed from the beginning as a source and textbook for both practicing and student toxico-logists with the central objective of equipping them for the regular statistical analysis of experimental data. Starting with the assumption of only basic mathematical skills and knowledge, it provides a complete and systematic yet practical introduction to the statistical methodologies that are available for, and utilized in, the discipline. For every technique that is presented, a worked example from toxicology is also presented. For those with computer skills, this edition has been enhanced with the addition of basic SAS.

Because toxicology is the study of the adverse responses of biological units to chemical and physical agents, any understanding and valid analysis must originate from an understanding of the underlying biological mechanisms and, from this understanding, selection of the best methods of analysis in accordance with their limitations and necessary assumptions. This book is written from the perspective of toxicologists, drawing on the author's combined experience of more than 30 years to provide insight and philosophical underpinnings for the overall structure of analysis and for the specific usage examples. It is the purpose of this book to provide both the methodological tools necessary to analyze experimental toxicology data and the insight to know when to use them.

Statistics and Experimental Design for Toxicologists and Pharmacologists, Fourth Edition is organized to provide an ordered development of skills and yet still facilitate ease of access to the information that is desired. The first section (Chapters 1 to 4) develops some basic principles, establishes (or reviews) introductory statistical concepts, presents an overview of automated computational hardware and software, and presents decision tree guidelines for selecting actual analytical techniques. The second section (Chapters 5 to 13) presents actual techniques, organized by their functional objectives. The third section (Chapters 14 to 18) reviews all the major possible data analysis applications in toxicology (LD_{50}/LC_{50} calculation, clinical chemistry, reproduction studies, behavioral studies, risk assessment, SARs, etc.) and in each case provides examples and guidelines for analysis. Each of these chapters not only should enable a person to perform a type of analysis, but should also ensure sufficient understanding of its assumptions and limitations to prevent misuse. The last two chapters stand apart and present Chapter 19, an introduction to good laboratory practices and Chapter 20, an overview of the areas of controversy in "statistical toxicology" such as the question of the existence of thresholds.

Chapters 13 ("Bayesian Analysis") and 19 ("Good Laboratory Practices") are new to this edition and Chapter 12 ("Meta-Analysis") has been rewritten and expanded. Bayesian analysis and meta analysis are rapidly expanding fields applicable to many toxicological analyses and good laboratory practices now influence all biomedical research.

About the Author

Shayne C. Gad, Ph.D., DABT, has been the Principal of Gad Consulting Services since 1994. He has more than 30 years of broad-based experience in toxicology, drug, and device development, and document preparation, statistics, and risk assessment, having previously been Director of Toxicology and Pharmacology for Synergen (Boulder, CO), Director of Medical Affairs Technical Support System Services for Becton Dickinson (RTP, NC), and Senior Director of Product Safety and Pharmacokinetics for G.D. Searle (Skokie, IL). A past president of the American College of Toxicology, a board-certified toxicologist (DABT), and Fellow of the Academy of Toxicological Sciences, he is also a member of the Society of Toxicology, Teratology Society, Society of Toxicological Pathologies, Biometrics Society, and the American Statistical Association. Dr. Gad has previously published 29 books and more than 300 chapters, papers, and abstracts in the above fields. He has contributed to and has personal experience with IND (he has successfully filed more than 64 of these), NDA, PLA, ANDA, 510(k), IDE, and PMS preparation, and has broad experience with the design, conduct, and analysis of preclinical and clinical safety and pharmacokinetic studies for drugs, devices, and combination products. He is also a retired Navy Captain with extensive operational experience at sea and overseas.

Acknowledgment

My greatest appreciation, for his diligent labor and critical review of drafts, to Russell Barbare, my son-in-law and friend, without whom this volume would be nonexistent.

Contents

1

Introduction

This book has been written for both practicing and student toxicologists, as both a basic text and a practical guide to the common statistical problems encountered in toxicology and the methodologies that are available to solve them. It has been enriched by the inclusion of discussions of why a particular procedure or interpretation is suggested and by the clear enumeration of the assumptions that are necessary for a procedure to be valid, and by worked-through examples and problems drawn from the actual practice of toxicology. A set of tables has been included to make this volume complete.

Since 1960, the field of toxicology has become increasingly complex and controversial in both its theory and practice. Much of this change is due to the evolution of the field. As in all other sciences, toxicology started as a descriptive science. Living organisms, be they human or otherwise, were dosed with or exposed to chemical or physical agents and the adverse effects that followed were observed. But as a sufficient body of descriptive data was accumulated, it became possible to infer and study underlying mechanisms of action — to determine in a broader sense why adverse effects occurred. As with all mature sciences, toxicology thus entered a later state of development, the mechanistic stage, where active contributions to the field encompass both descriptive and mechanistic studies. Frequently, in fact, present-day studies are a combination of both approaches.

As a result of this evolution, studies are being designed and executed to generate increased amounts of data, which are then utilized to address various areas of concern. The resulting problems of data analysis have then become more complex and toxicology has drawn more deeply from the well of available statistical techniques. At the same time, however, the field of statistics has also been very active and growing during the last 35 years — to some extent, at least, because of the very growth of toxicology. These simultaneous changes have led to an increasing complexity of data and, unfortunately, to the introduction of numerous confounding factors that severely limit the utility of the resulting data in all too many cases.

One (and perhaps the major) difficulty is that there is a very real necessity to understand the biological realities and implications of a problem, as well as to understand the peculiarities of toxicological data before procedures are selected and employed for analysis. Some of these characteristics include the following:

1. The need to work with a relatively small sample set of data collected from the members of a population (laboratory animals, cultured cells, and bacterial cultures), which are not actually our populations of interest (that is, people or a wildlife population).

2. Dealing frequently with data resulting from a sample that was censored on a basis other than the investigator's design. By censoring, of course, we mean that not all data points were collected as might be desired. This censoring could be the result of either a biological factor (the test animal was dead or too debilitated to manipulate) or a logistic factor (equipment was inoperative or a tissue was missed in necropsy).

3. The conditions under which our experiments are conducted are extremely varied. In pharmacology (the closest cousin to at least classical toxicology), the possible conditions of interaction of a chemical or physical agent with a person are limited to a small range of doses via a single route over a short course of treatment to a defined patient population. In toxicology, however, all these variables (dose, route, time span, and subject population) are determined by the investigator.

4. The time frames available to solve our problems are limited by practical and economic factors. This frequently means that there is not time to repeat a critical study if the first attempt fails. So a true iterative approach is not possible.

Unfortunately, there are very few toxicologists who are also statisticians, or vice versa. In fact, the training of most toxicologists in statistics remains limited to a single introductory course that concentrates on some theoretical basics. As a result, the armamentarium of statistical techniques of most toxicologists is limited and the tools that are usually present (t-tests, chi-square, analysis of variance, and linear regression) are neither fully developed nor well understood. It is hoped that this book will help change this situation.

As a point of departure toward this objective, it is essential that any analysis of study results be interpreted by a professional who firmly understands three concepts: the difference between biological significance and statistical significance, the nature and value of different types of data, and causality.

For the first concept, we should consider the four possible combinations of these two different types of significance, for which we find the relationship shown below:

		Statistical Significance	
		No	Yes
Biological	No	Case I	Case II
Significance	Yes	Case III	Case IV

Cases I and IV give us no problems, for the answers are the same statistically and biologically. But Cases II and III present problems. In Case II (the "false positive"), we have a circumstance where there is a statistical significance in the measured difference between treated and control groups, but there is no true biological significance to the finding. This is not an uncommon happening, for example, in the case of clinical chemistry parameters. This is called type I error by statisticians, and the probability of this happening is called the (alpha) level. In Case III (the "false negative"), we have no statistical significance, but the differences between groups are biologically/toxicologically significant. This is called type II error by statisticians, and the probability of such an error happening by random chance is called the (beta) level. An example of this second situation is when we see a few of a very rare tumor type in treated animals. In both of these latter cases, numerical analysis, no matter how well done, is no substitute for professional judgment. Here, perhaps, is the cutting edge of what really makes a practicing toxicologist. Along with this, however, must come a feeling for the different types of data and for the value or relative merit of each.

We will more fully explore the types of data (the second major concept) and their value (and the implications of value of data to such things as animal usage) in the next chapter of this book.

The reasons that biological and statistical significance are not identical are multiple, but a central one is certainly causality. Through our consideration of statistics, we should keep in mind that just because a treatment and a change in an observed organism are seemingly or actually associated with each other does not "prove" that the former caused the latter. Although this fact is now widely appreciated for correlation (for example, the fact that the number of storks' nests found each year in England is correlated with the number of human births that year does not mean that storks bring babies), it is just as true in the general case of significance. Timely establishment and proof that treatment causes an effect requires an understanding of the underlying mechanism and proof of its validity. At the same time, it is important that we realize that not finding a good correlation or suitable significance associated with a treatment and an effect likewise does not prove that the two are not associated — that a treatment does not cause an effect. At best, it gives us a certain level of confidence that under the conditions of the current test these items are not associated.

These points will be discussed in greater detail in the "Assumptions" section for each method, along with other common pitfalls and shortcomings

TABLE 1.1

Some Frequently Used Terms and Their General Meanings

Term	Meaning
95% confidence interval	A range of values (above, below or above and below) the sample (mean, median, mode, etc.) has a 95% chance of containing the true value of the population (mean, median, mode). Also called the fiducial limit; equivalent to Pc 0.05
Bias	Systemic error as opposed to a sampling error. For example, selection bias may occur when each member of the population does not have an equal chance of being selected for the sample
Degrees of freedom	The number of independent deviations and is usually abbreviated df
Independent variables	Also known as predictors or explanatory variables
p-Value	Another name for significance level; usually 0.05
Power	The effect of the experimental conditions on the dependent variable relative to sampling fluctuation. When the effect is maximized, the experiment is more powerful. Power can also be defined as the probability that there will not be a type II error (1-Beta). Conventionally, power should be at least 0.07
Random	Each individual member of the population has the same chance of being selected for the sample
Robust	Having inferences or conclusions little affected by departure from assumptions
Sensitivity	The number of subjects experiencing each experimental condition divided by the variance of scores in the sample.
Significance level	The probability that a difference has been erroneously declared to be significant, typically 0.05 and 0.01 corresponding respectively to a 5% and 1% chance of error.
Type I error	Concluding that there is an effect when there really is not an effect
Type II error	Concluding there is no effect when there really is an effect

(Adapted from Marriott, 1991.)

associated with the method. To help in better understanding the chapters to come, terms frequently used in discussion throughout this book should first be considered. These are presented in Table 1.1.

References

Marriott, F.H.C. (1991) *Dictionary of Statistical Terms*. Longman Scientific and Technical, Essex, England.

2

Basic Principles

Let us start by reviewing a few simple terms and concepts that are fundamental to an understanding of statistics.

Each measurement we make — each individual piece of experimental information we gather — is called a datum. It is extremely unusual, however, to either obtain or attempt to analyze a datum. Rather, we gather and analyze multiple pieces at one time, the resulting collection being called data.

Data are collected on the basis of their association with a treatment (intended or otherwise) as an effect (a property) that is measured in the experimental subjects of a study, such as body weights. These identifiers (that is, treatment and effect) are termed "variables." Our treatment variables (those that the researcher or nature control, and which can be directly controlled) are termed "independent," while our effect variables (such as weight, life span, and number of neoplasms) are termed "dependent" variables — their outcome is believed to depend on the "treatment" being studied.

All possible measures of a given set of variables in all possible subjects that exist are termed the "population" for those variables. Such a population of variables cannot be truly measured; for example, one would have to obtain, treat, and measure the weights of all of the Fischer 344 rats that were, are, or ever will be. Instead, we deal with a representative group — a "sample." If our sample of data is appropriately collected and of sufficient size, it serves to provide good estimates of the characteristics of the parent population from which it was drawn.

Two terms refer to the quality and reproducibility of our measurements of variables. The first, accuracy, is an expression of the closeness of a measured or computed value to its actual or "true" value in nature. The second, precision, reflects the closeness or reproducibility of a series of repeated measurements of the same quantity.

If we arrange all of our measurements of a particular variable in order as a point on an axis marked as to the values of that variable, and if our sample were large enough, the pattern of distribution of the data in the sample would begin to become apparent. This pattern is a representation of the frequency distribution of a given population of data — that is, of the incidence of different measurements, their central tendency, and dispersion.

The most common frequency distribution — and one we will talk about throughout this book — is the normal (or Gaussian) distribution. This distribution is the most common in nature and is such that two thirds of all values are within one standard deviation (to be defined in Chapter 2) of the mean (or average value for the entire population) and 95% are within 1.96 standard deviations of the mean. The mathematical equation for the normal curve is

$$y = \frac{e^{-\frac{(x-\mu)^2}{2\sigma^2}}}{\sigma\sqrt{2\pi}}$$

where μ is the mean and σ is the standard deviation.

There are other frequency distributions, such as the binomial, Poisson, and chi-square.

In all areas of biological research, optimal design and appropriate interpretation of experiments require that the researcher understand both the biological and technological underpinnings of the system being studied and of the data being generated. From the point of view of the statistician, it is vitally important that the experimenter both know and be able to communicate the nature of the data and understand its limitations. One classification of data types is presented in Table 2.1.

The nature of the data collected is determined by three considerations. These are the biological source of the data (the system being studied), the instrumentation and techniques being used to make measurements, and the design of the experiment. The researcher has some degree of control over each of these — least over the biological system (he/she normally has a choice of only one of several models to study) and most over the design of the experiment or study. Such choices, in fact, dictate the type of data generated by a study.

Statistical methods are based on specific assumptions. Parametric statistics — those that are most familiar to the majority of scientists — have

TABLE 2.1

Types of Variables (Data) and Examples of Each Type

Classified By	Type	Example
Scale		
Continuous	Scalar	Body weight
	Ranked	Severity of a lesion
Discontinuous	Scalar	Weeks until the first observation of a tumor in a carcinogenicity study
	Ranked	Clinical observations in animals
	Attribute	Eye colors in fruit flies
	Quantal	Dead/alive or present/absent
Frequency distribution	Normal	Body weights
	Bimodal	Some clinical chemistry parameters
	Others	Measures of time-to-incapacitation

more stringent underlying assumptions than do nonparametric statistics. Among the underlying assumptions for many parametric statistical methods (such as the analysis of variance) is that the data are continuous. The nature of the data associated with a variable (as described above) imparts a "value" to that data, the value being the power of the statistical tests that can be employed.

Continuous variables are those that can at least theoretically assume any of an infinite number of values between any two fixed points (such as measurements of body weight between 2.0 and 3.0 kg). Discontinuous variables, meanwhile, are those that can have only certain fixed values, with no possible intermediate values (such as counts of 5 and 6 dead animals, respectively).

Limitations on our ability to measure constrain the extent to which the real-world situation approaches the theoretical, but many of the variables studied in toxicology are in fact continuous. Examples of these are lengths, weights, concentrations, temperatures, periods of time, and percentages. For these continuous variables, we may describe the character of a sample with measures of central tendency and dispersion that we are most familiar with — the mean, denoted by the symbol $\bar{x}$ and called the arithmetic average, and the standard deviation SD, which is denoted by the symbol σ, and is calculated as being equal to

$$\sigma = \sqrt{\frac{\sum X^2 - \frac{(\sum x)^2}{N}}{N-1}}$$

where X is the individual datum and N is the total number of data in the group.

Contrasted with these continuous data, however, we have discontinuous (or discrete) data, which can only assume certain fixed numerical values. In these cases our choice of statistical tools or tests is, as we will find later, more limited.

Probability

Probability is simply the frequency with which, in a sufficiently large sample, a particular event will occur or a particular value will be found. Hypothesis testing, for example, is generally structured so that the likelihood of a treatment group being the same as a control group (the so-called null hypothesis) can be assessed as being less than a selected low level (very frequently 5%), which implies that we are $1.0 - \alpha$ (that is, $1.0 - 0.05 = 0.95$, or 95%) sure that the groups are *not* equivalent.

Functions of Statistics

Statistical methods may serve to do any combination of three possible tasks. The one we are most familiar with is hypothesis testing — that is, determining if two (or more) groups of data differ from each other at a predetermined level of confidence. A second function is the construction and use of models that may be used to predict future outcomes of chemical–biological interactions. This is most commonly seen in linear regression or in the derivation of some form of correlation coefficient. Model fitting allows us to relate one variable (typically a treatment or independent variable) to another. The third function, reduction of dimensionality, continues to be less commonly utilized than the first two. This final category includes methods for reducing the number of variables in a system while only minimally reducing the amount of information, therefore making a problem easier to visualize and to understand. Examples of such techniques are factor analysis and cluster analysis. A subset of this last function, discussed later under descriptive statistics, is the reduction of raw data to single expressions of central tendency and variability (such as the mean and standard deviation).

There is also a special subset of statistical techniques that is part of both the second and third functions of statistics. This is data transformation, which includes such things as the conversion of numbers to log or probit values.

As a matter of practicality, the contents of this book are primarily designed to address the first of the three functions of statistical methods that we presented (hypothesis testing). The second function, modeling — especially in the form of risk assessment — is becoming increasingly important as the science continues to evolve from the descriptive phase to a mechanistic phase (i.e., the elucidation of mechanisms of action), and as such is addressed in some detail. Likewise, because the interrelation of multiple factors is becoming a real concern, a discussion of reduction of dimensionality has been included.

Descriptive Statistics

Descriptive statistics are used to convey, in summary, the general nature of the data. As such, the parameters describing any single group of data have two components. One of these describes the location of the data, while the other gives a measure of the dispersion of the data in and about this location. Often overlooked is the fact that the choice of which parameters are used to give these pieces of information implies a particular type of distribution for the data.

Most commonly, location is described by giving the (arithmetic) mean and dispersion by giving the standard deviation (SD) or the standard error of the mean (SEM).

The use of the mean with either the SD or SEM implies, however, that we have reason to believe that the data being summarized are from a population that is at least approximately normally distributed. If this is not the case, then we should rather use a set of statistical quantities that do not require a normal distribution. These are the median, for location, and the semi-quartile distance, for a measure of dispersion.

Other descriptive statistics that are commonly used in toxicology include the geometric mean and the coefficient of variation (CV).

Calculation and discussion of the mean and standard deviation have already been covered. The formulae and notes for the other statistics mentioned follow:

Standard Error of the Mean

If we again denote the total number of data in a group as N, then the SEM would be calculated as

$$\text{SEM} = \frac{\text{SD}}{\sqrt{N}}$$

The standard deviation and the standard error of the mean are related to each other yet are quite different. To compare these two, let us first demonstrate their calculation from the same set of 15 observations.

		Sum (Σ)
Data Points (X_i):	1, 2, 3, 4, 4, 5, 5, 5, 6, 6, 6, 7, 7, 8, 9	78
Squares (X_i^2):	1, 4, 9, 16, 16, 25, 25, 25, 36, 36, 36, 49, 49, 64, 81	472

The standard deviation can then be calculated as:

$$\text{SD} = \sqrt{\frac{472 - \frac{(78)^2}{15}}{15 - 1}} = \sqrt{\frac{472 - \frac{(6084)}{15}}{14}} = \sqrt{\frac{472 - 405.6}{14}} = \sqrt{4.7428571} = 2.1778$$

with a mean ($\bar{x}$) of $78/15 = 5.2$ for the data group. The SEM for the same set of data, however, is

$$\text{SEM} = \frac{2.1778}{\sqrt{15}} = \frac{2.1778}{3.8730} = 0.562303$$

The SEM is quite a bit smaller than the SD, making it very attractive to use in reporting data. This size difference is because the SEM is actually an estimate of the error or variability involved in measuring the means of samples, and not an estimate of the error (or variability) involved in measuring the data from which means are calculated. This is implied by the Central Limit Theorem, which tells us three major things:

The distribution of sample means will be approximately normal regardless of the distribution of values in the original population from which the samples were drawn.

The mean value of a set of samples from the collection will tend toward the mean of the collection with a large number of samples; i.e., the mean of the collection of all possible means of samples of a given size is equal to the population mean.

The SD of the collection of all possible means of samples of a given size, called the SEM, depends on both the SD of the original population and the size of the sample.

Since the sample means are normally distributed regardless of the population distribution, a probable range for the population mean can be calculated based on the sample mean and the SEM. If the population mean is represented by $\bar{x}_P$ and the sample mean is represented by $\bar{x}_S$, then the range $\bar{x}_S \pm (1.96)(\text{SEM})$ has a 95% probability of containing $\bar{x}_P$, which is comparable to the earlier assertion that a normally distributed population has 95% of its values within $\bar{x}_P \pm (1.96)(\text{SD})$. Put simply, the SD is a measure of the variability of the data, while the SEM is a measure of the variability of the mean of samples of the data.

The SEM should be used only when the uncertainty of the estimate of the mean is of concern — which is almost never the case in toxicology. Rather, we are concerned with an estimate of the variability of the population — for which the SD is appropriate.

Median

When all the numbers in a group are arranged in a ranked order (that is, from smallest to largest), the median is the middle value. If there is an odd number of values in a group, then the middle value is obvious (in the case of 13 values, for example, the seventh largest is the median). When the number of values in the sample is even, the median is calculated as the midpoint between the $(N/2)$th and the $([N/2] + 1)$th number. For example, in the series of numbers 7, 12, 13, 19 the median value would be the midpoint between 12 and 13, which is 12.5. Note that for $N = 2$ the mean and median will be identical.

Semi-Quartile Distance

When all the data in a group are ranked, a quartile of the data contains one ordered quarter of the values. Typically, we are most interested in the borders of the middle two quartiles, Q_1 and Q_3, which together represent the semi-quartile distance and which contain the median as their center. Note this is the same as finding the distance between the 25th and 75th percentile. Given that there are N values in an ordered group of data, the upper limit of the jth quartile (Q_j) may be computed as being equal to the $[j(N+1)/4]$th value. Once we have used this formula to calculate the upper limits of Q_1 and Q_3, we can then compute the semi-quartile distance (which is also called the quartile deviation, and as such is abbreviated as the QD) with the formula $QD = (Q_3 - Q_1)/2$.

For example, for the fifteen-value data set 1, 2, 3, 4, 4, 5, 5, 5, 6, 6, 6, 7, 7, 8, 9, we can calculate the upper limits of Q_1 and Q_3 as

$$\text{Position of } Q_1 = \frac{1(15+1)}{4} = \frac{16}{4} = 4 \qquad \text{Position of } Q_3 = \frac{3(15+1)}{4} = \frac{48}{4} = 12$$

The 4th and 12th values in this data set are 4 and 7, respectively. The semi-quartile distance can then be calculated as

$$QD = (Q_3 - Q_1)/2 = (7 - 4)/2 = 1.5$$

In the case that the calculated lower and upper bounds are not integers, the values of Q_3 and Q_1 should be calculated using interpolation. For example, if we add the value 1 to the above data set, bring N up to 16, the position of $Q_1 = 4.25$ and $Q_3 = 12.25$. The value of Q_1 becomes 3 (the 4th data point) plus 0.25 times the difference between the fourth data point and the fifth, whose value is 4:

$$Q_1 = 3 + (4 - 3)*(0.25) = 3.25 \quad \text{and} \quad Q_3 = 7 + (7 - 7)*(0.25) = 7$$

therefore

$$QD = (Q_3 - Q_1)/2 = (7 - 3.25)/2 = 1.875$$

The value $Q_3 - Q_1$ is also known as the quartile distance or the interquartile range.

Geometric Mean

One final sample parameter which sees some use in toxicology (primarily in inhalation studies) is the geometric mean, denoted by the term $\overline{X}_g$. This is calculated as

$$\overline{X}_g = (X_1^* X_2^* X_3^* \dots X_N)^{1/N} = \sqrt[N]{X_1^* X_2^* X_3^* \dots X_N}$$

and has the attractive feature that it does not give excessive weight to extreme values (or "outliers"), such as the mass of a single very large particle in a dust sample. In effect, it "folds" extreme values in toward the center of the distribution, decreasing the sensitivity of the parameter to the undue influence of the outlier. This is particularly important in the case of aerosol samples where a few very large particles would cause the arithmetic mean of particle diameters to present a misleading picture of the nature of the "average" particle.

Coefficient of Variation

There are times when it is desired to describe the relative variability of one or more sets of data. The most common way of doing this is to compute the coefficient of variation (CV), which is calculated simply as the ratio of the SD to the mean, or

$$CV = SD/\overline{X}$$

A CV of 0.2 or 20% thus means that the SD is 20% of the mean. In toxicology the CV is frequently between 20 and 50% and may at times exceed 100%.

Outliers and Rounding of Numbers

These two considerations in the handling of numerical data can be, on occasion, of major concern to the toxicologist because of their pivotal nature in borderline cases. Outliers should also be of concern for other reasons, however. On the principle that one should always have a plan to deal with all reasonably likely contingencies in advance of their happening, early decisions should be made to select a policy for handling both outliers and the rounding of numbers.

Outliers are extreme (high or low) values that are widely divergent from the main body of a group of data and from our common experience. They may arise from an instrument (such as a balance) being faulty, the apparently natural urge of some animals to frustrate research, or be indicative of a "real" value. Outlying values can be detected by visual inspection of the data, use of a scattergram (described later), or (if the data set is small enough, which is usually the case in toxicology) by a large increase in the parameter estimating the dispersion of data, such as the SD.

When we can solidly tie one of the above error-producing processes (such as a balance being faulty) to an outlier, we can safely delete it from consideration. But if we cannot solidly tie such a cause to an outlier (even if we have strong suspicions), we have a much more complicated problem, for then such a value may be one of several other things. It could be the result of a particular case that is the grounds for the entire study — that is, the very "effect" that we are looking for — or it could be because of the collection of legitimate effects that constitute sample error. As will be discussed later (under exploratory data analysis), and is now more widely appreciated, outliers can be an indication of a biologically significant effect that is not yet statistically significant. Variance inflation can be the result of such outliers and can be used to detect them. Outliers, in fact, by increasing the variability within a small group, decrease the sensitivity of our statistical tests and actually preclude our having a statistically significant result (Beckman and Cook, 1983).

Alternatively, the outlier may be the result of, for example, an unobserved technician error, and may be such as to change the decisions made from a set of data. In this case we want to reject the data point — to exclude it from consideration with the rest of the data. But how can one identify these legitimate statistical rejection cases?

There are a wide variety of techniques for data rejection. Their proper use depends on one's having an understanding of the nature of the distribution of the data. For normally distributed data with a single extreme value, a simple method such as Chauvenet's Criterion (Meyer, 1975) may legitimately be employed. This states that if the probability of a value deviating from the mean is greater than $\frac{1}{2} N$, one should consider that there are adequate grounds for its rejections.

In practice, this approach is demonstrated below.

Use of Chauvenet's Criterion

Having collected 20 values as a data set, we find they include the following values: 1, 6, 7, 8, 8, 9, 9, 9, 10, 10, 10, 10, 10, 11, 11, 11, 12, 12, 13, and 14. Was the lowest value (1) erroneous and should it be rejected as an outlier? Some simple calculations are performed, as

$$\text{Mean} = 9.55$$

$$\text{SD} = 2.80$$

$$\text{Chauvenet's Criterion Value} = \frac{1}{2} N = 20/2 = 10$$

So we would reject the value of "1" if its probability of occurrence were less than 10%. Going to a table of Z scores (such as Table H in Appendix 1), we see that 10% of the values in a normal distribution are beyond ± 1.645 SD of the mean. Multiplying this by the SD for the sample, we get $(1.645)(2.80) = 4.606$. This means we would reject values beyond this range from the mean; that is, less than $(9.55 - 4.606) = 4.944$ or greater than $(9.55 + 4.606) = 14.156$. We therefore reject the value of "1."

One should note that as the sample size gets bigger, the rejection zone for Chauvenet's Criterion will also increase. Indeed, an N of 20 is about as large as this method is useful for.

A second, relatively straightforward approach for use when the data are normally distributed but contain several extreme values is to Winsorize the data. Although there are a number of variations to this approach, the simplest (called the G-1 method) calls for replacing the highest and lowest values in a set of data. In a group of data consisting of the values 54, 22, 18, 15, 14, 13, 11, and 4, we would replace 54 with a second 22, and 4 with a replicate 11. This would give us a group consisting of 22, 22, 18, 15, 14, 13, 11, and 11, which we would then treat as our original data. Winsorizing should not be performed, however, if the extreme values constitute more than a small minority of the entire data set.

Another approach is to use Dixon's Test (Dixon and Massey, 1969) to determine if extreme values should be rejected. In Dixon's test, the set of observations is first ordered according to their magnitude (as we did earlier for the data set used to demonstrate Chauvenet's Criterion, although there this step was simply to make the case clearer). The ratio of the difference of an extreme value from one of its nearest neighbor values in the range of values in the sample is then calculated, using a formula that varies with sample size. This ratio is then compared to a table value, and, if found to be equal or greater, is considered to be an outlier at the $p \leq 0.05$ level. The formula for the ratio varies with sample size and according to whether it is the smallest or largest value that is suspect.

If we have more information as to the nature of the data or the type of analysis to be performed, there are yet better techniques to handle outliers. Extensive discussions of these may be found elsewhere (Barnett and Lewis, 1994; Grubbs, 1969; Beckman and Cook, 1983; Snedecor and Cochran, 1989).

When the number of digits in a number is to be reduced (due to limitations of space or to reflect the extent of significance of a number) we must carry out the process of rounding off a number. Failure to have a rule for performing this operation can lead to both confusion and embarrassment for a facility (during such times as study audits). One common rule follows.

A digit to be rounded is not changed if it is followed by a digit less than 5 — the digits following it are simply dropped off ("truncated"). If the number is followed by a digit greater than 5 or by a 5 followed by other nonzero digits, it is increased to the next highest number. When the digit to be rounded is followed by 5 alone or by 5 followed by zeros, it is unchanged if it is even but increased by 1 if it is odd. Examples of this rule in effect are (in a case where we must reduce to whole digits):

137.4	becomes	137
137.6	becomes	138
138.52	becomes	139
137.5	becomes	138
138.5	becomes	138

The rationale behind this procedure is that over a period of time, the results should even out — as many digits will be increased as are decreased.

For sets of data, whatever rounding rule is used must generate consistent results. For example, if the values 0.233, 0.034, and 0.746 are rounded to two digits as separate values, the result is 0.23, 0.034, and 0.75 but if they are considered a set the values round to 0.23, 0.03, and 0.75 for consistency.

Sampling

Sampling — the selection of which individual data points will be collected, whether in the form of selecting which animals to collect blood from or to remove a portion of a diet mix for analysis — is an essential step upon which all other efforts toward a good experiment or study are based.

There are three assumptions about sampling that are common to most of the statistical analysis techniques that are used in toxicology. These are that the sample is collected without bias, that each member of a sample is collected independently of the others, and that members of a sample are collected with replacements. Precluding bias, both intentional and unintentional, means that at the time of selection of a sample to measure, each portion of the population from which that selection is to be made has an equal chance of being selected. Ways of precluding bias are discussed in detail in the chapter on experimental design.

Independence means that the selection of any portion of the sample is not affected by and does not affect the selection or measurement of any other portion.

Finally, sampling with replacement means that, in theory, after each portion is selected and measured, it is returned to the total sample pool and thus has the opportunity to be selected again. This is a corollary of the assumption of independence. Violation of this assumption (which is almost always the case in toxicology and all the life sciences) does not have serious consequences if the total pool from which samples are sufficiently large (say 20 or greater) is such that the chance of reselecting that portion is small anyway.

There are four major types of sampling methods — random, stratified, systematic, and cluster. Random is by far the most commonly employed method in toxicology. It stresses the fulfillment of the assumption of avoiding bias. When the entire pool of possibilities is mixed or randomized (procedures for randomization are presented in a later chapter), then the members of the group are selected in the order they are drawn from the pool.

Stratified sampling is performed by first dividing the entire pool into subsets or strata, then doing randomized sampling from each strata. This method is employed when the total pool contains subsets that are distinctly different but in which each subset contains similar members. An example is a large batch of a powdered pesticide in which it is desired to determine the nature of the particle size distribution. Larger pieces or particles are on the top, while progressively smaller particles have settled lower in the

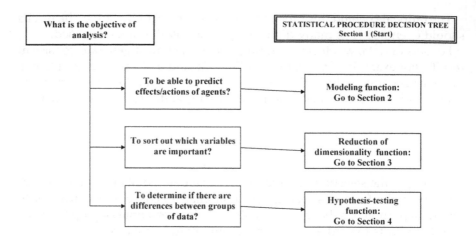

FIGURE 2.1
Overall decision tree for selecting statistical procedures.

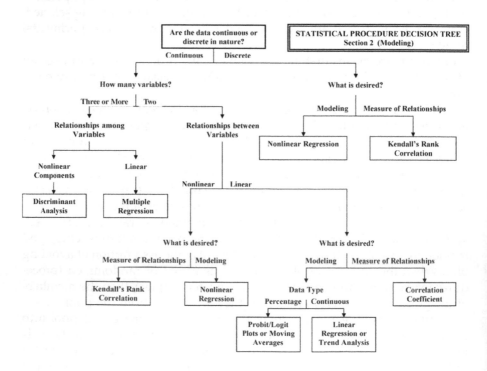

FIGURE 2.2
Decision tree for selecting hypothesis-testing procedures.

container, and, at the very bottom, the material has been packed and compressed into aggregates. To determine a timely representative answer, proportionally sized subsets from each layer or strata should be selected, mixed, and randomly sampled. This method is used more commonly in diet studies.

In systematic sampling, a sample is taken at set intervals (such as every fifth container of reagent or taking a sample of water from a fixed sample point in a flowing stream every hour). This is most commonly employed in quality assurance or (in the clinical chemistry lab) in quality control.

In cluster sampling, the pool is already divided into numerous separate groups (such as bottles of tablets), and we select small sets of groups (such as several bottles of tablets), then select a few members from each set. What one gets then is a cluster of measures. Again, this is a method most commonly used in quality control or in environmental studies when the effort and expense of physically collecting a small group of units is significant.

In classical toxicology studies sampling arises in a practical sense in a limited number of situations. The most common of these are as follows:

Selecting a subset of animals or test systems from a study to make some measurement (which either destroys or stresses the measured system, or is expensive) at an interval during a study. This may include such cases as doing interim necropsies in a chronic study or collecting and analyzing blood samples from some animals during a subchronic study.

Analyzing inhalation chamber atmospheres to characterize aerosol distributions with a new generation system.

Analyzing diet in which test material has been incorporated.

Performing quality control on an analytical chemistry operation by having duplicate analyses performed on some materials.

Selecting data to audit for quality assurance purposes.

Generalized Methodology Selection

One approach for the selection of appropriate techniques to employ in a particular situation is to use a decision-tree method. Figure 2.1 is a decision tree that leads to the choice of one of three other trees to assist in technique selection, with each of the subsequent trees addressing one of the three functions of statistics that was defined earlier in this chapter. Figure 2.2 is for the selection of hypothesis-testing procedures, Figure 2.3 for modeling procedures, and Figure 2.4 for reduction of dimensionality procedures. For the vast majority of situations, these trees will guide the user into the choice of the proper technique. The tests and terms in these trees will be explained subsequently.

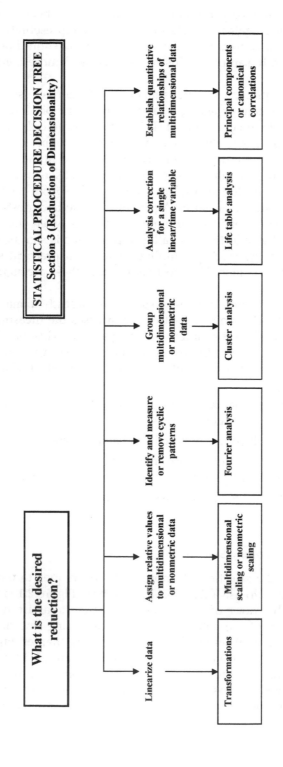

FIGURE 2.3
Decision tree for selecting modeling procedures.

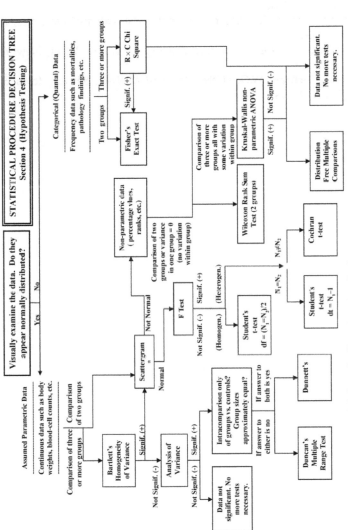

FIGURE 2.4
Decision tree for selecting reduction of dimensionality procedures.

References

Barnett, V. and Lewis, T. (1994) *Outliers in Statistical Data*. John Wiley, New York.

Beckman, R.J. and Cook, R.D. (1983) Outliers. *Technometrics*, 25, 119–163.

Dixon, W.J. and Massey, F.J., Jr. (1969) *Introduction to Statistical Analysis*, 3rd ed., McGraw-Hill, New York.

Grubbs, F.E. (1969) Procedure for detecting outlying observations in samples, *Technometrics*, 11, 1–21.

Meyer, S.L. (1975) *Data Analysis for Scientists and Engineers*. John Wiley, New York, pp. 17–18.

Snedecor, G.W. and Cochran, W.G. (1989) *Statistical Methods*, 8th ed., Iowa State University Press, Ames.

3

Experimental Design

Toxicological experiments generally have a twofold purpose. The first question is whether or not an agent results in an effect on a biological system. The second question, never far behind, is how much of an effect is present. Both the cost to perform research to answer such questions and the value that society places upon the results of such efforts have continued to increase rapidly. Additionally, it has become increasingly desirable that the results and conclusions of studies aimed at assessing the effects of environmental agents be as clear and unequivocal as possible. It is essential that every experiment and study yield as much information as possible, and that (more specifically) the results of each study have the greatest possible chance of answering the questions it was conducted to address. The statistical aspects of such efforts, so far as they are aimed at structuring experiments to maximize the possibilities of success, are called experimental design.

We have now become accustomed to developing exhaustively detailed protocols for an experiment or study prior to its conduct. But, typically, such protocols do not include or reflect a detailed plan for the statistical analysis of the data generated by the study and, certainly even less frequently, reflect such considerations in their design. *A priori* selection of statistical methodology (as opposed to the *post hoc* approach) is as significant a portion of the process of protocol development and experimental design as any other and can measurably enhance the value of the experiment or study. Prior selection of statistical methodologies is essential for proper design of other portions of a protocol such as the number of animals per group or the sampling intervals for body weight. Implied in such a selection is the notion that the toxicologist has both an in-depth knowledge of the area of investigation and an understanding of the general principles of experimental design, for the analysis of any set of data is dictated to a large extent by the manner in which the data are obtained.

The four basic statistical principles of experimental design are replication, randomization, concurrent ("local") control, and balance. In abbreviated form, these may be summarized as follows.

Replication

Any treatment must be applied to more than one experimental unit (animal, plate of cells, litter of offspring, etc.). This provides more accuracy in the measurement of a response than can be obtained from a single observation, since underlying experimental errors tend to cancel each other out. It also supplies an estimate of the experimental error derived from the variability among each of the measurements taken (or "replicates"). In practice, this means that an experiment should have enough experimental units in each treatment group (that is, a large enough N) so that reasonably sensitive statistical analysis of data can be performed. The estimation of sample size is addressed in detail later in this chapter.

Randomization

This is practiced to ensure that every treatment shall have its fair share of extreme high and extreme low values. It also serves to allow the toxicologist to proceed as if the assumption of "independence" is valid; that is, there is not avoidable (known) systematic bias in how one obtains data.

Concurrent Control

Comparisons between treatments should be made to the maximum extent possible between experimental units from the same closely defined population. Therefore, animals used as a "control" group should come from the same source, lot, age, etc., as test group animals. Except for the treatment being evaluated, test and control animals should be maintained and handled in exactly the same manner.

Balance

If the effect of several different factors is being evaluated simultaneously, the experiment should be laid out in such a way that the contributions of the different factors can be separately distinguished and estimated. There are several ways of accomplishing this using one of several different forms of design, as will be discussed below.

Experimental Design

There are four basic experimental design types used in toxicology. These are the randomized block, Latin square, factorial design, and nested design. Other designs that are used are really combinations of these basic designs and are very rarely employed in toxicology. Before examining these four basic types, however, we must first examine the basic concept of blocking. Blocking is, simply put, the arrangement or sorting of the members of a population (such as all of an available group of test animals) into groups based on certain characteristics that may (but are not sure to) alter an experimental outcome. Such characteristics that may cause a treatment to give a differential effect include genetic background, age, sex, overall activity levels, and so on. The process of blocking then acts (or attempts to act) so that each experimental group (or block) is assigned its fair share of the members of each of these subgroups.

Randomized Block

We should now recall that randomization is aimed at spreading out the effect of undetectable or unsuspected characteristics in a population of animals or some portion of this population. The merging of the two concepts of randomization and blocking leads to the first basic experimental design, the randomized block. This type of design requires that each treatment group have at least one member of each recognized group (such as age), the exact members of each block being assigned in an unbiased (or random) fashion.

Latin Square

The second type of experimental design assumes that we can characterize treatments (whether intended or otherwise) as belonging clearly to separate sets. In the simplest case, these categories are arranged into two sets, which may be thought of as rows (for, say, source litter of test animal, with the first litter as row 1, the next as row 2, etc.) and the secondary set of categories as columns (for, say, our ages of test animals, with 6 to 8 weeks as column 1, 8 to 10 weeks as column 2 and so on). Experimental units are then assigned so that each major treatment (control, low dose, intermediate dose, etc.) appears once and only once in each row and each column. If we denote our test groups as A (control), B (low), C (intermediate), and D (high), such an assignment would appear as in the table below.

	Age			
Source Litter	6–8 Weeks	8–10 Weeks	10–12 Weeks	12–14 Weeks
1	A	B	C	D
2	B	C	D	A
3	C	D	A	B
4	D	A	B	C

Factorial Design

The third type of experimental design is the factorial design, in which there are two or more clearly understood treatments, such as exposure level to test chemical, animal age, or temperature. The classical approach to this situation (and to that described under the Latin square) is to hold all but one of the treatments constant, and at any one time to vary just that one factor. Instead, in the factorial design all levels of a given factor are combined with all levels of every other factor in the experiment. When a change in one factor produces a different change in the response variable at one level of a factor than at other levels of this factor, there is an interaction between these two factors that can then be analyzed as an interaction effect.

Nested Design

The last of the major varieties of experimental design are the nested designs, where the levels of one factor are nested within (or are subsamples of) another factor. That is, each subfactor is evaluated only within the limits of its single larger factor.

Censoring

A second concept and its understanding are essential to the design of experiments in toxicology, that of censoring. Censoring is the exclusion of measurements from certain experimental units, or indeed of the experimental units themselves, from consideration in data analysis or inclusion in the experiment at all. Censoring may occur either prior to initiation of an experiment (where, in modern toxicology, this is almost always a planned procedure), during the course of an experiment (when they are almost universally unplanned, resulting from such as the death of animals on test), or after the conclusion of an experiment (when usually data are excluded because of being identified as some form of outlier).

In practice, *a priori* censoring in toxicology studies occurs in the assignment of experimental units (such as animals) to test groups. The most familiar example is in the common practice of assignment of test animals to acute, subacute, subchronic, and chronic studies, where the results of otherwise random assignments are evaluated for body weights of the assigned members. If the mean weights are found not to be comparable by some pre-established criterion (such as a 90% probability of difference by analysis of variance) then members are reassigned (censored) to achieve comparability in terms of starting body weights. Such a procedure of animal assignment to groups is known as a "censored randomization."

The first precise or calculable aspect of experimental design encountered is determining sufficient test and control group sizes to allow one to have an adequate level of confidence in the results of a study (that is, in the ability of the study design with the statistical tests used to detect a true difference — or effect — when it is present). The statistical test contributes a level of power to such a detection. Remember that the power of a statistical test is the probability that a test results in rejection of a hypothesis, H_0, say, when some other hypothesis, H, is valid. This is termed the power of the test "with respect to the (alternative) hypothesis H."

If there is a set of possible alternative hypotheses, the power, regarded as a function of H, is termed the "power function" of the test. When the alternatives are indexed by a single parameter θ, simple graphical presentation is possible. If the parameter is a vector θ, one can visualize a *power surface*.

If the power function is denoted by $\beta(\theta)$ and H_0 specifies $\theta = \theta_0$, then the value of $\beta(\theta)$ — the probability of rejecting H_0 when it is in fact valid — is the significance level. A test's power is greatest when the probability of a type II error is the least. Specified powers can be calculated for tests in any specific or general situation.

Some general rules to keep in mind are:

The more stringent the significance level, the greater the necessary sample size. More subjects are needed for a 1% level test than for a 5% level test.

Two-tailed tests require larger sample sizes than one-tailed tests. Assessing two directions at the same time requires a greater investment.

The smaller the critical effect size, the larger the necessary sample size. Subtle effects require greater efforts.

Any difference can be significant if the sample size is large enough.

The larger the power required, the larger the necessary sample size. Greater protection from failure requires greater effort. The smaller the sample size, the smaller the power, i.e., the greater the chance of failure.

The requirements and means of calculating necessary sample sizes depend on the desired (or practical) comparative sizes of test and control groups.

This number (N) can be calculated, for example, for equal sized test and control groups, using the formula

$$N = \frac{(t_{1-\alpha/2} + t_{1-\beta})}{\Delta^2} s^2$$

where $t_{1-\alpha/2}$ is the two-tailed t-value with $N - 1$ degrees of freedom corresponding to the desired level of confidence, $t_{1-\beta}$ is the one-tailed t-value with $N - 1$ degrees of freedom corresponding to the probability that the sample size will be adequate to achieve the desired precision, Δ is the acceptable range of variation in the variable of interest, and s is the sample standard deviation, derived typically from historical data and calculated as

$$s = \sqrt{\frac{1}{N-1}\Sigma(\bar{V} - V_i)^2}$$

with V being the variable of interest. If a one-sided test is used instead, substitute $t_{1-\alpha}$ instead of $t_{1-\alpha/2}$. The calculated value of N must be rounded up to the next whole number.

A good approximation can be generated by substituting the t-values (from a table such as Table F in Appendix 1) for an infinite number of degrees of freedom — this is equivalent to using the z-value instead of the t-value. This entire process is demonstrated in Example 3.1.

Example 3.1

In a subchronic dermal study in rabbits, the principal point of concern is the extent to which the compound causes oxidative damage to erythrocytes. To quantitate this, the laboratory will be measuring the numbers of reticulocytes in the blood. What then would be an adequate sample size to allow the question at hand to be addressed with reasonable certitude?

To do this, we use the one-tailed t-value for an infinite number of degrees of freedom at the 95% confidence level (that is, $\alpha = 0.05$). Going to a set of t tables, we find $t_{0.975}$ to be 1.96. For this we wish the chance of a false negative to be 10% or less; i.e., $\beta = 0.10$, for a $t_{0.90}$ of 1.28. From prior experience, we know that the usual values for reticulocytes in rabbit blood are from 0.5 to 1.9×10^6/ml. The acceptable range of variation, d, is therefore equal to half the span of this range, or 0.7. Likewise, examining the control data from previous rabbit studies, we find our sample standard deviation (SD) to be 0.825. When we insert all of these numbers into the equation (presented above) for sample size, we can calculate the required sample size (N) to be

$$N = \frac{(1.96 + 1.28)^2}{(0.7)^2}(0.825)^2 = \frac{10.50}{0.49}(0.68) = 14.58$$

In other words, in this case where there is little natural variability, measuring the reticulocyte counts of groups of fifteen animals each should be sufficient.

There are many formulas for calculating sample sizes, as demonstrated by Table 3.1, which provides various formulae for sample sizes for proportions in experiments involving groups of equal sizes. All formulae assume a two-sided test of significance; for a one-sided test, replace $Z_{\alpha/2}$ with Z_α where Z is the exact unconditional Z statistic (Sahai and Khurshid, 1996).

TABLE 3.1

Sample Size Formula for Proportions with Equal Group Sizes

Method	Formula	Source
Arcsine	$$N = \frac{(Z_{1-\alpha/2} + Z_{1-\beta})^2}{2\left[\arcsin(\sqrt{p_1}) - \arcsin(\sqrt{p_2})\right]^2}$$	Walter (1977) Sahai and Khurshid (1996)
Poisson	$$N = \frac{(Z_{1-\alpha/2} + Z_{1-\beta})^2(p_1 + p_2)}{\sigma^2}$$	Gail (1975) Sahai and Khurshid (1996)
Chi-squared with continuity correction	$$N = \frac{N}{4}\left[1 + \sqrt{\left\{\frac{2\sigma}{(Z_{1-\alpha/2} + Z_{1-\beta})^2 \overline{p}\overline{q}}\right\}}\right]^2 *$$	Fleiss et al., (1980) Sahai and Khurshid (1996)
Simple normal assuming homogeneity	$$N = \frac{2\overline{p}\overline{q}(Z_{1-\alpha/2} + Z_{1-\beta})^2}{\sigma^2} *$$	Cochran and Cox (1992) Sahai and Khurshid (1996)
Simple normal assuming heterogeneity	$$N = \frac{(p_1 q_1 + p_2 q_2)(Z_{1-\alpha/2} + Z_{1-\beta})^2}{\sigma^2}$$	Pocock (1979) Sahai and Khurshid (1996)
Difference method	$N \geq 2 \ (\sigma/\Delta)^2 \times (t_\alpha + t_\beta)^2$	McCance (1989)

$$* N = \frac{\left[Z_{1-\alpha/2}\sqrt{\{2\overline{p}(1-\overline{p})\}} + Z_{1-\beta}\sqrt{(p_1 q_2)}\right]^2}{\sigma^2}$$

where $q_1 = 1 - p_1$, $q_2 = 1 - p_2$, $\overline{p} = (p_1 + p_2)/2$, and $\overline{q} = 1 - \overline{p}$.

Table 3.2 gives formulae for proportions in experiments involving unequal groups. All formulae assume a two-sided test of significance; for a one-sided test, replace $Z_{\alpha/2}$ with Z_α.

There are a number of aspects of experimental design that are specific to the practice of toxicology. Before we look at a suggestion for step-by-step development of experimental designs, these aspects should first be considered as follows.

Frequently, the data gathered from specific measurements of animal characteristics are such that there is wide variability in the data. Often, such wide variability is not present in a control or low dose group, but in an intermediate dosage group variance inflation may occur. That is, there may be a large SD associated with the measurements from this intermediate group. In the face of such a set of data, the conclusion that there is no biological effect based on a finding of no statistically significant effect might well be erroneous (Gad, 1982).

In designing experiments, a toxicologist should keep in mind the potential effect of involuntary censoring on sample size. In other words, although the study described in Example 3.1 might start with five rabbits per group, this provides no margin should any die before the study is ended and blood samples are collected and analyzed. Just enough experimental units per group frequently leaves too few at the end to allow meaningful statistical analysis, and allowances should be made accordingly in establishing group sizes.

TABLE 3.2

Sample Size Formula for Proportions for Unequal Group Sizes

Method	Formula	Source
Arcsine	$$N = \frac{(Z_{1-\alpha/2} + Z_{1-\beta})^2}{\frac{4k}{k+1}\left[\arcsin(\sqrt{p_1}) - \arcsin(\sqrt{p_2})\right]^2}$$	Sahai and Khurshid (1996)
Poisson	$$N = \frac{(Z_{1-\alpha/2} + Z_{1-\beta})^2 (p_1 + p_2/k)}{\sigma^2}$$	Gail (1975) Sahai and Khurshid (1996)
Chi-squared	$$N = \frac{n'}{4}\left[1 + \sqrt{\left\{\frac{2\sigma}{(Z_{1-\alpha/2} + Z_{1-\beta})^2 \bar{p}\bar{q}}\right\}}\right]^2 *$$	Fleiss et al., (1980) Sahai and Khurshid (1996)
Simple normal assuming homogeneity	$$N = \frac{(1+1/k)(\bar{p}\bar{q})(Z_{1-\alpha/2} + Z_{1-\beta})^2}{\sigma^2}$$	Cochran and Cox (1992) Sahai and Khurshid (1996)
Simple normal assuming heterogeneity	$$N = \frac{(p_1 q_1 + p_2 q_2/k)(Z_{1-\alpha/2} + Z_{1-\beta})^2}{\sigma^2}$$	Pocock (1979) Sahai and Khurshid (1996)

$$* N = \frac{\left[Z_{1-\alpha/2}\sqrt{\{\bar{p}\bar{q}(1+1/k)\}} + Z_{1-\beta}\sqrt{\{p_1 q_1 + p_2 q_2/k\}}\right]^2}{\sigma^2}$$

where $q_1 = 1 - p_1$, $q_2 = 1 - p_2$, $\bar{p} = (p_1 + p_2)/2$, and $\bar{q} = 1 - \bar{p}$

It is certainly possible to pool the data from several identical toxicological studies. For example, after first having performed an acute inhalation study where only three treatment group animals survived to the point at which a critical measure (such as analysis of blood samples) was performed, we would not have enough data to perform a meaningful statistical analysis. We could then repeat the protocol with new control and treatment group animals from the same source. At the end, after assuring ourselves that the two sets of data are comparable, we could combine (or pool) the data from survivors of the second study with those from the first. The costs of this approach, however, would then be both a greater degree of effort expended (than if we had performed a single study with larger groups) and increased variability in the pooled samples (decreasing the power of our statistical methods). Note that metanalysis uses this approach.

Another frequently overlooked design option in toxicology is the use of an unbalanced design; that is, of different group sizes for different levels of treatment. There is no requirement that each group in a study (control, low dose, intermediate dose, and high dose) have an equal number of experimental units assigned to it. Indeed, there are frequently good reasons to assign more experimental units to one group than to others, and, as we shall see later in this book, all the major statistical methodologies have provisions to adjust for such inequalities, within certain limits. The two most common uses of the unbalanced design have larger groups assigned to either the highest dose to compensate for losses due to possible deaths during the study, or to the lowest dose to give more sensitivity in detecting effects at levels close to an effect threshold — or more confidence to the assertion that no effect exists.

We are frequently confronted with the situation where an undesired variable is influencing our experimental results in a nonrandom fashion. Such a variable is called a confounding variable — its presence, as discussed earlier, makes the clear attribution and analysis of effects at best difficult, and at worst impossible. Sometimes such confounding variables are the result of conscious design or management decisions, such as the use of different instruments, personnel, facilities, or procedures for different test groups within the same study. Occasionally, however, such confounding variables are the result of unintentional factors or actions, and are called lurking variables. Examples of such variables are almost always the result of standard operating procedures being violated— water not connected to a rack of animals over a weekend, a set of racks not cleaned as frequently as others, or a contaminated batch of feed used.

Finally, some thought must be given to the clear definition of what is meant by experimental unit and concurrent control.

The experimental unit in toxicology encompasses a wide variety of possibilities. It may be cells, plates of microorganisms, individual animals, litters of animals, etc. The importance of clearly defining the experimental unit is that the number of such units per group is the N, which is used in statistical calculations or analyses and critically affects such calculations.

The experimental unit is the unit that receives treatments and yields a response that is measured and becomes a datum. What this means in practice is that, for example, in reproduction or teratology studies where we treat the parental generation females and then determine results by counting or evaluating offspring, the experimental unit is still the parent. Therefore, the number of litters, not the number of offspring, is the N (Weil, 1970).

A true concurrent control is one that is identical in every manner with the treatment groups except for the treatment being evaluated. This means that all manipulations, including gavaging with equivalent volumes of vehicle or exposing to equivalent rates of air exchanges in an inhalation chamber, should be duplicated in control groups just as they occur in treatment groups.

The goal of the four principles of experimental design is statistical efficiency and the economizing of resources. It is possible to think of design as a logic flow analysis. Such an analysis is conducted in three steps and should be performed every time any major study or project is initiated or, indeed, at regular periods during the course of conduct of a series of "standard" smaller studies. These steps are detailed below.

Define the objective of the study — get a clear statement of what questions are being asked.

Can the question, in fact, be broken down into a set of subquestions?

Are we asking one or more of these questions repeatedly? For example, does "X" (an event or effect) develop at 30, 60, 90+ days and/or does it progress/regress or recover?

What is our model to be in answering this/these questions? Is it
   appropriate and acceptably sensitive?

For each subquestion (i.e., separate major variable to be studied):

How is the variable of interest to be measured?

What is the nature of data generated by the measure? Are we getting
   an efficient set of data? Are we buying too little information,
   (would another technique improve the quality of the information
   generated to the point that it becomes a higher "class" of data?),
   or too much information (i.e., does some underlying aspect of
   the measure limit the class of data obtainable within the bounds
   of feasibility of effort?).

Are there possible interactions between measurements? Can they be
   separated/identified?

Is our N (sample size) both sufficient and efficient?

What is the control — formal or informal? Is it appropriate?

Are we needlessly adding confounding variables (asking inadvert-
   ent or unwanted questions)?

Are there "lurking variables" present? These are undesired and not
   readily recognized differences that can affect results, such as
   different technicians observing different groups of animals.

How large an effect will be considered biologically significant? This
   is a question that can only be resolved by reference to experience
   or historical control data.

What are the possible outcomes of the study; i.e., what answers are
   possible to both our subquestions and to our major question?

How do we use these answers?

Do the possible answers offer a reasonable expectation of achieving
   the objectives that caused us to initiate the study?

What new questions may these answers cause us to ask? Can the
   study be redesigned, before it is actually started, so that these
   "revealed" questions may be answered in the original study?

A practical example of the application of this approach can be demonstrated in the process of designing a chronic inhalation study. Although in
such a situation the primary question being asked is usually "does the
chemical result in cancer by this route?" even at the beginning there are a
number of other questions the study is expected to answer. Two such questions are (1) if cancer is caused, what is the relative risk associated with it,
and (2) are there other expressions of toxicity associated with chronic exposure? Several, if not all, of these questions should be asked repeatedly
during the course of the study. Before the study starts, a plan and arrange-
ments must be formed to make measurements to allow us to answer these
questions.

When considering the last portion of our logic analysis, however, we must start by considering each of the things that may go wrong during the study. These include the occurrence of an infectious disease, the finding that extreme nasal and respiratory irritation was occurring in test animals, or the uncovering of a hidden variable. Do we continue to stop exposures? How will we now separate those portions of observed effects that are due to the chemical under study and those portions that are due to the disease process? Can we preclude (or minimize) the possibility of a disease outbreak by doing a more extensive health surveillance and quarantine on our test animals prior to the start of the study? Could we select a better test model — one that is not as sensitive to upper respiratory or nasal irritation?

For the reader who would like to further explore experimental design, there are a number of more detailed texts available that include more extensive treatments of the statistical aspects of experimental design. Among those recommended are Cochran and Cox (1992), Diamond (1981), Federer (1955), Hicks (1982), Kraemer and Thiemann (1987), and Myers (1972).

References

Cochran, W.G. and Cox, G.M. (1992) *Experimental Designs*, 2nd ed., John Wiley, New York.

Diamond, W.J. (1981) *Practical Experimental Designs*. Lifetime Learning Publications, Belmont, CA.

Federer, W.T. (1955) *Experimental Design*. Macmillan, New York.

Fleiss, J.L., Tytun, A., and Ury, H.K. (1980) A simple approximation for calculating sample size for comparing independent proportions. *Biometrics*, 36, 43–346.

Gad, S.C. (1982) Statistical analysis of behavioral toxicology data and studies. *Arch. Toxicol. Suppl.*, 5, 256–266.

Gail, M.H. (1975) A review and critique of some methods in competing risk analysis. *Biometrics*, 31, 209–222.

Hicks, C.R. (1982) *Fundamental Concepts in the Design of Experiments*. Holt, Rinehart, and Winston, New York.

Kraemer, H.C. and Thiemann, G. (1987) *How Many Subjects? Statistical Power Analysis in Research*. Sage Publications, Newbury Park, CA.

McCance, I. (1989) The number of animals. *NewsPhysiol. Sci.*, 4, 172–176.

Myers, J.L. (1972) *Fundamentals of Experimental Designs*. Allyn and Bacon, Boston.

Pocock, S.J. (1979) Allocation of patients to treatment in clinical trials. *Biometrics*, 35, 183–197.

Sahai, H. and Khurshid, A. (1996) Formulae and tables for the determination of sample sizes and power in clinical trials for testing differences in proportions for the two-sample design: A review. *Stat. Med.*, 15, 1–21.

Walter, S.D. (1977) Determination of significant relative risks and optimal sampling procedures in prospective and retrospective comparative studies of various sizes. *Am. J. Epidemiol.*, 105, 387–397.

Weil, C.S. (1970) Selection of the valid number of sampling units and a consideration of their combination in toxicological studies involving reproduction, teratogenesis or carcinogenesis. *Food Chem. Toxicol.*, 8, 177–182.

4

Computational Devices and Software

The range, scope, and availability of aids for the calculation of mathematical techniques in general and for statistical techniques in particular have increased at an almost geometric rate since the mid-1970s. There is no longer any reason to use paper and pencil to perform such calculations; the capabilities of electronic systems are sufficiently developed at each level (as discussed below) and the cost, compared to labor savings, is minimal.

There are now two tiers of computational support available for statistical analysis, and this chapter will attempt an overview of the major systems available within these tiers and the general characteristics and limitations of each. The three tiers range from programmable calculators (which represent Tier I, and include such devices as the Texas Instruments TI-83 and the Hewlett-Packard HP-41) to complete statistical packages for most desktop computers (the Tier II systems, which include such packages as SAS, SPSS, and Minitab).

As a general rule, as one goes from the systems in Tier I to those in Tier II, the cost, power, and capabilities of the systems increase, while both ease of use ("user friendliness") and flexibility decrease.

There are conventions associated with each system or instrument in these tiers, and these conventions should be known because they can affect both results and the ways in which results are reported.

The first two conventions, which apply to both tiers, have to do with the length of numbers. The first is that there is a preset number of digits that a machine (or software system) will accept as input, handle in calculations, and report as a result (either on a screen or as a printout). The most common such limit is 13 digits, that is, a number such as 12345.67890123. If, in the course of performing calculations, a longer string of digits is produced (for example, by dividing our example number by four, which should produce the actual number 3086.4197253075, a 14-digit number), a 13-digit system would then handle the last ("extra") digit in one of two ways. It would either truncate it (that is, just drop off the last digit) or it would round it by some rule (rounding was discussed in an earlier chapter). Truncation, particularly on the result of a long series of calculations (as is often the case in statistics) is more likely to produce erroneous results. Rounding is much less likely to produce errors, but knowledge of the rounding method utilized can be helpful.

The second convention, which applies to all systems, is that there is also a limit to how long a series of digits any system will report out; and there are different ways in which systems report out, and there are different ways in which systems report longer digit series. The first, less common, is truncation. The second is rounding combined with presentation of the results as an exponential. Exponentials are a series of digits followed by an expression of the appropriate powers of ten. Examples of such exponential expressions are

(a) 1.234567×10^{16}

(b) 1.234567×10^{-16}

(c) $1.234567 \ E^{16}$

(d) $1.234567 \ E^{-16}$

where (a) and (c) are the same number repressed two different ways, as are (b) and (d).

Tier I Systems (Programmable and Preprogrammed Statistical Calculators)

The two major Tier I systems (described more fully below) can readily perform all of the univariate (that is, two-variable) procedures described in this book. Libraries of programs, stored on magnetic cards, are available and well documented. Data sets may also be scored on magnetic cards, and by clever coding, programs and data sets of extreme length may be processed. But longer programs or larger data sets run extremely slowly. Accordingly, use of these instruments is limited, in practice, to univariate procedures and small data sets.

Two example Tier I systems are the Texas Instruments TI-89 and the Hewlett-Packard HP-49G+, both of which have graphing capabilities. These differ mainly in that the HP programs and data entry systems are programmable either in algebraic form or in reverse Polish notation (RPN) form, while the TI is in algebraic form. Both have USB ports to connect to desktop computers or other devices. Both have unique but simple programming languages.

Tier II (Desktop Computer Packages)

At levels above Tier I there are additional conventions as to the way systems operate and handle data. Additionally, there are a larger number and options for both machines and software systems because only a few common programming languages are involved.

The first common convention that must be understood and considered is that the systems actually perform a statistical operation in one of two modes, batch or interactive. In batch mode, the entire desired sequence of analysis is specified, and then the entire set of data is processed in accordance with this specified sequence. The drawback to this approach is that if a result early in the sequence indicates that an alternative latter set of procedures should be performed, there is no opportunity to change to these alternative (as opposed to the originally specified) procedures. In the interactive mode, procedures are specified (and results calculated) one step at a time, usually by the use of a series of menus.

The second convention concerns the structure of commonly used software systems. They are almost always divided into three separate parts or modules. The first module is a database manager, which allows the person using the system to store data in (and recall it from) a desired format, to perform a transformation(s) or arithmetic manipulation(s), and to pass the data along to either one of the other two modules or to another software system.

The second module performs the actual analysis of the data. Such modules also allow one to tailor reports to one's specifications.

The last module (which is not always present) is a graphics module, which presents (on a display screen) and prints out in any of a wide range of graphics and charts. The range may be limited to line, bar, and pie charts, or may extend to contour, cluster, and more exotic plots.

There are now two major microcomputer operating systems that support extensive statistical packages. These are the Macintosh and MS DOS ("IBM") systems. Each of these have virtually unlimited operating and storage memories, and printers and plotters are available. Each of these machines also has its own peculiarities.

Before reviewing software packages, several considerations should be presented. First, systems perform in one of two modes — either as libraries of programs, each of which can be selected to perform an individual procedure, or as an integrated system, where a single loading of the file allows access to each and every procedure. In the first mode, each step in an analysis (such as Bartlett's test, analysis of variance, and Duncan's) requires loading a separate file and then executing the procedure. Second, software may load from either CD-ROMs or diskettes.

Third, the available range of transformations should be carefully considered. At least a small number (log, reciprocal, probit, addition, subtraction, multiplication, division, and absolute value) are essential. If unusual data sets are to be handled or exploratory data analysis (discussed later) performed regularly, a more extensive set is required. Some of the packages (such as STATPRO) offer literally hundreds of transformations.

Fourth, all the packages listed in Table 4.1 have at least a basic set of capabilities. Besides database management and basic transformation and graphic functions, they can each perform the following simple tests (all discussed in later chapters):

TABLE 4.1

Popular Desktop Computer Statistical Packages

Title	Operating System(s)	EDA Performed	Graphics	Spreadsheet Import/Export	Menu (M) or Command (C) Driven
BMD8	Windows, Unix	No	Largely	Limited	C
DATA DESK	Windows, Mac	Yes	Complete	Full	M
E CHIP	Windows	No	Largely	Limited	M
JMP	Windows, Mac, Linux	Yes	Largely	Largely	M
MINITAB	Windows	No	Complete	Full	M/C
NCSS	Windows	Yes	Complete	Full	M
SAS/STAT	Windows, Unix, Linux	Yes	Requires SAS/GRAPH	Requires SAS/ACCESS	M/C
SPSS/PC	Windows	No	Complete	Full	C
STAT/MOST	Windows	Yes	Complete	Full	M
STATA	Windows, Mac, Unix	Yes	Complete	Full	M/C
STATISTICA	Windows, Mac	Yes	Complete	Full	M
SYSTAT	Windows	No	Complete	Full	C
XLSTAT	Windows (requires Excel)	No	Complete	Full	M

- Analysis of variance (ANOVA)
- 2 × 2 chi-square
- Linear regressions
- Student's *t*-test

Table 4.1 presents an overview of thirteen widely available commercially statistical packages for microcomputers, which the authors are familiar with. There are at least 120 additional packages available. Woodward et al. (1985) present an overview of many of these. For each package, the following information is presented:

Title	The Name of the Package
Operating system(s)	Which systems the package will operate on.
EDA	Does the system perform exploratory data analysis?
Graphics	How extensive are the graphic functions that the package performs?
Spreadsheet import/export	Can the system accept data and give output to popular spreadsheets such as EXCEL and LOTUS?
Menu (M) or command (C) driven	How does the user primarily interface with the system?

The top level of desktop system programs, which are summarized in Table 4.2, are all large commercial software packages that run on large computer systems with a time-sharing basis available. By definition, this means that these programs operate in a batch mode and use a unique (for each package) code language. Four of these libraries are briefly described below. Throughout this volume, many example problems are followed by a short review of the SAS code required to perform the operation, with the resulting output illustrated in a few examples. The other three packages function similarly.

The difficulty with the recently achieved, wide availability of automated analysis systems is that it has become increasingly easy to perform the wrong tests on the wrong data and from there to proceed to the wrong conclusions. This makes at least a basic understanding of the procedures and discipline of statistics a vital necessity for the research toxicologist.

TABLE 4.2

Top-Level Desktop System Programs

Package	Reference	Description
SPSS	Nie et al. (1995)	With manipulation, SPSS will perform al the procedures described in this book and the full range of graphics
BMD	Dixon (1994)	Has generally wider capabilities than SPSS (which are constantly being added to) and easily manipulated
SAS	SAS Institute (1999)	Widely available. Easier to format and very strong on data summarization. Has its own higher level programming language.
MINITAB	Ryan et al. (1996)	Easiest to use and least expensive of these four, but does not have the full range of capabilities

References

Dixon, W.J. (1994) *BMD-Biomedical Computer Programs*. University of California Press, Berkeley, CA, 1994.

Nie, N.H., Hall, C.H., Jenkins, J.G., Steinbrenner, K., and Bent, D.H. (1995) *Statistical Package for the Social Sciences*. McGraw-Hill, New York.

Ryan, T.A., Joyner, B.L., and Ryan, B.F. (1996) *Minitab Reference Manual*. Duxbury Press, Boston.

SAS Institute (1999) *SAS OnlineDoc, Version 8*. http://v8doc.sas.com/sashtml/ (Accessed August 2004) SAS Institute, Cary, NC.

Woodward, W.A., Elliott, A.C., and Gray, H.L. (1985) *Directory of Statistical Microcomputer Software*. Marcel Dekker, New York.

5

Methods for Data Preparation and Exploration

The data from toxicology studies should always be examined before any formal analysis is performed. Such examinations should be directed to determining if the data are suitable for analysis, and if so what form the analysis should take (see Figure 2.2, Chapter 2). If the data as collected are not suitable for analysis, or if they are only suitable for low-powered analytical techniques, one may wish to use one of many forms of data transformation to change the data characteristics so that they are more amenable to analysis.

The above two objectives, data examination and preparation, are the primary focus of this chapter. For data examination, two major techniques are presented — the scattergram and Bartlett's test. Likewise, for data preparation (with the issues of rounding and outliers having been addressed in a previous chapter) two techniques are presented — randomization (including a test for randomness in a sample of data) and transformation.

Finally, at the end of this chapter the concept of exploratory data analysis (EDA) is presented and briefly reviewed. This is a broad collection of techniques and approaches to "probe" data — that is, to both examine and to perform some initial, flexible analysis of the data.

Scattergram

Two of the major points to be made throughout this volume are (a) the use of the appropriate statistical tests, and (b) the effects of small sample sizes (as is often the case in toxicology) on our selection of statistical techniques. Frequently, simple examination of the nature and distribution of data collected from a study can also suggest patterns and results that were unanticipated and for which the use of additional or alternative statistical methodology is warranted. It was these three points that caused the author to consider a section on scattergrams and their use, essential for toxicologists.

Bartlett's test may be used to determine if the values in groups of data are homogeneous. If they are, this (along with the knowledge that they are from a continuous distribution) demonstrates that parametric methods are applicable.

But, if the values in the (continuous data) groups fail Bartlett's test (i.e., are heterogeneous), we cannot be secure in our belief that parametric methods are appropriate until we gain some confidence that the values are normally distributed. With large groups of data, we can compute parameters of the population (kurtosis and skewness, in particular) and from these parameters determine if the population is normal (with a certain level of confidence). If our concern is especially marked, we can use a chi-square goodness-of-fit test for normality. But when each group of data consists of 25 or fewer values, these measures or tests (kurtosis, skewness, and chi-square goodness-of-fit) are not accurate indicators of normality. Instead, in these cases we should prepare a scattergram of the data, then evaluate the scattergram to estimate if the data are normally distributed. This procedure consists of developing a histogram of the data, then examining the histogram to get a visual appreciation of the location and distribution of the data.

The abscissa (or horizontal scale) should be in the same scale as the values and should be divided so that the entire range of observed values is covered by the scale of the abscissa. Across such a scale we then simply enter symbols for each of our values. Example 5.1 shows such a plot.

Example 5.1 is a traditional and rather limited form of scatterplot but such plots can reveal significant information about the amount and types of association between the two variables, the existence and nature of outliers, the clustering of data, and a number of other two-dimensional factors (Anscombe, 1973; Chambers et al., 1983).

Current technology allows us to add significantly more graphical information to scatterplots by means of graphic symbols (letters, faces, different shapes, squares, colors, etc.) for the plotted data points. One relatively simple example of this approach is shown in Figure 5.1, where the simple case of dose (in a dermal study), dermal irritation, and white blood cell count are presented. This graph quite clearly suggests that as dose (variable x) is increased, dermal irritation (variable y) also increases; and as irritation becomes more severe, white blood cell count (variable z), an indicator of immune system involvement, suggesting infection or persistent inflammation, also increases. There is no direct association of variables x and z, however.

Cleveland and McGill (1984) presented an excellent, detailed overview of the expanded capabilities of the scatterplot, and the interested reader should refer to that article. Cleveland later (1985) expanded this to a book. Tufte (1983, 1990, 1997) has expanded on this.

Example 5.1

Suppose we have the two data sets below:

Group 1: 4.5, 5.4, 5.9, 6.0, 6.4, 6.5, 6.9, 7.0, 7.1, 7.0, 7.4, 7.5, 7.5, 7.5, 7.6, 8.0, 8.1, 8.4, 8.5, 8.6, 9.0, 9.4, 9.5 and 10.4.

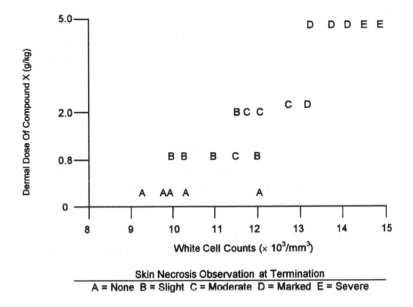

FIGURE 5.1
Exploratory data analysis; correlative plots.

Group 2: 4.0, 4.5, 5.0, 5.1, 5.4, 5.5, 5.6, 6.5, 6.5, 7.0, 7.4, 7.5, 7.5 8.0, 8.1, 8.5, 8.5, 9.0, 9.1, 9.5, 9.5, 10.1, 10.0 and 10.4.

Both of these groups contain 24 values and cover the same range. From them we can prepare the following scattergrams.

Group 1:

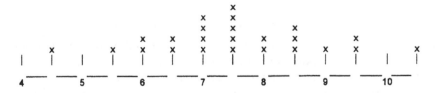

Group 2:

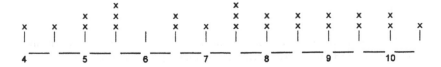

Group 1 can be seen to approximate a normal distribution (bell-shaped curve); we can proceed to perform the appropriate parametric tests with such data. But Group 2 clearly does not appear to be normally distributed. In this case, the appropriate nonparametric technique must be used.

Bartlett's Test for Homogeneity of Variance

Bartlett's test (see Sokal and Rohlf, 1994, pp. 403–407) is used to compare the variances (values reflecting the degree of variability in data sets) among three or more groups of data, where the data in the groups are continuous sets (such as body weights, organ weights, red blood cell counts, or diet consumption measurements). It is expected that such data will be suitable for parametric methods (normality of data is assumed) and Bartlett's is frequently used as a test for the assumption of equivalent variances.

Bartlett's is based on the calculation of the corrected χ^2 (chi-square) value by the formula:

$$\chi^2_{corr} = \frac{\sum df \left(\ln \left[\frac{\sum [df(S^2)]}{\sum df} \right] \right) - \sum \left[df(\ln S^2) \right]}{1 + \frac{1}{3(K-1)} \left[\sum \frac{1}{df} - \frac{1}{\sum df} \right]}$$

where S^2 = variance = $\dfrac{\sum X^2 - \frac{(\sum x)^2}{N}}{N-1}$

X = individual datum within each group
N = number of data within each group
K = number of groups being compared
df = degrees of freedom for each group = $(N-1)$

The corrected χ^2 value yielded by the above calculations is compared to the values listed in the chi-square table according to the numbers of degrees of freedom (such as found in Snedecor and Cochran, 1989, pp. 470–471).

If the calculated value is smaller than the table value at the selected p level (traditionally 0.05) the groups are accepted to be homogeneous and the use of ANOVA is assumed proper. If the calculated χ^2 is greater than the table value, the groups are heterogeneous and other tests (as indicated in Figure 2.3, the decision tree) are necessary. This is demonstrated in Example 5.2.

Example 5.2

If the monocytes in a sample of rat blood taken in the course of an inhalation study were counted, the results might appear as follows:

400 ppm		200 ppm		0 ppm	
(X_1)	$(X_1)^2$	(X_2)	$(X_2)^2$	(X_3)	$(X_3)^2$
9	81	5	25	7	49
5	25	5	25	6	36
5	25	4	16	5	25
4	16	6	36	7	49
		7	49		
$\Sigma X_1 = 25$	$\Sigma X_1^2 = 147$	$\Sigma X_2 = 27$	$\Sigma X_2^2 = 151$	$\Sigma X_3 = 25$	$\Sigma X_3^2 = 159$

$$S_1^2 = \frac{147 - \frac{(23)^2}{4}}{4-1} = 4.9167$$

$$S_2^2 = \frac{151 - \frac{(27)^2}{5}}{5-1} = 1.3000$$

$$S_3^2 - \frac{159 - \frac{(25)^2}{4}}{4-1} - 0.9167$$

In continuing the calculations, it is helpful to set up a table such as follows:

Concentration	N	df = (N − 1)	S^2	(df)(S^2)	ln(S^2)
400 ppm	4	3	4.9167	14.7501	1.5926
200 ppm	5	4	1.3000	5.2000	0.2624
0 ppm	4	3	0.9167	2.7501	−0.0870
Sums (Σ)	13	10		22.7002	

Concentration	(df)(ln S^2)	1/df
400 ppm	4.7778	0.3333
200 ppm	1.0496	0.2500
0 ppm	−0.2610	0.3333
Sums (Σ)	5.5664	0.9166

Now we substitute into our original formula for corrected χ^2

$$\chi_{corr}^2 = \frac{10\left(\ln\left[\frac{22.7002}{10}\right]\right) - 5.5664}{1 + \frac{1}{3(3-1)}\left[0.9166 - \frac{1}{10}\right]}$$

$$= \frac{10(0.8198) - 5.5664}{1 + 0.1667[0.8166]}$$

$$= 2.32$$

The table value for two degrees of freedom at the 0.05 level is 5.99. As our calculated value is less than this, the corrected χ^2 is not significant and the variances are accepted as homogeneous. We may thus use parametric methods (such as ANOVA) for further comparisons.

Assumptions and Limitations

Bartlett's test does not test for normality, but rather homogeneity of variance (also called equality of variances or homoscedasticity).

Homoscedasticity is an important assumption for Student's *t*-test, analysis of variance, and analysis of covariance.

The *F*-test (covered in the next chapter) is actually a test for the two sample (that is, control and one test group) case of homoscedasticity. Bartlett's is designed for three or more samples.

Bartlett's is very sensitive to departures from normality. As a result, a finding of a significant chi-square value in Bartlett's may indicate nonnormality rather than heteroscedasticity. Such a finding can be brought about by outliers, and the sensitivity to such erroneous findings is extreme with small sample sizes.

Statistical Goodness-of-Fit Tests

A goodness-of-fit test is a statistical procedure for comparing individual measurements to a specified type of statistical distribution. For example, a normal distribution is completely specified by its arithmetic mean and variance (square of the standard deviation — SD). The null hypothesis, that the data represent a sample from a single normal distribution, can be tested by a statistical goodness-of-fit test. Various goodness-of-fit tests have been devised to determine if the data deviate significantly from a specified distribution. If a significant departure occurs, it indicates only that the specified distribution can be rejected with some assurance. This does not necessarily mean that the true distribution contains two or more subpopulations. The true distribution may be a single distribution, based upon a different mathematical relationship, e.g., log-normal. In the latter case, logarithms of the measurement would not be expected to exhibit by a goodness-of-fit test a statistically significant departure from a log-normal distribution.

Everitt and Hand (1981) recommended use of a sample of 200 or more to conduct a valid analysis of mixtures of populations. Hosmer and Lemershow (1989) stated that even the maximum likelihood method, the best available method, should be used with extreme caution, or not at all, when separation between the means of the subpopulations is less than three standard deviations and sample sizes are less than 300. None of the available methods conclusively establishes bimodality, which may, however, occur when separation between the two means (modes) exceeds 2 SD. Conversely, inflections in probits or separations in histograms *less than* 2 SD apart may arise from genetic differences in test subjects.

Mendell et al. (1993) compared eight tests of normality to detect a mixture consisting of two normally distributed components with different means but equal variances. Fisher's skewness statistic was preferable when one component comprised less than 15% of the total distribution. When the two components comprised more nearly equal proportions (35 to 65%) of the total distribution, the Engelman and Hartigan test (1969) was preferable. For other mixing proportions, the maximum likelihood ratio test was best. Thus, the maximum likelihood ratio test appears to perform very well, with only small loss from optimality, even when it is not the best procedure.

The method of *maximum likelihood* provides estimators that are usually quite satisfactory. They have the desirable properties of being consistent, asymptotically normal, and asymptotically efficient for large samples under quite general conditions. They are often biased, but the bias is frequently removable by a simple adjustment (Examples 5.3 and 5.4). Other methods of obtaining estimators are also available, but the maximum likelihood method is the most frequently used.

Maximum likelihood estimators (MLEs) also have another desirable property: invariance. Let us denote the maximum likelihood estimator of the parameter θ by $\hat{\theta}$. Then, if $f(\theta)$ is a single-valued function of θ, the maximum likelihood estimator of $f(\theta)$ is $f(\hat{\theta})$. Thus, for example, $\hat{\sigma} = (\hat{\sigma}^2)^{1/2}$.

The principle of maximum likelihood tells us that we should use as our estimate that value which maximizes the likelihood of the observed event. The following examples demonstrate the technique.

Example 5.3

Derive the maximum likelihood estimator $\hat{p}$ of the binomial probability p for a coin-tossing experiment in which a coin is tossed n times and r heads are obtained.

We know that the likelihood of the observed event is

$$L = \binom{n}{r} p^r (1-p)^{n-r}$$

According to the principle of maximum likelihood, we choose the value of p that maximizes L. This value also maximizes

$$\ln(L) = \ln\binom{n}{r} + r(\ln p) + (n-r)\ln(1-p)$$

At the maximum, the derivative with respect to p must be zero. That is,

$$\frac{\partial}{\partial p}\ln(L) = \frac{r}{p} + \frac{n-4}{1-p}$$

where

$$\hat{p} = r/n$$

Example 5.4

A sample of n observations $x_1, \ldots, x_n$ is drawn from a normal population. Derive the maximum likelihood estimators $\hat{\mu}$ and $\hat{\sigma}^2$ for the mean and variance of the population.

The likelihood of the observed event is

$$L = \prod \frac{1}{\sqrt{2\pi\sigma^2}} \exp\left\{ -\frac{(x_i - \mu)^2}{2\sigma^2} \right\}$$

which can also be written as

$$L = \frac{1}{(2\pi)^{\pi/2}\sigma^n} \exp\left\{ -\frac{1}{2\sigma^2} \sum (x_i - \mu)^2 \right\}$$

According to the principle of maximum likelihood, we choose those values of μ and s^2 that maximize L. The same values maximize $\ln L$. So we equate the partial derivatives of $\ln L$ with respect to $\hat{\mu}$ and $\hat{\sigma}^2$ to zero and find that

$$\hat{\mu} = \sum x_i/n = \bar{x}$$

and

$$\hat{\sigma}^2 = \frac{1}{n} \sum (x_i - \bar{x})^2$$

In other words, the MLE for the population mean equals the sample mean and the MLE for the population variance is the sample variance. The latter estimator is slightly biased, but the bias can be removed by multiplying by $n/(n-1)$ and using the estimator

$$\hat{\sigma}^2 = \frac{1}{n-1} \sum (x_i - \bar{x})^2$$

These maximum likelihood methods can be used to obtain point estimates of a parameter, but we must remember that a point estimator is a random variable distributed in some way around the true value of the parameter. The true parameter value may be higher or lower than our estimate. It is often useful therefore to obtain an interval within which we are reasonably confident the true value will lie, and the generally accepted method is to construct what are known as *confidence limits*.

The following procedure will yield upper and lower 95% confidence limits with the property that when we say that these limits include the true value of the parameter, 95% of all such statements will be true, and 5% will be incorrect.

> Choose a (test) statistic involving the unknown parameter and no other unknown parameter.
>
> Place the appropriate sample values in the statistic.
>
> Obtain an equation for the unknown parameter by equating the test statistic to the upper 2 ½ % point of the relevant distribution.
>
> The solution of the equation gives one limit.
>
> Repeat the process with the lower 2 ½ % point to obtain the other limit.

We can also construct 95% confidence intervals using unequal tails (for example, using the upper 2% point and the lower 3% point). We usually want our confidence interval to be as short as possible, however, and with a symmetric distribution such as the normal or t, this is achieved using equal tails. The same procedure very nearly minimizes the confidence interval with other nonsymmetric distributions (for example, chi-square) and has the advantage of avoiding rather tedious computation.

When the appropriate statistic involves the square of the unknown parameter, both limits are obtained by equating the statistic to the upper 5% point of the relevant distribution. The use of two tails in this situation would result in a pair of non-intersecting intervals. When two or more parameters are involved, it is possible to construct a region within which we are reasonably confident the true parameter values will lie. Such regions are referred to as confidence regions. The implied interval for p_1 does not form a 95% confidence interval, however. Nor is it true that an 85.7375% confidence region for p_1, p_2, and p_3 can be obtained by considering the intersection of the three separate 95% confidence intervals, because the statistics used to obtain the individual confidence intervals are not independent. This problem is obvious with a multiparameter distribution such as the multinomial, but it even occurs with the normal distribution because the statistic that we use to obtain a confidence interval for the mean and the statistic that we use to obtain a confidence interval for the variance are not independent. The problem is not likely to be of great concern unless a large number of parameters is involved, as illustrated in Example 5.5.

Example 5.5

A sample of nine is drawn from a normal population with unknown mean and variance. The sample mean is 4.2 and the sample variance 1.69. Obtain a 95% confidence interval for the mean μ.

The confidence limits are obtained from the equation

$$\frac{\sqrt{9}(4.2-\mu)}{\sqrt{1.69}} = \pm 2.306$$

so

$$\mu = 4.2 \pm 2.30\sqrt{1.69/9} = 4.2 \pm 1.0$$

From this we determine that the 95% confidence interval is from 3.2 to 5.2.

Randomization

Randomization is the act of assigning a number of items (plates of bacteria or test animals, for example) to groups in such a manner that there is an equal chance for any one item to end up in any one group. This is a control against any possible bias in assignment of subjects to test groups. A variation on this is censored randomization, which insures that the groups are equivalent in some aspect after the assignment process is complete. The most common example of a censored randomization is one in which it is insured that the body weights of test animals in each group are not significantly different from those in the other groups. This is done by analyzing group weights both for homogeneity of variance and by analysis of variance after animal assignment, then re-randomizing if there is a significant difference at some nominal level, such as $p \leq 0.10$. The process is repeated until there is no difference.

There are several methods for actually performing the randomization process. The three most commonly used are card assignment, use of a random number table, and use of a computerized algorithm.

For the card-based method, individual identification numbers for items (plates or animals, for example) are placed on separate index cards. These cards are then shuffled, placed one at a time in succession into piles corresponding to the required test groups. The results are a random group assignment.

The random number table method requires only that one have unique numbers assigned to test subjects and access to a random number table. One simply sets up a table with a column for each group to which subjects are to be assigned. We start from the head of any one column of numbers in the random table (each time the table is used, a new starting point should be utilized). If our test subjects number less than 100, we utilize only the last two digits in each random number in the table. If they number more than 99 but less than 1000, we use only the last three digits. To generate group assignments, we read down a column, one number at a time. As we come across digits which correspond to a subject number, we assign that subject to a group (enter its identifying number in a column) proceeding to assign subjects to groups from left to right filling one row at a time. After a number is assigned to an animal, any duplication of its unique number is ignored. We use as many successive columns of random numbers as we may need to complete the process.

The third (and now most common) method is to use a random number generator that is built into a calculator or computer program. Procedures for generating these are generally documented in user manuals.

One is also occasionally required to evaluate whether a series of numbers (such as an assignment of animals to test groups) is random. This requires the use of a randomization test, of which there are a large variety. The chi-square test, described later, can be used to evaluate the goodness-of-fit to a random assignment. If the result is not critical, a simple sign test will work. For the sign test, we first determine the middle value in the numbers being checked for randomness. We then go through a list of the numbers assigned to each group, scoring each as a "+" (greater than our middle number) or "−" (less than our middle number). The number of pluses and minuses in each group should be approximately equal. This is demonstrated in Example 5.6.

Example 5.6

In auditing a study performed at a contract lab, we wish to ensure that their assignment of animals to test groups was random. Animals numbered 1 to 33 were assigned to groups of eleven animals each. Using the middle value in this series (17) as our check point, we assign signs as below.

Control		Test Group A		Test Group B	
Animal Number	Sign	Animal Number	Sign	Animal Number	Sign
17	0	18	+	11	−
14	−	1	−	2	−
7	−	12	−	22	+
26	+	9	−	28	+
21	+	5	−	19	+
15	−	20	+	3	−
16	−	33	+	29	+
6	−	27	+	10	−
25	+	8	−	23	+
32	+	24	+	30	+
4	−	31	+	13	−
Sum of Signs	−2		+1		+1

Note that 17 is scored as zero, insuring (as a check on results) that the sum of the sums of the three columns would be zero. The results in this case clearly demonstrate that there is no systematic bias in animal number assignments.

Transformations

If our initial inspection of a data set reveals it to have an unusual or undesired set of characteristics (or to lack a desired set of characteristics), we have a choice of three courses of action. We may proceed to select a method or test appropriate to this new set of conditions, or abandon the entire exercise, or transform the variable(s) under consideration in such a manner that the resulting transformed variates (X' and Y', for example, as opposed to the original variates X and Y) meet the assumptions or have the characteristics that are desired.

The key to all this is that the scale of measurement of most (if not all) variables is arbitrary. That is, although we are most familiar with a linear scale of measurement, there is nothing that makes this the "correct" scale on its own as opposed to a logarithmic scale (familiar logarithmic measurements are those of pH values, or earthquake intensity [Richter scale]). Transforming a set of data (converting X to X') is really as simple as changing a scale of measurement.

There are at least four good reasons to transform data:

1. To normalize the data, making them suitable for analysis by our most common parametric techniques such as analysis of variance (ANOVA). A simple test of whether a selected transformation will yield a distribution of data that satisfies the underlying assumptions for ANOVA is to plot the cumulative distribution of samples on probability paper (that is, a commercially available paper that has the probability function scale as one axis). One can then alter the scale of the second axis (that is, the axis other than the one that is on a probability scale) from linear to any other (logarithmic, reciprocal, square root, etc.) and see if a previously curved line indicating a skewed distribution becomes linear to indicate normality. The slope of the transformed line gives us an estimate of the standard deviation. And if the slopes of the lines of several samples or groups of data are similar, we accordingly know that the variances of the different groups are homogeneous.

2. To linearize the relationship between a paired set of data, such as dose and response. This is the most common use in toxicology for transformations and is demonstrated in the section under probit/logit plots.

3. To adjust data for the influence of another variable. This is an alternative in some situations to the more complicated process of analysis

TABLE 5.1

Common Data Transformations

Transformation	How Calculated[a]	Example of Use[b]
Arithmetic	$x' = x/y$ or $x' = x + c$	Organ weight/body weight
Reciprocals	$x' = 1/x$	Linearizing data, particularly rate phenomena
Arcsine (also called "angular")	$x' = \text{arcsine}(\sqrt{x})$	Normalizing dominant lethal and mutation rate data
Logarithmic	$x' = \log(x)$	pH values
Probability (probit)	$x' = \text{probability}(x)$	Percentage responding
Square roots	$x' = \sqrt{x}$	Surface area of animal from body weights
Box Cox	$x' = (x^v - 1)v$ for $v \neq 0$ $x' = \ln(x)$ for $v = 0$	For use when one has no prior knowledge of the appropriate transformation to use
Rank transformation	Depends on nature of samples	As a bridge between parametric and nonparametric statistics (Conover and Inman, 1981)

[a] x and y are original variables, x' is the transformed value. "c" represents a constant.

[b] Plotting a double reciprocal (i.e., $1/x$ vs. $1/y$) will linearize almost any data set. So will plotting the log transforms of a set of variables.

of covariance. A ready example of this usage is the calculation of organ weight to body weight ratios in *in vivo* toxicity studies, with the resulting ratios serving as the raw data for an analysis of variance performed to identify possible target organs. This use is discussed in detail later in this chapter.

4. Finally, to make the relationships between variables clearer by removing or adjusting for interactions with third, fourth, etc. uncontrolled variables that influence the pair of variables of interest. This case is discussed in detail under time series analysis.

Common transformations are presented in Table 5.1.

Exploratory Data Analysis

Over the past 20 years, an entirely new approach has been developed to get the most information out of the increasingly larger and more complex data sets that scientists are faced with. This approach involves the use of a very diverse set of fairly simple techniques which comprise exploratory data analysis (EDA). As expounded by Tukey (1977), there are four major ingredients to EDA:

Displays — These visually reveal the behavior of the data and suggest a framework for analysis. The scatterplot (presented earlier) is an example of this approach.

Residuals — These are what remain of a set of data after a fitted model (such as a linear regression) or some similar level of analysis has been removed.

Reexpressions — These involve questions of what scale would serve to best simplify and improve the analysis of the data. Simple transformations, such as those presented earlier in this chapter, are used to simplify data behavior (for example, linearizing or normalizing) and clarify analysis.

Resistance — This is a matter of decreasing the sensitivity of analysis and summary of data to misbehavior, so that the occurrence of a few outliers, for example, will not complicate or invalidate the methods used to analyze the data; in summarizing the location of a set of data, the median (but not the arithmetic mean) is high resistant.

These four ingredients are utilized in a process falling into two broad phases: an exploratory phase and a confirmatory phase. The exploratory phase isolates patterns in, and features of, the data and reveals them, allowing an inspection of the data before there is any firm choice of actual hypothesis testing or modeling methods has been made.

Confirmatory analysis allows evaluation of the reproducibility of the patterns or effects. Its role is close to that of classical hypothesis testing, but also often includes steps such as (a) incorporating information from an analysis of another, closely related set of data, (b) validating a result by assembling and analyzing additional data. These techniques are, in general, beyond the scope of this text. However, Velleman and Hoaglin (1981) and Hoaglin et al. (1982) present a clear overview of the more important methods, along with codes for their execution on a microcomputer (they have also now been incorporated into Minitab). A short examination of a single case of the use of these methods, however, is in order.

Toxicology has long recognized that no population — animal or human — is completely uniform in its response to any particular toxicant. Rather, a population is composed of a (presumably normal) distribution of individuals — some resistant to intoxication (hyporesponders), the bulk that respond close to a central value (such as an LD_{50}), and some that are very sensitive to intoxication (hyperresponders). This population distribution can, in fact, result in additional statistical techniques. The sensitivity of techniques such as ANOVA is reduced markedly by the occurrence of outliers (extreme high or low values — including hyper- and hyporesponders) which, in fact, serve to markedly inflate the variance (SD) associated with a sample.

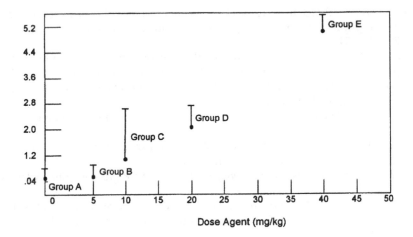

FIGURE 5.2
Variance inflation. (Points are means; error bars are +1 SD.)

Such variance inflation is particularly common in small groups that are exposed or dosed at just over or under a threshold level, causing a small number of individuals in the sample (who are more sensitive than the other members) to respond markedly. Such a situation is displayed in Figure 5.2, which plots the mean and SDs of methemoglobin levels in a series of groups of animals exposed to successively higher levels of a hemolytic agent.

Although the mean level of methemoglobin in group C is more than double that of the control group (A), no hypothesis test will show this difference to be significant because it has such a large SD associated with it. Yet this "inflated" variance exists because a single individual has such a marked response. The occurrence of the inflation is certainly an indicator that the data need to be examined closely. Indeed, all tabular data in toxicology should be visually inspected for both trend and variance inflation.

A concept related (but not identical) to resistance and exploratory data analysis is that of robustness. Robustness generally implies insensitivity to departures from assumptions surrounding an underlying model, such as normality.

In summarizing the location of data, the median, although highly resistant, is not extremely robust. But the mean is both nonresistant and nonrobust.

References

Anscombe, F.J. (1973) Graphics in statistical analysis. *American Statistician*, 27, 17–21.

Chambers, J.M., Cleveland, W.S., Kleiner, B., and Tukey, P.A. (1983) *Graphical Methods for Data Analysis*. Wadsworth, Belmont, CA.

Cleveland, W.S. (1985) *The Elements of Graphing Data*. Wadsworth Advanced Books, Monterey, CA.

Cleveland, W.S. and McGill, R. (1984) Graphical perception: theory, experimentation, and application to the development of graphical methods. *J. Am. Stat. Assoc.*, 79, 531–554.

Conover, J.W. and Inman, R.L. (1981) Rank transformations as a bridge between parametric and nonparametric statistics. *American Statistician*, 35, 124–129.

Engelman, L. and Hartigan, J.A. (1969) Percentage points of a test for clusters. *J. Am. Stat. Assoc.*, 64, 1647–1648.

Everitt, B.S. and Hand, D.J. (1981) *Finite Mixture Distributions*. Chapman and Hall, New York.

Hoaglin, D.C., Mosteller, F., and Tukey, J.W. (1982) *Understanding Robust and Explanatory Data Analysis*. John Wiley, New York.

Hosmer, D.W. and Lemershow, S. (1989) *Applied Logistic Regression*. John Wiley, New York.

Mendell, N.R., Finch, S.J., and Thode, H.C., Jr. (1993) Where is the likelihood ratio test powerful for detecting two component normal mixtures? *Biometrics*, 49, 907–915.

Snedecor, G.W. and Cochran, W.G. (1989) *Statistical Methods*, 8th ed., Iowa State University Press, Ames, Iowa, 1989.

Sokal, R.R. and Rohlf, F.J. (1994) *Biometry*, 3rd ed., W.H. Freeman, San Francisco.

Tufte, E.R. (1983) *The Visual Display of Quantitative Information*. Graphics Press, Cheshire, CT.

Tufte, E.R. (1990) *Envisioning Information*. Graphics Press, Cheshire, CT.

Tufte, E.R. (1997) *Visual Explanations*. Graphics Press, Cheshire, CT.

Tukey, J.W., (1977) *Exploratory Data Analysis*. Addison-Wesley, Reading, MA.

Velleman, P.F. and Hoaglin, D.C. (1981) *Applications, Basics and Computing of Exploratory Data Analysis*. Duxbury Press, Boston.

6

Nonparametric Hypothesis Testing of Categorical and Ranked Data

The test methods presented in this chapter are designed to maximize the analysis of low-information-value data while also maintaining acceptable levels of resistance. In general, the assumptions necessary for the use of these methods are less rigorous than those underlying the methods discussed in Chapter 7.

Categorical (or contingency table) presentations of data can contain any single type of data, but generally the contents are collected and arranged so that they can be classified as belonging to treatment and control groups, with the members of each of these groups then classified as belonging to one of two or more response categories (such as tumor/no tumor or normal/hyperplastic/neoplastic) (Agresti, 1996). For these cases, two forms of analysis are presented — Fisher's exact test (for the 2 × 2 contingency table) and the R × C chi-square test (for large tables). It should be noted, however, that there are versions of both of these tests that permit the analysis of any size of contingency table.

The analysis of rank data — what is generally called nonparametric statistical analysis — is an exact parallel of the more traditional (and familiar) parametric methods. There are methods for the single comparison case (just as Student's t-test is used) and for the multiple comparison case (just as analysis of variance is used) with appropriate *post hoc* tests for exact identification of the significance with a set of groups. Four tests are presented for evaluating statistical significance in rank data — the Wilcoxon rank sum test, distribution-free multiple comparisons, Mann–Whitney U test, and the Kruskall–Wallis nonparametric analysis of variance. For each of these tests, as for those in the next chapter, tables of distribution values for the evaluations of results are presented in Appendix 1.

It should be clearly understood that for data that do not fulfill the necessary assumptions for parametric analysis, these nonparametric methods are either as powerful or, in fact, more powerful than the equivalent parametric test.

Fisher's Exact Test

Fisher's exact test should be used to compare two sets of discontinuous, quantal (all or none) data. Small sets of such data can be checked by contingency data tables, such as those of Finney et al. (1963). Larger sets, however, require computation. These include frequency data such as incidences of mortality or certain histopathological findings, etc. Thus, the data can be expressed as ratios. These data do not fit on a continuous scale of measurement but usually involve numbers of responses classified as either negative or positive; that is, contingency table situation (Sokal and Rohlf, 1994, pp. 738–743).

The analysis is started by setting up a 2×2 contingency table to summarize the numbers of "positive" and "negative" responses as well as the totals of these, as follows:

	Positive	Negative	Total
Group I	A	B	A + B
Group II	C	D	C + D
Totals	A + C	B + D	$A + B + C + D = N_{total}$

Using the above set of symbols, the formula for P appears as follows:

$$P = \frac{(A+B)!(C+D)!(A+C)!(B+D)!}{N!A!B!C!D!}$$

(Note: A! is A factorial. For 4! — as an example, this would be $(4)\,(3)\,(2)\,(1) = 24$.)

The exact test produces a probability (P), which is the sum of the above calculation repeated for each possible arrangement of the numbers in the above cells (that is, A, B, C, and D) showing an association equal to or stronger than that between the two variables.

The P resulting from these computations will be the exact one- or two-tailed probability depending on which of these two approaches is being employed. This value tells us if the groups differ significantly (with a probability less than 0.05, say) and the degree of significance. This is demonstrated in Example 6.1.

Example 6.1

The pathology reports from 35 control and 20 treated rats show that 2 control and 5 treated animals have tumors of the spleen. Setting this up as a contingency table we see:

	Tumor-Bearing	No Tumors	Total
Control	2	33	35
Treated	5	15	20
Totals	7	48	55

The probability for this case calculates as:

$$\text{prob}_1 \frac{(35)!(20)!(7)!(48)!}{(55)!(2)!(5)!(33)!(15)!} = 0.046$$

In order to find the lesser probabilities, consider all tables whose interior numbers will add up to the same column and row totals but be of lesser probability (column and row headers are dropped for space):

0	35	35	One of the two most extreme possibilities.
7	13	20	Fisher's probability = 0.00038
7	48	55	

1	34	35	Fisher's probability = 0.00669
6	14	20	
7	48	55	

2	33	35	Results obtained. Fisher's probability = 0.04546
5	15	20	
7	48	55	

(More probable results are left out.)

7	28	35	Other extreme possibility. Fisher's
0	20	20	probability = 0.03314
7	48	55	

The exact one-tailed p-value is the sum of the p-value for the observed frequencies plus all of the less likely values in the same direction, so $p_{1\text{-tailed}} = 0.04546 + 0.00669 + 0.00038 = 0.05252$. Since this is greater than 0.05, we would not reject the one-tailed hypothesis of equal proportions at a significance level of 0.05. (This is close to 0.05, however, and may give the researcher encouragement to conduct a larger study). The two-tailed p-value is found by also adding the individual probabilities in the other direction that are less likely than the observed probability. In this case, there is only one, so $p_{2\text{-tailed}} = p_{1\text{-tailed}} + 0.03314 = 0.08566$.

SAS Analysis of Example 6.1

Fisher's exact test can be carried out with SAS (2000) using EXACT option with PROC FREQ, as shown in the sample code that follows. This example uses the WEIGHT statement since the summary results are input. If the data set containing the individual patient observations were used, the WEIGHT statement would be omitted.

Fisher's exact probability is printed under "Prob" on the output with a one-tailed value of 0.053. Note that SAS prints out a warning against use of the chi-test when cell sizes are too small.

SAS Code for Example 6.1

```
DATA TUMOR
INPUT GRP TRT $ CNT @@;
CARDS;
1 YES 2 1 _NO 33 2 YES 5 2 _NO 15;
RUN;
PROC FREQ DATA = TUMOR;
     TABLES TRT*GRP/EXACT NOPERCENT NOROW;
     WEIGHT CNT;
TITLE1 'Fisher's Exact Test';
TITLE2 'Example 6.1: Tumor incidence in rats after TRT;
RUN;
```

SAS Output for Example 6.1

```
Fisher's Exact Test
Example 6.1: Tumor incidence in rats after TRT
TABLE OF RESP BY GRP
```

RES Frequency Col. Pct.	GRP		Total
	1	2	
Yes	2	5	7
	5.71	25.00	7
No	33	15	48
	94.29	75.00	48
Total	35	20	55

STATISTICS FOR TABLE OF RESP BY GRP

Statistic	DF	Value	Prob
Chi-square	1	4.262	0.0439
Likelihood ratio chi-square	1	4.103	0.043
Continuity adj. chi-square	1	2.702	0.100
Mantel–Haenszel chi-square	1	4.184	0.041
Fisher's Exact Test			0.053
(Left)			
(Right)			0.993
(Two-Tail)			0.086
Phi coefficient		−0.278	
Contingency coefficient		0.268	
Cramer's V		−0.278	
Sample size = 70			

```
WARNING: 50% of the cells have expected counts less than
5. Chi-square may not be a valid test.
```

Assumptions and Limitations

Tables are available that provide individual exact probabilities for small sample size contingency tables. (See Zar, 1974, pp. 518–542.)

Fisher's exact test must be used in preference to the chi-square test when there are small cell sizes.

Ghent has developed and proposed a good (although, if performed by hand, laborious) method extending the calculation of exact probabilities to 2×3, 3×3, and $R \times C$ contingency tables (Ghent, 1972).

Fisher's probabilities are not necessarily symmetric. Although some analysts will double the one-tailed p-value to obtain the two-tailed result, this method is usually overly conservative.

2 × 2 Chi-Square

Although Fisher's exact test is preferable for analysis of most 2×2 contingency tables in toxicology, the chi-square test is still widely used and is preferable in a few unusual situations (particularly if cell sizes are large yet only limited computational support is available).

The general chi-square formula is:

$$X^2 = \sum_{i=1}^{n_1} \sum_{j=1}^{n_2} \frac{(O_{ij} - E_{ij})^2}{E_{ij}}$$

which, when simplified for the 2×2 case and expanded is:

$$= \frac{(O_{1A} - E_{1A})^2}{E_{1A}} + \frac{(O_{2A} - E_{2A})^2}{E_{2A}} + \frac{(O_{1B} - E_{1B})^2}{E_{1B}} + \frac{(O_{2B} - E_{2B})^2}{E_{2B}}$$

where *Os* are observed numbers (or counts) and *Es* are expected numbers. For cases where there are only two columns, i.e., $2 \times N$ cases, and $N < 50$, the Yates formula for correction is sometimes used:

$$X^2 = \sum_{i=1}^{2} \sum_{j=1}^{n} \frac{(|O_{ij} - E_{ij}| - 1/2)^2}{E_{ij}}$$

The common practice in toxicology is for the observed figures to be test or treatment group counts. The expected figure is calculated as:

$$E = \frac{(\text{column total})(\text{row total})}{\text{grand total}}$$

for each box or cell in a contingency table. Example 6.2 illustrates this.

Example 6.2

In a subacute toxicity study, there were 25 animals in the control group and 25 animals in the treatment group. Seven days after dosing, 5 control and 12 treatment animals were observed to be exhibiting fine muscle tremors. All other animals had no tremors. Do significantly more treatment animals have tremors?

	Tremors		No Tremors		
	Observed	(Expected)	Observed	(Expected)	Σ
Control	5	(8.5)	20	(16.5)	25
Treated	12	(8.5)	13	(16.5)	25
Σ	17		33		50

$$X^2 = \frac{(5-8.5)^2}{8.5} + \frac{(12-8.5)^2}{8.5} + \frac{(20-16.5)^2}{16.5} + \frac{(13-16.5)^2}{16.5}$$

$$= \frac{12.25}{8.5} + \frac{12.25}{8.5} + \frac{12.25}{16.5} + \frac{12.25}{16.5}$$

$$= 1.441 + 1.441 + 0.742 + 0.742$$

$$= 4.366$$

Our degrees of freedom are $(R-1)(C-1) = (2-1)(2-1) = 1$. Looking at a chi-square table (such as in Table C of Appendix 1) for one degree of freedom we see that this is greater than the test statistic at 0.05 (3.84) but less than that at 0.01 (6.64) so that $0.05 > p > 0.01$.

SAS Code for Example 6.2

```
DATA TREM
INPUT GRP RESP $ CNT @@;
CARDS;
1 YES 5 1 _NO 20
2 YES 12 2 _NO 13
;RUN;
```

```
*GRP 1 = Control, GRP 2 = Treated;
PROC FREQ DATA = TREM;
     TABLES RESP*GRP/CHISQ EXACT NOPERCENT NOROW;
WEIGHT CNT;
TITLE1 'The Chi-Square Test';
TITLE2 'Example 6.2; Tremors with treatment
RUN;
```

Assumptions and Limitations

ASSUMPTIONS:	Data are univariate and categorical
	Data are from a multinomial population
	Data are collected by random, independent sampling
	Groups being compared are of approximately same size, particularly for small group sizes
WHEN TO USE:	When the data are of a categorical (or frequency) nature
	When the data fit the assumptions above
	To test goodness-to-fit to a known form of distribution
	When cell sizes are large
WHEN NOT TO USE:	When the data are continuous rather than categorical
	When sample sizes are small and very unequal
	When sample sizes are too small (for example, when total N is less than 50 of if any expected value is less than 5)

R × C Chi-Square

The R × C chi-square test can be used to analyze discontinuous (frequency) data as in the Fisher's exact or 2 × 2 chi-square tests. However, in the R × C test (R = row, C = column) we wish to compare three or more sets of data. An example would be comparison of the incidence of tumors among mice on three or more oral dosage levels. We can consider the data as "positive" (tumors) or "negative" (no tumors). The expected frequency for any box is equal to: (row total)(column total)/(N_{total}).

As in the Fisher's exact test, the initial step is setting up a table (this time an R × C contingency table). This table would appear as follows:

	"Positive"	"Negative"	Total
Group I	A_1	B_1	$A_1 + B_1 = N_1$
Group II	A_2	B_2	$A_2 + B_2 = N_2$
Group R	A_R	B_R	$A_R + B_R = N_R$
Totals	N_A	N_B	N_{total}

Using these symbols, the formula for chi-square (X^2) is:

$$X^2 = \frac{N_{total}^2}{N_A N_B \dots N_K} \left(\frac{A_1^2}{N_1} + \frac{A_2^2}{N_2} + \dots + \frac{A_K^2}{N_K} - \frac{N_A^2}{N_{total}} \right)$$

The resulting X^2 value is compared to table values (as in Snedecor and Cochran, 1980, pp. 470–471 or Table C) according to the number of degrees of freedom, which is equal to $(R - 1)(C - 1)$. If X^2 is smaller than the table value at the 0.05 probability level, the groups are not significantly different. If the calculated X^2 is larger, there is some difference among the groups and $2 \times R$ chi-square or Fisher's exact tests will have to be compared to determine which group(s) differ from which other group(s). Example 6.3 demonstrates this.

Example 6.3

The R × C chi square can be used to analyze tumor incidence data gathered during a mouse feeding study as follows:

Dosage (mg/kg)	No. of Mice with Tumors	No. of Mice without Tumors	Total No. of Mice
2.00	19	16	35
1.00	13	24	37
0.50	17	20	37
0.25	22	12	34
0.00	20	23	43
Totals	91	95	186

$$X^2 = \frac{(186)^2}{(91)(95)} \left(\frac{19^2}{35} + \frac{13^2}{37} + \frac{17^2}{37} + \frac{22^2}{34} + \frac{20^2}{43} - \frac{91^2}{186} \right)$$

$$= (4.00)(1.71)$$

$$= 6.84$$

The smallest expected frequency would be $(91)(34)/186 = 16.6$; well above 5.0. The number of degrees of freedom is $(5 - 1)(2 - 1) = 4$. The chi-square table value for 4 degrees of freedom is 9.49 at the 0.05 probability level. Therefore, there is no significant association between tumor incidence and dose or concentration.

Assumptions and Limitations

Data is organized in a table (as shown below) with the independent variable placed vertically and the dependent variable extending horizontally. In this illustration, A, B, C, and D are the individual data in the "cells." The data must be raw; i.e., not converted to percentages or otherwise transformed.

		Columns (Dependent Variable)		
		No Effect	Effect	Totals
Rows (independent variable)	Control	A	B	A + B
	Treated	C	D	C + D
	Totals	A + C	B + D	A + B + C + D

The following minimum frequency thresholds should be obeyed:

For a 1×2 or 2×2 table, expected frequencies in each cell should be at least 5.

For a 2×3 table, expected frequencies should be at least 2.

For a 2×4 or 3×3 or larger table, if all expected frequencies but one are at least 5 and if the one small cell is at least 1, chi-square is still a good approximation.

The chi-square test as shown is always one-tailed.

Without the use of some form of correction, the test becomes less accurate as the differences between group size increases.

The results from each additional column (group) are approximately additive. Due to this characteristic, chi-square can be readily used for evaluating any $R \times C$ combination.

The results of the chi-square calculation must be a positive number.

Test is weak with either small sample sizes or when the expected frequency in any cell is less than the minimum frequency thresholds. This latter limitation can be overcome by "pooling" — combining cells. For large sample sizes, dropping data compromises the test.

Test results are independent of order of cells, unlike Kolmogorov–Smirnov results.

Can be used to test the probability of validity of any distribution.

Related tests include Fisher's exact test, G of likelihood ratio, and Kolmogorov–Smirnov.

Wilcoxon Rank Sum Test

The Wilcoxon rank sum test is commonly used for the comparison of two groups of nonparametric (interval or not normally distributed) data, such as those which are not measured exactly but rather as falling within certain limits (for example, how many animals died during each hour of an acute study).

The test is also used when there is no variability (variance = 0) within one or more of the groups we wish to compare (Sokal and Rohlf, 1994, pp. 432–437).

The data in both groups being compared are initially arranged and listed in order of increasing value. Then each number in the two groups must receive a rank value. Beginning with the smallest number in either group (which is given a rank of 1.0), each number is assigned a rank. If there are duplicate numbers (ties), then each value of equal size will receive the median rank for the entire identically sized group. Therefore, if the lowest number appears twice, both figures receive a rank of 1.5. This, in turn, means that the ranks of 1.0 and 2.0 have been used and that the next highest number has a rank of 3.0. If the lowest number appears three times, then each is ranked as 2.0 and the next number has a rank of 4.0. Thus, each tied number gets a "median" rank. This process continues until all the numbers are ranked. Each of the two columns of ranks (one for each group) is totaled giving the "sum of ranks" for each group being compared. As a check, we can calculate the value $N(N + 1)/2$, where N is the total number of data in both groups. The result should be equal to the sum of the sum of ranks for both groups.

The sum of rank values are compared to table values (such as Beyer, 1982, pp. 409–413, or Table J in Appendix 1) to determine the degree of significant differences, if any. These tables include two limits (an upper and a lower) that are dependent upon the probability level. If the number of data is not the same in both groups ($N_1 \neq N_2$), then the lesser sum of ranks (smaller N) is compared to the table limits to find the degree of significance. Normally the comparison of the two groups ends here and the degree of significant difference can be reported.

The values in Table J are calculated assuming normal distribution, so the Z statistic can also be used to determine significance:

$$Z = \frac{|R_1 - \mu_{R_1}| - 0.5}{\sigma_{R_1}}$$

where

$$\mu_{R_1} = \frac{N_1(N_1 + N_2 + 1)}{2}$$

and

$$\sigma_{R_1} = \sqrt{\frac{N_1 N_2 (N_1 + N_2 + 1)}{12}}$$

The factor of 0.5 in the formula is a continuity correction employed for smaller sample sizes. As the sample size increases, the distribution approaches normal and the correction is not needed, but there is some debate as to what level it should be applied. Note also that once the ranks are calculated, R_1 and R_2 are dependent on each other, so the Z statistic can be calculated using either one.

This calculation is demonstrated in Example 6.4.

Example 6.4

If we recorded the approximate times of death (in hours) of rats dosed with 5.0 g/kg (Group B) or 2.5 g/kg (Group A) of a given material, we might obtain the following results:

Hours to Death (Group A)		Hours to Death (Group B)	
4	3	7	4
6	6	3	3
7	1	6	1
7	7	7	2
2	7	2	5
5	4	5	4
7	6		
7	3		

So $N_1 = 16$, $N_2 = 12$, and $N = 28$. The ranked values of the responses are as shown in parenthesis below.

Hours to Death (Group A)				Hours to Death (Group B)			
4	(11.5)	3	(7.5)	7	(24.5)	4	(11.5)
6	(18.5)	6	(18.5)	3	(7.5)	3	(7.5)
7	(24.5)	1	(1.5)	6	(18.5)	1	(1.5)
7	(24.5)	7	(24.5)	7	(24.5)	2	(4.0)
2	(4.0)	7	(24.5)	2	(4.0)	5	(15.0)
5	(15.0)	4	(11.5)	5	(15.0)	4	(11.5)
7	(24.5)	6	(18.5)				
7	(24.5)	3	(7.5)				
Rank Sums		$R_1 = (261)$				$R_2 = (145)$	

As a check, $R_1 + R_2 = 406$ and $(N)(N + 1)/2 = (28)(29)/(2) = 406$.

The test statistic, based on a significance level of $\mu = 0.05$ in a normal approximation, becomes:

Null hypothesis (H_0):	$\theta_1 = \theta_2$
Alternative hypothesis (H_A):	$\theta_1 \neq \theta_2$
Test statistic:	$Z = \dfrac{\mid 261 - 232 \mid - 0.5}{\sqrt{448.38}} = 1.346$

Decision rule:	Reject H_0 if $\lvert Z \rvert > 1.96$
Conclusion:	Since 1.346 is not > 1.96, we do not reject H_0, concluding that there is insufficient evidence of a difference between the two doses in terms of time of death.

SAS Analysis of Example 6.4

Interpolation of Appendix H results in a two-tailed p-value of 0.178 for the Z statistic of 1.346. (The p-value can also be found from the SAS function PROBNORM).

The Wilcoxon rank sum test is performed in SAS using the NPAR1WAY procedure with the option WILCOXON, as shown in the SAS code that follows. The sum of ranks for either group can be used to compute the test statistic. Our manual calculations use $R_1 = 145$. When the smaller is used, a negative Z-value will result. In either case, for a two-tailed test, $\lvert Z \rvert = 1.346$ with a p-value of 0.178.

The output of this procedure also gives the results of the analysis using the Kruskal–Wallis chi-square approximation, which is discussed in the next chapter.

*SAS Code for Example 6.4

```
DATA RNKSM;
INPUT TRT $ PAT SCORE @@;
CARDS;
    A       4           A       6           A       7
    A       7           A       2           A       5
    A       7           A       7           A       3
    A       6           A       1           A       7
    A       7           A       4           A       6
    A       3           B       7           B       3
    B       6           B       7           B       2
    B       5           B       4           B       3
    B       1           B       2           B       5
    B       4
; RUN;
PROC NPAR1WAY WILCOXON DATA = RNKSM;
CLASS GRP; VAR SCORE;
TITLE1 'The Wilcoxon Rank-Sum Test';
TITLE2 'Example 6.4: Time to Death';
RUN;
```

SAS Output for Example 6.4

```
The Wilcoxon Rank Sum Test
Example 11: Time to Death
NPAR1WAY PROCEDURE
Wilcoxon Scores (Rank Sums) for Variable SCORE
Classified by Variable GRP
```

GRP	N	Sum of Scores	Expected Under H_0	Standard Deviation Under H_0	Mean Score
A	16	261.0	232.0	21.1750077	16.3125000
B	12	145.0	174.0	21.1750077	12.0833333

Average scores were used for ties Wilcoxon two-sample test.

(Normal approximation) (with continuity correction of 0.5).

(1) $S = 145.000 Z = -1.34593$ Prob $> |Z| = 0.1783$

(2) t-Test approx. significance $= 0.1895$

(3) Kruskal–Wallis test (chi-square approximation)

```
CHISQ = 1.8756 DF = 1 Prob > CHISQ = 0.1708
TABLE OF RESP BY GRP
```

RES Frequency Col. Pct.	GRP 1	GRP 2	Total
Yes	2	5	7
	5.71	25.00	
No	33	15	48
	94.29	75.00	
Total	35	20	55

Assumptions and Limitations

This is a two-sample test only and operates the same as the Mann–Whitney U test.

The model assumes no set distribution or form of data, but is in fact most appropriate for discontinuous or rank-type data.

The occurrence of too many ties (that is, of ties in 10% or more of the cases in the sample) causes this test to overestimate alpha (that is, to have an inflated false-positive rate). The test should not be used in such cases.

Distribution-Free Multiple Comparison

The distribution-free multiple comparison test should be used to compare three or more groups of nonparametric data. These groups are then analyzed two at a time for any significant differences (Hollander and Wolfe, 1973, pp. 124–129). The test can be used for data similar to those compared by the rank sum test. We often employ this test for reproduction and mutagenicity studies (such as comparing survival rates of offspring of rats fed various amounts of test materials in the diet).

As shown in Example 6.5, two values must be calculated for each pair of groups; the difference in mean ranks, and the probability level value against which the difference will be compared. To determine the difference in mean ranks we must first arrange the data within each of the groups in order of increasing values. Then we must assign rank values, beginning with the smallest overall figure. Note that this ranking is similar to that in the Wilcoxon test except that it applies to more than two groups.

The ranks are then added for each of the groups. As a check, the sum of these should equal $(N_{tot})(N_{tot} + 1)/2$, where N_{tot} is the total number of figures from all groups. Next, we can find the mean rank (R) for each group by dividing the sum of ranks by the numbers in the data (N) in the group. These mean ranks are then taken in those pairs that we want to compare (usually each test group vs. the control) and the differences are found $(|R_1 - R_2|)$. This value is expressed as an absolute figure; that is, it is always a positive number.

The second value for each pair of groups (the probability value) is calculated from the expression:

$$Z_{\alpha/[K(K-1)]}\sqrt{\frac{N_{tot}(N_{tot}+1)}{12}\left(\frac{1}{N_1}+\frac{1}{N_2}\right)}$$

where α is the level of significance for the comparison (usually 0.05, 0.01, 0.001, etc.), K is the total number of groups, and Z is a figure obtained from a normal probability table and determining the corresponding "Z-score" from Appendix 1, Table H.

The result of the probability value calculation for each pair of groups is compared to the corresponding mean difference $|R_1 - R_2|$. If $|R_1 - R_2|$ is smaller, there is no significant difference between the groups. If it is larger, the groups are different and $|R_1 - R_2|$ must be compared to the calculated probability values for $a = 0.01$ and $a = 0.001$ to find the degree of significance.

Example 6.5

Consider the following set of data (ranked in increasing order), which could represent the proportion of rats surviving given periods of time during diet inclusion of a test chemical at four dosage levels (survival index).

I		II		III		IV	
5.0 mg/kg		2.5 mg/kg		1.25 mg/kg		0.0 mg/kg	
% Value	Rank	% Value	Rank	% Value	Rank	% Value	Rank
40	2.0	40	2.0	50	5.5	60	9.0
40	2.0	50	5.5	50	5.5	60	9.0
50	5.5	80	12.0	60	9.0	80	12.0
100	17.5	80	12.0	100	17.5	90	14.0
		100	17.5	100	17.5	100	17.5
						100	17.5
Sum of Ranks	27.0		49.0		55.0		79.0
$N_I = 4$		$N_{II} = 5$		$N_{III} = 5$		$N_{IV} = 6$	
						$N_{tot} = 20$	

Note: Check sum of sums = 210; 20(21)/2 = 210

Mean ranks (R): $R_1 = 27.0/4 = 6.75$ $R_2 = 49.0/5 = 9.80$
                    $R_3 = 55.0/5 = 11.00$ $R_4 = 79.0/6 = 13.17$

Comparison Groups	$R_1 - R_2$	Probability Test Values
5.0 vs. 0.0	6.42	$= Z_{0.05/[4(3)]} \sqrt{\dfrac{20(21)}{12} \left(\dfrac{1}{4} + \dfrac{1}{6} \right)} = 2.637 \sqrt{14.589} = 10.07$
2.5 vs. 0.0	3.37	$= Z_{0.05/[4(3)]} \sqrt{\dfrac{20(21)}{12} \left(\dfrac{1}{5} + \dfrac{1}{6} \right)} = 2.637 \sqrt{12.833} = 9.45$
1.25 vs. 0.0	2.17	$= Z_{0.05/[4(3)]} \sqrt{\dfrac{20(21)}{12} \left(\dfrac{1}{5} + \dfrac{1}{6} \right)} = 2.637 \sqrt{12.833} = 9.45$

Since each of the $|R_1 - R_2|$ values is smaller than the corresponding probability calculation, the pairs of groups compared are not different at the 0.05 level of significance.

Assumptions and Limitations

As with the Wilcoxon rank sum, too many tied ranks inflate the false positive.
    Generally, this test should be used as a *post hoc* comparison after Kruskall–Wallis.

Mann–Whitney U Test

This is a nonparametric test in which the data in each group are first ordered from lowest to highest values, then the entire set (both control and treated values) is ranked, with the average rank being assigned to tied values. The ranks are then summed for each group and U is determined according to

$$U_t = n_c n_t + \frac{n_t(n+1)}{2} - R_t$$

$$U_c = n_c n_t + \frac{n_c(n_c+1)}{2} - R_c$$

where n_c, n_t = sample size for control and treated groups; and R_c, R_t = sum of ranks for the control and treated groups.

For the level of significance for a comparison of the two groups, the larger value of U_c or U_t is used. This is compared to critical values as found in tables (such as in Siegel, 1956, or in Appendix 1, Table E).

With the above discussion and methods in mind, we may now examine the actual variables that we encounter in teratology studies. These variables can be readily divided into two groups — measures of lethality and measures of teratogenic effect (Gaylor, 1978). Measures of lethality include: (a) corpora lutea per pregnant female, (b) implants per pregnant female, (c) live fetuses per pregnant female, (d) percentage of preimplantation loss per pregnant female, (e) percentage of resorptions per pregnant female, and (f) percentage of dead fetuses per pregnant female. Measures of teratogenic effect include: (a) percentage of abnormal fetuses per litter, (b) percentage of litters with abnormal fetuses, and (c) fetal weight gain. As demonstrated in Example 6.6, the Mann–Whitney U test is employed for the count data, but which test should be employed for the percentage variables should be decided on the same grounds as described later under reproduction studies in the applications chapter.

Example 6.6

In a two-week study, the levels of serum cholesterol in treatment and control animals are successfully measured and assigned ranks as below:

Treatment		Control	
Value	Rank	Value	Rank
10	1.0	19	4.0
18	3.0	28	13.0
26	10.5	29	14.5
31	16.0	26	10.5
15	2.0	35	19.0
24	8.0	23	7.0

	22	6.0	29	14.5
	33	17.0	34	18.0
	21	5.0	38	20.0
	25	9.0	27	12.0
Sum of ranks		77.5		132.5

The critical value for one-tailed $p \leq 0.05$ is $U \geq 73$. We then calculate:

$$U_t = (10)(10) + \frac{10(10+1)}{2} - 77.5 = 77.5$$

$$U_c = (10)(10) + \frac{10(10+1)}{2} - 132.5 = 2.5$$

As 77.5 is greater than 73, these groups are significantly different at the 0.05 level.

Assumptions and Limitations

It does not matter whether the observations are ranked from smallest to largest or vice versa.

This test should not be used for paired observations.

The test statistics from a Mann–Whitney are linearly related to those of Wilcoxon. The two tests will always yield the same result. The Mann–Whitney is presented here for historical completeness, as it has been much favored in reproductive and developmental toxicology studies. However, it should be noted that the authors do not include it in the decision tree for method selection (Chapter 2, Figures 2.1 to 2.4).

Kruskal–Wallis Nonparametric ANOVA

The Kruskal–Wallis nonparametric one-way analysis of variance should be the initial analysis performed when we have three or more groups of data that are by nature nonparametric (either not a normally distributed population, or of a discontinuous nature, or all the groups being analyzed are not from the same population) but not of a categorical (or quantal) nature. Commonly these will be either rank-type evaluation data (such as behavioral toxicity observation scores) or reproduction study data. The analysis is initiated (Pollard, 1977, pp. 170–173) by ranking all the observations from the combined groups to be analyzed. Ties are given the average rank of the tied values (that is, if two values which would tie for 12th rank — and therefore would be ranked 12th and 13th — both would be assigned the average rank of 12.5).

The sum of ranks of each group (r_1, r_2, ... r_k) is computed by adding all the rank values for each group. The test value H is then computed as:

$$H = \frac{12}{n(n+1)}\left[\sum_{i=1}^{k}\frac{r_i^2}{n_i}\right] - 3(n+1)$$

where n_i is the number of observations in group i and n is the total number of observations. The test statistic is then compared with a table of H values (such as in Appendix 1, Table D). If the calculated value of H is greater than the table value for the appropriate number of observations in each group, there is significant difference between the groups, but further testing (using the distribution-free multiple comparisons method) is necessary to determine where the difference lies (as demonstrated in Example 6.7).

Example 6.7

As part of a neurobehavioral toxicology study, righting reflex values (whole numbers ranging from 0 to 10) were determined for each of five rats in each of three groups. The values observed, and their ranks, are as follows:

Control Group Reflex Score	Rank	5 mg/kg Group Reflex Score	Rank	10 mg/kg Group Reflex Score	Rank
0	2.0	1	5.0	4	11.0
0	2.0	2	7.5	4	11.0
0	2.0	2	7.5	5	13.0
1	5.0	3	9.0	8	14.5
1	5.0	4	11.0	8	14.5
Sum of Ranks	16.0		40.0		64.0

From these the H value is calculated as:

$$H = \frac{12}{15(15+1)}\left[\frac{16^2}{5} + \frac{40^2}{5} + \frac{64^2}{5}\right] - 3(15+1)$$

$$= \frac{12}{240}\left[\frac{256 + 1600 + 4096}{5}\right] - 48$$

$$= 11.52$$

Consulting a table of values for H, we find that for the case where we have groups of observations each, the test values are 4.56 (for $p = 0.10$),

5.78 (for $p = 0.05$), and 7.98 (for $p = 0.01$). As our calculated H is greater than the $p = 0.01$ test value, we have determined that there is a significant difference between the groups at the level of $p < 0.01$ and would now have to continue to a multiple comparisons test to determine where the difference is.

SAS Analysis of Example 6.7

The Kruskal–Wallis test can be performed in SAS using the NPAR1WAY procedure with the option WILCOXON, as shown in the SAS code that follows. The output provides the rank sums for each group and the chi-square statistic, which corroborate the manual calculations demonstrated. The p-value is also printed out by SAS.

*SAS Code for Example 6.7

```
DATA REFL
INPUT DOSE $ RT REFLEX @@;
CARDS;
CNT 0          CNT 0          CNT 0          CNT 1          CNT 1
5.0 1          5.0 2          5.0 2          5.0 3          5.0 4
10.0 4         10.0 4         10.0 5         10.0 8         10.0 8
RUN;
PROC NPAR1WAY WILCOXON DATA = PSOR;
CLASS DOSE; VAR SCORE;
TITLE1 'The Kruskal-Wallis Test';
TITLE2 'Example 6.7.' Righting Reflex
RUN;
PROC RANK DATA = PSOR OUT = RNK;
VAR SCORE; RANKS RNKSCORE;
RUN;
PROC GLM DATA = RNK; CLASSES DOSE;
MODEL RNKSCORE = DOSE/SS3;
RUN;
```

Assumptions and Limitations

The test statistic H is used for both small and large samples.

When we find a significant difference, we do not know which groups are different. It is not correct to then perform a Mann–Whitney U test on all possible combinations; rather, a multiple comparison method must be used, such as the distribution-free multiple comparison.

Data must be independent for the test to be valid.

Too many tied ranks will decrease the power of this test and also lead to increased false-positive levels.

When $k = 2$, the Kruskal–Wallis chi-square value has 1 df. This test is identical to the normal approximation used for the Wilcoxon rank sum test. As noted in previous sections, a chi-square with 1 df. can be represented by the square of a standardized normal random variable. In the case of $k = 2$, the h-statistic is the square of the Wilcoxon rank sum Z-test (without the continuity correction).

The effect of adjusting for tied ranks is to slightly increase the value of the test statistic, h. Therefore, omission of this adjustment results in a more conservative test.

Log-Rank Test

The log-rank test is a statistical methodology for comparing the distribution of time until the occurrence of the event in independent groups. In toxicology, the most common event of interest is death or occurrence of a tumor, but it could also just as well be liver failure, neurotoxicity, or any other event that occurs only once in an individual. The elapsed time from initial treatment or observation until the event is the event time is often referred to as "survival time," even when the event is not "death."

The log-rank test provides a method for comparing "risk-adjusted" event rates, useful when test subjects in a study are subject to varying degrees of opportunity to experience the event. Such situations arise frequently in toxicology studies due to the finite duration of the study, early termination of the animal, or interruption of treatment before the event occurs.

Examples where use of the log-rank test might be appropriate include comparing survival times in carcinogenicity bioassay animals that are given a new treatment with those in the control group or comparing times to liver failure, several doses of a new NSAID where the animals are treated for 10 weeks or until cured, whichever comes first.

If every animal were followed until the event occurrence, the event times could be compared between two groups using the Wilcoxon rank sum test. However, some animals may die or complete the study before the event occurs. In such cases, the actual time of the event is unknown since the event does not occur while under study observation. The event times for these animals are based on the last known time of study observation and are called "censored" observations since they represent the lower bound of the true, unknown event times. The Wilcoxon rank sum test can be highly biased in the presence of the censored data.

The null hypothesis tested by the log-rank test is that of equal event time distributions among groups. Equality of the distributions of event times implies similar event rates among groups not only for the clinical trial as a whole, but also for any arbitrary time point during the trial. Rejection of the null hypothesis indicates that the event rates differ among groups at one or more time points during the study.

The idea behind the log-rank test for comparison of two life tables is simple: if there were no difference between the groups, the total deaths occurring at any time should split between the two groups at that time. So if the numbers at risk in the first and second groups in (say) the sixth month were 70 and 30, respectively, and 10 deaths occurred in that month we would expect:

$$10 \times \frac{70}{70+30} = 7$$

of these deaths to have occurred in the first group, and

$$10 \times \frac{30}{70+30} = 3$$

of these deaths to have occurred in the second group.

A similar calculation can be made at each time of death (in either group). By adding together for the first group the results of all such calculations, we obtain a single number, called the extent of exposure (E_1), which represents the "expected" number of deaths in that group if the two groups had the distribution of survival time. An extent of exposure (E_2) can be obtained for the second group in the same way. Let O_1 and O_2 denote the actual total numbers of deaths in the two groups. A useful arithmetic check is that the total number of deaths $O_1 + O_2$ must equal the sum $E_1 + E_2$ of the extents of exposure.

The discrepancy between the Os and Es can be measured by the quantity

$$x^2 = \frac{(|O_1 - E_1| - 1/2)^2}{E_1} + \frac{(|O_2 - E_2| - 1/2)^2}{E_2}$$

For rather obscure reasons, 2 is known as the log-rank statistic. An approximate significance test of the null hypothesis of identical distributions of survival time in the two groups is obtained by referring 2 to a chi-square distribution on 1 degree of freedom. The use of the adjustments $-\int$ in the equation above is a continuity correction for conservatism, which may be dropped when sample sizes are moderate or large.

Example 6.8

In a study of the effectiveness of a new monoclonal antibody to treat specific cancer, the times to reoccurrence of the cancer in treated animals in weeks were as follows:

Control Group			Treatment Group		
1	5	11	6	10	22
1	5	12	6	11	23
2	8	12	6	13	25
2	8	15	6	16	32
3	8	17	7	17	32
4	8	22	9	19	34
4	11	23	10	20	35

The table below presents the calculations for the log-rank test applied to these times. A chi-square value of 13.6 is significant at the $p < 0.001$ level.

Log-Rank Calculation for Tumor Data

Time	At Risk			Relapses			Extent of Exposure		
(t)	T	C	Total	T	C	Total	T	C	Total
1	21.0	21	42.0	0	2	2	1.0000	1.0000	2
2	21.0	19	40.0	0	2	2	1.0500	0.9500	2
3	21.0	17	38.0	0	1	1	0.5526	0.4474	1
4	21.0	16	37.0	0	2	2	1.1351	0.8649	2
5	21.0	14	35.0	0	2	2	1.2000	0.8000	2
6	20.5	12	32.5	3	0	3	1.8923	1.1077	3
7	17.0	12	29.0	1	0	1	0.5862	0.4138	1
8	16.0	12	28.0	0	4	4	2.2857	1.7143	4
10	14.5	8	22.5	1	0	1	0.6444	0.3556	1
11	12.5	8	20.5	0	2	2	1.2295	0.7705	2
12	12.0	6	18.0	0	2	2	1.3333	0.6667	2
13	12.0	4	16.0	1	0	1	0.7500	0.2500	1
15	11.0	4	15.0	0	1	1	0.7333	0.2667	1
16	11.0	3	14.0	1	0	1	0.7857	0.2143	1
17	9.5	3	12.5	0	1	1	0.7600	0.2400	1
22	7.0	2	9.0	1	1	2	1.5556	0.4444	2
23	6.0	1	7.0	1	1	2	1.7143	0.2857	2
Totals				9	21	30	19.2080	10.7920	30

Illustration: $T = 23$, $2(6/7) = 1.7143$, $2(1/7) = 0.2857$

Test of significance: $x^2 = \dfrac{(|9-19.2|-1/2)^2}{19.2} + \dfrac{(|21-10.8|-1/2)^2}{10.8} = 13.6$

Estimate of relative risk: $\hat{\theta} = \dfrac{9/19.2}{21/10.8} = 0.24$

SAS Analysis of Example 6.8

We use the SAS procedure LIFETEST to apply the log-rank test to the data set, as shown in the SAS code. The TIME statement following the PROC LIFETEST statement designates the event times (WKS). Censored times are designated by an indicator variable (CENS). In the example, SAS statements are used to generate CENS from the censored times which are input as negative values.

The output shows the log-rank test resulting in a chi-square value of 13.62, confirming our manual calculations. This is a significant result with a p-value of 0.001.

*SAS Code for Example 6.8

```
DATA MONOCLONAL
INPUT VAC $ WKS X @@;
        CENS = (WKS < 10);
        WKS = ABS (WKS);
        IF VAC = 'MONO' THEN TRT = 1;
        IF VAC = 'CONT' THEN TRT = 0;
CARDS;
```

CNT	1	MONO	6
CNT	1	MONO	6
CNT	2	MONO	6
CNT	2	MONO	6
CNT	3	MONO	7
CNT	4	MONO	9
CNT	4	MONO	10
CNT	5	MONO	10
CNT	5	MONO	11
CNT	8	MONO	13
CNT	8	MONO	16
CNT	8	MONO	17
CNT	8	MONO	19
CNT	11	MONO	20
CNT	11	MONO	22
CNT	12	MONO	23
CNT	12	MONO	25
CNT	15	MONO	32
CNT	17	MONO	32

```
CNT                    22              MONO                    34
CNT                    23              MONO                    35;
RUN;
PROC SORT DATA = MONOCLONAL; BY VAC PAT; RUN;
PROC PRINT DATA = MONO;
VAR VAC PAT WKS CENS X;
TITLE1 'The Log-Rank Test';
TITLE2 'Example': Time to tumor recurrence with
monoclonal
RUN;
PROC LIFETEST PLOT=(S) DATA = MONO;
TIME WKS*CENS (1)
STRATA VAC
RUN;
```

The log-rank test as presented by Peto et al. (1977) uses the product-limit life-table calculations rather than the actuarial estimators shown above. The distinction is unlikely to be of practical importance unless the grouping intervals are very coarse.

Peto and Pike (1973) suggest that the approximation in treating the null distribution of χ^2 as a chi-square is conservative, so that it will tend to understate the degree of statistical significance. In the formula for χ^2 we have used the continuity correction of subtracting $H = \frac{1}{2}$ from $|O_1 - E_1|$ and $|O_2 - E_2|$ before squaring. This is recommended by Peto et al. (1977) when, as in non-randomized studies, the permutational argument does not apply. Peto et al. (1977) give further details of the log-rank test and its extension to comparisons of more than two treatment groups and to tests that control for categorical confounding factors.

Assumptions and Limitations

The endpoint of concern is, or is defined, so that it is "right censored" — once it happens, it does not recur. Examples are death or a minimum or maximum value of an enzyme or physiologic function (such as respiration rate).

The method makes no assumptions on distribution.

Many variations of the log-rank test for comparing survival distributions exist. The most common variant is the same as the one presented previously but without the continuity correction:

$$x^2 = \frac{(O_1 - E_1)^2}{E_1} + \frac{(O_2 - E_2)^2}{E_2}$$

where O_i and E_i are computed for each group, as in the formulas given previously. This statistic also has an approximate chi-square distribution with 1 degree of freedom under H_0.

The Wilcoxon rank sum test could be used to analyze the event times in the absence of censoring. A "generalized Wilcoxon" test, sometimes called the Gehan test, based on an approximate chi-square distribution, has been developed for use in the presence of censored observations.

Both the log-rank and the generalized Wilcoxon tests are nonparametric tests and require no assumptions regarding the distribution of event times. When the event rate is greater earlier in the trial rather than toward the end, the generalized Wilcoxon test is the more appropriate test since it gives greater weight to the earlier differences.

Survival and failure times often follow the exponential distribution. If such a model can be assumed, a more powerful alternative to the log-rank test is the likelihood ratio test.

This parametric test assumes that event probabilities are constant over time. That is, the chance that a patient becomes event positive at time t given that he is event negative up to time t does not depend on t. A plot of the negative log of the event vs. distribution showing a linear trend through the origin is consistent with exponential event times.

Life tables can be constructed to provide estimates of the event time distributions. Estimates commonly used are known as the Kaplan–Meier estimates.

References

Agresti, A. (1996) *An Introduction to Categorical Data Analysis*. John Wiley & Sons, New York.

Beyer, W.H. (1982) *CRC Handbook of Tables for Probability and Statistics*. CRC Press, Boca Raton, FL.

Finney, D.J., Latscha, R., Bennet, B.M., and Hsu, P. (1963) *Tables for Testing Significance in a 2 × 2 Contingency Table*. Cambridge University Press, Cambridge.

Gaylor, D.W. (1978) Methods and concepts of biometrics applied to teratology, in *Handbook of Teratology*, Vol. 4, Wilson, J.G. and Fraser, F.C., Eds., Plenum Press, New York, pp. 429–444.

Ghent, A.W. (1972) A method for exact testing of 2 × 2, 2 × 3, 3 × 3 and other contingency tables, employing binomiate coefficients. *The American Midland Naturalist*, 88, 15–27.

Hollander, M. and Wolfe, D.A. (1973) *Nonparametric Statistical Methods*. John Wiley, New York, pp. 124–129.

Peto, R. and Pike, M.C. (1973) Conservatism of approximation $(0 - E)^2/E$ in the log rank test for survival data on tumour incidence data. *Biometrics*, 29, 579–584.

Peto, R., Pike, M.C., Armitage, P., Breslow, N.E., Cox, D.R., Howard, S.V., Kantel, N., McPherson, K., Peto, J., and Smith, P.G. (1977) Design and analysis of randomized clinical trials requiring prolonged observations of each patient, II. Analyses and examples. *Br. J. Canc.*, 35, 1–39.

Pollard, J.H. (1977) *Numerical and Statistical Techniques*. McGraw-Hill, New York.

SAS Institute. 2000 *SAS/STAT Users Guide*, Ver. 8, Cary, NC.

Siegel, S. (1956) *Nonparametric Statistics for the Behavioral Sciences*. McGraw-Hill, New York.

Snedecor, G.W. and Cochran, W.G. (1980) *Statistical Methods*, 7th ed., Iowa State University Press, Ames, Iowa.

Sokal, R.R. and Rohlf, F.J. (1994) *Biometry: The Principles and Practice of Statistics in Biological Research*, 3rd ed., W.H. Freeman, San Francisco.

Zar, J.H. (1974) *Biostatistical Analysis*. Prentice-Hall, Englewood Cliffs, NJ.

7

Hypothesis Testing: Univariate Parametric Tests

In this chapter we consider test methods for univariate case data from normally distributed populations. Such data generally have a higher information value associated with them but the traditional hypothesis testing techniques (which include all the methods described in this chapter) are generally neither resistant nor robust. All the data analyzed by these methods are also effectively continuous; that is, at least for practical purposes, the data may be represented by any number and each such data number has a measurable relationship to other data numbers.

Student's t-Test (Unpaired t-Test)

Pairs of groups of continuous, randomly distributed data are compared via this test. We can use this test to compare three or more groups of data, but they must be intercompared by examination of two groups taken at a time and are preferentially compared by analysis of variance (ANOVA). Usually this means comparison of a test group vs. a control group, although two test groups may be compared as well. To determine which of the three types of t-tests described in this chapter should be employed, the F-test is usually performed first. This will tell us if the variances of the data are approximately equal, which is a requirement for the use of the parametric methods. If the F-test indicates homogeneous variances and the numbers of data within the groups (N) are equal, then the Student's t-test is the appropriate procedure (Sokal and Rohlf, 1994, pp. 226–231). If the F is significant (the data are heterogeneous) and the two groups have equal numbers of data, the modified Student's t-test is applicable (Diem and Lentner, 1975).

The value of t for Student's t-test is calculated using the formula:

$$t = \frac{\bar{x}_1 - \bar{x}_2}{\sum D_1^2 + \sum D_2^2} \sqrt{\frac{N_1 N_2}{N_1 + N_2} (N_1 + N_2 - 2)}$$

where

$$\sum D^2 = \frac{(N\sum x^2) - (\sum x)^2}{N}$$

The value of t obtained from the above calculations is compared to the values in a t distribution table (such as in Appendix 1, Table F) according to the appropriate number of degrees of freedom (df). If the F-value is not significant (i.e., variances are homogeneous), the df $= N_1 + N_2 - 2$. If the F was significant and $N_1 = N_2$, then the df $= N - 1$. Although this case indicates a nonrandom distribution, the modified t-test is still valid. If the calculated value is larger than the table value at $p = 0.05$, it may then be compared to the appropriate other table values in order of decreasing probability to determine the degree of significance between the two groups. Example 7.1 demonstrates this methodology.

Example 7.1

Suppose we wish to compare two groups (a test and control group) of dog weights following inhalation of a vapor. First, we would test for homogeneity of variance using the F-test. Assuming that this test gave negative (homogeneous) results, we would perform the t-test as follows:

Dog #	Test Weight		Control Weight	
	X_1 (kg)	X_1^2	X_2 (kg)	X_2^2
1	8.3	68.89	8.4	70.56
2	8.8	77.44	10.2	104.04
3	9.3	86.49	9.6	92.16
4	9.3	86.49	9.4	88.36
Sums	$\sum X_1 = 35.7$	$\sum X_1^2 = 319.31$	$\sum X_2 = 37.6$	$\sum X_2^2 = 355.12$
Means	8.92		9.40	

The difference in means $= 9.40 - 8.92 = 0.48$

$$\sum D_1^2 = \frac{4(319.31) - (35.7)^2}{4} = \frac{2.75}{4} = 0.6875$$

$$\sum D_2^2 = \frac{4(355.12) - (37.6)^2}{4} = \frac{6.72}{4} = 1.6800$$

$$t = \frac{0.48}{\sqrt{0.6875 + 1.6800}} \sqrt{\frac{4(4)}{4+4}(4+4-2)} = 1.08$$

The table value for t at the 0.05 probability level for $(4 + 4 - 2)$, or 6 degrees of freedom, is 2.447. Therefore, the dog weights are not significantly different at $p = 0.05$.

SAS Analysis of Example 7.1

The SAS code and output for analyzing this data set are shown on the next page. The program first computes the body weights, then prints a listing of the data using PROC PRINT (1). The *t*-test is conducted using PROC MEANS, although PROC UNIVARIATE could also be used. Body weight is also included in the VAR statement to obtain the mean and standard deviation of the body weight data. The test is significant if the *p*-value is less than the significance level of the test. In this case, we accept the null hypothesis at a two-tailed significance level of 0.05, since the *p*-value is greater than 0.05.

* SAS Code for Example 7.1:

```
DATA DOG;
INPUT WT KG @@;
BMI = WT KG
LABEL Test = 1, Cont. = 2
CARDS;
1      Test 8.33          Cont.  8.4
2      Test 8.86          Cont. 10.2
3      Test 9.37          Cont.  9.6
4      Test 9.38          Cont.  9.4
; RUN;
PROC PRINT LABEL DATA = DOGS
VAR WT KG Test Cont.
FORMAT BMI BMI0 5.1;
TITLE1 'One-Sample t-Test';
TITLE2 'EXAMPLE 7.1: Body Weight
RUN;
PROC MEANS MEAN STD N T PRT DATA = DOGS
VAR TEST CONT.
RUN;
```

Assumptions and Limitations

The test assumes that the data are univariate, continuous, and normally distributed.

Data are collected by random sampling.

The test should be used when the assumptions in 1 and 2 are met and there are only two groups to be compared.

Do not use when the data are ranked, when the data are not approximately normally distributed, or when there are more than two groups to be compared. Do not use for paired observations.

This is the most commonly misused test method, except in those few cases where one is truly only comparing two groups of data and the group sizes are roughly equivalent. It is not valid for multiple comparisons (because of resulting additive errors) or where group sizes are very unequal.

Test is robust for moderate departures from normality and, when N_1 and N_2 are approximately equal, robust for moderate departures from homogeneity of variances.

The main difference between the Z-test and the *t*-test is that the Z statistic is based on a known standard deviation, σ, while the *t* statistic uses the sample standard deviation, *s*, as an estimate of σ. With the assumption of normally distributed data, the variance s^2 is more closely estimated by the sample variance s^2 as *n* gets large. It can be shown that the *t*-test is equivalent to the Z-test for infinite degrees of freedom. In practice, a "large" sample is usually considered $n \geq 30$.

Cochran *t*-Test

The Cochran test should be used to compare two groups of continuous data when the variances (as indicated by the *F*-test) are heterogeneous and the numbers of data within the groups are not equal ($N_1 \neq N_2$). This is the situation, for example, when the data, although expected to be randomly distributed, were found not to be (Cochran and Cox, 1992, pp. 100–102).

Two *t*-values are calculated for this test, the "observed" t (t_{obs}) and the "expected" t (t'). The observed *t* is obtained by:

$$t_{obs} = \frac{\overline{X}_1 - \overline{X}_2}{\sqrt{W_1 + W_2}}$$

where W = SEM² (standard error of the mean squared) = S^2/N and S^2 (variance) is calculated as:

$$s^2 = \frac{\sum X^2 - \frac{(\sum x)^2}{N}}{N-1}$$

The value for t' is obtained from:

$$t' = \frac{t_1 W_1 + t_2 W_2}{W_1 + W_2}$$

where t_1 and t_2 are values for the two groups taken from the t distribution table corresponding to $N - 1$ degrees of freedom (for each group) at the 0.05 probability level (or such level as one may select).

The calculated t_{obs} is compared to the calculated t' value (or values, if t' values were prepared for more than one probability level). If t_{obs} is smaller than a t', the groups are not considered to be significantly different at that probability level. This procedure is shown in Example 7.2.

Example 7.2

Using the red blood cell count comparison from the F-test on page 86 (with $N_1 = 5$, $N_2 = 4$), the following results were determined:

$$\bar{X}_1 = \frac{37.60}{5} = 7.52 \qquad W_1 = \frac{0.804}{5} = 0.1608$$

$$\bar{X}_2 = \frac{29.62}{4} = 7.40 \qquad W_2 = \frac{0.025}{4} = 0.0062$$

(Note that S^2 values of 0.804 and 0.025 are calculated in Example 6.7)

$$t_{obs} = \frac{7.52 - 7.40}{\sqrt{0.1608 + 0.0062}}$$

From the t distribution table we use $t_1 = 2.776$ (df = 4) and $t_2 = 3.182$ (df = 3) for the 0.05 level of significance; there is no statistical difference at $p = 0.05$ between the two groups.

Assumptions and Limitations

The test assumes that the data are univariate, continuous, normally distributed and that group sizes are unequal.

The test is robust for moderate departures from normality, and very robust for departures from equality of variances.

F-Test

This is a test of the homogeneity of variances between two groups of data (Sokal and Rohlf, 1994, pp. 187–188). It is used in two separate cases. The first is when Bartlett's indicates heterogeneity of variances among three or more groups (i.e., it is used to determine which pairs of groups are heterogeneous). Second, the F-test is the initial step in comparing two groups of continuous data that we would expect to be parametric (two groups not usually being compared using ANOVA), the results indicating whether the data are from the same population and whether subsequent parametric comparisons would be valid.

The F is calculated by dividing the larger variance (S_1^2) by the smaller one (S_2^2). S^2 is calculated as:

$$s^2 = \frac{\sum X^2 - \frac{(\sum x)^2}{N}}{N-1}$$

where N is the number of data in the group and X represents the individual values within the group. Frequently, S^2 values may be obtained from ANOVA calculations. Use of this is demonstrated in Example 7.3.

The calculated F-value is compared to the appropriate number in an F-value table (such as Appendix 1, Table G) for the appropriate degrees of freedom ($N - 1$) in the numerator (along the top of the table) and in the denominator (along the side of the table). If the calculated value is smaller, it is not significant and the variances are considered homogeneous (and the Student's t-test would be appropriate for further comparison). If the calculated F-value is greater, F is significant and the variances are heterogeneous (and the next test would be modified Student's t-test if $N_1 = N_2$ or the Cochran t-test if $N_1 \neq N_2$; see Figure 2.3 to review the decision tree).

Example 7.3

If we wished to compare the red blood cell counts (RBC) of dogs receiving test materials in their diets with the RBCs of control dogs we might obtain the following results:

Test Weight		Control Weight	
X_1 (kg)	X_1^2	X_2 (kg)	X_2^2
8.3	68.89	8.4	70.56
8.8	77.44	10.2	104.04
9.3	86.49	9.6	92.16
9.0	81.00	9.4	88.36
9.3	86.49		
$\sum X_1 = 44.7$	$\sum X_1^2 = 400.31$	$\sum X_2 = 37.6$	$\sum X_2^2 = 355.12$
Mean = 8.94		Mean = 9.40	

Test RBC		Control RBC	
X_1	X_1^2	X_2	X_2^2
8.23	67.73	7.22	52.13
8.59	73.79	7.55	57.00
7.51	56.40	7.53	56.70
6.60	46.56	7.32	53.58
6.67	44.49		
$\Sigma X_1 = 37.60$	$\Sigma X_1^2 = 285.97$	$\Sigma X_2 = 29.62$	$\Sigma X_2^2 = 219.41$

$$\text{Variance for } X_1 = S_1^2 = \frac{285.95 - \frac{(37.60)^2}{5}}{5-1} = 0.804$$

$$\text{Variance for } X_2 = S_1^2 = \frac{219.41 - \frac{(29.62)^2}{4}}{4-1} = 0.025$$

$$F = \frac{0.804}{0.025} = 32.16$$

From the table for F values, for 4 (numerator) vs. 3 (denominator) df, we read the limit of 9.12 (from Appendix 1, Table G) at the 0.05 level. As our calculated value is larger (and, therefore, significant), the variances are heterogeneous and the Cochran t-test would be appropriate for comparison of the two groups of data.

Assumptions and Limitations

This test could be considered as a two-group equivalent of the Bartlett's test. If the test statistic is close to 1.0, the results are (of course) not significant. The test assumes normality and independence of data.

Analysis of Variance (ANOVA)

ANOVA is used for comparison of three or more groups of continuous data when the variances are homogeneous and the data are independent and normally distributed.

A series of calculations are required for ANOVA, starting with the values within each group being added (X) and then these sums being added (X). Each figure within the groups is squared, and these squares are then summed (X^2) and these sums added (X^2). Next the "correction factor" (CF) can be calculated from the following formula:

$$CF = \frac{\left(\Sigma_1^k \Sigma_1^N X\right)^2}{N_1 + N_2 + \cdots + N_k}$$

where N is the number of values in each group and K is the number of groups. The total sum of squares (SS) is then determined as follows:

$$SS_{total} = \sum_{1}^{k} \sum_{1}^{N} X^2 - CF$$

In turn, the sum of squares between groups (bg) is found from:

$$SS_{bg} = \frac{(\sum X_1)^2}{N_1} + \frac{(\sum X_2)^2}{N_2} + \cdots + \frac{(\sum X_k)^2}{N_k} - CF$$

The sum of squares within groups (wg) is then the difference between the last two figures, or:

$$SS_{wg} = SS_{total} - SS_{bg}$$

Now, there are three types of degrees of freedom to determine. The first, total df, is the total number of data within all groups under analysis minus one $(N_1 + N_2 + \ldots N_k - 1)$. The second figure (the df between groups) is the number of groups minus one $(K - 1)$. The last figure, the df within groups or "error df," is the difference between the first two figures $(df_{total} - df_{bg})$.

The next set of calculations requires determination of the two mean squares (MS_{bg} and MS_{wg}). These are the respective SS values divided by the corresponding df figures ($MS = SS/df$). The final calculation is that of the F ratio. For this, the MS bg is divided by the MS wg ($F = MS_{bg}/MS_{wg}$).

A table of the results of these calculations (using data from Example 7.4 at the end of this section) would appear as follows:

	df	SS	MS	F
Bg	3	0.04075	0.01358	4.94
Wg	12	0.03305	0.00275	
Total	15	0.07380		

For interpretation, the F ratio value obtained in the ANOVA is compared to a table of F-values (Appendix 1, Table G). If $F \leq 1.0$, the results are not significant and comparison with the table values is not necessary. The df for the greater MS (MS_{bg}) are indicated along the top of the table. Read down the side of the table to the line corresponding to the df for the lesser MS (MS_{wg}). The figure shown at the desired significance level (traditionally 0.05) is compared to the calculated F-value. If the calculated number is smaller, there is no significant difference among the groups being compared. If the calculated value is larger, there is some difference but further (*post hoc*) testing will be required before we know which groups differ significantly.

Example 7.4

Suppose we want to compare four groups of dog kidney weights, expressed as percentage of body weights, following an inhalation study. Assuming homogeneity of variance (from Bartlett's test), we could complete the following calculations:

	400 ppm	200 ppm	100 ppm	0 ppm	
	0.43	0.49	0.34	0.34	
	0.52	0.48	0.40	0.32	
	0.43	0.40	0.42	0.33	
	0.55	0.34	0.40	0.39	
ΣX	1.93	1.71	1.56	1.38	$\Sigma\Sigma X = 6.58$

Next, the preceding figures are squared:

	400 ppm	200 ppm	100 ppm	0 ppm	
	0.1849	0.2401	0.1156	0.1156	
	0.2701	0.2304	0.1600	0.1024	
	0.1849	0.1600	0.1764	0.1089	
	0.3025	0.1156	0.1600	0.1521	
ΣX	0.9427	0.7461	0.6120	0.4790	$\Sigma\Sigma X = 2.7798$

$$CF = \frac{(6.82)^2}{4+4+4+4} = 2.7060$$

$$SS_{total} = 2.7798 - 2.7060 = 0.0738$$

$$SS_{bg} = \frac{(1.93)^2}{4} + \frac{(1.71)^2}{4} + \frac{(1.56)^2}{4} + \frac{(1.38)^2}{4} - 2.7060 = 0.04075$$

The total df = 4 + 4 + 4 + 4 − 1 = 15

The other df are: $df_{bg} = 4 - 1 = 3$; $df_{wg} = 15 - 3 = 12$

$$MS_{bg} = \frac{0.04075}{3} = 0.01358$$

$$MS_{wg} = \frac{0.03305}{12} = 0.00275$$

$$F = \frac{0.01358}{0.00275} = 4.94$$

Going to an F value we find that the 3 df_{bg} (greater MS) and 12 df_{wg} (lesser MS), the 0.05 value of F is 3.49. As our calculated value is greater, there is a difference among groups at the 0.05 probability level. To determine where the difference is, further comparisons by a *post hoc* test will be necessary.

SAS Analysis of Example 7.4

These data can be analyzed with the GLM procedure of SAS as shown in the following SAS code:

*SAS Code for Example 7.4

```
DATA KIDNEY
INPUT DOSEGRP $ WEIGHT @@;
CARDS;
400        0.43
400        0.52
400        0.43
400        0.55
200        0.49
200        0.48
200        0.40
200        0.34
100        0.34
100        0.40
100        0.42
100        0.40
  0        0.34
  0        0.32
  0        0.33
  0        0.39
;RUN;
PROC SORT DATA = KIDNEY; CLASS DOSEGRP;
MODEL WEIGHTS = DOSEGRP;
MEANS DOSEGRP/T DUNCAN;
RUN;
```

Assumptions and Limitations

What is presented here is the workhorse of toxicology — the one-way analysis of variance. Many other forms exist for more complicated experimental designs.

The test is robust for moderate departures from normality if the sample sizes are large enough. Unfortunately, this is rarely the case in toxicology.

ANOVA is robust for moderate departures from equality of variances (as determined by Bartlett's test) if the sample sizes are approximately equal.

It is not appropriate to use a *t*-test (or version of ANOVA for two groups at a time) to identify where significant differences are within the design group. A multiple-comparison *post hoc* method must be used.

Post Hoc Tests

There is a wide variety of *post hoc* tests available to analyze data after finding significant result in an ANOVA. Each of these tests has advantages and disadvantages, proponents and critics. Four of the tests are commonly used in toxicology and will be presented or previewed here. These are Dunnett's *t*-test and Williams' *t*-test (Snedecor and Cochran, 1989). Two other tests that are available in many of the mainframe statistical packages are Tukey's method and the Student–Newman– Keuls method (Zar, 1974, pp. 151–161).

If ANOVA reveals no significance it is not appropriate to proceed to perform a *post hoc* test in the hope of finding differences. To do so would only be another form of multiple comparisons, increasing the type I error rate beyond the desired level.

Duncan's Multiple Range Test

Duncan's multiple range test (Duncan, 1955) is used to compare groups of continuous and randomly distributed data (such as body weights, organ weights, etc.). The test normally involves three or more groups taken one pair at a time. It should only follow observation of a significant *F*-value in the ANOVA and can serve to determine which group (or groups) differs significantly from which other group (or groups).

There are two alternative methods of calculation. The selection of the proper one is based on whether the number of data (N) are equal or unequal in the groups.

Groups with Equal Number of Data ($N_1 = N_2$)

Two sets of calculations must be carried out: first, the determination of the difference between the means of pairs of groups; second, the preparation of a probability rate against which each difference in means is compared (as shown in the first of the two examples in this section).

The means (averages) are determined (or taken from the ANOVA calculation) and ranked in either decreasing or increasing order. If two means are the same, they take up two equal positions (thus, for four means we could have ranks of 1, 2, 2, and 4 rather than 1, 2, 3, and 4). The groups are then taken in pairs and the differences between the means $\left(\bar{X}_1 - \bar{X}_2\right)$ expressed as positive numbers are calculated. Usually, each pair consists of a test group and the control group, although multiple test groups may be intracompared if so desired. The relative rank of the two groups being compared must be considered. If a test group is ranked "2" and the control group is ranked "1," then we say that there are two places between them, while if the test group were ranked "3," then there would be three places between it and the control.

To establish the probability table, the standard error of the mean (SEM) must be calculated. This can be done as presented in Chapter 2 or shown as:

$$\text{SEM} = \sqrt{\frac{\text{error mean square}}{N}} = \sqrt{\frac{\text{mean square within group}}{N}}$$

where N is the number of animals or replications per dose level. The mean square within groups (MS_{wg}) can be calculated from the information given in the ANOVA procedure (refer to the earlier section on ANOVA). The SEM is then multiplied by a series of table values (as in Harter, 1960, or Beyer, 1982, pp. 369–378) to set up a probability table. The table values used for the calculations are chosen according to the probability levels (note that the tables have sections for 0.05, 0.01, and 0.001 levels) and the number of means apart for the groups being compared and the number of "error" df. The "error" df is the number of df within the groups. This last figure is determined from the ANOVA calculation and can be taken from ANOVA output. For some values of df, the table values are not given and should thus be interpolated. Example 7.5 demonstrates this case.

Example 7.5

Using the data given in Example 7.4 (4 groups of dogs, with 4 dogs in each group), we can make the following calculations:

Ranks	1	2	3	4
Concentration:	0 ppm	100 ppm	200 ppm	400 ppm
Mean kidney weight	0.345	0.390	0.428	0.482
Groups compared	$(\bar{X}_1 - \bar{X}_2)$	No. of means apart		Probability
2 vs. 1 (100 vs. 0 ppm)	0.045	2		$p > 0.05$
3 vs. 1 (200 vs. 0 ppm)	0.083	3		$p > 0.05$
4 vs. 1 (400 vs. 0 ppm)	0.137	4		$0.01 > p > 0.001$
4 vs. 2 (400 vs. 100 ppm)	0.092	3		$0.05 > p > 0.01$

The MS_{wg} from the ANOVA example was 0.00275. Therefore, the SEM = $(0.00275/4)^{1/2}$ = 0.02622. The "error" df (df_{wg}) was 12 so the following table values are used.

No. of means apart	Probability Levels		
	0.05	0.01	0.001
2	3.082	4.320	6.106
3	3.225	4.504	6.340
4	3.313	4.662	6.494

When these are multiplied by the SEM we get the following probability table:

No. of means apart	Probability Levels		
	0.05	0.01	0.001
2	0.0808	0.1133	0.1601
3	0.0846	0.1161	0.1662
4	0.0869	0.1212	0.1703

Groups with Unequal Numbers of Data ($N_1 \neq N_2$)

This procedure is very similar to that discussed above. As before, the means are ranked and the differences between the means are determined $(\bar{X}_1 - \bar{X}_2)$. Next, weighing values ("a_{ij}" values) are calculated for the pairs of groups being compared in accordance with

$$a_u = \sqrt{\frac{2N_i N_j}{(N_i + N_j)}} = \sqrt{\frac{2N_1 N_2}{(N_1 + N_2)}}$$

This weighting value for each pair of groups is multiplied by $(\bar{X}_1 - \bar{X}_2)$ for each value to arrive at a "t" value. It is the "t" that will later be compared to a probability table.

The probability table is set up as above, except that instead of multiplying the appropriate table values by SEM, SEM² is used. This is equal to $(MS_{wg})^{1/2}$.

For the desired comparison of two groups at a time, either the $(\bar{X}_1 - \bar{X}_2)$ value (if $N_1 = N_2$) is compared to the appropriate probability table. Each comparison must be made according to the number of places between the means. If the table value is larger at the 0.05 level, the two groups are not considered to be statistically different. If the table value is smaller, the groups are different and the comparison is repeated at lower levels of significance. Thus, the degree of significance may be determined. We might have significant differences at 0.05 but not at 0.01, in which case the probability would be represented at $0.05 > p > 0.01$. Example 7.6 demonstrates this case.

Example 7.6

Suppose that the 400 ppm level from the above example had only 3 dogs, but that the mean for the group and the mean square within groups were the same. To continue Duncan's we would calculate the weighing factors as follows:

$$100 \text{ ppm vs. } 0 \text{ ppm} \quad N_1 = 4; N_2 = 4 \quad a_u = \sqrt{\frac{2(4)(4)}{4+4}} = 2.00$$

$$200 \text{ ppm vs. } 0 \text{ ppm} \quad N_1 = 4; N_2 = 4 \quad a_u = \sqrt{\frac{2(4)(4)}{4+4}} = 2.00$$

$$400 \text{ ppm vs. } 0 \text{ ppm} \quad N_1 = 3; N_2 = 4 \quad a_u = \sqrt{\frac{2(3)(4)}{3+4}} = 1.85$$

$$400 \text{ ppm vs. } 100 \text{ ppm} \quad N_1 = 3; N_2 = 4 \quad a_u = \sqrt{\frac{2(3)(4)}{3+4}} = 1.85$$

Using the $(\bar{X}_1 - \bar{X}_2)$ from the above example we can set up the following table:

Concentrations (ppm)	No. of Means Apart	$(\bar{X}_1 - \bar{X}_2)$	a_u	$(\bar{X}_1 - \bar{X}_2)$
100 vs. 0	2	0.045	2.000	2.000(0.045) = 0.090
200 vs. 0	3	0.083	2.000	2.000(0.083) = 0.166
400 vs. 0	4	0.137	1.852	1.852(0.137) = 0.254
400 vs. 100	3	0.092	1.852	1.852(0.092) = 0.170

Next we calculate SEM as being $(0.00275)^{1/2} = 0.05244$. This is multiplied by the appropriate table values chosen for 11 df (df_{wg} for this example). This gives the following probability table.

No. of means apart	Probability Levels		
	0.05	0.01	0.001
2	0.1632	0.2303	0.3291
3	0.1707	0.2401	0.3417
4	0.1753	0.2463	0.3501

Comparing the *t*-values with the probability table values we get the following:

Groups Compared	Probability
100 vs. 0 ppm	$p > 0.05$
200 vs. 0 ppm	$p > 0.05$
400 vs. 0 ppm	$0.01 > p > 0.001$
400 vs. 100 ppm	$0.05 > p > 0.01$

Assumptions and Limitations

Duncan's test assures a set alpha level or type 1 error rate for all tests when means are separated by no more than ordered step increases. Preserving this alpha level means that the test is less sensitive than some others, such as the Student–Newman–Keuls. The test is inherently conservative and not resistant or robust.

Scheffé's Multiple Comparisons

Scheffé's is another *post hoc* comparison method for groups of continuous and randomly distributed data. It also normally involves three or more groups (Scheffé's, 1999, and Harris, 1975). It is widely considered a more powerful significance test than Duncan's.

Each *post hoc* comparison is tested by comparing an obtained test value (F_{contr}) with the appropriate critical F-value at the selected level of significance (the table F-value multiplied by $K - 1$ for an F with $K - 1$ and $N - K$ df). F_{contr} is computed as follows:

Compute the mean for each sample (group)

Denote the residual mean square by MS_{wg}

Compute the test statistic as:

$$F_{contr} = \frac{(C_1\bar{X}_1 + C_2\bar{X}_2 + \ldots + C_k\bar{X}_k)^2}{(k-1)MS_{wg}\left(\frac{C_1^2}{N_1} + \frac{C_2^2}{N_2} + \ldots + \frac{C_k^2}{N_k}\right)}$$

where C_k is the comparison number such that $C_1 + C_2 + \ldots + C_k = 0$. (See Example 7.7).

Example 7.7

At the end of a short-term feeding study the following body weight changes were recorded:

	Group 1	Group 2	Group 3
	10.2	12.2	9.2
	8.2	10.6	10.5
	8.9	9.9	9.2
	8.0	13.0	8.7
	8.3	8.1	9.0
	8.0	10.8	
		11.5	
Totals	51.6	76.1	46.6
Means	8.60	10.87	9.32
$MS_{wg} = 1.395$			

To avoid logical inconsistencies with pair-wise comparisons, we compare the group having the largest sample mean (group 2) with that having the smallest sample mean (group 1), then with the group having the next smallest sample mean, and so on. As soon as we find a nonsignificant comparison in this process (or no group with a smaller sample mean remains), we replace the group having the largest sample mean with that having the second largest sample mean and repeat the comparison process.

Accordingly, our first comparison is between groups 2 and 1. We set $C_1 = -1$, $C_2 = 1$, and $C_3 = 0$ and calculate our test statistic:

$$F_{contr} = \frac{(10.87 - 8.60)^2}{(3-1)1.395(1/6 + 1/7)} = 5.97$$

The critical region for F at $p \leq 0.05$ for 2 and 11 df is 3.98. Therefore, these groups are significantly different at this level. We next compare groups 2 and 3, using $C_1 = 0$, $C_2 = 1$, and $C_3 = -1$.

$$F_{contr} = \frac{(10.87 - 9.32)^2}{(3-1)1.395(1/7 + 1/5)} = 2.51$$

This is less than the critical region value, so these groups are not significantly different.

Assumptions and Limitations

The Scheffé procedure is robust to moderate violations of the normality and homogeneity of variance assumptions.

It is not formulated on the basis of groups with equal numbers (as one of Duncan's procedures is), and if $N_1 \neq N_2$ there is no separate weighing procedure.

It tests all linear contrasts among the population means (the other three methods confine themselves to pairwise comparison, except they use a Bonferroni-type correlation procedure).

The Scheffé procedure is powerful because of it robustness, yet it is very conservative. Type 1 error (the false-positive rate) is held constant at the selected test level for each comparison.

Dunnett's *t*-Test

Dunnett's *t*-test (Dunnett, 1955, 1964) has as its starting point the assumption that what is desired is a comparison of each of several means with one other mean and only one other mean; in other words, that one wishes to compare each and every treatment group with the control group, but not compare treatment groups with each other. The problem here is that, in toxicology, one is frequently interested in comparing treatment groups with other treatment groups. However, if one wants only to compare treatment groups vs. a control group, Dunnett's is a useful approach. In a study with K groups (one of them being the control) we will wish to make $K - 1$ comparisons. In such a situation, we want to have a P level for the entire set of $K - 1$ decisions (not for each individual decision). The Dunnett's distribution is predicated on this assumption. The parameters for utilizing a Dunnett's table, such as found in his original article, are K (as above) and the number of df for MS_{wg}. The test value is calculated as:

$$t = \frac{|T_i - T_j|}{\sqrt{2MS/n}}$$

where n is the number of observations in each of the groups. The MS_{wg} is as we have defined it previously; T_j is the control group mean; and T_i is the mean of, in order, each successive test group observation. Note that one uses the absolute value of the positive number resulting from subtracting T_i from T_j. This is to ensure a positive number for our final t.

Example 7.8 demonstrates this test, again with the data from Example 7.4.

Example 7.8

The means, Ns, and sums for the groups previously presented in Example 6.8, are

	Control	100 ppm	200 ppm	400 ppm
Sum (ΣX)	1.38	1.56	1.71	1.93
N	4	4	4	4
Mean	0.345	0.39	0.4275	0.4825

The MS_{wg} was 0.00275, and our test t for 4 groups and 12 df is 2.41. Substituting in the equation, we calculate our t for the control vs. the 400 ppm to be

$$t = \frac{|0.345 - 0.4825|}{\sqrt{2(0.00275)/4}} = \frac{0.1375}{\sqrt{0.001375}} = 3.708$$

which exceeds our test value of 2.41, showing that these two groups are significantly different at $p \leq 0.05$. The values for the comparisons of the control vs. the 200 and 100 ppm groups are then found to be, respectively, 2.225 and 1.214. Both of these are less than our test value, and therefore the groups are not significantly different.

Assumptions and Limitations

Dunnett's seeks to ensure that the type I error rate will be fixed at the desired level by incorporating correction factors into the design of the test value table. Treated group sizes must be approximately equal.

Williams' *t*-Test

Williams' *t*-test (Williams, 1971, 1972) has also become popular, although the following discussion will show that its use is quite limited in toxicology. It is designed to detect the highest level (in a set of dose/exposure levels) at which there is no significant effect. It assumes that the response of interest (such as change in body weights) occurs at higher levels, but not at lower levels, and that the responses are monotonically ordered so that $X_0 \leq X_1 \ldots \leq X_k$. This is, however, frequently not the case. The Williams' technique handles the occurrence of such discontinuities in a response series by replacing the offending value and the value immediately preceding it with weighted average values. The test is also adversely affected by any mortality at high dose levels. Such mortalities "impose a severe penalty, reducing the power of detecting an effect not only at level K but also at all lower doses" (Williams, 1972, p. 529). Accordingly, it is not generally applicable in toxicology studies.

Analysis of Covariance

Analysis of covariance (ANCOVA) is a method for comparing sets of data which consist of two variables (treatment and effect, with our effect variable being called the "variate") when a third variable (called the "covariate") exists that can be measured but not controlled and that has a definite effect on the variable of interest. In other words, it provides an indirect type of statistical control, allowing us to increase the precision of a study and to remove a potential source of bias. One common example of this is in the analysis of organ weights in toxicity studies. Our true interest here is the effect of our dose or exposure level on the specific organ weights, but most organ weights also increase (in the young, growing animals most commonly used in such studies) in proportion to increases in animal body weight. As we are not interested here in the effect of this covariate (body weight), we measure it to allow for adjustment. We must be careful before using ANCOVA, however, to ensure that the underlying nature of the correspondence between the variate and covariate is such that we can rely on it as a tool for adjustments (Anderson et al., 1980; Kotz and Johnson, 1982).

Calculation is performed in two steps. The first is a type of linear regression between the variate Y and the covariate X.

This regression, performed as described under the linear regression section, gives us the model

$$Y = a_1 + BX + e$$

which in turn allows us to define adjusted means $(\bar{Y} \text{ and } X)$ such that

$$\bar{Y}_{1a} = \bar{Y}_1 - (\bar{X}_1 - X)$$

If we consider the case where K treatments are being compared such that $K = 1, 2, \ldots k$, and we let X_{ik} and Y_{ik} represent the predictor and predicted values for each individual i in group k, we can let X_k and Y_k be the means. Then, we define the between-group (for treatment) sum of squares and cross products as:

$$T_{xx} = \sum_{k=1}^{K} n_k (\bar{X}_k - \bar{X})^2$$

$$T_{yy} = \sum_{k=1}^{K} n_k (\bar{Y}_k - \bar{Y})^2$$

$$T_{zz} = \sum_{k=1}^{K} n_k (\bar{Z}_k - \bar{Z})^2$$

In a like manner, within-group sums of squares and cross products are calculated as:

$$\sum xx = \sum_{k=1}^{K} \sum_{i=1}^{n_i} (X_{ik} - X_k)^2$$

$$\sum yy = \sum_{k=1}^{K} \sum_{i=1}^{n_i} (Y_{ik} - Y_k)^2$$

$$\sum xy = \sum_{k=1}^{K} \sum_{i=1}^{n_i} (X_{ik} - X_k)(Y_{ik} - Y_k)$$

where i indicates the sum from all the individuals within each group and f = total number of subjects minus number of groups.

$$S_{xx} = T_{xx} + \sum xx$$

$$S_{yy} = T_{yy} + \sum yy$$

$$S_{xy} = T_{xy} + \sum xy$$

With these in hand, we can then calculate the residual mean squares of treatments (St^2) and error (Se^2)

$$St^2 = \frac{T_{yy} - \frac{(S_{xy})^2}{S_{xx}} + \frac{(\Sigma xy)^2}{\Sigma xx}}{f-1}$$

$$Se^2 = \frac{\Sigma yy - \frac{(\Sigma y)^2}{\Sigma xx}}{f-1}$$

These can be used to calculate an F statistic to test the null hypothesis that all treatment effects are equal.

$$F = \frac{St^2}{Se^2}$$

The estimated regression coefficient of Y over X is

$$B = \frac{\Sigma xy}{\Sigma xx}$$

The estimated standard error for the adjusted difference between two groups is given by:

$$Sd = \sqrt{Se^2 \left(\frac{1}{n_i} + \frac{1}{n_j} + \frac{(X_i - X_j)^2}{\Sigma xx} \right)}$$

where n_0 and n_1 are the sample sizes of the two groups. A test of the null hypothesis that the adjusted differences between the groups is zero is provided by:

$$t = \frac{Y_1 - Y_0 - B(X_1 - X_0)}{Sd}$$

The test value for the t is then looked up in the t table with $f - 1$ df.

Computation is markedly simplified if all the groups are of equal size, as demonstrated in Example 7.9.

Example 7.9

An ionophore was evaluated as a potential blood pressure-reducing agent. Early studies indicated that there was an adverse effect on blood cholesterol and hemoglobin levels, so a special study was performed to evaluate this specific effect. The hemoglobin (Hgb) level covariate was measured at study start along with the percent changes in serum triglycerides between study start and the end of the 13-week study. Was there a difference in effects of the two ionopheres?

Ionophore A		Ionophore B	
Hgb (X)	Serum Triglyceride % Change (Y)	Hgb (X)	Serum Triglyceride % Change (Y)
7.0	5	5.1	10
6.0	10	6.0	15
7.1	−5	7.2	−15
8.6	−20	6.4	5
6.3	0	5.5	10
7.5	−15	6.0	−15
6.6	10	5.6	−5
7.4	−10	5.5	−10
5.3	20	6.7	−20
6.5	−15	8.6	−40
6.2	5	6.4	−5
7.8	0	6.0	−10
8.5	−40	9.3	−40
9.2	−25	8.5	−20
5.0	25	7.9	−35
		5.0	0
		6.5	10

To apply ANCOVA using Hgb as a covariate, we first obtain some summary results from the data as follows:

	Ionophore A (Group 1)	Ionophore B (Group 2)	Combined
Σx	112.00	119.60	231.60
Σx^2	804.14	821.64	1625.78
Σy	−65.00	−185.00	−250.00
Σy^2	4575.00	6475.00	11050.00
Σxy	−708.50	−1506.50	−2215.00
Mean of x	7.000	6.6444	6.8118
Mean of y	−4.625	−10.2778	−7.3529
N	16	18	34

We compute for the ionophore A group ($i = 1$):

$$S_{xx(1)} = 804.14 - (112)^2/16 = 20.140$$

$$S_{yy(1)} = 4575.00 - (-65)^{-2}/16 = 4310.938$$

$$S_{xy(1)} = -708.50 - (112)(-65)/16 = -253.500$$

Similarly, for the ionophore B group ($i = 2$), we obtain:

$$S_{xx(2)} = 26.964$$

$$S_{yy(2)} = 4573.611$$

$$S_{xy(2)} = -277.278$$

Finally, for the combined data (ignoring groups), we compute:

$$S_{xx} = 48.175$$

$$S_{yy} = 9211.765$$

$$S_{xy} = -512.059$$

The sums of squares can now be obtained as:

$$TOT(SS) = 9211.8$$

$$SSE = \frac{(20.140 + 26.964)(4310.938 + 4573.611) - [-253.5 - 277.28]^2}{(20.140 + 26.964)} = 2903.6$$

$$SSG = \frac{(48.175)(9211.765) - (-512.059)^2}{(48.175)} - 2903.6 = 865.4$$

$$SSX = (4310.938 + 473.611) - 2903.6 = 5980.9$$

and the ANCOVA summary table can be completed as follows:

Source	df	SS	MS	F
TREATMENT	1	865.4	865.4	9.2[a]
X (Hgb)	1	5980.9	5980.9	63.8
Error	31	2903.7	93.7	
Total	33	9211.8		

[a] Significant ($p < 0.05$); critical F-value $= 4.16$.

The F statistics are formed as the ratios of effect MS to the MSE (93.7). Each F statistic is compared with the critical F-value with 1 upper and 31 lower df. The critical F-value for $a = 0.05$ is 4.16.

The significant covariate effect ($F = 63.8$) indicates that the triglyceride response has a significant linear relationship with HbA_{1c}. The significant

F-value for TREATMENT indicates that the mean triglyceride response adjusted for hemoglobin effect differs between treatment groups.

SAS Analysis of Example 7.9

*SAS Code for Example 7.9

```
DATA IONO; HGB SERTRI
INPUT TRT $ @@;
CARDS;
A       7.0     5      A      6.0       10
A       7.1    -5      A      8.6      -20
A       6.3     0      A      7.5      -15
...
B       7.9   -35      B      7.4        0
B       5.0     0      B      6.5      -10
; RUN;
PROC SORT DATA = IONO
BY TRT HGB SERTRI; RUN;
      ***Print data set;
PROC PRINT DATA = IONO
VAR TRT PAT HGB SETRI;
TITLE1 'Analysis of Covariance';
TITLE2 'Example 7.9: Serum Triglyceride Hemoglobin';
RUN;
      **Plot data, by group;
PROC PLOT VPERCENT=45 DATA = IONO;
PLOT SERTRI * HGB=TRT; RUN;
      ** Obtain summary statistics for each group;
PROC MEANS MEAN STD N DATA = IONO;
BY TRT; VAR HGD SERTRI; RUN;
      ** Use hemoglobin levels control as covariate;
PROC GLM DATA = IONO; CLASSES TRT;
MODEL SERTRI = TRT HGB/SS3 SOLUTION;
LSMEANS TRT/PDIFF STDERR;
RUN;
```

```
        ** Compare groups with ANOVA, ignoring the
   covariate;
   PROC GLM DATA = IONO; CLASSES TRT;
   MODEL SERTRI = TRT/SS3; RUN;
```

SAS Output for Example 7.9

```
Analysis of Covariance
Example 7.9 Triglyceride Changes with Ionophores
```

OBS	TRT	HGB	SERTRI	
1	A	5.0	25	
2	A	5.3	20	
3	A	6.0	10	
4	A	6.2	5	
5	A	6.3	0	
6	A	6.5	−15	
7	A	6.6	10	
8	A	7.0	−10	
9	A	7.0	5	
10	A	7.1	−5	
11	A	7.4	−10	
12	A	7.5	−15	
13	A	7.8	0	
14	A	8.5	−40	
15	A	8.6	−20	
16	A	9.2	−25	(1)
17	B	5.0	0	
18	B	5.1	10	
19	B	5.5	−10	
20	B	5.5	10	
21	B	5.6	−5	
22	B	6.0	−15	
23	B	6.0	−10	
24	B	6.0	15	
25	B	6.4	−5	
26	B	6.4	5	

27	B	6.5	−10
28	B	6.7	−20
29	B	7.2	−15
30	B	7.4	0
31	B	7.9	−35
32	B	8.5	−20
33	B	8.6	−40
34	B	9.3	−40

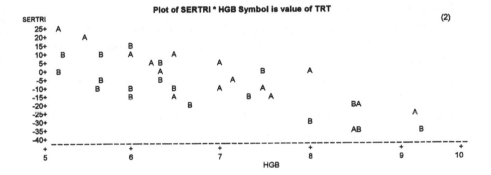

Plot of SERTRI * HGB Symbol is value of TRT   (2)

Analysis of Covariance
Example 7.9: Triglyceride Changes with Ionophores
General Linear Models Procedure
Class Level Information

Class	Levels	Values
TRT	2	A B

Number of observations in data set = 34
Dependent Variable: SERTRI

Source	DF	Sum of Squares	Mean Square	F-Value	Pr > F
Model	1	327.21609	327.21609	1.18	0.2858
Error	32	8884.54861	277.64214		
Corrected Total	33	9211.76471			

R-Square	C.V.	Root MSE	TRICHG Mean
0.035522	−266.6113	16.663	−7.3529

Source	DF	Type III SS	Mean Square	F Value	Pr > F
TRT	1	327.21609	327.21609	1.18	0.2858

Assumptions and Limitations

The underlying assumptions for ANCOVA are fairly rigid and restrictive. The assumptions include:

That the slopes of the regression lines of a Y and X are equal from group to group. This can be examined visually or formally (i.e., by a test). If this condition is not met, ANCOVA cannot be used.

That the relationship between X and Y is linear.

That the covariate X is measured without error. Power of the test declines as error increases.

That there are no unmeasured confounding variables.

That the errors inherent in each variable are independent of each other. Lack of independence effectively (but to an immeasurable degree) reduces sample size.

That the variances of the errors in groups are equivalent between groups.

That the measured data that form the groups are normally distributed. ANCOVA is generally robust to departures from normality.

Of the seven assumptions above, the most serious are the first four.

References

Anderson, S., Auquier, A., Hauck, W.W., Oakes, D., Vandaele, W., and Weisburg, H.I. (1980) *Statistical Methods for Comparative Studies*. John Wiley & Sons, New York.

Beyer, W.H. (1982) *CRC Handbook of Tables for Probability and Statistics*. CRC Press, Boca Raton, FL.

Cochran, W.G. and Cox, G.M. (1992) *Experimental Designs*, 2nd ed., John Wiley & Sons, New York.

Diem, K. and Lentner, C. (1975) *Documenta Geigy Scientific Tables*. Geigy, New York, pp. 158–159.

Duncan, D.B. (1955) Multiple range and multiple F tests. *Biometrics*, 11, 1–42.

Dunnett, C.W. (1955) A multiple comparison procedure for comparing several treatments with a control. *J. Am. Stat. Assoc.*, 50, 1096–1121.

Dunnett, C.W. (1964) New tables for multiple comparison with a control. *Biometrics*, 16, 671–685.

Harris, R.J. (1975) *A Primer of Multivariate Statistics*. Academic Press, New York, pp. 96–101.

Harter, A.L. (1960) Critical values for Duncan's new multiple range test. *Biometrics*, 16, 671–685, 1960.

Kotz, S. and Johnson, N.L. (1982) *Encyclopedia of Statistical Sciences*, vol. 1., John Wiley & Sons, New York, pp. 61–69.

Scheffé, H. (1999) *The Analysis of Variance*. Wiley-InterScience, New York.
Snedecor, G.W. and Cochran, W.G. (1989) *Statistical Methods*, 8th ed., Iowa State University Press, Ames, Iowa.
Sokal, R.R. and Rohlf, F.J. (1994) *Biometry: The Principles and Practice of Statistics in Biological Research*, 3rd ed., W.H. Freeman, San Francisco.
Williams, D.A. (1971) A test for differences between treatment means when several dose levels are compared with a zero dose control. *Biometrics*, 27, 117.
Williams, D.A. (1972) The comparison of several dose levels with a zero dose control. *Biomet.*, 28, 519–531.
Zar, J.H. (1974) *Biostatistical Analysis*. Prentice-Hall, Englewood Cliffs, NJ p. 50.

8

Modeling and Extrapolation

The mathematical modeling of biological systems, restricted even to the field of toxicology, is an extremely large and vigorously growing area. Broadly speaking, modeling is the principal conceptual tool by which toxicology seeks to develop as a mechanistic science. In an iterative process, models are developed or proposed, tested by experiment, reformulated, and so on, in a continuous cycle. Such a cycle could also be described as two related types of modeling: explanatory (where the concept is formed) and correlative (where data are organized and relationships derived). The outcomes of the cycle are predictions (with cautious extrapolation beyond the range of existing data) and inferences (about analogous situations that are difficult or impossible to measure directly, such as risk of carcinogenesis at extremely low levels of exposure to a material). It is the special area of these last two (prediction and inference) that will be addressed here. An excellent introduction to the broader field of modeling of biological systems can still be found in Gold (1977).

In toxicology, modeling is of prime interest in seeking to relate a treatment variable with an effect variable and, from the resulting model, to predict effects at exact points where no experiment has been done (but in the range where we have performed experiments, such as "determining" LD_{50}s), to estimate how good our prediction is, and occasionally, simply to determine if a pattern of effects is related to a pattern of treatment. Methods are presented for all three of these objectives under each of several conditions.

For use in prediction, the techniques of linear regression, probit/logit analysis (a special case of linear regression), moving averages (an efficient approximation method), and nonlinear regression (for doses where data cannot be made to fit a linear pattern) are presented. For evaluating the predictive value of these models, both the correlations coefficient (for parametric data) and Kendall's rank correlation (for nonparametric data) are given. Finally, the concept of trend analysis is introduced and a method presented.

Not included here are methods for modeling when there are three or more variables. Rather, two of these methods (discriminant analysis and multiple regression) and a related method (canonical correlation analysis) are discussed in the later chapter on multivariate methods.

When we are trying to establish a pattern between several data points (whether this pattern is in the form of a line or a curve), what we are doing is interpolating. It is possible for any given set of points to produce an infinite set of lines or curves that pass near (for lines) or through (for curves) the data points. In most cases, we cannot actually know the "real" pattern so we apply a basic principle of science — Occam's razor. We use the simplest explanation (or, in this case, model) that fits the facts (or data). A line is, of course, the simplest pattern to deal with and describe, so fitting the best line (linear regression) is the most common form of model in toxicology.

Linear Regression

Foremost among the methods for interpolating within a known data relationship is regression — the fitting of a line or curve to a set of known data points on a graph, and the interpolation ("estimation") of this line or curve in areas where we have no data points. The simplest of these regression models is that of linear regression (valid when increasing the value of one variable changes the value of the related variable in a linear fashion, either positively or negatively). This is the case we will explore here, using the method of least squares (Table 8.1).

Given that we have two sets of variables, x (say mg/kg of test material administered) and y (say percentage of animals so dosed that die), what is required is solving for a and b in the equation $Y_i = a + bx_i$ (where the uppercase Y_i is the fitted value of y_i at x_i, and we wish to minimize $[y_i - Y_i]^2$). So we solve the equations

$$b = \frac{\sum x_1 y_1 - n\bar{x}\bar{y}}{\sum x_1^2 - n\bar{x}^2}$$

and

$$a = \bar{y} - b\bar{x}$$

where a is the y intercept, b is the slope of the time, and n is the number of data points. Use of this is demonstrated in Example 8.1.

Note that in actuality, dose–response relationships are often not linear and instead we must use either a transformation (to linearize the data) or a nonlinear regression method (a good discussion of which may be found in Gallant, 1975, and later in this chapter).

Note also that we can use the correlation test statistic (to be described in the correlation coefficient section) to determine if the regression is significant

TABLE 8.1

Linear Regression Analysis of Variance

Source of Variation	Sum of Squares	Degrees of Freedom	Mean Square
Regression	$b_1^2(\Sigma x_1^2 - nx^2)$	1	By division
Residual	By difference	$n - 2$	By division
Total	$\Sigma y_1^2 - ny^2$	$n - 3$	

(and, therefore, valid at a defined level of certainty). A more specific test for significance would be the linear regression analysis of variance (Pollard, 1977). To do so we start by developing the appropriate ANOVA table, as demonstrated in Example 8.1.

Example 8.1

From a short-term toxicity study we have the following results:

	Dose Administered (mg/kg)		% Animals Dead	
	x_1	x_1^2	Y_1	x_1y_1
	1	1	10	10
	3	9	20	60
	4	16	18	72
	5	25	20	100
Sums	13	51	68	242
Means	3.25		17	

$$b = \frac{242 - (4)(3.25)(17)}{51 - (4)(10.5625)}$$

$$a = 17 - (2.4)(3.25) = 9.20$$

We therefore see that our fitted line is $Y = 9.2 + 2.4X$.

These ANOVA table data are then used as shown in Example 7.4. We then calculate

$$F = \frac{\text{regression mean square}}{\text{residual mean square}}$$

SAS Analysis of Example 8.1

The SAS code and output for analyzing the data set of this example are shown on the following pages. A printout of the data set is first obtained using PROC PRINT, followed by the summary statistics for the x and y values using PROC MEANS.

For the regression analysis, we utilize the SAS GLM procedure, although PROC REG may also be used. When GLM is used without a CLASS statement, SAS assumes the independent variables in the MODEL statement are quantitative and performs a regression analysis.

The regression estimates, a and b, are printed in the "Estimate" column of the output.

The P and CLM options specified after the MODEL statement request the predicted values based on the regression equation for each data point, along with 95% confidence intervals for the mean response at the corresponding x values.

* SAS Code for Example 8.1

```
DATA LETHALITY;
INPUT AN X_DOSE Y_LETHALITY@@;
CARDS;
1     10
3     20
4     18
5     20
RUN;
PROC SORT DATA = LETHALITY; BY X_DOSE Y_LETHAL; RUN;
PROC PRINT DATA = LETHALITY;
      VAR AN X_DOSE Y_LETHAL;
TITLE1 'Linear Regression & Correlation';
TITLE2 'Example 8.1 Lethality Curve'
RUN;
PROC MEANS MEAN STD N; VAR X_DOSE Y_LETHAL;
RUN;
PROC GLM DATA = LETHALITY;
MODEL Y_LETH = X_DOSE/P CLM SS3;
QUIT;
RUN;
```

Example 8.2

We desire to test the significance of the regression line in Example 7.6.

$$\Sigma y_i^2 = 10^2 + 20^2 + 18^2 + 20^2 = 1224$$

$$\text{regression SS} = (2.4)^2[51 - 4(3.25)^2] = 50.4$$

$$\text{total SS} = 1224 - 4(17)^2 = 68.0$$

$$\text{residual SS} = 68.0 - 50.4 = 17.6$$

$$F = 50.4/8.8 = 5.73$$

This value is not significant at the 0.05 level; therefore, the regression is not significant. A significant F-value (as found in an F distribution table for the appropriate degrees of freedom) indicates that the regression line is an accurate prediction of observed values at that confidence level. Note that the portion of the total sum of squares explained by the regression is called the coefficient of correlation, which in the above example is equal to 0.862 (or 0.74). Calculation or the correlation coefficient is described later in this chapter.

Finally, we might wish to determine the confidence intervals for our regression line; that is, given a regression line with calculated values for Y_i given x_i, within what limits may we be certain (with say a 95% probability) what the real value of Y_i is?

If we denote the residual mean square in the ANOVA by s^2, the 95% confidence limits for a (denoted by A, the notation for the true, as opposed to the estimated, value for this parameter) are calculated as:

$$t_{n-2} = \frac{a - A}{\sqrt{\frac{s^2\left(\Sigma x^2\right)}{n\left(\Sigma x^2\right) - n^2\bar{x}^2}}}$$

So, for this example:

$$t_{n-2} = \frac{9.2 - A}{\sqrt{\frac{8.8(51)}{4(51) - (16)(10.562)}}} = \frac{9.2 - A}{\sqrt{\frac{448}{35.008}}}$$

$$= \frac{9.2 - A}{3.58} = -4.303$$

$$A = 9.2 - 15.405$$

Assumptions and Limitations

All the regression methods are for interpolation, not extrapolation. That is, they are valid only in the range that we have data — the experimental region. Not beyond.

The method assumes that the data are independent and normally distributed and it is sensitive to outliers. The x-axis (or horizontal) component plays an extremely important part in developing the least square fit. All points have equal weight in determining the height of a regression line, but extreme x-axis values unduly influence the slope of the line.

A good fit between a line and a set of data (that is, a strong correlation between treatment and response variables) does not imply any causal relationship.

It is assumed that the treatment variable can be measured without error, that each data point is independent, that variances are equivalent, and that a linear relationship does not exist between the variables.

There are many excellent texts on regression, which is a powerful technique. These include Draper and Smith (1981) and Montgomery, Peck, and Vining (2001), which are not overly rigorous mathematically.

Probit/Log Transforms and Regression

As we noted in the preceding section, dose–response problems (among the most common interpolation problems encountered in toxicology) rarely are straightforward enough to make a valid linear regression directly from the raw data. The most common valid interpolation methods are based upon probability ("probit") and logarithmic ("log") value scales, with percentage responses (death, tumor incidence, etc.) being expressed on the probit scale, whereas doses (Y_i) are expressed on the log scale. There are two strategies for such an approach. The first is based on transforming the data to these scales, then doing a weighted linear regression on the transformed data. (If one does not have access to a computer or a high-powered programmable calculator, the only practical strategy is not to assign weights.) The second requires the use of algorithms (approximate calculation techniques) for the probit value and regression process and is extremely burdensome to perform manually.

Our approach to the first strategy requires that we construct a table with the pairs of values of x_i and y_i listed in order of increasing values of Y_i (percentage response). Beside each of these columns a set of blank columns should be left so that the transformed values may be listed. We then simply add the columns described in the linear regression procedure. Log and probit values may be taken from any of a number of sets of tables (such as provided in Appendix 1) and the rest of the table is then developed from these transformed x_i and y_i values (denoted as x_i' and y_i'). A standard linear regression is then performed (see Example 8.3).

The second strategy we discussed has been broached by a number of authors (Bliss, 1935; Finney, 1977; Litchfield and Wilcoxon, 1949; Prentice, 1976). All of these methods, however, are computationally cumbersome. It is possible to approximate the necessary iterative process using the algorithms developed by Abramowitz and Stegun (1964), but even this merely reduces the complexity to a point where the procedure may be readily programmed on a small computer or programmable calculator.

Example 8.3

% of Animals Killed [x_1]	Probit of x_1 [x_1']	Dose in mg/kg [y_1]	Log of y_1 [y_1']	$(x_1')^2$	$x_1'y_1'$
2	2.9463	3	0.4771	8.6806	1.40568
10	3.7184	5	0.6990	13.8264	2.59916
42	4.7981	10	1.0000	23.0217	4.79810
90	6.2816	20	1.3010	39.4585	8.17223
98	7.2537	30	1.4771	52.6162	10.4190
Sums	$\Sigma\, x_1' =$ 24.9981		$\Sigma\, y_1' =$ 4.9542	$\Sigma\, (x_1')^2 =$ 137.6034	$\Sigma\, x_1' \Sigma\, y_1' =$ 27.68974

Our interpolated log of the LD_{50} (calculated by using $Y = -0.200591 + 0.240226x$ where $x = 5.000$ — the probit of 50% — in the regression equation) is 1.000539. When we convert this log value to its linear equivalent, we get an LD_{50} of 10.0 mg/kg.

Finally, our calculated correlation coefficient is $r = .997$. A goodness-of-fit of the data using chi-square may also be calculated.

Assumptions and Limitations

The probit distribution is derived from a common error function, with the midpoint (50% point) moved to a score of 5.00.

The underlying frequency distribution becomes asymptotic as it approaches the extremes of the range. That is, in the range of 16 to 84%, the corresponding probit values change gradually — the curve is relatively linear. But beyond this range, they change ever more rapidly as they approach either 0 or 100%. In fact, there are no values for either of these numbers.

A normally distributed population is assumed, and the results are sensitive to outliers.

Moving Averages

An obvious drawback to the interpolation procedures we have examined to date is that they do take a significant amount of time (although they are simple enough to be done manually), especially if the only result we desire is an LD_{50}, LC_{50}, or LT_{50}.

The method of moving averages (Thompson and Weil, 1952; Weil, 1952; Dykstra and Robertson, 1983;) gives a rapid and reasonably accurate estimate of this "median effective dose" (*m*) and the estimated standard deviation of its logarithm.

Such methodology requires that the same number of animals be used per dosage level and that the spacing between successive dosage exposure levels be geometrically constant (i.e., levels of 1, 2, 4, and 8 mg/kg or 1, 3, 9, and 27 ppm). Given this and access to a table for the computation of moving averages (such as found in Appendix 1) one can readily calculate the median effective dose with the formula (illustrated for dose):

$$\log m = \log D + d(K - 1)/2 + df$$

where
   m = median effective dose or exposure
   D = the lowest dose tested
   d = the log of the ratio of successive doses/exposures
and *f* = a table value taken from Weil (1952; on Table 1 in Appendix 1) for
   the proper *K* (the total number of levels tested minus 1)

Included in Appendix 1 (and in Beyer, 1982) are a complete set of tables covering the full range of possibilities at *N*s up to 10 (where *N* is the number of animals per dosage level). Also included are simplified formulas and calculated values useful if the factor between dosage levels is 2.0 (the logarithm of which is 0.30103 = *d* in this case).

Example 8.4 demonstrates the use of this method and the new tables.

Example 8.4

As part of an inhalation study we exposed groups of 5 rats each to levels of 20, 40, 80, and 160 ppm of a chemical vapor. These exposures killed 0, 1, 3, and 5 animals, respectively. From the $N = 5$, $K = 3$ tables on the *r* value 0, 1, 3, 5 line we get an *f* of 0.7 and an σ_f of 0.31623. We can then calculate the LC_{50} to be:

$$\text{Log } LC_{50} = 1.30130 + 0.30103(2)/2 + 0.30103(0.7)$$

$$= 1.30103 + 0.51175$$

$$= 1.81278$$

$$LC_{50} = 65.0 \text{ ppm with 95\% confidence intervals of}$$

$$\pm 2.179 * d * \sigma_f \text{ or } \pm 2.179(0.30103)(0.31623)$$

$$= \pm 0.20743$$

Therefore, the log confidence limits are 1.81278 ± 0.20743 = 1.60535 to 2.02021; on the linear scale 40.3 to 104.8 ppm.

Assumptions and Limitations

A common misconception is that the moving average method cannot be used to determine the slope of the response curve. This is not true — Weil has published a straightforward method for determining slope in conjunction with a moving average determination of the LD_{50} (Weil, 1983).

The method also provides confidence intervals.

Nonlinear Regression

More often than not in toxicology (and, in fact, in the biological sciences in general) we find that our data demonstrate a relationship between two variables (such as age and body weight) that is not linear. That is, a change in one variable (e.g., age) does not produce a directly proportional change in the other (e.g., body weight). But some form of relationship between the variables is apparent. If understanding such a relationship and being able to predict unknown points is of value, we have a pair of options available to us. The first, which was discussed and reviewed earlier, is to use one or more transformations to linearize our data and then to make use of linear regression. This approach, although most commonly used, has a number of drawbacks. Not all data can be suitably transformed; sometimes the transformations necessary to linearize the data require a cumbersome series of calculations; and the resulting linear regression is not always sufficient to account for the differences among sample values — there are significant deviations around the linear regression line (that is, a line may still not give us a good fit to the data or do an adequate job of representing the relationship between the data). In such cases, we have available a second option — the fitting of data to some nonlinear function such as some form of the curve. This is, in general form, nonlinear regression and may involve fitting data to an infinite number of possible functions. But most often we are interested in fitting curves to a polynomial function of the general form.

$$Y = a + bx + cx^2 + dx^3 + \cdots$$

where x is the independent variable. As the number of powers of x increases, the curve becomes increasingly complex and will be able to fit a given set of data increasingly well.

Generally in toxicology, however, if we plot the log of a response (such as body weight) vs. a linear scale of our dose or stimulus, we get one of four types

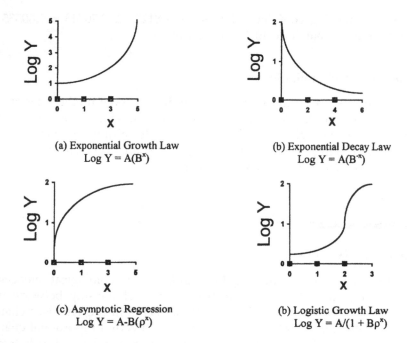

FIGURE 8.1
Common curvilinear curves.

of nonlinear curves. These are (Snedecor and Cochran, 1989):

Exponential growth, where $\log Y = A(B^x)$, such as the growth curve for the log phase of a bacterial culture.

Exponential decay, where $\log Y = A(B^{-x})$, such as a radioactive decay curve.

Asymptotic regression, where $\log Y = A - B(p^x)$, such as a first order reaction curve.

Logistic growth curve, where $\log Y = A/(1 + Bp^x)$, such as a population growth curve.

In all these cases, A and B are constant while p is a log transform. These curves are illustrated in Figure 8.1.

All four types of curves are fit by iterative processes; that is, best guess numbers are initially chosen for each of the constants and, after a fit is attempted, the constants are modified to improve the fit. This process is repeated until an acceptable fit has been generated. Analysis of variance or covariance can be used to objectively evaluate the acceptability of it. Needless to say, the use of a computer generally accelerates such a curve-fitting process.

Assumptions and Limitations

The principle of using least squares may still be applicable in fitting the best curve, if the assumptions of normality, independence, and reasonably error-free measurement of response are valid.

Growth curves are best modeled using a nonlinear method.

Correlation Coefficient

The correlation procedure is used to determine the degree of linear correlation (direct relationship) between two groups of continuous (and normally distributed) variables; it will indicate whether there is any statistical relationship between the variables in the two groups. For example, we may wish to determine if the liver weights of dogs on a feeding study are correlated with their body weights. Thus, we will record the body and liver weights at the time of sacrifice and then calculate the correlation coefficient between these pairs of values to determine if there is some relationship.

A formula for calculating the linear correlation coefficient (r_{xy}) is as follows:

$$r_{xy} = \frac{N\Sigma XY - (\Sigma X)(\Sigma Y)}{\sqrt{N\Sigma X^2 - (\Sigma X)^2}\sqrt{N\Sigma Y^2 - (\Sigma Y)^2}}$$

where X is each value for one variable (such as the dog body weights in the above example), Y is the matching value for the second variable (the liver weights), and N is the number of pairs of X and Y. Once we have obtained r_{xy} it is possible to calculate t_r, which can be used for more precise examination of the degree of significant linear relationship between the two groups. This value is calculated as follows:

$$t_r = \frac{r_{xy}\sqrt{N-2}}{\sqrt{1-r_{xy}^2}}$$

This calculation is also equivalent to r = sample covariance/($Sx\ Sy$), as was seen earlier under ANCOVA.

The value obtained for r_{xy} can be compared to table values (such as Snedecor and Cochran, 1989, p. 477) for the number of pairs of data involved minus two. If the r_{xy} is smaller (at the selected test probability level, such as 0.05), the correlation is not significantly different from zero (no correlation). If r_{xy} is larger than the table value, there is a positive statistical relationship between the groups. Comparisons are then made at lower levels of probability to determine the degree of relationship (note that if r_{xy} = either 1.0 or −1.0, there is complete correlation between the groups). If r_{xy} is a negative number and the absolute is greater than the table value, there is an inverse relationship between the groups; that is, a change in one group is associated with a change in the opposite direction in the second group of variables. Both computations are demonstrated in Example 8.5.

Since the comparison of r_{xy} with the table values may be considered a somewhat weak test, it is perhaps more meaningful to compare the t_r value with values in a t distribution table for $N - 2$ degrees of freedom (df), as is done for the Student's t-test. This will give a more exact determination of the degree of statistical correlation between the two groups.

It should be noted that this method examines only possible linear relationships between sets of continuous, normally distributed data. Other mathematical relationships (log, log/linear, exponential, etc.) between data sets exist that require either the use of another correlation testing method or that one or more of the data sets be transformed so that they are of linear nature. This second approach requires, of course, that one know the nature of the data so that an appropriate transform may be used. A few transforms were discussed earlier in the sections on linear regression and probit/log analysis.

Example 8.5

If we computed the dog body weight vs. dog liver weight for a study we could have the following results:

Dog #	Body Weight (kg) X	X^2	Liver Weight (g) Y	Y^2	XY
1	8.4	70.56	243	59049	2041.2
2	8.5	72.25	225	50625	1912.5
3	9.3	86.49	241	58081	2241.3
4	9.5	90.25	263	69169	2498.5
5	10.5	110.25	256	65536	2688.0
6	8.6	73.96	266	70756	2287.6
Sums	$\Sigma X = 54.8$	$\Sigma X^2 =$ 503.76	$\Sigma Y = 1494$	$\Sigma Y^2 =$ 373216	$\Sigma XY =$ 13669.1

$$r_{xy} = \frac{6(13669.1) - (54.8)(1494)}{\sqrt{6(503.76) - (54.8)^2}\sqrt{6(373216) - (1494)^2}} = 0.381$$

The table value for six pairs of data (read beside the $N - 2$ value, or $6 - 2 = 4$) is 0.811 at a 0.05 probability level. Thus, there is a lack of statistical correlation (at $p = 0.05$) between the body weights and liver weights for this group of dogs.

The t_r value for these data would be calculated as follows:

$$t_r = \frac{0.381\sqrt{6-2}}{\sqrt{1-(0.381)^2}} = 0.824$$

The value for the t-distribution table for 4 df at the 0.05 level is 2.776; therefore, this again suggests a lack of significant correlation at $p = .05$.

Assumptions and Limitations

A strong correlation does not imply that a treatment causes an effect.

The distances of data points from the regression line are the portions of the data not "explained" by the model. These are called residuals. Poor correlation coefficients imply high residuals, which may be due to many small contributions (variations of data from the regression line) or a few large ones. Extreme values (outliers) greatly reduce correlation.

X and Y are assumed to be independent.

Feinstein (1979) has provided a fine discussion of the difference between correlation (or association of variables) and causation.

Kendall's Coefficient of Rank Correlation

Kendall's rank correlation, represented by τ(tau), should be used to evaluate the degree of association between two sets of data when the nature of the data is such that the relationship may not be linear. Most commonly, this is when the data are not continuous and/or normally distributed. An example of such a case is when we are trying to determine if there is a relationship between the length of hydra and their survival time in a test medium in hours, as is presented in Example 8.6. Both of our variables here are discontinuous, yet we suspect a relationship exists. Another common use is in comparing the subjective scoring done by two different observers.

Tau is calculated at $\tau = N/n(n-1)$ where n is the sample size and N is the count of ranks, calculated as $N = 4(C_i) - n(n-1)$, with the computing of C_i being demonstrated in the example.

If a second variable Y_2 is exactly correlated with the first variable Y_1, then the variates Y_2 should be in the same order as the Y_1 variates. However, if the correlation is less than exact, the order of the variates Y_2 will not correspond entirely to that of Y_1. The quantity N measures how well the second variable corresponds to the order of the first. It has maximum value of $n(n-1)$ and a minimum value of $-n(n-1)$.

A table of data is set up with each of the two variables being ranked separately. Tied ranks are assigned as demonstrated earlier under the Kruskall–Wallis test. From this point, disregard the original variates and deal only with the ranks. Place the ranks of one of the two variables in rank order (from lowest to highest), paired with the rank values assigned for the other variable. If one (but not the other) variable has tied ranks, order the pairs by the variables without ties (Sokal and Rohlf, 1994, p. 601–607).

The most common way to compute a sum of the counts is also demonstrated in Example 8.6.

The resulting value of τ will range from −1 to +1, as does the familiar parametric correlation coefficient, r.

Example 8.6

During the validation of an in vitro method, it was noticed that larger hydra seem to survive longer in test media than do small individuals. To evaluate this, 15 hydra of random size were measured (mm), then placed in test media. How many hours each individual survives was recorded over a 24-h period. These data are presented below, along with ranks for each variable.

Length	Rank (R_1)	Survival	Rank (R_2)
3	6.5	19	9
4	10	17	7
6	15	11	1
1	1.5	25	15
3	6.5	18	8
3	6.5	22	12
1	1.5	24	14
4	10	16	6
4	10	15	5
2	3.5	21	11
5	13	13	3
5	13	14	4
3	6.5	20	10
2	3.5	23	13
5	13	12	2

We then arrange this based on the order of the rank of survival time (there are no ties here). We then calculate our counts of ranks. The conventional method is to obtain a sum of the counts c_i, as follows: examine the first value in the column of ranks paired with the ordered column. In the case below, this is rank 15. Count all ranks subsequent to it that rank greater than 15. There are 14 ranks following the 2 and all of them are less than 15. Therefore, we count a score of $C_1 = 0$. We repeat this process for each subsequent rank of r_1, giving us a final score of 1 (by this point it is obvious that our original hypothesis — that larger hydra live longer in test media than do small individuals — was in error).

R_2	R_1	Following (R_2) Ranks Greater than R_1	Counts (C_i)
1	15	—	$C_1 = 0$
2	13	—	$C_2 = 0$
3	13	—	$C_3 = 0$
4	13	—	$C_4 = 0$
5	10	—	$C_5 = 0$
6	10	—	$C_6 = 0$
7	10	—	$C_7 = 0$
8	6.5	—	$C_8 = 0$
9	6.5	—	$C_9 = 0$
10	6.5	—	$C_{10} = 0$
11	3.5	6.5	$C_{11} = 1$
12	6.5	—	$C_{12} = 0$
13	3.5	—	$C_{13} = 0$
14	1.5	—	$C_{14} = 0$
15	1.5	—	$C_{15} = 0$
			$C_i - 1$

Our count of ranks, N, is then calculated as

$$N = 4(1) - 15(15 - 1) = -206$$

We can then calculate τ as

$$\tau = -206 / [15(15 - 1)] = -0.9810$$

In other words, there is a strong negative correlation between our variables.

Assumptions and Limitations

A very robust estimator that does not assume normality, linearity, or minimal error of measurement.

References

Abramowitz, M. and Stegun, I.A. (1964) *Handbook of Mathematical Functions.* National Bureau of Standards, Washington, DC, pp. 925–964.

Beyer, W.H. (1982) *CRC Handbook of Tables for Probability and Statistics.* CRC Press, Boca Raton, FL.

Bliss, C.I. (1935) The calculation of the dosage-mortality curve. *Ann. Appl. Biol.*, 22, 134–167.

Draper, N.R. and Smith, H. (1981) *Applied Regression Analysis.* John Wiley, New York.

Dykstra, R.L. and Robertson, T. (1983) On testing monotone tendencies. *J. Am. Stat. Assoc.*, 78, 342–350.

Feinstein, A.R. (1979) Scientific standards vs. statistical associations and biological logic in the analysis of causation. *Clin. Pharmacol. Therapeut.*, 25, 481–492.

Finney, D.K. (1977) *Probit Analysis*, 3rd ed., Cambridge University Press, Cambridge.

Gallant, A.R. (1975) Nonlinear regression. *American Statistican*, 29, 73–81.

Gold, H.J. (1977) *Mathematical Modeling of Biological System: An Introductory Guidebook.* John Wiley, New York.

Litchfield, J.T. and Wilcoxon, F. (1949) A simplified method of evaluating dose effect experiments. *J. Pharmacol. Exp. Therapeut.*, 96, 99–113.

Montgomery, D.C., Peck, E.A, and Vining, G.G. (2001) *Introduction to Linear Regression Analysis*, 3rd ed., Wiley-InterScience, New York.

Pollard, J.H. (1977) *Numerical and Statistical Techniques.* Cambridge University Press, New York.

Prentice, R.L. (1976) A generalization of the probit and logit methods for dose response curves. *Biometrics*, 32, 761–768.

Snedecor, G.W. and Cochran, W.G. (1989) *Statistical Methods*, 8th ed., Iowa State University Press, Ames.

Sokal, R.R. and Rohlf, F.J. (1994) *Biometry: The Principles and Practice of Statistics in Biological Research*, 3rd ed., W.H. Freeman, San Francisco.

Thompson, W.R. and Weil, C.S. (1952) On the construction of tables for moving average interpolation. *Biometrics*, 8, 51–54.

Weil, C.S. (1952) Tables for convenient calculation of median-effective dose (LD_{50} or ED_{50}) and instructions in their use. *Biometrics*, 8, 249–263.

Weil, C.S. (1983) Economical LD_{50} and slope determinations. *Drug Chem. Toxicol.*, 6, 595–603.

9

Trend Analysis

Trend analysis is a collection of techniques that have been "discovered" by toxicology over the last 25 years (Tarone, 1975; Tarone and Gart, 1980), but which has only gained popularity in the last 15 years. The actual methodology dates back to the mid-1950s (Cox and Stuart, 1955).

Trend analysis methods are a variation on the theme of regression testing. In the broadest sense, the methods are used to determine whether a sequence of observations taken over an ordered range of a variable (most commonly time) exhibit some form of pattern of change (e.g., an increase-upward trend) associated with another variable of interest (in toxicology, some form or measure of dosage of an exposure).

There are a number of tests available to evaluate data sets to determine if a trend exists. These can be generalized or specialized for particular types of data, such as counts, preparations, or discrete, continuous values.

Trend corresponds to sustained and systematic variations over a long period of time. It is associated with the structural causes of the phenomenon in question; for example, population growth, technological progress, new ways of organization, or capital accumulation.

The identification of trend has always posed a serious statistical problem. The problem is not one of mathematical or analytical complexity but of conceptual complexity. This problem exists because the trend as well as the remaining components of a time series are latent (nonobservables) variables and, therefore, assumptions must be made on their behavioral pattern. The trend is generally thought of as a smooth and slow movement over a long term. The concept of "long" in this connection is relative and what is identified as trend for a given series span might well be part of a long cycle once the series is considerably augmented. Often, a long cycle is treated as a trend because the length of the observed time series is shorter than one complete face of this type of cycle.

The ways in which data are collected in toxicology studies frequently serve to complicate trend analysis, as the length of time for the phenomena underlying a trend to express themselves is frequently artificially censored.

To avoid the complexity of the problem posed by a statistically vague definition, statisticians have resorted to two simple solutions. One consists of estimating trend and cyclical fluctuations together, calling this combined movement "trend-cycle." The other consists of defining the trend in terms of the series length, denoting it as the longest nonperiodic movement.

Trend Models

Within the large class of models identified for trend, we can distinguish two main categories: deterministic trends and stochastic trends.

Deterministic trend models are based on the assumption that the trend of a time series can be approximated closely by simple mathematical functions of time over the entire span of the series. The most common representation of a deterministic trend is by means of polynomials or of transcendental functions. The time series from which the trend is to be identified is assumed to be generated by a nonstationary process where the nonstationarity results from a deterministic trend. A classical model is the regression or error model (Anderson, 1971) where the observed series is treated as the sum of a systematic part, or trend, and a random part, or irregular. This model can be written as:

$$Z_t = Y_t + U_t$$

where U_t is a purely random process; that is, $U_t \sim$ i.i.d.(O, σ_U^2), independent and identically distributed with expected value O and variance σ_U^2.

Trend tests are generally described as "k-sample tests of the null hypothesis of identical distribution against an alternative of linear order," i.e., if sample I has distribution function F_i, $i = 1,\ldots,k$, then the null hypothesis

$$H_0: F_1 = F_1 = \ldots = F_k$$

is tested against the alternative

$$H_1: F_1\ F_2 \geq \ldots \geq F_k$$

(or its reverse), where at least one of the inequalities is strict. These tests can be thought of as special cases of tests of regression or correlation in which association is sought between the observations and its ordered sample index. They are also related to analysis of variance except that the tests are tailored to be powerful against the subset of alternatives H_1 instead of the more general set $\{F_i \neq F_j$, some $i \neq j\}$.

Different tests arise from requiring power against specific elements or subsets of this rather extensive set of alternatives.

The most popular trend test in toxicology is currently presented by Tarone (1975) because it is used by the National Cancer Institute (NCI) in the analysis of carcinogenicity data. A simple but efficient alternative is the Cox and Stuart test (Cox and Stuart, 1955), which is a modification of the sign test. For each point at which we have a measure (such as the incidence of animals observed with tumors) we form a pair of observations — one from each of the groups we wish to compare. In a traditional NCI bioassay, this would mean pairing control with low dose and low dose with high dose (to explore a dose-related trend) or each time period observation in a dose group (except the first) with its predecessor (to evaluate a time-related trend). When the second observation in a pair exceeds the earlier observation, we record a plus sign for that pair. When the first observation is greater than the second, we record a minus sign for that pair. A preponderance of plus signs suggests a downward trend, whereas an excess of minus signs suggests an upward trend. A formal test at a preselected confidence level can then be performed.

More formally put, after having defined what trend we want to test for, we first match pairs as (X_1, X_{1+c}), (X_2, X_{2+c}), ... $(X_{n'-c}, X_{n'})$ where $c = n'/2$ when n' is even and $c = (n' + 1)/2$ when n' is odd (where n' is the number of observations in a set). The hypothesis is then tested by comparing the resulting number of excess positive or negative signs against a sign test table such as found in Beyer (1982).

We can, of course, combine a number of observations to allow ourselves to actively test for a set of trends, such as the existence of a trend of increasing difference between two groups of animals over a period of time. This is demonstrated in Example 9.1.

Example 9.1

In a chronic feeding study in rats, we tested the hypothesis that, in the second year of the study, there was a dose-responsive increase in tumor incidence associated with the test compound. We utilize below a Cox–Stuart test for trend to address this question. All groups start the second year with an equal number of animals.

	Control			Low Doses					
Month of Study	Total X Animals with Tumors	Change (X_{B-A})	Total Y Animal Tumors	Change (Y_{B-A})	Compared to Control $(Y - X)$	Total Z Animals with Tumors	Change (Z_{B-A})	Compared to Control $(Z - X)$
---	---	---	---	---	---	---	---	---
12 (A)	1	NA	0	NA	NA	5	NA	NA
13 (B)	1	1	0	0	0	7	2	(+)2
14 (C)	3	2	1	1	(–)1	11	4	(+)2
15 (D)	3	0	1	0	0	11	0	0
16 (E)	4	1	1	0	(–)1	13	2	(+)1
17 (F)	5	1	3	2	(+)1	14	1	0
18 (G)	5	0	3	0	0	15	1	(+)1
19 (H)	5	0	5	2	(+)2	18	3	(+)3
20 (I)	6	1	6	1	0	19	1	0
21 (J)	8	2	7	1	(–)1	22	3	(+)1
22 (K)	12	4	9	2	(–)2	26	4	0
23 (L)	14	2	12	3	(+)1	28	2	0
24 (M)	18	4	17	5	(+)1	31	3	(–)1
			Sum of signs Y – X	(+)4 (–)4 = 0 (No trend)			Sum of signs Z – X	(+)6 (–)1 = +5

Note: Reference to the sign table is not necessary for the low dose comparison (where there is no trend) but clearly shows the high dose to be significant at the $p \leq 0.05$ level.

Assumptions and Limitations

Trend tests seek to evaluate whether there is a monotonic tendency in response to a change in treatment. That is, the dose response direction is absolute — as dose goes up, the incidence of tumors increases. Thus the test loses power rapidly in response to the occurrences of "reversals," for example, a low dose group with a decreased tumor incidence. There are methods (such as those of Dykstra and Robertson, 1983) that "smooth the bumps" of reversals in long data series. In toxicology, however, most data series are short (that is, there are only a few dose levels).

Tarone's trend test is most powerful at detecting dose-related trends when tumor onset hazard functions are proportional to each other. For more power against other dose-related group differences, weighted versions of the statistic are also available; see Breslow (1984) or Crowley and Breslow (1984) for details.

In 1985, the United States *Federal Register* recommended that the analysis of tumor incidence data be carried out with a Cochran–Armitage (Armitage, 1955; Cochran, 1954) trend test. The test statistic of the Cochran–Armitage test is defined as this term:

$$T_{CA} = \sqrt{\frac{N}{(N-r)}} \frac{\sum_{i=1}^{k}\left(r_i - \frac{n_i}{N}r\right)d_i}{\sqrt{\sum_{i=1}^{k}\frac{n_i}{N}d_i^2 - \left(\sum_{i=1}^{k}\frac{n_i}{N}d_i\right)^2}}$$

where $r = \Sigma r_i$, $N = \Sigma n_i$, and dose scores d_i ($d_1 \leq d_2 \leq \dots \leq d_k$). Armitage's test statistic is the square of this term (T_{CA}^2). As one-sided tests are carried out for an increase of tumor rates, the square is not considered. Instead, the above-mentioned test statistic, which is presented by Portier and Hoel (1984), is used. This test statistic is asymptotically standard normal distributed. The Cochran–Armitage test is asymptotically efficient for all monotone alternatives (Tarone, 1975), but this result only holds asymptotically. And tumors are rare events, so the binominal proportions are small. In this situation approximations may become unreliable.

Therefore, exact tests that can be performed using two different approaches, conditional and unconditional, are considered. In the first case, the total number of tumors r is regarded as fixed. As a result, the null distribution of the test statistic is independent of the common probability p. The exact conditional null distribution is a multivariate hypergeometric distribution.

The unconditional model treats the sum of all tumors as a random variable. Then the exact unconditional null distribution is a multivariate binomial distribution. The distribution depends on the unknown probability.

References

Anderson, T.W. (1971) *The Statistical Analysis of Time Series*. Wiley, New York.
Armitage, P. (1955) Tests for linear trends in proportions and frequencies. *Biometrics*, 11, 375–386.

Beyer, W.H. (1982) CRC Handbook for Probability and Statistics, CRC Press, Boca Raton, FL.

Breslow, N. (1984) Comparison of survival curves, in *Cancer Clinical Trials: Methods and Practice*, Buse, M.F., Staguet, M.J., and Sylvester, R.F., Eds., Oxford University Press, Oxford, pp. 381–406.

Cochran, W.F. (1954) Some models for strengthening the common x^2 tests. *Biometrics*, 10, 417–451.

Cox, D.R. and Stuart, A. (1955) Some quick tests for trend in location and dispersion. *Biometrics*, 42, 80–95.

Crowley, J. and Breslow, N. (1984) Statistical analysis of survival data. *Annu. Rev. Publ. Health*, 5, 385–411.

Dykstra, R.L. and Robertson, T. (1983) On testing monotone tendencies. *J. Am. Stat. Assoc.*, 78, 342–350.

Federal *Register.* (1985) 50(50), .

Portier, C. and Hoel, D. (1984) Type I error of trend tests in proportions and the design of cancer screens. *Comm. Stat. Theor. Meth.*, A13, 1–14.

Tarone, R.E. (1975) Tests for trend in life table analysis. *Biometrika*, 62, 679–682.

Tarone, R.E. and Gart, J.J. (1980) On the robustness of combined tests for trends in proportions. *J. Am. Stat. Assoc.*, 75, 110–116.

10

Methods for the Reduction of Dimensionality

Techniques for the reduction of dimensionality are those that simplify the understanding of data, either visually or numerically, while causing only minimal reductions in the amount of information present. These techniques operate primarily by pooling or combining groups of variables into single variables, but may also entail the identification and elimination of low-information-content (or irrelevant) variables.

In an earlier chapter we presented the descriptive statistics (calculations of means, standard deviations, etc.) that are the simplest and most familiar form of reduction of dimensionality. The first topic we will address in this chapter is classification, which provides the general conceptual tools for identifying and quantifying similarities and differences between groups of things that have more than a single linear scale of measurement in common (for example, that have both been determined to have or lack a number of enzyme activities).

Two common methodology sets will then be presented and reviewed. Both table preparation and statistical graphics are familiar activities for most toxicologists, but are much misused. Guidelines for the proper preparation of tables and graphs are presented, along with some of the more recent improvements in presentation techniques that are available. Both tables and graphs can address two objectives: presentation (communication of information) and analysis.

We will then move on to two collections of methodologies that combine graphic and computational methods, multidimensional/nonmetric scaling, and cluster analysis. Multidimensional scaling (MDS) is a set of techniques for quantitatively analyzing similarities, dissimilarities, and distances between data in a display-like manner. Nonmetric scaling is an analogous set of methods for displaying and relating data when measurements are non-quantitative (the data are described by attributes or ranks). Cluster analysis is a collection of graphic and numerical methodologies for classifying things based on the relationships between the values of the variables that they share.

The final pair of methods for reduction of dimensionality that will be tackled in this chapter are Fourier analysis and the life table analysis. Fourier analysis seeks to identify cyclic patterns in data and then either analyze the patterns or the residuals after the patterns are taken out. Life table analysis techniques are directed to identifying and quantitating the time course of risks (such as death, or the occurrence of tumors).

Three purely multivariate techniques for the reduction of dimensionality (principal components analysis, factor analysis, and canonical correlation analysis) are discussed in a later chapter on multivariate analysis.

Classification

Classification is both a basic concept and a collection of techniques that are necessary prerequisites for further analysis of data when the members of a set of data are (or can be) each described by several variables. At least some degree of classification (which is broadly defined as the dividing of the members of a group into smaller groups in accordance with a set of decision rules) is necessary prior to any data collection. Whether formally or informally, an investigator has to decide which things are similar enough to be counted as the same and develop rules for governing collection procedures. Such rules can be as simple as "measure and record body weights only of live animals on study," or as complex as demonstrated by the expanded weighting classification presented in Example 10.1. Such a classification also demonstrates that the selection of which variables to measure will determine the final classification of data.

Example 10.1

Is animal of desired species?	Yes/No
Is animal member of study group?	Yes/No
Is animal alive?	Yes/No
Which group does animal belong to?	
Control	
Low dose	
Intermediate dose	
High dose	
What sex is animal?	Male/Female
Is the measured weight in acceptable range?	Yes/No

Classifications of data have two purposes (Hartigan, 1983; Gordon, 1981): data simplification (also called a descriptive function) and prediction. Simplification is necessary because there is a limit to both the volume and complexity of data that the human mind can comprehend and deal with conceptually. Classification allows us to attach a label (or name) to each group of data, to summarize the data (that is, assign individual elements of data to groups and to characterize the population of the group), and to define the relationships between groups (that is, develop a taxonomy).

Prediction, meanwhile, is the use of summaries of data and knowledge of the relationships between groups to develop hypotheses as to what will happen when further data are collected (as when more animals or people are exposed to an agent under defined conditions) and as to the mechanisms that cause such relationships to develop. Indeed, classification is the prime device for the discovery of mechanisms in all of science. A classic example of this was Darwin's realization that there were reasons (the mechanisms of evolution) behind the differences and similarities in species that had caused Linnaeus to earlier develop his initial modern classification scheme (or taxonomy) for animals.

To develop a classification, one first sets bounds wide enough to encompass the entire range of data to be considered, but not unnecessarily wide. This is typically done by selecting some global variables (variables every piece of data have in common) and limiting the range of each so that it just encompasses all the cases on hand. Then one selects a set of local variables (characteristics that only some of the cases have; e.g., the occurrence of certain tumor types, enzyme activity levels, or dietary preferences) that serve to differentiate between groups. Data are then collected, and a system for measuring differences and similarities is developed. Such measurements are based on some form of measurement of distance between two cases (x and y) in terms of each single variable scale. If the variable is a continuous one, then the simplest measure of distance between two pieces of data is the Euclidean distance, $[d(x, y)]$ defined as:

$$d(x,y) = \sqrt{\sum (x_i - y_i)^2}$$

For categorical or discontinuous data, the simplest distance measure is the matching distance, defined as:

$$d(x, y) = \text{number of times } x_i \neq y_i$$

After we have developed a table of such distance measurements for each of the local variables, some weighting factor is assigned to each variable. A weighting factor seeks to give greater importance to those variables that are believed to have more relevance or predictive value. The weighted variables are then used to assign each piece of data to a group. The actual act of developing numerically based classifications and assigning data members to them is the realm of cluster analysis, which will be discussed later in this chapter. Classification of biological data based on qualitative factors has been well discussed by Glass (1975), and Gordon (1981) does an excellent job of introducing the entire field and mathematical concepts.

Relevant examples of the use of classification techniques range from the simple to the complex. Schaper et al. (1985) developed and used a very simple classification of response methodology to identify those airborne chemicals that alter the normal respiratory response induced by CO_2. At the other end of the spectrum, Kowalski and Bender (1972) developed a more mathematically based system to classify chemical data (a methodology they termed pattern recognition).

Table Preparation

Tables are the most common means of presenting data (in both raw and summarized forms) in toxicology. There are some general principals that should be adhered to in the construction and use of tables. However, these are not commonly followed, or even understood. Some of these principles also apply to graphs and their use, which is a major reason for placing this discussion of tables in this section of the book.

Tables should be used when some form of repetitive data must be presented. They should be explained in the accompanying text, but their contents should not be repeated in the text.

If only a few (two or three) pieces of data are to be presented, place them in the text (not as a table).

Tables should be self-contained but should not try to repeat the materials and methods portions of a report in the table's contents.

Column headings in tables should always specify the units of measure. The contents should be arranged like elements, which are read down in a table, not across. The contents should be arranged so that there is an order to the elements (such as ascending dose or increase in time or exposure). This helps in mentally summarizing and understanding the table contents.

Standard conditions that apply to all the data elements presented in the table or the columns or rows of tables should be presented in footnotes, and not in the columns of a table. The footnotes should be denoted by superscript symbols that are clear and consistent.

Symbols (such as ascending numbers, letters, or sets of asterisks) should be assigned in the order they will be encountered in reading the table. For example, the first footnote encountered should be denoted by 1, the second by 2, etc.

There are common conventions that set aside certain groups of footnote symbols for particular uses. In toxicology, statistical significance is very commonly denoted by the use of superscript as, bs, cs, denoting $p \leq 0.05$, $p \leq 0.01$, and $p \leq 0.001$, respectively. The use of these same symbols for other footnotes should be avoided, as it may lead to confusion.

Any table that presents a large number of identical entries as results should be reconsidered. For example, it's wasteful to present a table of the effects of 15 different dose levels on body weight where 6 of the doses had no effect (a 0% change), 8 cut weight gain in half (a 50% response), and 1 dose did something in between this result could be presented very clearly by a single sentence in the text.

The table legend (title) should clearly label (summarize) the contents or intent of the table.

Do not use more significant figures than are necessary or appropriate for the precision of the data — to do so misleads the reader. And the number of significant figures presented in a column (also called the level of decimalization) should be constant for the entire column and all equivalent data sets presented in the column. If three columns all present mean rabbit weights, for example, then all should report weights to the gram (or tenths of grams, tens of grams, etc.). If some of the members of the column were actually measured to a different level of precision (which happens when different scales are used to weigh different groups of animals), those with more precision should be rounded for inclusion in the table.

Do not present descriptive statistics (such as means and standard deviations) that are labeled in a misleading manner. Examples of this are presenting the arithmetic average of two numbers as a mean and calculating and presenting a standard deviation for these two numbers. Similarly, presenting SEMs (standard error of measurements) without also presenting the size of the groups that they are calculated for is incorrect and makes evaluation of the presented data difficult.

Statistical Graphics

The use of graphics in one form or another in statistics is the single most effective and robust statistical tool and, at the same time, one of the most poorly understood and improperly used.

Graphs are used in statistics (and in toxicology) for one of four major purposes. Each of the four is a variation on the central theme of making complex data easier to understand and use. These four major functions are exploration, analysis, communication and display of data, and graphical aids. Exploration (which may be simply summarizing data or trying to expose relationships between variables) is determining the characteristics of data sets and deciding on one or more appropriate forms of further analysis, such as the scatterplot. Analysis is the use of graphs to formally evaluate some aspect of the data, such as whether there are outliers present or if an underlying assumption of a population distribution is fulfilled. As long ago as 1960 (Anderson, 1960), some 18 graphical methods for analyzing multivariate data relationships were developed and proposed.

Communication and display of data are the most commonly used functions of statistical graphics in toxicology, whether used for internal reports, presentations at meetings, or formal publications in the professional literature. In communicating data, graphs should not be used to duplicate data that are presented in tables, but rather to show important trends and/or relationships in the data. Although such communication is most commonly of a quantitative compilation of actual data, it can also be used to summarize and present the results of statistical analysis. The fourth and final function of graphics is one that is largely becoming outdated as microcomputers become more widely available. Graphical aids to calculation include nomograms (the classic example in toxicology of a nomogram is presented by Litchfield and Wilcoxon, 1949, for determining median effective doses) and extrapolating and interpolating data graphically based on plotted data.

There are many forms of statistical graphics (a partial list, classified by function, is presented in Table 10.1), and a number of these (such as scatterplots and histograms) can be used for each of a number of possible functions. Most of these plots are based on a Cartesian system (that is, they use a set of rectangular coordinates), and our review of construction and use will focus on these forms of graphs.

TABLE 10.1

Forms of Statistical Graphics (by Function)

EXPLORATION		
Data Summary	**Two Variables**	**Three or More Variables**
Box and whisker plot	Autocorrelation plot	Biplot
Histogram[a]	Cross-correlation plot	Cluster trees[a]
Dot-array diagram	Scatter plot[a]	Labeled scatter plot[a]
Frequency polygon	Sequence plot	Glyphs and metroglyphs
Ogive		Face plots
Stem and leaf diagram		Fourier plots
		Similarity and preference maps
		Multidimensional scaling displays
		Weathervane plot

ANALYSIS		
Distribution Assessment	**Model Evaluation and Assumption Verification**	**Decision Making**
Probability plot	Average vs. standard deviation	Control chart
Q–Q plot	Component-plus-residual plot	Custom chart
P–P plot	Partial-residual plot	Half-normal plot
Hanging histogram	Residual plots	Ridge trace
Rootagram		Youden plot
Poissonness plot		

COMMUNICATION AND DISPLAY OF DATA		
Quantitative Graphics	**Summary of Statistical Analyses**	**Graphical Aids**
Line chart[a]	Means plot	Confidence limits
Pictogram	Sliding reference distribution	Graph paper
Pie chart[a]	Notched box plot	Power curves
Contour plot[a]	Factor space/response	Nomographs
Stereogram	Interaction plot	Sample-size curves
Color Map	Contour plot	Trilinear coordinates
Histogram[a]	Predicted response plot	
	Confidence region plot	

[a] Reviewed in the text of this book.

Construction of a rectangular graph of any form starts with the selection of the appropriate form of graph followed by the laying out of the coordinates (or axes). Even graphs that are going to encompass mulivariate data (that is, more than two variables) generally have as their starting point two major coordinates. The vertical axis, or ordinate (also called the Y axis), is used to present an independent variable. Each of these axes is scaled in the units of measure that will most clearly present the trends of interest in the data. The range covered by the scale of each axis is selected to cover the entire region for which data is presented. The actual demarking of the measurement scale along an axis should allow for easy and accurate assessment of the coordinates of any data point, yet should not be cluttered.

Actual data points should be presented by symbols that present the appropriate indicators of location and, if they represent a summary of data from a normal data population, it would be appropriate to present a symbol for the mean and some indication of the variability (or error) associated with that population, commonly by using "error bars," which present the standard deviation (or standard error) from the mean. If, however, the data are not normal or continuous it would be more appropriate to indicate location by the median and present the range or semi-quartile distance for variability estimates. The symbols that are used to present data points can also be used to present a significant amount of additional information. At the simplest level a set of clearly distinct symbols (circles, triangles, squares, etc.) are very commonly used to provide a third dimension of data (most commonly treatment group). But by clever use of symbols, all sorts of additional information can be presented. Using a method such as Chernoff's faces (Chernoff, 1973), in which faces are used as symbols of the data points (and various aspects of the faces present additional data, such as the presence or absence of eyes denoting presence or absence of a secondary pathological condition), it is possible to present a large number of different variables on a single graph.

Already presented in this book are the basics of constructing and using simple line (or curve) plots and scatterplots. Separate parts of this chapter will address biplots and cluster analysis. There are three other forms of graphs that are commonly used in toxicology that we will now look at. These are histograms, pie charts, and contour plots.

Histograms are graphs of simple frequency distribution. Commonly, the abscissa is the variable of interest (such as lifespan or litter size) and is generally shown as classes or intervals of measurements (such as age ranges of 0 to 10, 10 to 20 weeks, etc.). The ordinate, meanwhile, is the incidence or frequency of observations. The result is a set of vertical bars, each of which represents the incidence of a particular set of observations. Measures of error or variability about each incidence are reflected by some form of error bar on top of or in the frequency bars, as shown in Figure 10.1. The size of class intervals may be unequal (in effect, one can combine or pool several small class intervals), but it is proper in such cases to vary the width of the bars to indicate differences in interval size.

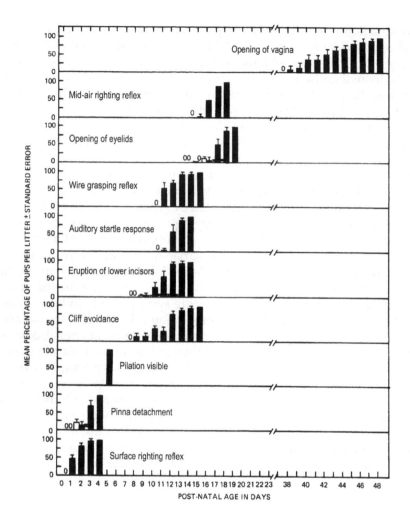

FIGURE 10.1
Acquisitions of postnatal development landmarks in rats.

Pie charts are the only common form of quantitative graphic technique that is not rectangular. Rather, the figure is presented as a circle out of which several "slices" are delimited. The only major use of the pie chart is in presenting a breakdown of the components of a group. Typically, the entire set of data under consideration (such as total body weight) constitutes the pie while each slice represents a percentage of the whole (such as the percentages represented by each of several organs). The total number of slices in a pie should be small for the presentation to be effective. Variability or error can be readily presented by having a subslice of each sector shaded and labeled accordingly.

Finally, there is the contour plot, which is used to depict the relationships in a three-variable, continuous data system. That is, a contour plot visually portrays each contour as a locus of the values of two variables associated with a constant value of the third variable. An example would be a relief map that gives both latitude and longitude of constant altitude using contour lines.

Just as with tables, there are a number of practical guidelines for the construction and use of graphs.

Graphs should be used to present or communicate data when there are pronounced or interesting trends or data relationships. If these are not present, the data should be presented in the text or in a table.

The most common misuse of graphs is to either conceal or exaggerate the extent of the difference by using inappropriately scaled or ranged axis. Tufte (1983) has termed a statistic for evaluating the appropriateness of scale size — the lie factor — defined as:

$$\text{Lie factor} = \frac{\text{Size of effect shown in graph}}{\text{Size of effect in data}}$$

An acceptable range for the lie factor is from 0.95 to 1.05. Less means the size of an effect is being understated, more that the effect is being exaggerated.

There is rarely a reason for both graphs and tables of the same data to be presented. However, in some instances it is important to both present exact values and demonstrate a trend in data that is not apparent from an inspection of the tabular data.

The axes should be clearly labeled with the names of the variables and the units of measurement.

Scale breaks should be used where there are large gaps between data points. The breaks should be clearly visible.

Visual comparison of related graphs should be made as easy as possible. This can be done by such actions as using identical scales of measurement and placing the graphs next to each other.

The legends should make the graphs as freestanding as possible. All the guidelines presented earlier for footnotes and headings for tables should also be followed for graphs.

There are a number of excellent references available for those who would like to further pursue statistical graphics. Anscombe (1973) presents an excellent short overview, while books by Schmid (1983) and Tufte (1983, 1990, and 1997) provide a wealth of information. A final reference that is very worthwhile is Craver's *Make Graph Paper with Your Copier* (1991), which will provide the investigator with master forms to make over 200 different kinds of graphs.

Multidimensional and Nonmetric Scaling

Multidimensional scaling (MDS) is a collection of analysis methods for data sets that have three or more variables making up each data point. MDS displays the relationships of three or more dimensional extensions of the methods of statistical graphics.

MDS presents the structure of a set of objects from data that approximate the distances between pairs of the objects. The data, called similarities, dissimilarities, distances, or proximities, must be in such a form that the degree of similarities and differences between the pairs of the objects (each of which represents a real-life data point) can be measured and handled as a distance (remember the discussion of measures of distances under classifications). Similarity is a matter of degree — small differences between objects cause them to be similar (a high degree of similarity), while large differences cause them to be considered dissimilar (a small degree of similarity).

In addition to the traditional human conceptual or subjective judgments or similarity, data can be an "objective" similarity measure (the difference in weight between a pair of animals) or an index calculated from multivariate data (the proportion of agreement in the results of a number of carcinogenicity studies). However, the data must always represent the degree of similarity of pairs of objects.

Each object or data point is represented by a point in a multidimensional space. These plots or projected points are arranged in this space so that the distances between pairs of points have the strongest possible relation to the degree of similarity among the pairs of objects. That is, two similar objects are represented by two points that are close together, and two dissimilar objects are represented by a pair of points that are far apart. The space is usually a two- or three-dimensional Euclidean space, but may be non-Euclidean and may have more dimensions.

MDS is a general term that includes a number of different types of techniques. However, all seek to allow geometric analysis of multivariate data. The forms of MDS can be classified (Young, 1985) according to the nature of the similarities in the data. It can be qualitative (nonmetric) or quantitative (metric MDS). The types can also be classified by the number of variables involved and by the nature of the model used; for example, classical MDS (there is only one data matrix, and no weighting factors are used on the data), replicated MDS (more than one matrix and no weighting), and weighted MDS (more than one matrix and at least some of the data being weighted).

MDS can be used in toxicology to analyze the similarities and differences between effects produced by different agents, in an attempt to use an understanding of the mechanism underlying the actions of one agent to determine the mechanisms of the other agents. Actual algorithms and a good intermediate level presentation of MDS can be found in Davison (1983).

Nonmetric scaling is a set of graphic techniques closely related to MDS and is definitely useful for the reduction of dimensionality. Its major objective is to arrange a set of objects (each object, for our purposes, consisting of a number of related observations) graphically in a few dimensions, while retaining the maximum possible fidelity to the original relationships between members (that is, values that are most different are portrayed as most distant). It is not a linear technique and does not preserve linear relationships (i.e., A is not shown as twice as far from C as B, even though its "value difference" may be twice as much). The spacings (interpoint distances) are kept such that if the distance of the original scale between members A and B is greater than that between C and D, the distances on the model scale shall likewise be greater between A and B than between C and D. Figure 5.1 (in Chapter 5), uses a form of this technique in adding a third dimension by using letters to present degrees of effect on the skin.

This technique functions by taking observed measures of similarity or dissimilarity between every pair of M objects, then finding a representation of the objects as points in Euclidean space that the interpoint distances in some sense "match" the observed similarities or dissimilarities by means of weighting constants.

Cluster Analysis

Cluster analysis is a quantitative form of classification. It serves to help develop decision rules and then uses these rules to assign a heterogeneous collection of objects to a series of sets. This is almost entirely an applied methodology (as opposed to theoretical). The final result of cluster analysis is one of several forms of graphic displays and a methodology (set of decision classifying rules) for the assignment of new members into the classifications.

The classification procedures used are based on either density of population or distance between members. These methods can serve to generate a basis for the classification of large numbers of dissimilar variables such as behavioral observations and compounds with distinct but related structures and mechanisms (Gad et al., 1984, 1985), or to separate tumor patterns caused by treatment from those caused by old age (Salsburg, 1979).

There are five types of clustering techniques (Everitt, 1999).

Hierarchical Techniques

Classes are subclassified into groups, with the process being repeated at several levels to produce a tree that gives sufficient definition to groups.

Optimizing Techniques

Clusters are formed by optimization of a clustering criterion. The resulting classes are mutually exclusive; the objects are partitioned clearly into sets.

Density or Mode-Seeking Techniques

Clusters are identified and formed by locating regions in a graphic representation that contains concentrations of data points.

Clumping Techniques

A variation of density-seeking techniques in which assignment to a cluster is weighted on some variables, so that clusters may overlap in graphic projections.

Others

Methods that do not clearly fall into the previous categories. Romesburg (1984) provides an excellent step-by-step guide to cluster analysis.

Fourier or Time Analysis

Fourier analysis (Bloomfield, 1976) is most frequently a univariate method used for either simplifying data (which is the basis for its inclusion in this chapter) or for modeling. It can, however, also be a multivariate technique for data analysis.

In a sense, it is like trend analysis; it looks at the relationship of sets of data from a different perspective. In the case of Fourier analysis, the approach is to resolve the time dimension variable in the data set. At the most simple level, it assumes that many events are periodic in nature, and if we can remove the variation in other variables because of this periodicity (by using Fourier transforms), we can better analyze the remaining variation from other variables. The complications to this are (a) there may be several over-lying cyclic time-based periodicities, and (b) we may be interested in the time cycle events for their own sake.

Fourier analysis allows one to identify, quantitate, and (if we wish) remove the time-based cycles in data (with their amplitudes, phases, and frequencies) by use of the Fourier transform:

$$nJ_i = x_i \exp(-w_i t)$$

where
n = length
J = the discrete Fourier transform for that case
x = actual data
i = increment in the series
w = frequency
and t = time

A graphic example of the use of Fourier analysis in toxicology is provided in Figure 10.2.

Life Tables

Chronic *in vivo* toxicity studies are generally the most complex and expensive studies conducted by a toxicologist. Answers to a number of questions are

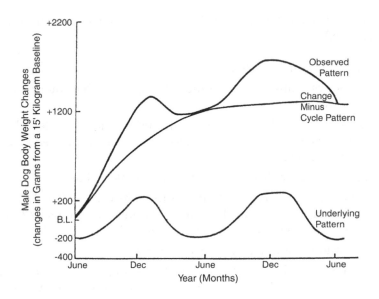

FIGURE 10.2
Use of time–series analysis.

sought in such a study — notably if a material results in a significant increase in mortality or in the incidence of tumors in those animals exposed to it. But we are also interested in the time course of these adverse effects (or risks). The classic approach to assessing these age-specific hazard rates is by using life tables (also called survivorship tables).

It may readily be seen that during any selected period of time (t_i) we have a number of competing risks affecting an animal. There are risks of (a) "natural death," (b) death induced by a direct or indirect action of the test compound, and (c) death due to such occurrences of interest of tumors (Hammond et al., 1978). And we are indeed interested in determining if (and when) the last two of these risks become significantly different than the "natural" risks (defined as what is seen to happen in the control group). Life table methods enable us to make such determinations as the duration of survival (or time until tumors develop) and the probability of survival (or of developing a tumor) during any period of time.

We start by deciding the interval length (t_i) we wish to examine within the study. The information we gain becomes more exact as the interval is shortened. But as interval length is decreased, the number of intervals increases and calculations become more cumbersome and less indicative of time-related trends because random fluctuations become more apparent. For a 2-year or lifetime rodent study, an interval length of 1 month is commonly employed. Some life table methods, such as the Kaplan–Meyer, have each new event (such as a death) define the start of a new interval.

Having established the interval length, we can tabulate our data (Cutler and Ederer, 1958). We start by establishing the following columns in each table (a separate table being established for each group of animals; i.e., by sex and dose level):

The interval of time selected (t_i)

The number of animals in the group that entered that interval of the study alive (l_i)

The number of animals withdrawn from study during the interval (such as those taken for an interim sacrifice or that may have been killed by a technician error) (w_i)

The number of animals that died during the interval (d_i)

The number of animals at risk during the interval, $l_i = l_i - 1/2\, w_i$, or the number on study at the start of the interval minus 1/2 of the number withdrawn during the interval

The proportion of animals that died $= D_i = d_i/l_i$

The cumulative probability of an animal surviving until the end of that interval of study, $Pi = 1 - D_i$, or 1 minus the number of animals that died during that interval divided by the number of animals at risk

The number of animals dying until that interval (M_i)

The number of animals found to have died during the interval (m_i)

The probability of dying during the interval of the study $c_i = 1 - (M_i + m_i/l_i)$, or the total number of animals dead until that interval plus the animals discovered to have died during that interval divided by the number of animals at risk through the end of that interval

The cumulative proportion surviving, pi, is equivalent to the cumulative product of the interval probabilities of survival (i.e., $p_i = p_1 \cdot p_2 \cdot p_3 \cdot \ldots \cdot p_x$)

The cumulative probability of dying, C_i, equal to the cumulative product of the interval probabilities to that point (i.e., $C_i = c_1 \cdot c_2 \cdot c_3 \cdot \ldots \cdot c_x$)

With such tables established for each group in a study (as shown in Example 10.2), we may now proceed to test the hypotheses that each of the treated groups has a significantly shorter duration of survival, or that of the treated groups died more quickly (note that plots of total animals dead and total animals surviving will give one an appreciation of the data, but can lead to no statistical conclusions).

Now, for these two groups, we wish to determine effective sample size and to compare survival probabilities in the interval months 14 to 15.

For the exposure group we compute sample size as:

$$I_{g14-15} = \frac{0.8400(1 - 0.8400)}{(0.0367)^2} = 99.7854$$

Likewise, we get a sample size of 98.1720 for the control group.

The standard error of difference for the two groups here is

$$SD = \sqrt{0.0367^2 + 0.0173^2} = 0.040573$$

The probability of survival differences is $P_D = 0.9697 - 0.8400 = 0.1297$. Our test statistic is then $0.1297/0.040573 = 3.196$. From our z-value table we see that the critical values are

$$p \le 0.05 = 1.960$$

$$p \le 0.01 = 2.575$$

$$p \le 0.001 = 3.270$$

As our calculated value is larger than all but the last of these, we find our groups to be significantly different at the 0.01 level ($0.01 > p > 0.001$).

Example 10.2

Interval (Months) t_i	Alive at the Beginning of Interval l_i	Animals Withdrawn ω_i	Died During Interval d_i	Animals at Risk l_i	Proportion of Animals Dead D_i	Probability of Survival P_i	Cumulative Proportion Surviving P_i	Standard Error of Survival S_i
			TEST LEVEL 1					
8–9	109	0	0	109	0	1.0000	1.0000	0.0000
9–10	109	0	2	109	0.0184	0.9816	0.9816	0.0129
10–11	107	0	0	107	0	1.0000	0.9816	0.0128
11–12	107	10	0	102	0	1.0000	0.9816	0.0128
12–13	97	0	1	97	0.0103	0.9897	0.9713	0.0162
13–14	96	0	1	96	0.0104	0.9896	0.9614	0.0190
14–15	95	0	12	95	0.1263	0.8737	0.8400	0.0367
15–16	83	0	2	83	0.0241	0.9759	0.8198	0.0385
16–17	81	0	3	81	0.0370	0.9630	0.7894	0.0409
17–18	78	20	1	68	0.0147	0.9853	0.7778	0.0419
18–19	57	0	2	57	0.0351	0.6949	0.7505	0.0446
			CONTROL LEVEL					
11–12	99	0	1	99	0.0101	0.9899	0.9899	0.0100
12–13	98	0	0	98	0	1.0000	0.9899	0.0100
13–14	98	0	0	98	0	1.0000	0.9899	0.0100
14–15	98	0	2	98	0.0204	0.9796	0.9697	0.0172
15–16	96	0	1	96	0.0104	0.9896	0.9596	0.0198
16–17	95	0	0	95	0	1.0000	0.9596	0.0198
17–18	95	20	2	85	0.0235	0.8765	0.9370	0.0249
18–19	73	0	2	73	0.0274	0.9726	0.9113	0.0302

There are a multiplicity of methods for testing significance in life tables, with (as is often the case) the power of the tests increasing as does the difficulty of computation (Breslow, 1975; Cox, 1972; Haseman, 1977; Tarone, 1975).

We begin our method of statistical comparison of survival at any point in the study by determining the standard error of the K interval survival rate as (Garrett, 1947).

$$S_k = P_k \sqrt{\sum_{i=1}^{k} \left(\frac{D_i}{l'_x - d_x} \right)}$$

We may also determine the effective sample size (l_1) in accordance with

$$l_1 = \frac{P(1-p)}{S^2}$$

We may now compute the standard error of difference for any two groups (1 and 2) as

$$S_D = \sqrt{S_1^2 + S_2^2}$$

The difference in survival probabilities for the two groups is then calculated as

$$P_D = P_1 - P_2$$

We can then calculate a test statistic as

$$t' = \frac{P_D}{S_D}$$

This is then compared to the z distribution table. If $t' > z$ at the desired probability level, it is significant at that level. Example 8.5 illustrates the life table technique for mortality data. With increasing recognition of the effects of time (both as age and length of exposure to unmeasured background risks), life table analysis has become a mainstay in chronic toxicology. An example is the reassessment of the ED_{01} study (SOT ED_{01} Task Force, 1981), which radically changed interpretation of the results and understanding of underlying methods when adjustment for time-on-study was made.

The increased importance and interest in the analysis of survival data has not been restricted to toxicology, but rather has encompassed all the life sciences. Those with further interest should consult Lee (1980) or Elandt-Johnson and Johnson (1980), both are general in their approach to the subject.

References

Anderson E. (1960) A semigraphical method for the analysis of complex problems. *Technomet*, 2, 387–391.

Anscombe, F.J. (1973) Graphs in statistical analysis. *Am. Stat.*, 27, 17–21.

Bloomfield, P. (1976) *Fourier Analysis of Time Series: An Introduction*. John Wiley, New York.

Breslow, N.E. (1975) Analysis of survival data under the proportional hazards model. *Int. Stat. Rev.*, 43, 45–58.

Chernoff, H. (1973) The use of faces to represent points in K-dimensional space graphically. *J. Am. Stat. Assoc.*, 68, 361–368.

Cox, D.R. (1972) Regression models and life-tables. *J. Roy. Stat. Soc.*, 34B, 187–220.

Craver, J.S. (1991) *Make Graph Paper with Your Copier*. Price Stern Sloan, New York.

Cutler, S.J. and Ederer, F. (1958) Maximum utilization of the life table method in analyzing survival. *J. Chron. Dis.*, 8, 699–712.

Davison, M.L. (1983) *Multidimensional Scaling*. John Wiley, New York.

Elandt-Johnson, R.C. and Johnson, N.L. (1980) *Survival Models and Data Analysis*. John Wiley, New York.

Everitt, B. (1999) *Cluster Analysis*, 4th ed., Arnold Publishers, New York.

Gad, S.C., Gavigan, F.A., Siino, K.M., and Reilly, C. (1984) The toxicology of eight ionophores: neurobehavioral and membrane and acute effects and their structure activity relationships (SARs). *Toxicologist*, 4, 702.

Gad, S.C., Reilly, C., Siino, K.M., and Gavigan, F.A. (1985) Thirteen cationic ionophores: neurobehavioral and membrane effects. *Drug Chem. Toxicol.*, 8(6), 451–468.

Garrett, H.E. (1947) *Statistics in Psychology and Education*. Longman, New York, pp. 215–218.

Glass, L. (1975) Classification of biological networks by their qualitative dynamics. *J. Theor. Biol.*, 54, 85–107.

Gordon, A.D. (1981) *Classification*. Chapman and Hall, New York.

Hammond, E.C., Garfinkel, L., and Lew, E.A. (1978) Longevity, selective mortality, and competitive risks in relation to chemical carcinogenesis. *Environ. Res.*, 16, 153–173.

Hartigan, J.A. (1983) Classification, in *Encyclopedia of Statistical Sciences*, vol. 2, Katz, S. and Johnson, N.L., Eds., John Wiley, New York.

Haseman, J.K. (1977) Response to use of statistics when examining life time studies in rodents to detect carcinogenicity. *J. Toxicol. Environ. Health*, 3, 633–636.

Kowalski, B.R. and Bender, C.F. (1972) Pattern recognition: a powerful approach to interpreting chemical data. *J. Am. Chem. Soc.*, 94, 5632–5639.

Lee, E.T. (1980) *Statistical Methods for Survival Data Analysis*. Lifetime Learning Publications, Belmont, CA.

Litchfield, J.T. and Wilcoxon, F. (1949) A simplified method of evaluating dose effect experiments. *J. Pharmacol. Exp. Therapeut.*, 96, 99–113.

Romesburg, H.C. (1984) *Cluster Analysis for Researchers*. Lifetime Learning Publications, Belmont, CA.

Salsburg, D. (1979) *The Effects of Life-Time Feeding Studies on Patterns of Senile Lesions in Mice and Rats* (unpublished manuscript).

Schaper, M., Thompson, R.D., and Alarie, Y. (1985) A method to classify airborne chemicals which alter the normal ventilatory response induced by CO_2. *Toxicol. Appl. Pharmacol.*, 79, 332–341.

Schmid, C.F. (1983) *Statistical Graphics*. John Wiley, New York.

SOT ED_{01} Task Force. (1981) Reexamination of the ED_{01} study-adjusting for time on study. *Fundam. Appl. Toxicol.*, 1, 8–123.

Tarone, R.E. (1975) Tests for trend in life table analysis. *Biometrika*, 62, 679–682.

Tufte, E.R. (1983) *The Visual Display of Quantitative Information*. Graphics Press, Cheshire, CT.

Tufte, E.R. (1990) *Envisioning Information*. Graphics Press, Cheshire, CT.

Tufte, E.R. (1997) *Visual Explanations*. Graphics Press, Cheshire, CT.

Young, F.W. (1985) Multidimensional scaling, in *Encyclopedia of Statistical Sciences*, vol. 5, Katz, S. and Johnson, N.L., Eds., John Wiley, New York, pp. 649–659.

11

Multivariate Methods

In a book of this kind, an in-depth explanation of the available multivariate statistical techniques is an impossibility. However, as the complexity of problems in toxicology increases, we can expect to confront more frequently data that are not univariate but rather multivariate (or multidimensional). For example, a multidimensional study might be one in which the animals are being dosed with two materials that interact. Suppose we measure body weight, tumor incidence, and two clinical chemistry values for test material effects and interaction. Our dimensions—or variables—are now: A = dose "x", B = dose "y", W = body weight, C = tumor incidence, D and E = levels of clinical chemistry parameters, and possibly also t (length of dosing).

These situations are particularly common in chronic studies (Schaffer et al., 1967). Although we can continue to use multiple sets of univariate techniques as we have in the past, there are significant losses of power, efficiency, and information when this is done, as well as an increased possibility of error (Davidson, 1972).

In this chapter we will try to communicate an understanding of the nature of multivariate data, its distributions and descriptive parameters, and how to collect and prepare these data. We will also look briefly at the workings and uses of each of the most commonly employed multivariate techniques, together with several examples from the literature of their employment in toxicology and the other biological sciences. We shall group the methods according to their primary function: hypothesis testing (are these significant or not?), model fitting (what is the relationship between these variables, or what would happen if a population were exposed to x?), and reduction of dimensionality (which variables are most meaningful?). It should be noted (and will soon be obvious), however, that most multivariate techniques actually combine several of these functions.

The most fundamental concept in multivariate analysis is that of a multivariate population distribution. By this point it is assumed that the reader is familiar with the univariate random variable and with such standard distributions as the normal distribution. This chapter extends these univariate ideas to the multivariate case. We will discuss the general properties of the multivariate case. We will discuss the general properties of multivariate distributions and consider the particular example of the multivariate normal distribution (Greenacre, 1981).

We will denote an n-dimensional random variable, sometimes called a random vector, by $\mathbf{X}$ where

$$\mathbf{X}^t = [\mathbf{X}_1,...,\mathbf{X}_n]$$

and $x_1,...,x_n$ are univariate random variables and $\mathbf{X}_1,...,\mathbf{X}_n$ are matrices.

We are primarily concerned with the case where the variables are continuous, but it is probably easier to understand the ideas if we start by considering the case where the components of $\mathbf{X}$ are discrete random variables. In this case, the (joint) distribution of $\mathbf{X}$ is described by the joint probability functions $P(x_1,...,x_n)$, where

$$P(x_1,...,\mathbf{X}_n) = \text{Prob}(\mathbf{X}_1,...,\mathbf{X}_n = \mathbf{X}_n)$$

We will abbreviate $P(x_1,...,x_n)$ as $P(x)$. Note that lowercase letters indicate particular values of the corresponding random variables.

The function $P(x)$ must satisfy two conditions similar to those required in the univariate case, namely that

$$P(x) \geq 0 \text{ for every } x$$

and

$$\sum P(x) = 1$$

where the summation is over all possible values of $\mathbf{X}$.

From the joint distribution, it is often useful to calculate two other types of distribution—the marginal distributions and the conditional distributions.

Suppose we are interested in the distribution of one component of $\mathbf{X}$, say X_1, without regard to the values of the other variables. The probability distribution of X_1 can then be obtained from the joint distribution by summing over all the other variables. Thus

$$\text{Prob } (X_1 = x_i) = \sum P(x_1, ..., X_n)$$

where the summation is for all x such that the ith component is fixed to be x_i; in other words, over $x_1,..., x_{i-1}, x_{i+1},...,x_n$.

When the distribution of a single variable is obtained from a joint distribution by summing over all other variables, then it usually called a marginal distribution. If the joint distribution is equal to the product of all the marginal distributions for every x, so that

$$P(x) = \prod_{i=1}^{n} P_i(X_i)$$

where $P_i(X_i)$ denotes the marginal distribution of X_i, then the random variables are said to be independent.

Also note that marginal joint distributions are possible by summing over less than $(p - 1)$ variables. For example, the marginal joint distribution of X_1 and X_2 is obtained by summing the joint distribution for all variables from X_3 to X_n.

If some of the variables are set equal to specified constant values, then the distribution of the remaining variables is called a conditional distribution. For a conditional probability, we should note that for two events A and B,

Prob$(A \mid B)$ = probability of event A given that B has occurred
$$= P(A\Omega B)/P(B)$$

where $A\Omega B$ denotes the intersection of events A and B, meaning that both events occur. By analogy, we find that the conditional distribution of a random variable is given by the ratio of the joint distribution to the appropriate marginal distribution.

In the case $p = 2$, the conditional distribution of X_1, given that X_2 take the particular values x_2, is given by

$$P(x_1 \mid x_2) = \text{Prob}(X_1 = x_1 \mid X_2 = x_2) = P(x_1, x_2)/P_2(x_2)$$

where $P_2(x_2)$ is the marginal distribution of X_2. More generally,

$$P(x_1, \ldots, x_k \mid x_{k+1}, \ldots, x_n) - P(x)/P_M(x_{k+1}, \ldots, x_n)$$

where $P_M(x_{k+1}, \ldots, x_n)$ denotes the marginal joint distribution of $X_{k+1}, \ldots, X_n$. These concepts are demonstrated in Example 11.1.

Example 11.1

Groups of four rats are each dosed one at a time, with a new chemical. If we let

X_1 = number of deaths in first two animals dosed
X_2 = number of deaths in last three animals dosed

| | The Joint Distribution | | | |
| | X_1 | | | Marginal Distribution |
X_2	0	1	2	of X_2
0	1/16	1/16	0	1/8
1	1/8	2/16	1/16	2/8
2	1/16	2/16	1/8	2/8
3	0	1/16	1/16	1/8
Marginal distribution of X_1	1/2	1/2	1/2	

then we can find the joint distribution of X_1 and X_2, the marginal distribution of X_1, and the conditional distribution of X_1, given that $X_2 = 2$.

Note that the sum of the joint probabilities is 1. The marginal distributions of X_1 and X_2 can be found by summing the column and row joint probabilities, respectively. Of course, in this simple case we could write down the (marginal) distribution of X_1 (and of X_2) directly. The conditional distribution of X_1, given that $X_2 = 2$, is obtained by looking at the row of joint probabilities where $X_2 = 2$ and normalizing them so that they sum to 1. This is done by dividing by the row total, which is the appropriate marginal probability. Thus

$$\text{Prob}(X_1 = 0 \mid X_2 = 2) = (1/16)/(3/8) = 1/6$$

$$\text{Prob}(X_1 = 1 \mid X_2 = 2) = 1/2$$

$$\text{Prob}(X_1 = 0 \mid X_2 = 2) = 1/3$$

Note that these three conditional probabilities sum to unity (that is, 1.0).

In the two-variable case, the distribution of a continuous random variable may be described by the cumulative distribution function (abbreviated c.d.f.) or by its derivative, called the probability density function (abbreviated p.d.f.). For continuous multivariate distributions, we may define suitable multivariate analogues of these functions. The joint c.d.f., which we will denote by $F(x_1,...,x_n)$, is defined by

$$F(x_1...,x_n) = \text{Prob}(x_1 \le x_2 \le ... \, x_{n-1} \le x_n)$$

The joint p.d.f., which we denoted by $f(x_1,...,x_n)$, or $f(x)$, is then given by the nth partial derivative

$$f(x) = \frac{\delta^P F(x_1,...,x_n)}{\delta x_1 \delta x_2 ... \delta x_n}$$

assuming that $F(x)$ is absolutely continuous.

This joint p.d.f. satisfies the assumptions:

$$f(x) \le 0 \text{ for every } x \text{ and}$$

$$\int_{-\infty}^{\infty} ... \int_{-\infty}^{\infty} f(x)dx_1 ... \, dx_n = 1$$

As in the univariate case, the joint p.d.f. is not a single probability, but rather the probabilities that can be found by integrating over the required subset of n-space. The above definitions can be related to the bivariate case where $n = 2$, and one can think of the two variables as being defined along perpendicular axes in a plane. For any point (x_1, y_1) in this plane, the joint p.d.f. gives the height of a surface above this plane. The total volume under this three-dimensional surface is defined to be 1 and is the total probability. For any area in the plane, the corresponding volume underneath the surface gives the probability of having a bivariate observation in the given area.

Marginal and conditional distributions are easily defined in the continuous case. The marginal p.d.f. of one component of **X**, say X_i, may be found from the joint p.d.f. by integrating out all the other variables. Thus

$$f_i(x_i) = \int_{-\infty}^{\infty} \ldots \int_{-\infty}^{\infty} f(x)dx_1 \ldots dx_{i+1} \ldots dx_n$$

The random variables are independent if the joint p.d.f. is equal to the product of all the marginal p.d.f.s for every x. We also note that marginal joint distributions are possible by integrating out less than $(n - 1)$ variables.

The density functions of conditional continuous distributions can be found by dividing the joint p.d.f. by the appropriate marginal p.d.f., which is clearly analogous to the discrete case. Thus, in the case $n = 2$, the conditional p.d.f. of X_1, given that X_2 takes the particular value x_n, will be denoted by $h(x_1 \mid x_2)$ and is given by

$$h(x_1 \mid x_2) = f(x_1 \mid x_2)/f_2(x_2)$$

More generally, the conditional joint p.d.f. of $X_1, \ldots, X_n$ given $X_{k+1} = x_{k+1}, \ldots, X_n = x_n$ is given by

$$h(x_1, \ldots, x_k \mid x_{k+1}, \ldots, x_n) = f(x)/f_M(x_{k+1}, \ldots, x_n)$$

where $f_M(x_{k+1}, \ldots, x_n)$ denotes the marginal joint p.d.f. of $X_{k+1}, \ldots, X_n$.

In the two-variable (or univariate) case, it is common to summarize a probability distribution's characteristics by giving two parameters, the mean and standard deviation. To summarize multivariate distributions, we need to find the mean and variance (the square of the standard deviation) of each of the n variables, together with a measure of the way each pair of variable is related. The latter target is achieved by calculating a set of quantities, called "covariances," or their standardized counterparts, called "correlations." These quantities are defined as below.

Means. The mean vector **T** = $[1, \ldots, n]$ is such that

$$E(X_i) = \int_{-\infty}^{\infty} x f_i(x)dx$$

is the mean of the ith component of X. This definition is given for the case where X_i is continuous. If X_i is discrete, then $E(X_i)$ is given by $\Sigma x P_i(x)$, where $P_i(x)$ is the (marginal) probability distribution of X_i.

Variances. The variance of the ith component of X is given by

$$\text{Var}(X_i) = E[(X_i - \mu_i)^2]$$

$$= EX_i^2 - \mu_i^2$$

This is usually denoted by σ_i^2 in the univariate case, but in order to fit in with the covariance notation provided below, we will usually denote it by σ_{ii} in the multivariate case.

Covariances. The covariance of two variable X_i and X_j is defined by

$$\text{Cov}(X_i, X_j) = E[(X_i - \mu_i)(X_j - \mu_j)]$$

Thus, it is the product moment of the two variables about their respective means. In particular, if $i = j$, we note that the covariance of a variable with itself is simply the variance of the variable. Thus, there is really no need to define variance separately in the multivariate case, as it is a special case of covariance.

The covariance of X_i and X_j is usually denoted by σ_{ij}. Thus, if $i = j$, the variance of X_i is denoted by σ_{ii}, as noted above.

The equation given above for the covariance is often presented as:

$$\sigma_{ij} = E[X_i X_j] - \mu_i \mu_j$$

The covariance matrix. With n variables, there are n variances and $\frac{1}{2} n(n - 1)$ covariances, and these quantities are all second moments. It is often useful to present these quantities in a $(n \times n)$ matrix, denoted by Σ, whose (i, j)th element is σ_{ij}. Thus,

$$\Sigma = \begin{bmatrix} \sigma_{11} & \sigma_{12} & \cdots & \sigma_{1n} \\ \sigma_{21} & \sigma_{22} & & \sigma_{2n} \\ \cdots & & & \cdots \\ \sigma_{n1} & \sigma_{n2} & \cdots & \sigma_{nn} \end{bmatrix}$$

This matrix is called the dispersion matrix, or the covariance matrix, and we will use the latter term. The diagonal terms are the variances, while the off-diagonal terms, the covariances, are such that $\sigma_{ij} = \sigma_{ji}$. Thus the matrix is symmetric.

Using the two covariance equations, we can express the matrix in two alternative and useful forms, namely,

$$\Sigma = E[(X - \mu)(X - \mu)^\tau]$$

$$= E[XX^\tau] - \mu\mu^\tau$$

Linear compounds. Perhaps the main use of covariances is as a stepping stone to the calculation of correlations (see below), but they are also useful for a variety of other purposes. Here we illustrate their use in finding the variance of any linear combination of the components of X. Such combinations arise in a variety of situations. Consider the general linear compound

$$Y = \mathbf{a}^\tau X$$

where $\mathbf{a}^\tau = [a_1, \ldots, a_n]$ is a vector of constants. Then Y is a univariate random variable. Its mean is clearly given by

$$E(Y) = \mathbf{a}^\tau \mu$$

while its variance is given by

$$\text{Var}(Y) = E[\mathbf{a}^\tau (X - \mu)]^2$$

As $\mathbf{a}^\tau (X - \mu)$ is a scalar and therefore equal to its transpose, we can express $\text{Var}(Y)$ in terms of Σ, using the formula above, as

$$\text{Var}(Y) = E[\mathbf{a}^\tau (X - \mu)(X - \mu)^\tau \mathbf{a}]$$

$$= \mathbf{a}^\tau E[(X - \mu)(X - \mu)^\tau] \mathbf{a}$$

$$= \mathbf{a}^\tau \Sigma \mathbf{a}$$

Correlation. Although covariances are useful for many mathematical purposes, they are rarely used as descriptive statistics. If two variables are related in a linear way, then the covariance will be positive or negative depending on whether the relationship has a positive or negative slope. But the size of the coefficient is difficult to interpret because it depends on the units in which the two variables are measured. Thus the covariance is often standardized by dividing by the product of the standard deviations of the two variables to give a quantity called the correlation coefficient. The correlation between variable X_i and X_j will be denoted by r_{ij} and is given by

$$r_{ij} = \sigma_{ij} / (\sigma_i \sigma_j)$$

where σ_i denotes the standard deviation of X_i.

It can be shown that r_{ij} must lie between -1 and $+1$, using the fact that $\text{Var}(aX_i + bX_j) \geq 0$ for every a and b, and putting $a = \text{Var}(X_j)$ and $b = \pm\text{Var}(X_i)$.

The correlation coefficient provides a measure of the linear association between two variables. The coefficient is positive if the relationship between the two variables has a positive slope; that is, the "high" values of one variable tend to go with the "high" values of the other variable. Conversely, the coefficient is negative if the relationship has a negative slope.

If two variables are independent, then their covariance, and hence their correlation, will be zero. But it is important to note that the converse of this statement is not true. It is possible to construct examples where two variables have zero correlation and yet are dependent on one another, often in a nonlinear way. This emphasizes the fact that the correlation coefficient is of no real use as a descriptive statistic (and may be positively misleading) if the relationship between two variables is of a nonlinear form. However, if the two variables follow a bivariate normal distribution, then a zero correlation implies independence (or no correlation).

The correlation matrix. With n variables, there are $n(n - 1)/2$ distinct correlations. The number n in this case is also often represented by ρ, the Greek letter "rho." It is often useful to present them in a $(\rho \times \rho)$ matrix whose (i, j)th element is defined to be r_{ij}. This matrix, called the correlation matrix, will be denoted by P, i.e., capital "rho." The diagonal terms of ρ are unity, and the off-diagonal terms are such that ρ is symmetric.

In order to relate the covariance and correlation matrices, let us define a $(\rho \times \rho)$ diagonal matrix, $\mathbf{D}$, whose diagonal terms are the standard deviations of the components of X, so that

$$\mathbf{D} = \begin{bmatrix} \sigma_1 & 0 & \cdots & 0 \\ 0 & \sigma_2 & & 0 \\ \cdots & & & \cdots \\ 0 & 0 & \cdots & \sigma_n \end{bmatrix}$$

Then the covariance and correlation matrices are related by

$$\Sigma = \mathbf{D} P\, \mathbf{D}$$

or

$$P = \mathbf{D}^{-1} \Sigma \mathbf{D}^{-1}$$

where the diagonal terms of the matrix $\mathbf{D}^{-1}$ are the reciprocals of the respective standard deviations.

We can now complete this section with a slightly more advanced discussion of the matrix properties of Σ and P, and in particular of their rank.

First, we show that both Σ and $\mathbf{P}$ are positive and semidefinite. As any variance must be nonnegative, we have that

$$\text{Var}(a^\tau X) \geq 0 \text{ for every } a$$

But $\text{Var}(a^\tau X) = a^\tau \Sigma a$, and so Σ must be positive semidefinite. We also note that Σ is related to $\mathbf{P}$, where $\mathbf{D}$ is nonsingular, and so it follows that $\mathbf{P}$ is also positive semidefinite.

Because Σ is nonsingular, we may also show that the rank of $\mathbf{P}$ is the same as the rank of Σ. This rank must be less than or equal to n.

If Σ (and hence **P**) is of full rank n, then Σ (and hence **P**) will be of positive definite. In this case, $\text{Var}(a^\tau X) = a^\tau \Sigma a$ is strictly greater than zero for every $a \neq 0$. But if rank $(\Sigma) < p$, then Σ (and hence **P**) will be singular, and this indicates a linear constraint on the components of X. This means that there exists a vector $\mathbf{a} \neq \mathbf{0}$ such that $\mathbf{a}^\tau X$ is identically equal to a constant. The most commonly assumed and used multivariate distribution is the multivariate normal distribution.

First we should remember that a univariate normal random variable X, with mean X and variance σ^2, has density function

$$f(x) = \frac{1}{\sigma\sqrt{2\pi}} e^{\frac{(x-\bar{x})^2}{2\sigma^2}}$$

and we write

$$X \sim N(\bar{X}, \sigma^2)$$

In the multivariate case, we say that an n-dimensional random variable X follows the multivariate normal distribution if its joint p.d.f. is of the form

$$f(x) = \frac{1}{(2\pi)^{n/2}\sqrt{|\Sigma|}} e^{\frac{(x-\bar{x})^\tau}{2\Sigma(x-\bar{x})}}$$

where Σ is any $(n \times n)$ symmetric positive definite matrix. If $X_1,\ldots,X_p$ are independent random variables where $X_i \approx N(\mu_i, \sigma_i^2)$, then their joint p.d.f. is simply the product of the appropriate (marginal) density functions, so that

$$f(x_1,\ldots,x_n) = \frac{1}{(2\pi)^{n/2}\prod\sigma_i} e^{-\frac{\Sigma(x_i-\bar{x}_i)^2}{2\sigma_i^2}}$$

In this case $X^\tau = [X_i,\ldots,X_n]$ has mean

$$\bar{X}^\tau = [\bar{X}_1 + \bar{X}_p]$$

and covariance matrix

$$\Sigma = \begin{bmatrix} \sigma_1^2 & 0 & \ldots & 0 \\ 0 & \sigma_2^2 & & 0 \\ \ldots & & & \ldots \\ 0 & 0 & \ldots & \sigma_n^2 \end{bmatrix}$$

There are other important multivariate distributions—the multivariate discrete distribution (where the data are rank order and not continuous) and the multivariate continuous distribution (where the data are continuous but not normally distributed).

Multivariate data are virtually never processed and analyzed other than by computer. One must first set up an appropriate data base file, then enter the data, coding some of them to meet the requirements of the software being utilized (for example, if only numerical data are analyzed, sex may have to be coded as 1 for male and 2 for females).

Having recorded the data, it is then essential to review for suspect values and errors of various kinds. There are many different types of suspect values and it is helpful to distinguish among them.

- Outliers. These are defined to be observations that appear to be inconsistent with the rest of the data. They may be caused by gross recording or entering errors. But it is important to realize that an apparent outlier may occasionally be genuine and indicate a non-normal distribution or valuable data point.

- Inversions. A common type of error occurs when two consecutive digits are interchanged at the recording, coding, or entering stage. The error may be trivial if, for example, 56.74 appears as 56.47, but it may generate an outlier if 56.74 appears as 65.74.

- Repetitions. At the coding or entering stage, it is quite easy to repeat an entire number in two successive rows or columns of a table, thereby omitting one number completely.

- Values in the wrong column. It is also easy to get numbers into the wrong columns.

- Other errors and suspect values. There are many other types of error, including possible misrecording of data of a minor nature.

The general term used to denote procedures for detecting and correcting errors is data editing. This includes checks for completeness, consistency, and credibility. Some editing can be done at the end of the data entry stage. In addition, many routine checks can be made by the computer itself, particularly those for gross outliers. An important class of such checks are range tests. For each variable an allowable range of possible values is specified, and the computer checks that all observed values lie within the given range. Bivariate and multivariate checks are also possible. For example, one may specify an allowable range for some functions of two of more variables. A set of checks called "if–then" checks are also possible. For example, if both age and date of birth are recorded for each animal, then one can check that the answers are consistent. If the date of birth is given, then one can deduce the corresponding age. In fact, in this example the age observation is redundant. It is sometimes a good idea to include one or two redundant variables as a check on accuracy. Various other general procedures for detecting out-liers are described by Barnett and Lewis (1981), as briefly discussed earlier.

In addition to the above procedure, another simple but very useful check is to examine a printout of the data visually. Although it may be impractical to check every digit, the eye is very efficient at picking out many types of obvious errors, particularly repetitions and gross outliers.

When a questionable value or error has been detected, the toxicologist must decide what to do about it. One may be able to go back to the original data source and check the observation. Inversions, repetitions, and values in the wrong column can often be corrected in this way. Outliers are more difficult to handle, particularly when they are impossible to check or have been misrecorded in the first place. It may be sensible to treat them as missing values and try to insert a value "guessed" in an appropriate way (e.g., by interpolation or by prediction from other variables). Alternatively, the value may have to be left as unrecorded; then either all observations for the given individual will have to be discarded or one will have to accept unequal numbers of observations for the different variables. With a univariate set of observations, the analysis usually begins with the calculation of two sum-mary statistics, namely the mean and standard deviation. In the multivariate case, the analysis usually begins with the calculation of the mean and stan-dard deviation for each variable and, in addition, the correlation coefficient for each pair of variables is usually calculated. Therein summary statistics are vital to having a preliminary look at the data.

The sample mean of the jth variable is given by

$$\bar{X}_j = \sum_{r=1}^{n} X_{rj}/n$$

and the sample mean vector, x, is given by $x^\tau = [x_1, x_2, \ldots, x_n]$. If the observa-tions are a random sample from a population with mean x, then the sample mean vector x^τ is usually the point estimate of x, and this estimate can easily be shown to be unbiased.

The standard deviation of the jth variable is given by

$$S_j = \sqrt{\sum_{r=1}^{n} (x_{rj} - \bar{x}_j)^2/(n-1)}$$

The correlation coefficient of variables i and j is given by

$$r_{ij} = \frac{\sum_{r=1}^{n} (X_{ri} - \bar{X}_j)(X_{rj} - \bar{X}_j)}{(n-1)s_i S_j}$$

These coefficients can be conveniently assembled in the sample correlation matrix, **R**, which is given by

$$\mathbf{R} = \begin{bmatrix} 1 & r_{12} & \cdots & r_{1n} \\ r_{21} & 1 & & r_{2n} \\ \cdots & & & \cdots \\ r_{n1} & r_{n2} & \cdots & 1 \end{bmatrix}$$

Note that the diagonal terms are all unity. This matrix provides an estimate of the corresponding population correlation matrix, **P**, which was defined earlier in this section. We note in passing that this estimate is generally not unbiased, but the bias is generally small and does not stop us from using the estimate. The virtues of lack of bias are sometimes overlooked.

The interpretation of means and standard deviations is straightforward. It is worth looking to see if, for example, some variables have much higher scatter than others. It is also worth looking at the form of the distribution of each variable and considering whether any of the variables need to be transformed. For example, the logarithmic transformation is often used to reduce positive skewness and produce a distribution that is closer to normal. One may also consider the removal of outliers at this stage.

There are three significant multivariate techniques that have hypothesis testing as their primary function: MANOVA, MANCOVA, and factor analysis.

MANOVA (multivariate analysis of variance) is the multidimensional extension of the ANOVA process we explored before. It can be shown to have grown out of Hotelling's T^2 (Hotelling, 1931), which provides a means of testing the overall null hypothesis that two groups do not differ in their means on any of p measures. MANOVA accomplishes its comparison of two (or more) groups by reducing the set of p measures on each group to a simple number applying the linear combining rule $W_i = w_j X_{ij}$ (where w_j is a weighting factor) and then computing a univariate F-ratio on the combined variables. New sets of weights (w_j) are selected in turn until that set which maximizes the F-ratio is found. The final resulting maximum F-ratio (based on the multiple discriminant functions) is then the basis of the significance test. As with ANOVA, MANOVA can be one-way or higher order, and MANOVA has as a basic assumption a multivariate normal distribution.

Gray and Laskey (1980) used MANOVA to analyze the reproductive effects of manganese in the mouse, allowing identification of significant effects at multiple sites. Witten et al. (1981) utilized MANOVA to determine the significance of the effects of dose, time, and cell division in the action of abrin on the lymphocytes.

Multivariate analysis of covariance (MANCOVA) is the multivariate analog of analysis of covariance. As with MANOVA, it is based on the assumption that the data being analyzed are from a multivariate normal population. The MANCOVA test utilizes the two residual matrices using the statistic and is an extension of ANCOVA with two or more uncontrolled variables (or covariables). A detailed discussion can be found in Tatsuoka (1971).

Factor analysis is not just a technique for hypothesis testing, it can also serve as a reduction of dimensionality function. It seeks to separate the variance unique to particular sets of values from that common to all members in that variable system and is based on the assumption that the intercorrelations among the n original variables are the result of there being some smaller number of variables ("factors") that explain the bulk of variation seen in the variables. There are several approaches to achieving the end results, but they all seek a determination of what percentage of the variance of each variable is explained by each factor (a factor being one variable or a combination of variables). The model in factor analysis is $y = A + x$, where y = n-dimensional vector of observable responses; A = factor loading an $n \times q$ matrix of unknown parameters; f = q-dimensional vector of common factor; and z = n-dimensional vector of unique factor.

Used for the reduction of dimensionality, factor analysis is said to be a linear technique because it does not change the linear relationships between the variables being examined.

Joung et al. (1979) used factor analysis to develop a generalized water quality index that promises suitability across the U.S., with appropriate weightings for ten parameters.

Factor analysis promises great utility as a tool for developing models in risk analysis where a number of parameters act and interact.

Now we move on to multivariate modeling techniques. We shall briefly discuss two of these: multiple regression and discriminant analysis.

Multiple regression and correlation seeks to predict one (or a few) variable from several others. It assumes that the available variables can be logically divided into two (or more) sets and serves to establish maximal linear (or some other scale) relationships among the sets.

The linear model for the regression is simply:

$$Y = b_0 + b_1 X_1 + b_2 X_2 + \ldots + b_p X_p$$

where Y = the predicted value, b = values set to maximize correlations between X and Y, and X = the actual observations (with Xs being independent of predictor variables and Ys being dependent variables or outcome measures). One of the outputs from the process will be the coefficient of multiple correlation, which is simply the multivariate equivalent of the correlation coefficient (r).

Schaeffer et al. (1982) have neatly demonstrated the utilization of multiple regression in studying the contribution of two components of a mixture to its toxicologic action, using quantitative results from an Ames test as an end point. Paintz et al. (1982) similarly used multiple regression to model the quantitative structure–activity relationships of a series of fourteen 1-benzoyl-3-methyl-pyrazole derivatives.

Discriminant analysis has for its main purpose to find linear combinations of variables that maximize the differences between the populations being studied, with the objective of establishing a model to sort objects into their appropriate populations with minimal error. At least four major questions are, in a sense, being asked of the data:

Are there significant differences among the K groups?

If the groups do exhibit statistical differences, how do the central masses (or centroids, the multivariate equivalent of means) of the populations differ?

What are the relative distances among the K groups?

How are now (or at this point unknown) members allocated to *establish* groups? How do you predict the set of responses of characteristics of an as yet untried exposure case?

The discriminant functions used to produce the linear combinations are of the form:

$$D_i = d_{i1}Z_1 + d_{i2}Z_2 + \dots + d_{ip}Z_p$$

where D_i = the score on the discriminant function i, ds = weighing coefficient(s), and Zs = standardized value(s) of the discriminating variables used in the analysis.

It should be noted that discriminant analysis can also be used for the hypothesis testing function by the expedient of evaluating how well it correctly classifies members into proper groups (e.g., control, treatment 1, treatment 2, etc.).

Taketomo et al. (1982) used discriminant analysis in a retrospective study of gentamycin nephrotoxicity to identify patient risk factors (that is, variables that contributed to a prediction of a patient being at risk).

Finally, we introduce four techniques whose primary function is the reduction of dimensionality—canonical correlation analysis, principal components analysis, biplot analysis, and correspondence analysis.

Canonical correlation analysis provides the canonical R, an overall measure of the relationship between two sets of variables (one set consisting of several outcome measures, the other of several predictor variables). The canonical R is calculated on two numbers for each subject:

$$W_i = \sum_i W_i X_{ij} \quad \text{and} \quad V_i = \sum_i V_i Y_{ij}$$

where Xs = predictor variable(s), Ys = outcome measure(s), and W_i and V_i = canonical coefficients.

MANOVA can be considered a special case of canonical correlation analysis. Canonical correlation analysis can be used in hypothesis testing also for testing the association of pairs of sets of weights, each with a corresponding coefficient of canonical correlation, each uncorrelated with any of the preceding sets of weights, and each accounting for successively less of the variation shared by the two sets of variables. For example, Young and Matthews (1981) used canonical correlation analysis to evaluate the relationship between plant growth and environmental factors at 12 different sites.

The main purpose of principal components analysis is to describe, as economically as possible, the total variance in a sample in a few dimensions: one wishes to reduce the dimensionality of the original data while minimizing the loss of information. It seeks to resolve the total variation of a set of variables into linearly independent composite variables that successively account for the maximum possible variability in the data. The fundamental equation is $Y = AZ$, where A = matrix of scales eigenvectors; Z = original data matrice; and Y = principal components.

The concentration here, as in factor analysis, is on relationships within a single set of variables. Note that the results of principal components analysis are affected by linear transformations.

Cremer and Seville (1982) used principal components to compare the difference in blood parameters resulting from each of two separate pyrethroids. Henry and Hidy (1979), meanwhile, used principal components to identify the most significant contributors to air quality problems.

The biplot display (Gabriel, 1981) of multivariate data is a relatively new technique, but promises wide applicability to problems in toxicology. It is, in a sense, a form of exploratory data analysis, used for data summarization and description.

The biplot is a graphical display of a matrix Y_{nm} of N rows and M columns by means of row and column marker. The display carries one marker for each row and each column. The "bi" in biplot refers to the joint display of rows and columns. Such plots are used primarily for inspection of data and for data diagnostics when such data are in the form of matrices.

Shy-Modjeska et al. (1984) illustrated this usage in the analysis of aminoglycoside renal data from beagle dogs, allowing the simultaneous display of relationships among different observed variables and presentation of the relationship of both individuals and treatment groups to these variables.

Correspondence analysis is a technique for displaying the rows and columns of a two-way contingency table as points in a corresponding low dimensional vector space (Greenacre, 1981). As such it is equivalent to simultaneous linear regression (for contingency table data, such as tumor incidences, which is a very common data form in toxicology). As such it can be considered a special case of canonical correlation analysis. The data are defined, described, and analyzed in a geometric framework. This is particularly attractive to such sets of observations in toxicology as multiple endpoint behavioral scores and scored multiple tissue lesions.

There are a number of good surveys of multivariate techniques available (Atchely and Bryant, 1975; Bryant and Atchely, 1975; Seal, 1964) that are not excessively mathematical. More rigorous mathematical treatments on an introductory level are also available (e.g., Gnanadesikan, 1997). It should be noted that most of the techniques we have described are available in the better computer statistical packages.

References

Atchely, W.R. and Bryant, E.H. (1975) *Multivariate Statistical Methods: Among Groups Covariation*. Dowden, Hutchinson and Ross, Stroudsburg, PA.

Barnett, V. and Lewis, T. (1981) *Outliers in Statistical Data*. John Wiley, New York.

Bryant, E.H. and Atchely, W.R. (1975) *Multivariate Statistical Methods: Within-Groups Covariation*. Dowden, Hutchinson and Ross, Stroudsburg, PA.

Cremer, J.E. and Seville, M.P. (1982) Comparative effects of two pyrethroids, dietamethrin and cismethrin, on plasma catecholamines and on blood glucose and lactate. *Toxicol. Appl. Pharmacol.*, 66, 124–133.

Davidson, M.L. (1972) Univariate vs. multivariate tests in repeated-measures experiments. *Psychol. Bull.*, 77, 446–452.

Gabriel, K.R. (1981) Biplot display of multivariate matrices for inspection of data and diagnosis, in *Interpreting Multivariate Data*, Barnett, V., Ed., John Wiley, New York, pp. 147–173.

Gnanadesikan, R. (1997) *Methods for Statistical Data Analysis of Multivariate Observations*, 2nd ed., John Wiley, New York.

Gray, L.E. and Laskey, J.W. (1980) Multivariate analysis of the effects of manganese on the reproductive physiology and behavior of the male house mouse. *J. Toxicol. Environ. Health*, 6, 861–868.

Greenacre, M.J. (1981) Practical correspondence analysis in interpreting multivariate data, in *Outhers in Statistical Data*, Barnett, V., Ed., John Wiley, New York.

Henry, R.D. and Hidy, G.M. (1979) Multivariate analysis of particulate sulfate and other air quality variables by principle components. *Atmos. Environ.*, 13, 1581–1596.

Hotelling, H. (1931) The generalization of Student's ratio. *Ann. Math. Stat.*, 2, 360–378.

Joung, H.M., Miller, W.M., Mahannah, C.N., and Guitjens, J.C. (1979) A generalized water quality index based on multivariate factor analysis. *J. Environ. Qual.*, 8, 95–100.

Paintz, M., Bekemeier, H., Metzner, J., and Wenzel, U. (1982) Pharmacological activities of a homologous series of pyrazole derivatives including quantitative structure–activity relationships (QSAR). *Agents Actions*, Suppl. 10, 47–58.

Schaeffer, D.J., Glave, W.R., and Janardan, K.G. (1982) Multivariate statistical methods in toxicology, III, specifying joint toxic interaction using multiple regression analysis. *J. Toxicol. Env. Health*, 9, 705–718.

Schaffer, J.W., Forbes, J.A., and Defelice, E.A. (1967) Some suggested approaches to the analysis of chronic toxicity and chronic drug administration data. *Toxicol. Appl. Pharmacol.*, 10, 514–522.

Seal, H.L. (1964) *Multivariate Statistical Analysis for Biologists*. Methuen, London.

Shy-Modjeska, J.S., Riviere, J.E., and Rawldings, J.O. (1984) Application of biplot methods to the multivariate analysis of toxicological and pharmacokinetic data. *Toxicol. Appl. Pharmacol.*, 72, 91–101.

Taketomo, R.T., McGhan, W.F., Fushiki, M.R., Shimada, A., and Gumpert, N.F. (1982) Gentamycin nephrotoxicity application of multivariate analysis. *Clin. Pharm.*, 1, 554–549.

Tatsuoka, M.M. (1971) *Multivariate Analysis.* John Wiley, New York.

Witten, M., Bennet, C.E., and Glassman, A. (1981) Studies on the toxicity and binding kinetics of abrin in normal and Epstein Barr virus–transformed lymphocyte culture—I: experimental results. *Exp. Cell. Biol.,* 49, 306–318.

Young, J.E. and Matthews, P. (1981) Pollution injury in Southeast Northumberland, England, UK; the analysis of field data using economical correlation analysis. *Environ. Pollut. B Chem. Phys.,* 2, 353–366.

12

Meta-Analysis

Meta-analysis ("analysis of analyses") is a review and re-analysis of data and methods from several studies on a topic, usually used in order to clarify or generalize results when no one study has provided a clear and convincing outcome (Sacks et al., 1987; Bailar, 1997). For example, Smith et al. (1993) used meta-analysis to analyze the safety and efficacy of the results of 34 separate trials of cholesterol-lowering drugs. The phrase itself was coined by Gene V. Glass in 1976. Meta-analytic methods are being used increasingly in biomedical and toxicological science to try to obtain a qualitative or quantitative synthesis of the available research data on a particular issue. The technique is most commonly applied to the synthesis of several randomized trials but has also been used to reconsider single studies.

Even though the term has only been around since 1976, the field existed much earlier — as early as 1931 L.H.C. Tippett proposed a test of statistical significance for combining experimental results. Despite this long history and the increasing number of meta-analytic studies published, there is still considerable controversy over the validity and uses of the technique. The basic issue that causes the debate is this: a meta-analysis not only includes whatever weaknesses, biases, or errors exist in its individual studies, but also potentially balloons the problem by introducing the flaws of the overall study, i.e., the meta-analysis itself. A weak or biased study selection plan could end up picking inadequate or skewed studies, perhaps leading to a result even further removed from the underlying trend.

Despite the increased potential for error with these methods and the sometimes vehement opposition to its use (Shapiro, 1994), there are several advantages that make meta-analysis attractive. Most of the advantages come from the continuous, tremendous increase in the volume and availability of data facilitated by the steady increase in computer processing power, storage ability, and connectivity. This accessibility makes it possible to conduct analyses that would be unethical, expensive, or inordinately lengthy to carry out *de novo*. For example, if one wants to analyze the effects of a substance known to be toxic for humans, it is unethical to actually dose people with it but one can analyze the data available from accidental exposure. Also, a thorough appraisal of previous work is often as effective as a new study but considerably faster and less expensive. The increased volume of data can also be used to increase statistical power, improve effect size estimates, or resolve conflicts between individual studies. There is also the possibility that an overview will reveal an effect or answer questions that individual trials did not. Finally, it should be noted that it is possible to perform meta-analysis using a Bayesian approach to incorporating study or trial data (Babapulle et al., 2004), which provides a rigorous and parsimonious approach to data set incorporation decisions.

It should be noted that the area where meta-analysis is not applicable is the study of new compounds or situations (for example, effects of a known compound on a species not exposed previously to that compound). Meta-analysis needs a volume of prior, appropriate data — it cannot create data where there is none.

Planning a Meta-Analysis

The idea of creating a statistical plan of analysis before the data is amassed is generally central to good experimental design, but meta-analysis raises possibilities that make prior planning virtually necessary from the quality and validity standpoints. In order to address the critical issue of study selection as well as the other potential pitfalls, the design of the meta-analysis should be carefully thought out beforehand to control whatever systemic error or bias the individual studies may have had while introducing none of its own.

There have been a number of sets of recommendations for planning of meta-analyses, such as those by Sacks et al. (1987), Thacker (1988), and Sutton et al. (2000). Most of these recommendations contain the same elements, occasionally in a different order. The list presented here of issues to consider in design is meant to be reasonably inclusive and is fairly long, but it is not meant to discourage the use of meta-analysis. It should be taken more as a view of the quality and extent of pre-planning needed for the generation of valid results rather than as a checklist and should also be a good overview of the kind of considerations needed to properly plan the selection and statistical analysis of any volume of data.

The meta-analysis should be designed to answer a specific question or questions and should base all successive decisions on the ability to answer the question as presented in as unbiased a way as possible.

All methods, decisions, and controls and the reasons for them should be documented with enough detail that the analysis could be replicated.

The literature or data search criteria should be set before the actual search begins and the criteria should be based on study methods or characteristics, not the actual results. While computer searches are increasingly powerful and popular, they should not be the only method used since they often miss studies, especially older ones. Also, remember to review the references of the studies found for further studies. All those that are found should be reported and then a log of rejected studies should be kept, noting what criteria they were rejected on. Possible criteria include:

1. Should studies be limited to those that are published? Negative studies (studies that report little or no benefit from following a particular course of action) are less likely to be published than are positive studies. Therefore, the published literature may be biased toward studies with positive results, and a synthesis of these studies would give a biased estimate of the impact of pursuing some courses of action. Unpublished studies, however, often are of lower quality than are published studies, and poor research methods often produce an underestimate of impact. Additionally, the unpublished studies are often difficult to discover.

2. Should studies be limited to those that meet additional quality control criteria? If investigators impose an additional set of criteria before including a study in meta-analysis, this may further improve the average quality of the studies used, but it introduces still greater concerns about selection bias. Moreover, different investigators might use different criteria for a "good" study and therefore select a different group of studies for meta-analysis.

3. Should studies be limited to GLP compliant studies? This is a variant of the above question concerning quality control. At one time, rigid quality standards were more likely to be met by GLP than by non-GLP studies. Increasingly, however, this is not necessarily the case.

4. Should studies be limited to those that appear in peer-reviewed publications? Peer review is considered the primary method for quality control in medical publishing. Some investigators recommend that only those studies that are published in peer-reviewed publications be considered in meta-analysis. Although this may seem an attractive option, it might produce an even more highly biased selection of studies.

5. Should studies be limited to those using identical methods? For practical purposes, this would mean using only separately published studies from multicenter trials, for which the methods were the same for all and the similarity of methods was monitored. This criterion is very difficult to achieve.

A critical analysis of the found data should be conducted to determine what data can be combined or considered together and what data should be considered as separate subgroups. Note that some of these criteria are similar to those used to decide on studies and, if they can be quantified before study selection, may be used for inclusion or exclusion of the studies. One way to make decisions of data pooling is to run an appropriate homogeneity test at a set confidence level.

Can data be combined from randomized and nonrandomized trials? Since nonrandomized trials are often of lower quality, the same issues arise as in the selection of studies.

Can data be used from non-placebo trials? There is some evidence that using historical controls may favor new treatments over the results of a properly controlled, randomized trial (Sacks et al., 1982).

Can data with different patient characteristics, diagnostic criteria, therapies, or endpoints be grouped?

Is the data considered to be from one population or from separate populations? Under the assumption of homogeneity (same population), experimental error is the difference. Under heterogeneity, experimental error and population variability are both reasons for differences. There are graphical and statistical methods that have been suggested to answer this question; such tests can often be used to answer a, b, and c of this section.

Based on the above criteria, and possibly other quality criteria, what data should be excluded from consideration?

Bias should be controlled as much as possible and accuracy should be maintained at the same time.

The data extractor should be blinded (as much as possible) to extraneous information such as study result summaries. Ideally, use more than one extractor and compare the degree of variance between them.

The data extraction should be standardized and cross-checked.

The source of money, resources, and other significant support should be reported.

The analyst(s) should report authorship or significant ties with the authors of the included studies.

The statistical methods for standardization of scale and combination of methods should be selected based on the type of data and the question to be answered. (Some specific methods will be covered in this chapter.)

The chance of making Type I and Type II errors should be reported for each study.

Confidence intervals for each type of error should also be reported.

Since all data may not be combined, the possibility of having subgroups needs to be planned for.

The meta-analysis should analyze and report the sensitivity of the included studies.

Each study should be evaluated for quality.

The methods of study assessment should be decided before evaluation.

Publication bias (see above) can skew sensitivity and should be controlled.

The applicability of the analysis and conclusions should be considered.

How definitive is the conclusion?

What can it be applied to and where should it not be applied?

What is the economic impact of the results?

What alternate explanations are possible for the results?

What guidelines for future research can be recommended from this study?

Methodologic Analysis (Qualitative)

Sometimes the question to be answered is not how much benefit is derived from the use of a particular treatment but whether there is any benefit at all. In this case, a qualitative meta-analysis may be done, in which the quality of the research concerning the treatment is scored according to a list of objective criteria. The meta-analyst then examines the methodologically superior studies to determine whether or not the question of adverse effects or risks is answered consistently by them. This qualitative approach has been called methodologic analysis (Gerbarg and Horwitz, 1988) or quality scores analysis (Greenland, 1994).

In some cases, the methodologically strongest studies agree with one another and disagree with the weaker studies, which may or may not be consistent with one another. An example was provided by a meta-analysis that showed that of the eight major controlled trials of the bacillus Calmette-Guérin (BCG) vaccine against tuberculosis, only three trials met all or almost all of the methodologic criteria and had precise statistical estimates (Clemens et al. 1983). These three trials agreed that BCG vaccine afforded a high level of protection against tuberculosis. The remaining five studies were methodologically weaker and had large confidence intervals. Their conclusions varied from showing a weak protective efficacy to actual harm from the vaccine. It should be noted that not everyone believes that this type of methodologic analysis is generally useful (Greenland, 1994).

Pooled Analysis (Quantitative)

Usually, the main purpose of meta-analysis is quantitative. The goal is to develop better overall estimates of the amount of adverse effects caused by particular treatments, based on the combining (pooling) of estimates found in the existing studies of the agents or treatments of interest. This type of meta-analysis is sometimes called a pooled analysis (Gerbarg and Horwitz, 1988) because the analysts pool the observations of many studies and then calculate parameters such as risk ratios or odds ratios from the pooled data.

Because of the many decisions regarding inclusion or exclusion of studies, different meta-analyses might reach very different conclusions on the same topic. Even after the studies are chosen, there are many other methodological issues in choosing how to combine means and variances (e.g., what weighting methods should be used). Pooled analyses should report both relative risks and alterations in risks or adverse events as well as absolute risks and increases in risks or adverse events (Sinclair and Bracken, 1994).

Standardizing Results

It is somewhat unlikely that sets of data brought together for meta-analysis will have common ways of reporting their conclusions. It follows that it is often necessary to transform the results of the individual studies to a common scale in order to properly compare them, whether the intent of the analysis is qualitative or quantitative. Even a graphical exploration of the data would be almost useless unless a unified scale is used.

This does not mean it is always possible to scale the results similarly. For example, if study AAA's primary toxicity endpoint is based on changes in food consumption and body weight and study BBB's primary toxicity endpoint is mortality, they are obviously not directly comparable. However, if AAA reported mortality also or BBB reported food consumption and body weight, a common metric is reachable. The scale should be based on the type of data being examined and the question the analysis is designed to answer.

Quantal Data

Quantal data are data that have only two possible values, for example dead/alive or present/absent. Common statistics used to standardize quantal data include risk ratios and risk reduction.

Risk Ratio (Relative Risk)

The relative risk is the ratio of the likelihood of effect in the exposed group to the likelihood of effect in the control group. For these calculations we will refer to groups of data A, B, C, and D such that:

	Effect	No Effect
Exposure	A	B
No exposure	C	D

so the risk ratio is calculated as:

$$RR = [A/(A+B)]/[C/(C+D)]$$

For calculation of the confidence interval around risk ratio and, indeed, for most calculations using odds or odds ratios, the interval is figured using the natural logarithm of the data and then converted back when reporting it. Since values of zero in the data are possible and log transformations are undefined for values of zero, it is common to add a factor of 0.5 to each cell before calculation. It has been suggested that this also reduces bias caused by small cell counts, so the calculations presented here will use the correction factor. The variance of ln(RR) is:

$$s^2_{\ln(RR)} = \frac{1}{A} - \frac{1}{A+B} + \frac{1}{C} - \frac{1}{C+D}$$

This makes the transformed confidence interval:

$$\ln(RR) \pm z_{\alpha/2}\sqrt{s^2_{\ln(RR)}} = \ln(RR) \pm z_{\alpha/2}s_{\ln(RR)}$$

where α is the confidence level, usually 95% (0.05), and $z_{\alpha/2}$ is the z statistic for $\alpha/2$. The result should then be exponentiated to get the true confidence interval. Note that this is the two-sided confidence interval—a one-sided interval would use z_α rather than $z_{\alpha/2}$. One-sided tests are rarely used, however.

Example 12.1

In a transgenic mouse study, there were 34 mice dosed subcutaneously with compound X in a vehicle and 20 dosed with the vehicle only. Out of these, 16 dosed animals developed tumors at the dosing site and 5 control animals did. Thus:

	Tumor	No Tumor
X + vehicle	16	18
Vehicle only	5	15

The confidence interval is calculated from:

$$\ln(RR) = \ln(16.5/(16.5+18.5))/(5.5/(5.5+15.5))$$

$$= \ln(1.8)$$

$$= 0.598$$

$$s^2_{\ln(RR)} = \frac{1}{16.5} - \frac{1}{16.5+18.5} + \frac{1}{5.5} - \frac{1}{5.5+15.5} = 0.1662$$

$$\ln(RR) \pm z_{\alpha/2}\sqrt{s^2_{\ln(RR)}} = 0.598 \pm (1.96)\sqrt{0.1662} = 0.190 \text{ to } 1.006$$

So the 95% confidence interval for the risk ratio is $e^{0.190}$ to $e^{1.006}$ or 1.209 to 2.735. Since this is above 1 (exactly equal risk), we can be 95% certain that the risk of injection site tumors is greater for compound X relative to using the vehicle alone.

Risk Reduction (Risk Difference)

The risk reduction is the risk in the exposed group minus the risk in the control group:

$$RD = p_E - p_C$$

where RD is used to differentiate this with risk ratio, p_E is the probability of an event in the exposed group, and p_C is the probability in the control group. This is different from the risk ratio in that risk ratio is a measure of correlation between treatment and outcome, where risk reductions are a measure of the effect difference between exposed and control populations. Using our previous terms, this makes the risk ratio a qualitative measure (is there an effect?) while the risk reduction is a qualitative measure (how much effect?). Using the same A, B, C, and D designations that were used for risk ratio:

$$p_E = [A/(A+B)] \quad \text{and} \quad p_C = [C/(C+D)]$$

Since this measure is not a ratio or a relative measure, log transformation is not used for the calculation of the confidence interval. The variance is calculated as:

$$s_{RD}^2 = \frac{p_E(1-p_E)}{n_E} + \frac{p_C(1-p_C)}{n_C}$$

where n_E is the total number in the exposed group and n_C is the total number in the control group. This makes the confidence interval:

$$RD \pm z_{\alpha/2}\sqrt{s_{RD}^2} = RD \pm z_{\alpha/2}s_{RD}$$

where α is the confidence level, usually 95% (0.05), and $z_{\alpha/2}$ is the z statistic for $\alpha/2$.

Example 12.2

Using the same data as Example 12.1, calculate the 95% confidence interval of the risk reduction.

	Tumor	No Tumor
X + vehicle	16	18
Vehicle only	5	15

$$RD = p_E - p_C$$

$$= [A/(A+B)] - [C/(C+D)]$$

$$= [16.5/(16.5+18.5)] - [5.5/(5.5+15.5)]$$

[once again with a correction factor]

$$= 0.4714 - 0.2619$$

$$= 0.210$$

$$s_{RD}^2 = \frac{0.4714(1-0.4714)}{34} + \frac{0.2619(1-0.2619)}{20} = 0.0170$$

$$RD \pm z_{\alpha/2}\sqrt{s_{RD}^2} = 0.210 \pm (1.96)(0.130) = -0.046 \text{ to } 0.466$$

Since this interval includes zero, we cannot say with 95% certainty that there is difference in risk between the exposed and the control populations.

Incidence Rate Data

Incidence rate data are computed from a number of events in a specific number of systems over a specific period of time. Incidence rates are often called rates of occurrence.

Rate of Occurrence

The rate of occurrence of an effect is a measure of both how many events have been noted and how often they happen; e.g., events per person per year. So, if E events have been seen in S subjects (animals, humans, etc.) over time t,

$$\text{rate of occurrence} = RO = E/(S * t)$$

Like the risk ratio, the confidence interval for this is calculated using the natural logarithm and then converted back for reporting. The variance of $\ln(RO)$ is complex but can be simplified for practical purposes to:

$$s_{\ln(RO)}^2 = 1/E$$

and the confidence interval is calculated just like the confidence interval for risk ratio.

Example 12.3

Drug Y has been given to 1202 persons over the course of 3 years. During this time there have been 74 reported cases of fainting that were not attributable to other sources. The fainting occurrence rate and 95% confidence interval for that rate are calculated as:

$$RO = 74/(1202 * 3) = 0.0205 \text{ episodes per person per year}$$

$$\ln(RO) = -3.886$$

$$s^2_{\ln(RO)} = 1/74 = 0.01351$$

$$\ln(RR) \pm z_{\alpha/2}\sqrt{s^2_{\ln(RR)}} = -3.886 \pm (1.96)\sqrt{0.01351} = -4.002 \text{ to } -3.770$$

So the 95% confidence interval for the rate of occurrence is $e^{-4.002}$ to $e^{-3.770}$ or 0.01823 to 0.02305.

Continuous Data

Continuous data can theoretically take any value within a range of values. A common statistic used to examine continuous data is the treatment effect, which can be computed with and without weighting the data.

Treatment Effect (Mean Difference)

The unweighted effect of treatment is easy to calculate:

$$TE = \mu_E - \mu_C$$

where μ_E is the mean of the exposed group and μ_C is the mean of the control group. The variance of this measure is:

$$s^2_{TE} = \sigma^2 \left(\frac{1}{n_E} + \frac{1}{n_C} \right)$$

where σ^2 is the population variance, n_E is the number in the exposed group, and n_C is the number in the control group. Since σ^2 is usually not available, it is often estimated from the sample variances of the exposed and control groups:

$$\sigma^2 \approx s^2_{Pooled} = \frac{(n_E - 1)s^2_E + (n_C - 1)s^2_C}{n_E + n_C - 2}$$

Note that the use of this statistic assumes homogeneity of population variance. Given the mean difference and the variance, the confidence interval can be estimated the same way the confidence interval for risk difference is.

Example 12.4

Often the toxicity of a drug is measured not against a control, but against the toxicity of a standard agent. This is especially the case for chemotherapy drugs, which often have noticeable adverse events at the therapeutic dosage level. A new drug, X001, and the reference standard, S001, were both given to rats for 2 weeks. At the end the body weights of the groups were summarized:

Compound	N	Mean Weight	SD
X001	23	12.74	2.62
S001	18	10.98	1.74

The treatment effect is:

$$TE = \mu_E - \mu_C = 12.74 - 10.98 = 1.76$$

To calculate the standard deviation, first calculate the estimated population variance.

$$\sigma^2 = \frac{(12-1)(2.62)^2 + (18-1)(1.74)^2}{23+18-2} = 5.19$$

$$s_{TE} = \sqrt{5.19\left(\frac{1}{23} + \frac{1}{18}\right)} = 0.717$$

Weighted Treatment Effect
(Standardized Mean Difference)

Better quality studies often have a lower standard deviation than those of poor quality. This happens because the more accurate studies measure the population deviation closely while the poorer quality studies measure both the population deviation and introduce variability. Therefore, one common way of weighting study outcomes is by the inverse of the standard deviation, giving the high variability studies a lesser weight. So:

$$TE_W = (\mu_E - \mu_C)/s_{TE}$$

where s_{TE} is calculated as shown above. The variance of this measure is complex but can be estimated for normal populations as:

$$s_W^2 = \frac{n_E + n_C}{n_E n_C} + \frac{TE_W^2}{2(n_E + n_C)}$$

and for large populations with equal variance this can be simplified to:

$$s_W^2 = \frac{n_E + n_C}{n_E n_C}$$

Once again, the confidence interval can be estimated from the weighted mean difference and its variance.

Example 12.5

An additional study similar to that from Example 12.4 was performed. The data from both are summarized as:

	X001			S001		
Study Number	N	Mean	SD	N	Mean	SD
12.5	23	12.74	2.62	18	10.98	1.74
12.6	76	11.43	3.47	42	11.33	3.86

The 95% confidence intervals for both studies require some intermediate calculations.

Study Number	T. Effect	Est.σ^2	s_{TE}	Weighted
12.5	1.76	5.19	0.717	2.455
12.6	0.10	13.05	0.695	0.144

$$s_{W12.5}^2 = \frac{23 + 18}{(23)(18)} + \frac{(2.455)^2}{2(23 + 18)} = 0.173$$

$$s_{W12.6}^2 = \frac{76 + 42}{(76)(42)} + \frac{(0.144)^2}{2(76 + 42)} = 0.037$$

Study Number	s_W	Interval (1.96)*s_W	Lower	Weighted Mean	Upper
12.5	0.416	0.815	1.640	2.455	3.270
12.6	0.192	0.376	−0.232	0.144	0.520

Ranked Data

Ranked data are ordered in a qualitative manner, i.e., it is possible to know whether one data point is greater, equivalent to, or less than another, but without a quantity that can reasonably be said to represent the difference. An example would be the Draize scale for eye irritation scores. Whitehead and Jones (1994) have detailed a method for combining data from different scales using a stratified proportional odds model that generates an odds ratio. The assumptions and an example set of calculations are detailed in their article but are beyond the scope of this book due to their complexity.

Graphical Explorations

Plotting a meta-analysis can serve several functions. Three of the more important are:

In conjunction with quantitative tests, diagrams can be used to check for bias or heterogeneity.

A well-designed plot will often show the overall shape and trends of the data better than a simple table of data.

Graphical explorations of the data can show results or suggest areas to explore that are not easily seen from the basic information.

Each visualization tool is useful for examining a different aspect of the studies included in a meta-analysis. The box-and-whisker plot is used for examination of confidence intervals, the standardized residual plot for examination of homogeneity/heterogeneity, the funnel plot for examination of bias, and the L'Abbé plot for examination of odds ratios.

Box-and-Whisker Plot for Examination of Confidence Intervals

In a box-and-whisker plot, the confidence intervals calculated for each individual study are presented. This diagram is a simple but effective tool for exploration and also for presenting a summary, since the pooled estimate can also be plotted on the same axis. This is a very flexible tool since it can be used to display odds, risk ratios, or means.

The marker size in Figure 12.1 is proportional to the inverse of the variance for the individual study. This is a common practice in that it helps emphasize the more precise studies.

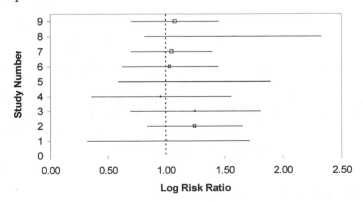

FIGURE 12.1
Example of box-and-whisker plot.

Standardized Residuals for Examination of Heterogeneity/Homogeneity

If the individual studies of the meta-analysis are heterogeneous, the differences in the results will be due to random variation and should, on an appropriate scale, approximate the normal curve. The appropriate scale in this case is:

$$SR_i = (\mu_i - \mu_T)/\text{var}_i)^{1/2}$$

where μ_i is the individual study mean, μ_T is the weighted mean treatment effect of all the studies, and var$_i$ is the variance of the individual study. The calculation of μ_T is detailed in the section on the standard χ^2 homogeneity test (as $\bar{T}$). This statistic is often called the z score, a name that is also sometimes applied to the standardized mean, i.e., $\mu_i/(\text{var}_i)^{1/2}$.

Example 12.6

The means and variances of 11 studies have been calculated as well as the overall weighted mean (4.49):

Study Number	Mean	Variance	$(\text{Var})^{1/2}$	SR_i
1	5.60	2.23	1.49	0.66
2	4.83	1.52	1.23	0.09
3	3.76	1.80	1.34	−0.65
4	2.74	1.20	1.09	−1.72
5	1.34	2.00	1.41	−2.32
6	5.93	1.20	1.10	1.19
7	4.89	0.88	0.94	0.29
8	4.46	1.68	1.30	−0.12
9	3.48	1.82	1.35	−0.85
10	7.85	1.84	1.36	2.38
11	4.18	1.80	1.34	−0.32

Then the histogram, with an interval size of 0.5 in this case, is graphed on the same interval as the normal curve.

The shape of the distribution of the standardized scores appears to be more spread out (variable) than expected from random chance as well as being asymmetric. Therefore it appears that this set of studies does not follow the normal curve closely and may be heterogeneous.

Funnel Plot for Examination of Bias

The funnel plot is an informal method for examining groups of studies for bias. The treatment effect is plotted on the x-axis and a measure of the study size is plotted on the y-axis. An unbiased set of studies will then roughly form the shape of an inverted cone or funnel.

In order to explain this apparent shape, remember that smaller studies will have more random error than larger ones. Smaller studies will then be low on the y-axis but have treatment effects spread out loosely along the x-axis on both sides of the "real" treatment effect level. Larger studies estimating the true effect better will be higher on the y-axis, and less likely to vary much from the accurate population effect. Note that the axis of the funnel should correspond reasonably closely to the true population treatment effect and therefore should only fall on the y-axis ($x = 0$) when the true effect is zero.

The x-axis scale can be based on any of the standardized measures that have already been presented. It is usual for risk ratios to be presented as ln(risk ratio) and this also applies to other measures transformed on the log scale. The y-axis is usually presented as 1/(standard error). Since standard error goes down as N increases, the inverse rises as N does, which makes it an appropriate measure. This scale is the most commonly used and it makes for a much more compact scale—study sizes can easily vary by factors of 100 or more but 1/(standard error) will usually fall within a small range.

Most bias will tend to exclude studies only in one direction, i.e., only the side with greater effects or only the side with lesser effects, and will show up as asymmetry in the shape of the funnel. Publication bias is usually the biggest concern but an asymmetrical shape can also result from many other factors including differing underlying populations, poor design of the smaller studies, unequal measurement scales, or even deliberate deception.

Quantitative methods for the examination of bias exist but often give contradictory results. So far there is not a consensus as to which quantitative method or methods are better (Sutton et al., 2000).

L'Abbé Plot for Examination of Risks

L'Abbé plots are useful in examining risks for studies producing quantal data (L'Abbé et al., 1987). For each study, the risk or incidence for the exposed group are plotted against the risk or incidence for the control group, with marker size proportional to the inverse of the variance. A line is added to show where risks are equal.

From Figure 12.2 it can be inferred that exposure generally causes higher incidence in populations that have a low baseline incidence rate (0.0 to about 0.25) but the effect reverses for populations having a higher baseline. Note that the incidence events in this case could be either desirable or undesirable.

If there are subpopulations, it is often useful to graph them on the same L'Abbé plot but differentiate them by symbol or color. This provides an informal but effective way to compare differences in effect between subpopulations.

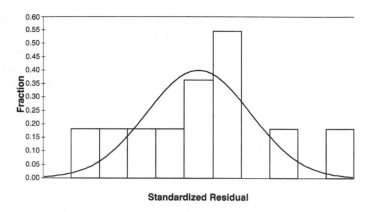

Standardized Residual

FIGURE 12.2
L'Abbé plot.

Standard χ^2 Homogeneity Test

Cochran (1954) proposed a homogeneity test that is still in common use. The hypothesis is that the underlying treatment effects are the same and the interstudy differences are due to random variation, which can be written as:

$$H_0: \theta_1 = \theta_2 = \quad = \theta_k$$

where θ_i is the underlying treatment effect for the ith study and there are k studies.

Cochran's test statistic, often represented as Q, has a χ^2 distribution with $k-1$ degrees of freedom, so the result can be looked up in standard χ^2 tables. This statistic is:

$$Q = \sum_{i=1}^{k} w_i (T_i - \bar{T})^2$$

where T_i is the treatment effect of study i. The weighted mean treatment effect is:

$$\bar{T} = \frac{\sum_{i=1}^{k} w_i T}{\sum_{i=1}^{k} w_i}$$

Each weight w_i is the inverse of the variance of the estimate for T_i. The calculations of variance for the different types of data were presented previously.

The number of studies in a meta-analysis will usually be low enough to lower the power of this test. As a result, heterogeneity is often rejected even when H_0 is true. Fleiss (1986) recommended that a cutoff level of 90% (0.10) instead of the more conventional 95% (0.05) and this has become customary (Sutton, et al. 2000). Sutton et al. also noted several other situations where the Q statistic may not approximate the underlying effect and listed several alternate tests available. Nevertheless, this test has survived critical examination (Hedges and Olkin, 1985) and remains in common use.

Example 12.7

Using the data from Example 12.6, we can test the homogeneity quantitatively:

Study	Mean	Variance	w_i	$(w_i)(T_i)$	Q_i
1	5.60	2.23	0.45	2.51	0.56
2	4.83	1.52	0.66	3.18	0.08
3	3.76	1.80	0.56	2.09	0.30
4	2.74	1.20	0.83	2.29	2.56
5	1.34	2.00	0.50	0.67	4.96
6	5.93	1.20	0.83	4.93	1.72
7	4.89	0.88	1.13	5.54	0.18
8	4.46	1.68	0.59	2.65	0.00
9	3.48	1.82	0.55	1.91	0.57
10	7.85	1.84	0.54	4.25	6.11
11	4.18	1.80	0.55	2.32	0.05
Sum			7.21	32.35	17.07

The critical value of χ^2 for $\alpha = 0.10$ and df $= 11 - 1 = 10$ is 15.987. Since the calculated value of 17.07 is greater than this, we reject H_0, i.e., the studies are heterogeneous at the $\alpha = 0.10$ level. This matches what our informal graphical analysis shows in Figure 12.3.

Pooling Results

In the ideal world, the individual data from the various studies are available and the studies are essentially equal in methods, quality, and homogeneity and measurement of results, allowing pooling of all of the data directly. In that case, meta-analytic methods are not needed further — standard statistical methods will suffice to generate an estimate of effect (such as Figure 12.4). From a practical and statistical standpoint, this is only likely when the studies come from different locations doing a single trial or test under the same conditions or different runs of a trial or test at a single location, again under

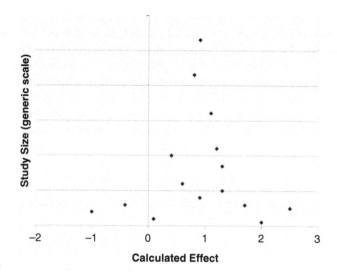

FIGURE 12.3
Standardized residual plot with the normal curve.

the same conditions. Even then, the person(s) doing the meta-analysis should be reluctant to do such pooling and should conduct the appropriate tests to verify that the methods, quality, and homogeneity and measurement of results are consistent enough that the data can reasonably be pooled.

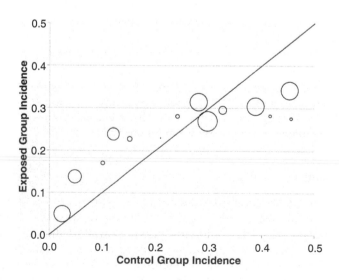

FIGURE 12.4
Funnel plot showing a generally symmetrical shape.

The most likely scenario is that the results are reported as means and standard deviations (or variances) or as tables for quantal data. The choice of methods thus usually depends on how the data are reported and whether or not it is heterogeneous. A general model that works for heterogeneous studies is the inverse-variance weighted method, which can be used for any of the measures previously calculated for either quantal or continuous data. In the case of heterogeneity, a random effects model can be used, which is also usable with the measures calculated previously. There are also two classic methods specific to odd ratios: the Mantel-Haenszel method and Peto's method.

There are some times when only a p-value is available and several formulas do exist to pool p-values. The most significant problem with this approach is that it does not generate any estimates of effect size. Hasselblad (1995) outlines two circumstances where such an approach is reasonable: first, if there are studies that should be included that only report p-values, and second, when the study designs (or situations) are significantly different and therefore combining measures of effects would not be reasonable.

Finally, there will be meta-analyses whose individual studies have measures of outcome that are so disparate that they cannot easily be reduced to a common scale. While methods do exist to handle these, they are beyond the scope of this book. Interested readers are referred to Sutton et al. (2000), pp. 205–228.

Inverse-Variance Weighted Method

As discussed previously, studies with better methodology and higher numbers will generally produce smaller variances than those with low rigor and/or low numbers. Therefore, it is reasonable to weigh results by the inverse of the variance to give larger weight to those studies with lower variance. This idea was proposed by Birge in 1932 and Cochran in 1937 and is still in common use.

Thus, each weight $w_i = 1/v_i$ where v_i is the variance for study i and the pooled estimate $\bar{T}$ is calculated exactly as discussed under the section on the standard χ^2 homogeneity test.

The variance for this estimate is calculated as:

$$\mathrm{var}(\bar{T}) = 1 \Big/ \sum_{i=1}^{k} w_i$$

Example 12.8

The time to maximum plasma concentration of drug X has been estimated in eight studies, each of which compared the standard capsule with a time-release capsule. Using the following mean differences and variances, generate an estimated treatment effect.

Study	Mean Difference (T_i)	Variance	Weight (w_i)	$(w_i)(T_i)$
1	0.83	1.10	0.91	0.76
2	1.84	0.72	1.38	2.55
3	1.48	1.06	0.95	1.40
4	1.44	0.54	1.85	2.66
5	2.09	0.87	1.15	2.41
6	1.12	1.20	0.83	0.93
7	1.20	0.67	1.50	1.80
8	1.66	0.90	1.11	1.84
Sum			9.69	14.39

So, estimated treatment mean is $14.39/9.69 = 1.48$, estimated variance is $1/9.69 = 0.10$, and the confidence interval is found using the usual formula for a normal distribution:

$$CI_\alpha = \bar{T} \pm Z_{\alpha/2}(\sqrt{\text{var}(\bar{T})})$$

which, in this case, comes to:

$$CI_{0.95} = 1.48 \pm 1.96(\sqrt{0.10}) = 0.86 \text{ to } 2.10$$

So, for this case, since the interval does not include zero, the effect of the time release capsule is statistically significant and is estimated as 1.48 hours. This is despite the fact that most of the individual studies themselves did not show significance (figuring $T_i \pm 1.96*[\text{var}(T_i)]^{1/2}$).

These calculations could be done identically for odds, except in that case the mean treatment effect would be calculated using the logarithm of the odds and the resultant confidence interval would need to be exponentiated to generate the real odds interval.

Weighted Random Effects Method

The inverse variance method above assumes that there is an underlying but unknown effect and that variability between studies is due to normal variations in measurement. Thus, it significantly underestimates the range of effects if the studies actually vary in "true" effect, i.e., under conditions of heterogeneity. In a weighted random effects model, we make the assumption that the effects themselves vary but those variations have a normal distribution and weight the individual estimates accordingly. By assuming heterogeneity, this model remains useful when the tests of homogeneity fail, but at the cost of a larger variance and therefore a broader confidence interval than those produced by other models.

Computationally, the weighted random effects model starts by calculating the test statistic Q just as shown in the standard χ^2 homogeneity test. If $Q \leq k - 1$, where k is the number of studies, then it is calculated identically to the inverse-variance weighted method. Otherwise, the adjusted weights must be calculated using several intermediate calculations:

$$\bar{w} = \sum_{i=1}^{k} w_i/k = \frac{1}{k}\sum_{i=1}^{k} w_i$$

$$s_w^2 = \frac{1}{k-1}\left[\left(\sum_{i=1}^{k} w_i^2\right) - k\bar{w}^2\right]$$

$$U = (k-1)\left(\bar{w} - \frac{s_w^2}{k\bar{w}}\right)$$

$$\hat{\tau}^2 = (Q-(k-1))/U$$

And, finally, the adjusted weights are:

$$w_i' = \frac{1}{[(1/w_i)+\hat{\tau}^2]}$$

The resultant adjusted weights are then used to recalculate the weighted mean, variance, and confidence intervals exactly like the previous calculations:

$$\bar{T}' = \frac{\sum_{i=1}^{k} w_i'T_i}{\sum_{i=1}^{k} w_i'}$$

$$\mathrm{var}(\bar{T}') = 1\bigg/\sum_{i=1}^{k} w_i'$$

$$CI_\alpha = \bar{T}' \pm Z_{\alpha/2}(\sqrt{\mathrm{var}(\bar{T}')})$$

Using these equations, it is possible to estimate the confidence intervals of either mean differences or odds by using the same computations shown in Example 12.8.

Mantel-Haenszel Method

The Mantel-Haenszel method (Mantel and Haenszel, 1959) is specific to combining odds ratios. For the following equations, a_i is the number A from study i, b_i is the number B from study i, etc., all referring to the following array:

	Effect	No Effect
Exposure	A	B
No exposure	C	D

For M-H, the pooled estimate is calculated as:

$$\bar{T}_{MH} = \frac{\sum_{i=1}^{k} a_i d_i / n_i}{\sum_{i=1}^{k} b_i c_i / n_i}$$

and the variance of the logarithm of the odds ratio, derived by Robins et al. (1986a,b), is

$$\text{var}_{MH} = \frac{\sum_{i=1}^{k} P_i R_i}{2\left(\sum_{i=1}^{k} R_i\right)^2} + \frac{\sum_{i=1}^{k} (P_i S_i + Q_i R_i)}{2\left(\sum_{i=1}^{k} R_i\right)\left(\sum_{i=1}^{k} S_i\right)} + \frac{\sum_{i=1}^{k} Q_i S_i}{2\left(\sum_{i=1}^{k} S_i\right)^2}$$

where $P_i = (a_i + d_i)/n_i$, $Q_i = (b_i + c_i)/n_i$, $R_i = a_i d_i/n_i$, and $S_i = b_i c_i/n_i$. Note that the odds ratio is calculated directly, but the variance is calculated for the logarithm of the odds ratio.

For either the Mantel-Haenszel method or Peto's method (next section) the confidence interval equation is

$$CI_\alpha = \exp(\ln(\bar{T}) \pm Z_{\alpha/2}\sqrt{\text{var}})$$

Assumptions and Limitations

This method assumes heterogeneity. If this is not the case, the random effects model is better.

For trials that have a zero in any arm (A, B, C, or D above), it can be seen from the equation for the pooled estimate that the M-H method ignores the trial either in the numerator or the denominator (or both). Such trials can obviously be significant, so either the M-H method should not be used for trials with a zero-event arm or a correction factor of 0.5 should be added to each cell.

This method is preferred if there are a large number of studies each with a small total number of subjects, samples, etc.

Peto's Method

Peto's method (Peto et al., 1977) is also designed only for combining odds ratios and may be considered a modification of Mantel-Haenszel.

Part of the calculations for Peto's method requires finding the difference between the observed events and the expected events, similar to the calculations used for the 2×2 and $R \times C$ chi-square:

$$\bar{T}_{\text{PETO}} = \exp\left[\left.\sum_{i=1}^{k}(O_i - E_i)\middle/\sum_{i=1}^{k}v_i\right.\right]$$

Referring to the array shown in the Mantel-Haenszel method

$$O_i = a_i$$

$$E_i = (\text{proportion affected})(\text{number exposed}) = \frac{a_i + c_i}{n_i}(a_i + b_i)$$

and $\quad v_i = \dfrac{(a_i + b_i)(c_i + d_i)(a_i + c_i)(b_i + d_i)}{n_i^2(n_i - 1)}$

It should be noted that the above equation for v_i is often represented differently so the equivalent equation may not match the one from other books. Once the above calculations are done, the variance (again for the logarithm) is simply

$$\text{var}_{\text{PETO}} = 1\middle/\sum_{i=1}^{k}v_i$$

Assumptions and Limitations

This method assumes heterogeneity. If this is not the case, the random effects model is better.

Unlike the M-H method, this method is reasonable to use when any arm of a study has a zero in it.

Similarly, Peto's method performs well where any arm has a relatively small value in it.

This method may considerably underestimate for high exposure effects, i.e., where the odds ratio is high.

Example 12.9

Five separate tests of developmental mutagenicity of compound B in frogs were conducted and the results are presented below. In this case, developmental mutagenicity was defined as any abnormal physiological signs.

	Exposed Frogs		Unexposed Frogs	
Study	Abnormal (A)	Normal (B)	Abnormal (C)	Normal (D)
1	3	14	1	15
2	8	39	5	44
3	14	65	11	70
4	12	40	4	42
5	5	19	3	23

Using the Mantel-Haenszel method:

Study	$P = (A+D)/N$	$Q = (B+C)/N$	$R = A*D/N$	$S = B*C/N$	$P*R$	$P*S$	$Q*R$	$Q*S$
1	0.5455	0.4545	0.3636	0.4242	0.7438	0.2314	0.6198	0.1928
2	0.5417	0.4583	3.6667	2.0313	1.9861	1.1003	1.6806	0.9310
3	0.5250	0.4750	6.1250	4.4688	3.2156	2.3461	2.9094	2.1227
4	0.5510	0.4490	5.1429	1.6327	2.8338	0.8996	2.3090	0.7330
5	0.5600	0.4400	2.3000	1.1400	1.2880	0.6384	1.0120	0.5016
SUM	2.7231	2.2769	18.5982	9.6969	10.0674	5.2158	8.5308	4.4811

$$T_{MH} = 18.5982/9.6969 = 1.9179$$

$$\text{var}_{MH} = \frac{10.0674}{2(18.5982)^2} + \frac{5.2158 + 8.5308}{2(18.5982)(9.6969)} + \frac{4.4811}{2(9.6969)^2}$$

$$= 0.0765$$

$$CI_{\alpha} = \exp(\ln(1.9179) \pm 1.96\sqrt{0.0765}) = 1.12 \text{ to } 3.30$$

Using Peto's method:

Study	$O = A$	$E = (A+C)(A+B)/N$	$O - E$	$V = (A+B)(C+D)(A+C)$ $(B+D)/[N^2(N-1)]$
1	3	2.0606	0.9394	0.9054
2	8	6.3646	1.6354	2.8382
3	14	12.3438	1.6563	5.3058
4	12	8.4898	3.5102	3.3688
5	5	3.8400	1.1600	1.7115
SUM		33.0987	8.9013	14.1297

$$T_p = \exp(8.9013/14.1297) = 1.875$$

$$\text{var}_p = 1/14.1297 = 0.0708$$

$$CI_{\alpha} = \exp(\ln(1.8775) \pm 1.96\sqrt{0.0708}) = 1.11 \text{ to } 3.16$$

For both calculations, the confidence interval does not include unity, so we estimate that the odds of a frog having a physiological abnormality are greater if it was exposed to compound B by a factor of approximately 1.9.

Methods for Combining *p*-Values

As previously mentioned, combining *p*-values is not the preferred method of performing a meta-analysis primarily since it generates no indication of effect size. Also, the power of such methods has been demonstrated to generally be low. The method Fisher gave in 1932 and Pearson independently derived in 1933, often referred to as the sum of logs, is generally preferred unless there is reason to believe another test will perform better (Hedges and Olkin, 1985). For a review of other methods, please refer to Hedges and Olkin (1985) or Sutton et al. (2000).

Fisher's Sum of Logarithms Method

The test statistic

$$-2\sum_{i=1}^{k} \log(p_i)$$

has a χ^2 distribution with 2*k degrees of freedom with a $1 - \alpha$ critical value.

Example 12.10

Given the *p*-values in the following table, is the pooled *p*-value significant at the $\alpha = 0.05$ level?

Study	p	ln(*p*)
1	0.53	−0.635
2	0.46	−0.777
3	0.68	−0.386
4	0.72	−0.329
5	0.99	−0.010
SUM		−2.136

The value to test is −2* −2.136 = 4.271 with a chi-square test with 2 * 5 = 10 degrees of significance. The resultant value is 0.93, so the test is not significant for $1 - \alpha = 0.95$ critical value.

Summary

There is still some controversy about the areas and methods for proper application of meta-analysis, but the advantages and the burgeoning volume of studies available are making it an increasingly popular technique. Of all statistical analyses discussed in this book, meta-analysis is the one that most needs pre-planning in order to assure the validity of the conclusion.

There are several tools to graphically examine various aspects of a meta-analysis, one common homogeneity test, and several well-characterized methods for pooling results. In each case, however, the analyst is advised to carefully study the weaknesses and strengths of the methods before choosing which ones to use.

References

Babapulle, M.N., Joseph, L., Bélisle, P., Brophy, J.M., and Eisenberg, M. (2004) A hierarchical Bayesian meta-analysis of randomized clinical trials of drug-eluting stents. *Lancet*, 364, 583–591.

Bailar, J.C. (1997) The promise and problems of meta-analysis. *N. Engl. J. Med.*, 337, 559–561.

Birge, R.T. (1932) The calculation of errors by the method of least squares. *Phys. Rev.*, 16, 1–32.

Clemens, J.D., Chuong, J.J., and Feinstein, A.R. (1983) The BCG controversy: a methodological and statistical reappraisal. *J. Am. Med. Assoc.*, 249, 2362–2369.

Cochran, W.G. (1937) Problems arising in the analysis of a series of similar experiments. *J. Roy. Stat. Soc.*, Suppl. 4, 102–118.

Cochran, W.G. (1954) The combination of estimates from different experiments. *Biometrics*, 10, 101–129.

Fisher, R.A. (1932) *Statistical Methods for Research Workers*, 4th ed., Oliver & Boyd, London.

Fleiss, J.L. (1986) Analysis of data from multiclinical trials. *Contr. Clin. Trials*, 7, 267–275.

Gerbarg, Z.B. and Horwitz, R.I. (1988) Resolving conflicting clinical trials: guidelines for meta-analysis. *J. Clin. Epidemiol.*, 41, 503–509.

Glass, G.V. (1976) Primary, secondary, and meta-analysis of research. *Educational Researcher*, 5, 3–8.

Greenland, S. (1994) Invited commentary: a critical look at some popular meta-analytic methods. *Am. J. Epidemiol.*, 140, 290–296.

Hasselblad, V. (1995) Meta-analysis of environmental health data. *Sci. Total Environ.*, 15, 160–161, 545–548.

Hedges, L.V. and Olkin, I. (1985) *Statistical Methods for Meta-Analysis*. Academic Press, Orlando, FL.

L'Abbé, K.A., Detsky, A.S., and O'Rourke, K. (1987) Meta-analysis in clinical research. *Ann. Intern. Med.*, 107(2), 224–233.

Mantel, N. and Haenszel, W. (1959) Statistical aspects of the analysis of data from retrospective studies of disease. *J. Nat. Canc. Inst.*, 22, 719–748.

Pearson, K. (1933) On a method of determining whether a sample of given size n supposed to have been drawn from a parent population having a known probability integral has probably been drawn at random. *Biometrika*, 25, 379–410.

Peto, R., Pike, M.C., Armitage, P., Breslow, N.E., Cos, D.R., Howard, S.V., Mantel, N., McPherson, K., Peto, J., and Smith, P.G. (1977) Design and analysis of randomized clinical trials requiring prolonged observation of each patient. II: analysis and examples. *Br. J. Canc.*, 35, 1–39.

Robins, J., Breslow, N., and Greenland, S. (1986a) Estimators of the Mantel-Haenszel variance consistent in both sparse data and large-strata limiting methods. *Biometrics*, 42, 311–323.

Robins, J., Greenland, S., and Breslow, N.E. (1986b) A general estimator for the variance of the Mantel-Haenszel odds ratio. *Am. J. Epidemiol.*, 124, 719–723.

Sacks, H., Chalmers, T.C., Smith, H., Jr. (1982) Randomized versus historical controls for clinical trials. *Am. J. Med.*, 72, 233–240.

Sacks, H.S., Berrier, J., Reitman, D., Ancona-Berk, V.A., and Chalmers, T.C. (1987) Meta-analyses of randomized controlled trials. *N. Engl. J. Med.*, 316(8), 450–455.

Shapiro, S. (1994) Meta-analysis/shmeta-analysis. *Am. J. Epidemiol.*, 140, 771–778.

Sinclair, J.C. and Bracken, M.B. (1994) Clinically useful measures of effect in binary analyses of randomized trials. *J. Clin. Epidemiol.*, 47, 881–889.

Smith, G.D., Song, F., and Sheldon, T.A. (1993) Cholesterol lowering and mortality: the importance of considering initial level of risk. *Br. Med. J.*, 306(6889):1367–73. Erratum in: *Br. Med. J.*, 306(6893), 1648.

Sutton, A.J., Abrams, K.R., Jones, D.R., Sheldon, T.A., and Song, F. (2000) *Methods for Meta-Analysis in Medical Research*. John Wiley & Sons, New York.

Thacker, S.B. (1988) Meta-analysis: a quantitative approach to research integration. *J. Am. Med. Ass.*, 259, 1685–1689.

Tippett, L.H.C. (1931) *The Method of Statistics*. Williams & Norgate, London.

Whitehead, A. and Jones, N.M.B. (1994) A meta-analysis of clinical trials involving different classifications of response into ordered categories. *Stat. Med.*, 13, 2503–2515.

13

Bayesian Analysis

The basic equations and foundations of Bayesian analysis come from Reverend Thomas Bayes, whose papers were published after his death in 1763. Simply put, Reverend Bayes developed and formalized an equation for estimating the probability of a result given the previous occurrence of an event or events and, using formal logic, showed that this formula works whether or not the result and event(s) were dependent on one another. The logical extension of this is that it is possible to generate an estimate of probability of a result based on any number of prior events by simply using Bayes' theorem iteratively; i.e., calculating the probabilities one after another. In other words, intermediate results can be integrated in calculating the probability of future outcomes.

There are some terms used in these statistics that have not been previously covered. First, statistics based on Bayes' formulas are called Bayesian statistics, and other statistics, which are based on estimates of frequency distributions, are called frequentist statistics. The calculations for Bayesian statistics require a previously calculated or enumerated probability distribution, which is called the "prior." When this prior is applied to the data, a new probability distribution is generated that is termed the "posterior." Finally, when the confidence interval is calculated from the posterior using Bayesian methods, it is called a (or the) "credible interval." It should be noted that some texts use the two terms confidence interval or credible interval interchangeably, but purists prefer the latter term and we will differentiate the two in this chapter.

Given how long ago the basic postulates were produced, it would initially seem that Bayesian statistics would be very prevalent. It can also be shown that, for some calculations, Bayesian methods often produce more defined estimates and smaller confidence intervals compared to frequentist methods. There are, however, two big factors that slowed the acceptance and use of Bayesian statistics: first, they are often computationally intensive and do not lend themselves to easy use of lookup tables. While frequentist statistics are also often computationally intensive, they can usually be estimated with lookup tables that have existed for decades or more. Second, in order for Bayes' theorem to be valid, the prior must be independent of the results, i.e., the prior must be decided before viewing the data. This has two ancillary consequences:

The analyst must do more work initially in order to use Bayesian statistics. Instead of deciding which test to use as in frequentist statistics, they must define the probability distribution to use. Since the use of Bayesian statistics in medicine and toxicology has been on the rise only relatively recently (Goodman, 2001), there may not be easy guidelines available. For guidelines for frequentist statistics please see Chapter 2, Figures 2.1 to 2.4.

Deciding ahead of time which distribution to use looks suspiciously like an attempt to "fix" or predetermine the results, which is completely contrary to good statistical practice.

Fortunately, for the use of Bayesian statistics the computational objection has been minimized because analysis has shown that there are some shortcuts for common types of distributions and the phenomenal development in computer speed and power in the last two decades has made the more complex calculations easy. Careful analysis has shown that the choice of prior does not influence Bayesian results as much as would be expected, and that simple adjustments to the prior can significantly lower any prejudice introduced by a poor choice (Bolstad, 2004, pp. 261–274). This effectively took care of the two most serious objections. The resulting rise in publication and acceptance of analyses done with Bayesian statistics has then made it easier for statisticians to find peer consensus on which methods are applicable.

Bayes' Formula

Reverend Bayes' formulas are complex and often initially daunting, but are built on a simple idea: if the occurrence of one event is known, then that may change the probability that another event happens. This idea should be intuitively obvious as illustrated in Example 13.1.

Example 13.1

Out of 100 mice, 60 are exposed to compound C. The number of adrenal tumors in the exposed and unexposed groups are:

	Effect (E)	No Effect ($\bar{E}$)
Exposure (X)	18	42
No exposure ($\bar{X}$)	4	36

So, looking at all 100 mice, the chance of E is $(18+4)/100 = 22\%$. However, if we know that X occurred, the chance is $18/60 = 30\%$.

More formally, the chance of effect E happening given that exposure X occurred equals the chance of X and E occurring together in the entire population divided by the probability of X occurring, so

$$P(E|X) = \frac{\text{Probability of } X \text{ and } E \text{ occurring together}}{P(X)}$$

where P(X) is read as "the probability of X occurring" and P(E | X) is read as "the probability of E occurring given that X occurred." Using some basic theorems of set and probability theory, Reverend Bayes restated this as:

$$P(E|X) = \frac{P(X|E) \times P(E)}{P(X|E) \times P(E) + P(X|E) \times P(E)}$$

We can see that this formula works for our example:

$$P(E|X) = \frac{[18/22] \times [22/100]}{[18/22] \times [22/100] + [42/78] \times [78/100]} = \frac{18/100}{18/100 + 42/100} = 30\%$$

It is worthwhile to note the base numbers were converted to probabilities by dividing by the population, which was 100 in this case, but due to the way formula works, that factor cancels out in calculation. This is not just due to this specific case—it is a general result of Bayesian statistics. Purists prefer that the prior probabilities add up to 1 (or integrate to 1 in the case of continuous probabilities) but any prior weighting at all will work mathematically.

Bayesian Applications for Assays

Assays are often defined by sensitivity and specificity. Sensitivity is the chance of having a positive result on the assay if the condition truly exists. This means that (1 − sensitivity) is the chance of making a Type II error, i.e., a false negative. Specificity is the chance of a negative result if the condition does not exist, so (1 − specificity) is the Type I error rate or false positive rate. If these values are known and the prevalence of the condition in the population is known, then the posterior probability can be calculated.

Example 13.2

(Adapted from Gad and Rousseaux, 2002). Suppose that a population has approximately 2% prevalence of hyperparathyroidism, which causes elevated blood calcium but is asymptomatic except in extreme cases. A blood calcium test detects 90% of the incidences of this disease (90% accurate) but the same elevated levels occur in normal individuals at a 5% rate, so the test is 100 − 5 = 95% specific. We want to find a group of subjects with asymptomatic hyperparathyroidism in order to test a new compound.

From a Bayesian standpoint:

Bayesian Term	Explanation	P
$P(E)$	Population having the condition	0.02
$P(E)$	Population not having the condition	0.98
$P(X|E)$	Chance of finding elevated Ca levels if the subject has the condition	0.90
$P(X|E)$	Chance of finding elevated Ca levels if the subject does not have the condition	0.05

so the posterior can be calculated as

$$P(E|X) = \frac{[0.90] \times [0.02]}{[0.90] \times [0.02] + [0.05] \times [0.98]} = 0.269 = 27\%$$

So, even with a positive screen on the calcium test, there is only a 27% chance that our screened subjects will have the condition.

Example 13.3

Referring to the previous example (Example 13.2), it is obviously ethically undesirable to treat a population with only a 27% chance of having the condition since approximately 73% of the subjects would be inappropriate for the treatment. This accuracy can be improved by the follow-up use of a parathyroid hormone radioimmune assay, which has a 95% sensitivity and a 98% specificity. If we use the screened population as our prior instead of the general population:

Bayesian Term	Explanation	P
$P(E)$	Population having the condition	0.27
$P(E)$	Population not having the condition	0.73
$P(X|E)$	Chance of a positive assay if the subject has the condition	0.95
$P(X|E)$	Chance of a positive assay if the subject does not have the condition	0.02

so the posterior is

$$P(E|X) = \frac{0.95 \times 0.27}{0.95 \times 0.27 + 0.02 \times 0.73} = 0.945$$

This time, including both our screens, there is a 94.5% chance of selecting a group of subjects with hyperparathyroidism. This is obviously a great improvement over the previous results and it is worthwhile to note how such an improvement can be obtained. Referring to Bayes' formula, the criteria for a highly probable posterior can be expressed as:

$$P(X|E) \times P(E) \gg P(X|E) \times P(E)$$

where >> is read as "is significantly greater than." In other words, for an assay to yield a high posterior probability, the sensitivity of the assay times the population actually having the condition should be much greater than one minus the specificity of the assay times the population not having the condition, or more simply, the true positive results should be much greater than the false positive results. In our example, the sensitivity and specificity went up noticeably but the real difference was the great increase in the proportion of the population having the disease. This is not merely an academic exercise—false positives in cancer screenings, for example, can lead to radiation or chemotherapy of a healthy individual.

Order of Bayesian Statistics

As shown in Examples 13.2 and 13.3 above, Bayesian analyses can be used iteratively to generate posterior values. If the calculations are done in reverse order:

$$P(E|X) = \frac{0.95 \times 0.02}{0.95 \times 0.02 + 0.02 \times 0.98} = 0.49$$

$$P(E|X) = \frac{[0.90] \times [0.49]}{[0.90] \times [0.49] + [0.05] \times [0.51]} = 0.945 = 94.5\%$$

The similar results are not an accident: it does not matter what order the calculations are done as long as the test results are independent of each other. It is also possible to calculate the posterior in one step rather than doing it iteratively. Looking at it in a simplified way the single-step analysis is

Prior * Results = Posterior

and the multi-step analysis is

Prior1 * Results1 = Posterior1 = Prior2

Prior2 * Results2 = Posterior2 = Prior3

Prior3 * Results3 = Posterior3

etc.

So the single-step calculation is

$$\text{Prior}_1 * \text{Results}_1 * \ldots * \text{Results}_N = \text{Posterior}_N$$

Therefore, Bayesian estimations can be done in steps or all at once. This is an advantage for tests or trials that take a long time in that intermediate results can be estimated as the data comes in rather than waiting for trial termination. There is also a significant consequence for the planning of assays or screenings. If you want to find all of the individuals in a population with a specific characteristic and the situation is such that more than one screen is needed, then it is more economical to use the less expensive tests first. The same proportion of individuals will be identified in the end no matter what the order of the tests, but the more expensive test(s) are required on a smaller number of individuals.

Assumptions and Limitations

Test results must be independent of each other. This also means that the population remaining after one test must have the same proportional response to the following test(s) as the original population did.

If the calculations are done on an iterative basis, care must be taken to correct for cumulative round-off errors.

Discrete Data

In some situations the initial event or exposure is composed of several exclusive (independent) subevents; for example, when using discrete data. Bayes' formula for the probability of discrete event i out of n discrete events is

$$P(E_i|X) = \frac{P(X|E_i) \times P(E_i)}{\sum_{j=1}^{n} P(X|E_j) \times P(E_j)}$$

Example 13.3 shows the calculation of subevent probabilities.

Example 13.4

One of the common measurements made in a urinalysis is the specific gravity of the sample, defined as the mass of a specific volume of the sample as a ratio to the mass of the same volume of water. While this is technically a continuous variable, the measurements are such that several discrete values are the only results. Assume that we have an estimate of the distribution of values in the general population, a subgroup of N individuals, and a measurement of samples from 75 randomly selected individuals in the subgroup, it is possible to estimate the posterior distribution of specific gravities from the subgroup.

The prior in this case is the usual population distribution, which is

Specific Gravity	Proportion (Prior)
1.00	0.0031
1.05	0.2855
1.10	0.3041
1.15	0.2074
1.20	0.1251
1.25	0.0626
1.30	0.0122

The data is the count of samples in the specific range, which is

Specific Gravity	Count
1.00	2
1.05	26
1.10	24
1.15	15
1.20	6
1.25	3
1.30	0

Putting these in terms of the previous equation gives:

Specific Gravity	Prior $P(X \mid E_i)$	Count $P(E_i)$	Individual $P(X \mid E_i) \cdot P(E_i)$	Scaled Individual $P(E_i \mid X)$
1.00	0.0031	2	0.0062	0.0062/18.7770 = 0.0003
1.05	0.2855	26	7.4320	7.4320/18.7770 = 0.3953
1.10	0.3041	24	7.2984	7.2984/18.7770 = 0.3887
1.15	0.2074	15	3.1110	3.1110/18.7770 = 0.1657
1.20	0.1251	6	0.7506	0.7506/18.7770 = 0.0400
1.25	0.0626	3	0.1878	0.1878/18.7770 = 0.0100
1.30	0.0122	0	0	0.0000/18.7770 = 0.0000
Sums	1.0000	76	18.7770 $\sum P(X \mid E_i) \cdot P(E_i)$	1.0000

Earlier we stated that the prior did not have to add up to unity for Bayesian calculations to be valid, but we can see from the example above that this is also true for how we represent the data distribution. Generally, neither the prior nor the data has to be converted to a probability scale—such conversion is done in the last step of calculation. The only strict requirement is that the prior and the data must each be internally proportional. Interpretation of these results is easier if we look at the prior, data, and posterior graphically, as shown in Figure 13.1.

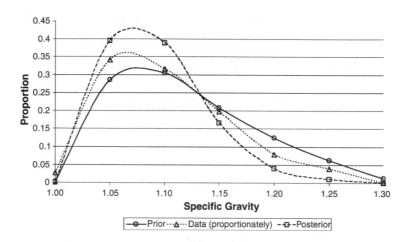

FIGURE 13.1
Analysis of urine-specific gravity by Bayesian methods.

Note that the posterior and the sample distribution should not be confused—they are obviously different shapes. Instead, the posterior is a better estimate of the specific gravity with the highest probability of occurrence in the selected subgroup.

Simplified Calculation of Bayesian Parameters

Example 13.3 and Figure 13.1 also illustrate a point made earlier about the complexity of calculation of Bayesian statistics. While the calculation of the individual posterior values was fairly straightforward, calculation of a maximum likelihood estimator or a credible interval for the curves shown would require fairly advanced modeling and estimation. This is even more true when dealing with continuous curves especially since posteriors are frequently nonstandard distributions. Add to this the possibility of large data sets with tens of thousands or even millions of values, e.g., data from late-stage drug trials, and it is easy to see why increased computer capacity has been a boom to Bayesian statistics. Fortunately, there are shortcuts for some distributions. The trivial case of this occurs if a flat proper prior is chosen—it can be shown that any distribution is unchanged if the value of the prior is the same at every point. Then the Bayesian calculations are the same as the frequentist calculations. Note that this assumes a *proper* prior, i.e., the prior probabilities sum to unity in the discrete case or the prior integrates to unity for continuous data. We will also examine two other cases; both the normal curve and the beta set of distributions have sets of update rules that avoid onerous calculations.

Normal Curves

If the population can be assumed to have a normal distribution and the prior also has a normal distribution, then Bayesian calculations generate another normal distribution with an updated mean and standard deviation. As a side note, it should be noted that a non-normal prior will generate a non-normal posterior even when used with a normal population. We will start by looking at a strictly defined case and then see what adjustments are required when various assumptions change. The assumptions of the strict case are:

The population is normal with mean μ_N and variance σ_N^2.

The prior is also normal with mean μ_P and variance σ_P^2.

Then the variance of the mean based on Bayesian mathematics is

$$\sigma_R^2 = \frac{\sigma_P^2 \sigma_N^2}{\sigma_P^2 + \sigma_N^2}$$

and the mean is

$$\mu_B = \frac{1/\sigma_P^2}{1/\sigma_N^2 + 1/\sigma_P^2} \mu_P + \frac{1/\sigma_N^2}{1/\sigma_N^2 + 1/\sigma_P^2} \mu_N$$

which can also be rewritten as

$$\mu_B = \frac{\sigma_B^2}{\sigma_P^2} \mu_P + \frac{\sigma_B^2}{\sigma_N^2} \mu_N$$

From this second version of the equation for the posterior mean it can be seen that it is a weighted combination of the prior and the population mean, with the weights inversely proportional to the variance of the respective populations. This inverse variance is often called the precision, so one way of looking at the posterior mean is that it is the combination of prior and population means weighted by the precision. Since the new Bayesian estimate is normal with a defined mean and variance, credible intervals, differences in means, and probabilities for this posterior can be calculated using the methods already enumerated for normal populations.

It is unlikely that the true population mean and variance will be known. If the population variance is known or well characterized but the mean is not, then the mean can be estimated based on a random sample of size n and mean μ_R drawn from the population. In this case the variance of the mean is

$$\sigma^2_{\mu B} = \frac{\sigma^2_N \sigma^2_P}{\sigma^2_N + n\sigma^2_P}$$

and the estimated mean is

$$\mu_B = \frac{1/\sigma^2_P}{n/\sigma^2_N + 1/\sigma^2_P}\mu_P + \frac{n/\sigma^2_N}{n/\sigma^2_N + 1/\sigma^2_P}\mu_R$$

which can also be rewritten as

$$\mu_B = \frac{\sigma^2_{\mu B}}{\sigma^2_P}\mu_P + n\frac{\sigma^2_{\mu B}}{\sigma^2_N}\mu_R$$

It is important to note here that we went from calculating the mean and variance of the population to estimating the mean and variance of the mean, so $\sigma^2_{\mu B}$ is the equivalent of standard error of the mean (SEM), not standard deviation. This makes the Bayesian credible interval for μ_B at the $(1 - \alpha)$ ∞ 100% level

$$\mu_B \pm z_{\alpha/2} \times \sigma_{\mu B}$$

Since the posterior is normal, then the relationship between the credible interval for the mean and the credible interval for the data is known and therefore the Bayesian credible interval for the range of data at the same level would be

$$\mu_B \pm z_{\alpha/2} \times \sqrt{n} \times \sigma_{\mu B}$$

It is unlikely that the true population variance will be known while the population mean is unknown. This makes it necessary to estimate the population variance from the known sample variance, σ_R^2, once again assuming that a sample of n values with mean μ_R is drawn from the population. Also, in cases where it is suspected that the population variance may not be easily estimable, e.g., in cases of known or probable variance inflation, it is better to estimate the population variance using the sample. In these cases the sample variance is calculated normally and used instead of σ_P^2 in the above equations. The additional uncertainty of estimating the population variance as the sample variance should be taken into account, however. This is done by using *Student's t* instead of z-values, which makes the $(1 - \alpha)$ ∞ 100%

interval for the mean

$$\mu_B \pm t_{\alpha/2} \times \sigma_{\mu B}$$

Since this case, i.e., where the prior is fully known but the population mean and variance are unknown, is the one most likely to be encountered when dealing with real data it is worthwhile to give the equations necessary for calculating credible intervals for the differences in means.

The credible interval for the difference between means depends on whether or not the variances of the populations the samples are drawn from are equal. The simplest case is when the variances are assumed to be equal. In that case the two-sided credible interval for the difference in means is

$$\mu_{1B} - \mu_{2B} \pm t_{\alpha/2} \sqrt{\sigma^2_{\mu1B} + \sigma^2_{\mu2B}}$$

where μ_{1B}, μ_{2B}, $\sigma^2_{\mu1B}$, and $\sigma^2_{\mu B}$ are the means and variances of the mean for the Bayesian priors calculated as shown above and the degrees of freedom used for the t-value is $n_1 + n_2 - 2$.

The assumption of equal variances may not be applicable, again perhaps due to variance inflation. Oddly enough, the two-sided credible interval for the difference in means is exactly the same as the previous equation, i.e.,

$$\mu_{1B} - \mu_{2B} \pm t_{\alpha/2} \sqrt{\sigma^2_{\mu1B} + \sigma^2_{\mu2B}}$$

Unfortunately, the degrees of freedom are not as straightforward as the previous equation. The estimation was proposed by Satterthwaite in 1946 and is still in common use today:

$$df_{est} = \frac{\left(\sigma_1^2/n_1 + \sigma_2^2/n_2 \right)^2}{\frac{\left(\sigma_1^2/n_1 \right)^2}{n_1 - 1} + \frac{\left(\sigma_2^2/n_2 \right)}{n_2 - 1}}$$

df_{est} is then rounded down to the nearest integer.

Example 13.5
Part 1:

The blood urea nitrogen (BUN) in a ferret population has previously been measured as normally distributed with mean 27 mg/dL and standard deviation 14 mg/dL, which will be taken as the prior. Assuming that the population was retested and the new values were a mean of 24 mg/dL and the standard deviation was 9 mg/dL, the posterior values for the population can be calculated.

From the equations above it is easier to calculate σ_B first and then use it to calculate μ_B.

$$\sigma_B = \sqrt{\frac{14^2 \times 9^2}{14^2 \times 9^2}} = \sqrt{57.314} = 7.57$$

So

$$\mu_B = \frac{57.314}{14^2} 27 + \frac{57.314}{9^2} 24 = 24.88$$

Part 2:

Assuming instead that the population mean was actually based on a sample size of 12 and the population standard deviation is known and still equal to 9 mg/dL then the calculations for the posterior mean and standard deviation are different. Once again it is easier to calculate σ_B first and then use it to calculate μ_B.

$$\sigma_{\mu B} = \sqrt{\frac{9^2 \times 14^2}{9^2 + 12(14^2)}} = \sqrt{6.525} = 2.55$$

So

$$\mu_B = \frac{6.525}{14^2} 27 + 12 \frac{6.525}{9^2} 24 = 24.10$$

and therefore the 95% two-sided credible interval for the mean is

$$24.88 \pm 1.96 \times 2.55 = 19.88 \text{ to } 29.88$$

The frequentist estimate for the same interval would be 24.00 ± 1.96 * 2.60 = 18.90 to 29.10. It is instructive to note that the frequentist confidence interval is centered differently than the Bayesian credible interval and that the credible interval is shorter. By our assumptions, it is impossible to "know" which is correct, but, in general, a shorter range will lead to a higher probability of finding statistical significance over repeated random tests.

Part 3:

If the population standard deviation is assumed instead to be *unknown* but the sample standard deviation remains at 9 mg/dL, then σ_B and σ_B are exactly the same as in Part 2, but the credible interval is based on t

(at $12 - 1 = 11$ degrees of freedom) instead of z and therefore broadens to

$$24.88 \pm 2.201 \times 2.55 = 19.27 \text{ to } 30.49$$

Part 4:

Assume a second random sample is taken where $n_2 = 11$, $\mu_2 = 33$, and $\sigma_2 = 13$. The difference in means (assuming equal variance) requires some intermediate steps to calculate. First it is necessary to calculate $\sigma_{\mu 2B}$ and μ_{2B}

$$\sigma_{\mu 2B} = \sqrt{\frac{13^2 \times 14^2}{13^2 + 11(14^2)}} = \sqrt{14.247} = 3.77$$

and

$$\mu_{2B} = \frac{14.247}{14^2} 27 + 11 \frac{14.247}{13^2} 33 = 32.56$$

which makes the credible interval for the difference in means

$$24.10 - 32.56 \pm 2.080 \sqrt{6.525 + 14.247}$$

or −17.94 to 1.02 with $= 0.05$ and df $= 12 + 11 - 2 = 21$. If the hypothesis H_0 was that the means were equal, then it has not been disproved since the interval includes zero. If the assumption instead is that the variances are unequal then the equation is the same except that the degrees of freedom should be

$$df_{est} = \frac{(6.525/12 + 14.247/11)^2}{\frac{(6.525/12)^2}{12-1} + \frac{(14.247/11)^2}{11-1}} \approx 17$$

which makes the credible interval −18.07 to 1.16, which is (not surprisingly) wider than the one based on the assumption of equal variances.

Beta Distributions

Another family of distributions that have simple update rules for Bayesian statistics is the Beta(a, b) distribution, which is the class of curves defined by the equation

$$y = k \times x^{n-1}(1-x)^{b-1}$$

TABLE 13.1

Calculating Statistics for Beta Distributions.

Statistic[a]	Formula
Mean	$= a/(a + b)$
Mode (maximum likelihood)	$= (a - 1)/(a + b - 2)$
Equivalent sample size (n_{eq})	$= a + b + 1$
Variance	$= \dfrac{ab}{(a+b)^2(a+b+1)}$

[a] All statistics are calculated for the zero to one interval.

where a, b, and k are constants and x takes values from zero to one. The values of a and b define the shape of the curve while the value of k can be adjusted to make the distribution proper (area equal to 1) in the interval from zero to one. The mathematics of this set of curves make for easy calculations, as shown in Table 13.1.

The equivalent sample size is the sample size that would produce the specified variance for the distribution. All of the calculated statistics for beta distributions are independent of the factor k, so no calculation of k is needed, making working with the distribution easier. In addition, as a and b rise, the beta distribution closely approximates normal curves with the mean and variance shown. If both a and b are ten or greater then the tables for the normal curve can be used to get good approximations of probabilities and confidence intervals (Woodworth, 2004).

Beta distributions are often used as priors for estimating proportions using quantal data. The first reason for this is the update rules to generate a posterior from a beta prior and quantal data are simple: for y successes out of n events, the posterior is a new beta distribution with

$$a' = a + y$$

and

$$b' = b + (n - y)$$

In other words, add the successes (or positive events) to the prior value of a to generate the posterior value of a (a') and add the number of failures (or negative events) to the prior value of b to generate the posterior value of b (b'). Another good reason to use beta functions for quantal data has to do with the probable distribution of proportions: if the number of positive events (y) out of n events is known, then the likelihood function for (the true population proportion) is

$$P(\pi|y,n) = \binom{n}{y}\pi^{y}(1-\pi)^{n-y}$$

which is obviously identifiable as a beta function. To put it another way, if a trial is done that produces quantal data and the number of data points and successes are known, then there is a calculable beta function that shows the probability curve of the true population proportion and that beta function can be used to get confidence intervals.

To construct a beta prior it is necessary to define the parameters a and b. Since these are somewhat abstract, it is more informative to define two of the listed parameters above and then calculate the a and b values necessary to produce them. The distribution could be defined by the prior mean and variance, the maximum likelihood and the variance, the maximum likelihood and the equivalent sample size, or any other combination of two of the statistics listed above. In practice, it is usually easiest to define the mean and the equivalent sample size (n_{eq}) since those equations are the easiest to solve. Predefining n_{eq} also gives an easy way to control how strong the belief in the prior is, e.g., if the belief is weak then n_{eq} should be small and if the belief is strong then n_{eq} should be large.

Example 13.6

A laboratory is testing substance C2 on 12 dogs for its ability to cause abnormalities in cardiac rhythm, defined simply as either presence or absence of abnormalities after administration of C2. The biostatistician noted that substance C1 was pharmacologically similar, had previously been tested at the same dosage on 15 animals, and showed 3 abnormalities, and therefore decided to use this as the prior. So

$$\text{Mean} = (a)/(a+b) = 3/15 \text{ and } n_{eq} = a+b+1 = 15$$

Solving gives $a = 2.8$ and $b = 11.2$. When the test was performed, 8 of the 12 dogs showed abnormalities.

Since the prior distribution was beta, the posterior distribution is beta and is defined by the update rules

$$a' = a+y = 2.8+8 = 10.8$$

and

$$b' = b+(n-y) = 11.2+(12-8) = 15.2$$

therefore

$$\text{Posterior mean} = m' = 10.8/(10.8 + 15.2) = 0.415$$

$$\text{Posterior mode} = M' = (10.8 - 1)/(10.8 + 15.2 - 2) = 0.408$$

$$n'_{eq} = a + b + 1 = 27$$

$$\text{Var}(y') = \frac{(10.8)(15.2)}{(10.5 + 15.2)^2 (10.8 + 15.2 + 1)} = 0.008994$$

Since both a' and b' are greater than 10, a normal distribution with mean 0.415 and variance 0.008994 can be used to approximate a 95% confidence interval instead of resorting to numerical integration.

$$m' \pm z_{\alpha/2} \times \sqrt{\text{Var}(y')} = 0.415 \pm 1.96 \times \sqrt{0.008994}$$

$$= 0.229 \text{ to } 0.601$$

Choosing a Good Prior

It is instructive to graph the prior, likelihood of π based on the data, and the posterior from Example 13.5, as shown in Figure 13.2. The a and b values for the prior and the posterior are given above. Solving for those values for

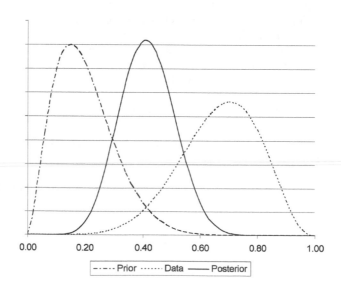

FIGURE 13.2
Prior, data, and posterior beta distributions from Example 13.5.

the data gives $a = 22/3$ and $b = 11/3$, which is the curve shown in the figure as "Data." Looking at the graph clearly shows that the posterior largely does not overlap either of the other data sets. The choice of prior has significantly shifted the posterior away from the likelihood distribution of the data itself. Given the situation described, it is somewhat unlikely (in a common sense manner, not necessarily in a statistical manner) that the prior was a good choice and it is also unlikely that the posterior is a good representation of the true population mean. As stated previously, going back and changing the prior afterward invalidates the results, so it is necessary to start every analysis by choosing a prior that is robust; i.e., one that will lead to a posterior close to the true population value without prejudicing the posterior too much.

The way to make a robust prior is the same no matter what distribution is chosen: flatten it. In other words, increase the probability of extreme values (thereby lowering the probability of the values near the mean). For a normal prior, this would mean increasing the estimated standard deviation (or variance). For a beta prior, setting the equivalent sample size to a smaller number is one easy way to achieve this. One technique that can be easily used with numerical integration techniques is to simply add a small, equal amount to the entire range of probabilities under consideration. The use of a robust beta is illustrated in Example 13.6

Example 13.7

The data from Example 13.6 is reanalyzed by a statistician who has not seen the results but has seen the data from substance C1. That person decides to use the same mean but decides that an equivalent sample size of five is more appropriate. So

$$\text{Mean} = (a)/(a + b) = 3/15 \text{ and } n_{eq} = a + b + 1 = 5$$

Solving gives

$$a = 4/3 \text{ and } b = 8/3$$

Since the prior distribution was beta, the posterior distribution is beta and is defined by the update rules

$$a' = a + y = 1.333 + 8 = 9.333$$

and

$$b' = b + (n - y) = 2.667 + (12 - 8) = 6.667$$

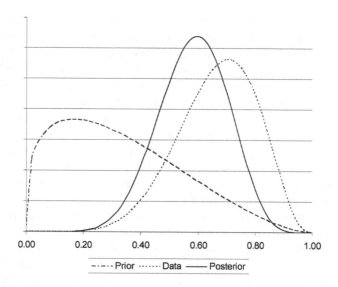

0.00   0.20   0.40   0.60   0.80   1.00

—·—·· Prior ······· Data —— Posterior

FIGURE 13.3
Prior, data, and posterior beta distributions from Example 13.6.

therefore

$$\text{Posterior mean} = m' = 9.333/(9.333+6.667) = 0.583$$

$$\text{Posterior mode} = M' = (9.333-1)/(9.333+6.667-2) = 0.595$$

$$n'_{eq} = a+b+1 = 17$$

$$\text{Var}(y') = \frac{(9.333)(6.667)}{(9.333+16.667)^2(9.333+16.667+1)} = 0.014297$$

Since both a' and b' are less than 10, numerical integration is used to approximate the 95% confidence interval, which is 0.37 to 0.77 (note that it is not symmetrical). The results are shown in Figure 13.3. The posterior now resembles the data more than it did in the previous example even though the prior mean is the same as before.

The preceding examples show that statisticians have to strike a careful balance when specifying a prior. A well-defined prior can greatly increase the confidence in the results, but a poor prior can skew them instead. It should also be noted that certain distributions of the prior can strongly prejudice the posterior in favor of acceptance of a null hypothesis H_θ that frequentist statistics would firmly reject. Basically, if the prior has a strong peak at the value θ and a generally flat but non-zero distribution otherwise, then the posterior has a much higher probability of including θ in its credible interval than the equivalent frequentist statistics would in their confidence intervals. This result is known as the Lindley Paradox from the work of Dennis Lindley in 1957. A short discussion can be found in Press (1989) and more details are available from the works of Bernardo (1980), Lindley (1965), and Shafer (1982).

In general, if there is already data with a high degree of confidence or another method of estimation that produces accurate results, it is reasonable to use it. Even then it is prudent to spread out the distribution of the prior. A truly conservative approach with a new or unique situation would be to use an uninformative (flat) prior, which leaves the estimations the same as the frequentist statistics.

Summary

Bayesian statistics can often, when properly applied, produce better estimates than frequentist statistics, but widespread use has been slowed because of their numerical complexity and the necessity of producing a prior before knowledge of the data. Rise in computing power and the calculation of simple update rules for common distribution families has led to greater ease of use. A flat prior produces the same results that frequentist statistics do. Both the normal family of distributions and the beta family of distributions have simple update rules. Proper selection of a prior is still critical for the calculated posterior to be applicable, but the rise in the use of Bayesian statistics has led to better understanding of the elements needed for a statistically useful prior.

References

Bayes, T. (1763) An essay towards solving a problem in the doctrine of chances. *Phil. Trans. Roy. Soc. Lond.*, 53, 370–418.

Bernardo, J.M. (1980) A Bayesian analysis of classical hypothesis testing, in *Bayesian Statistics*, Bernardo, J.M., DeGroot, M.H., Lindley, D.V., and Smith, A.F.M., Eds., University Press, Valencia, Spain.

Bolstad, W.M. (2004) *Introduction to Bayesian Statistics*. John Wiley & Sons, Englewood, NJ.

Gad, S.C. and Rousseaux, C.G. (2002) Use and misuse of statistics in the design and interpretation of studies, in *Handbook of Toxicologic Pathology*, Hascheck, W.M., Rousseaux, C.G., and Wallig, M.A., Eds., 2nd ed., Vol. 1, pp. 406–401.

Goodman, S. (2001) What can Bayesian analysis do for us? Presented to U.S.F.D.A. Oncologic Drugs Advisory Committee, Pediatric Subcommittee on Nov. 28, 2001. Accessed June 1, 2004, http://www.fda.gov/ohrms/dockets/ac/01/slides/3803s1_05A_Goodman/index.htm.

Lindley, D.V. (1957) A statistical paradox. *Biometrika*, 44, 187–192.
Lindley, D.V. (1965) *Introduction to Probability and Statistics*. Cambridge University Press, Boston, MA.
Press, S.J. (1989) *Bayesian Statistics*. John Wiley & Sons, New York.
Satterthwaite, F.E. (1946) An approximate distribution of estimates of variance components. *Biometrics Bull.*, 2, 110–114.
Shafer, G. (1982) Lindley's paradox. *J. Am. Stat. Assoc.*, 77(378), 325–351.
Woodworth, G.G. (2004) *Biostatistics: A Bayesian Introduction*. John Wiley & Sons, Englewood, NJ.

14

Data Analysis Applications in Toxicology

Having reviewed basic principles and provided a set of methods for statistical handling of data, the remainder of this book will address the practical aspects and difficulties encountered in day-to-day toxicological work.

As a starting point, we present in Table 14.1 an overview of data types actually encountered in toxicology, classified by type (as presented at the beginning of this book). It should be stressed, however, that this classification is of the most frequent measure of each sort of observation (such as body weight) and will not always be an accurate classification.

There are now common practices in the analysis of toxicology data, although they are not necessarily the best. These are discussed in the remainder of this chapter, which seeks to review statistical methods on a use-by-use basis and to provide a foundation for the selection of alternatives in specific situations.

Median Lethal and Effective Doses

For many years, the starting point for evaluating the toxicity of an agent was to determine its LD_{50} or LC_{50}, which are the dose or concentration of a material at which half of a population of animals would be expected to die. These figures are analogous to the ED_{50} (effective dose for half a population) used in pharmacologic activities and are derived by the same means.

To calculate either of these figures we need, at each of several dosage (or exposure) levels, the number of animals dosed and the number that died. If we seek only to establish the median effective dose in a range-finding test, then 4 or 5 animals per dose level, using Thompson's method of moving averages, is the most efficient methodology and will give a sufficiently accurate solution. With two dose levels, if the ratio between the high and low dose is two or less, even total or no mortality at these two dose levels will yield an acceptably accurate medial lethal dose, although a partial mortality is desirable. If, however, we wish to estimate a number of toxicity levels (LD_{10}, LD_{90}) and are interested in more precisely establishing the slope of the dose/lethality

TABLE 14.1

Classification of Data Commonly Encountered in Toxicology, by Type

Continuous normal:	Body weights
	Food consumption
	Organ weights: absolute and relative
	Mouse ear swelling test (MEST) measurements
	Pregnancy rates
	Survival rates
	Crown-rump lengths
	Hematology (some)
	Clinical chemistry (some)
Continuous but not normal:	Hematology (some WBC)
	Clinical chemistry (some)
	Urinalysis
Scalar data:	Neurobehavioral signs (some)
	PDI scores
	Histopathology (some)
Count data:	Resorption sites
	Implantation sites
	Stillborns
	Hematology
	(Some reticulocyte counts // Howel-Jolly // WBC differentials)
Categorical data:	Clinical signs
	Neurobehavioral signs (some)
	Ocular scores
	GP sensitization scores
	MEST sensitization
	Counts
	Fetal abnormalities
	Dose/mortality data
	Sex ratios
	Histopathology data (most)

curve, the use of at least 10 animals per dosage level with the log/probit regression technique described in Chapter 8 would be the most common approach. Note that in the equation $Y_i = a + bx_i$, b is the slope of the regression line, and that our method already allows us to calculate 95% confidence intervals about any point on this line. Note that the confidence interval at any one point will be different from the interval at other points and must be calculated separately. Additionally, the nature of the probit transform is such that toward the extremes—LD_{10} and LD_{90}, for example—the confidence intervals will "balloon." That is, they become very wide. Because the slope of the fitted line in these assays has a very large uncertainty in relation to the uncertainty of the LD_{50} itself (the midpoint of the distribution), much caution must be used with calculated LD_xs other than LD_{50}s. The imprecision of the LD_{35}, a value close to the LD_{50}, is discussed by Weil (1972), as is that of the slope of the log dose–probit line (Weil, 1975). Debanne and Haller (1985) recently reviewed the statistical aspects of different methodologies for estimating a median effective dose.

There have been questions for years as to the value of LD_{50} and the efficiency of the current study design (which uses large numbers of animals) in determining it. As long ago as 1953, Weil et al. presented forceful arguments that an estimate having only minimally reduced precision could be made using significantly fewer animals. More recently, the last few years have seen an increased level of concern over the numbers and uses of animals in research and testing, and additional arguments against existing methodologies for determining the LD_{50}, or even the need to make the determination at all, have been made (Zbinden and Flury-Roversi, 1981). In response, a number of suggestions for alternative methodologies have been advanced (DePass et al., 1984; Gad et al., 1984; Bruce, 1985).

Body and Organ Weights

Among the sets of data commonly collected in studies where animals are dosed with (or exposed to) a chemical are body weight and the weights of selected organs. In fact, body weight is frequently the most sensitive indication of an adverse effect. How to best analyze this and in what form to analyze the organ weight data (as absolute weights, weight changes, or percentages of body weight) have been the subjects of a number of articles (Jackson, 1962; Weil, 1962, 1970; Weil and Gad, 1980).

Both absolute body weights and rates of body weight change (calculated as changes from a baseline measurement value, which is traditionally the animal's weight immediately prior to the first dosing with or exposure to test material) are almost universally best analyzed by ANOVA followed, if called for, by a *post hoc* test. Even if the groups were randomized properly at the beginning of a study (no group being significantly different in mean body weight from any other group, and all animals in all groups within two standard deviations of the overall mean body weight), there is an advantage to performing the computationally, slightly more cumbersome (compared to absolute body weights) changes in body weight analysis. The advantage is an increase in sensitivity, because the adjustment of starting points (the setting of initial weights as a "zero" value) acts to reduce the amount of initial variability. In this case, Bartlett's test is performed first to ensure homogeneity of variance and the appropriate sequence of analysis follows.

With smaller sample sizes, the normality of the data becomes increasingly uncertain and nonparametric methods such as Kruskal–Wallis may be more appropriate (see Zar, 1998).

The analysis of relative (to body weight) organ weights is a valuable tool for identifying possible target organs (Gad et al., 1984). How to perform this analysis is still a matter of some disagreement, however.

Weil (1962) presented evidence that organ weight data expressed as percentages of body weight should be analyzed separately for each sex. Furthermore, since the conclusions from organ weight data of males differed so often from those of females, data from animals of each sex should be used in this measurement. Also, Weil (1970, 1973), Boyd and Knight (1963), and Boyd (1972) have discussed in detail other factors that influence organ weights and that must be taken into account.

The two competing approaches to analyzing relative organ weights call for either calculating organ weights as a percentage of total body weight (at the time of necropsy) and analyzing the results by ANOVA or ANCOVA, with body weights as the covariates as discussed previously by the authors (Weil and Gad, 1980).

A number of considerations should be kept in mind when these questions are addressed. First, one must keep a firm grasp on the difference between biological significance and statistical significance. In this particular case, we are especially interested in examining organ weights when an organ weight change is not proportional to changes in whole body weights. Second, we are now required to detect smaller and smaller changes while still retaining a similar sensitivity (i.e., the $p < 0.05$ level).

There are several devices to attain the desired increase in power. One is to use larger and larger sample sizes (number of animals) and the other is to utilize the most powerful test we can. However, the use of even currently employed numbers of animals is being vigorously questioned and the power of statistical tests must, therefore, now assume an increased importance in our considerations.

The biological rationale behind analyzing both absolute body weight and the organ weight–to–body weight ratio (this latter as opposed to a covariance analysis of organ weights) is that in the majority of cases, except for the brain, the organs of interest in the body change weight (except in extreme cases of obesity or starvation) in proportion to total body weight. We are particularly interested in detecting cases where this is not so. Analysis of actual data from several hundred studies (unpublished data) has shown no significant difference in rates of weight change of target organs (other than the brain) compared to total body weight for healthy animals in those species commonly used for repeated dose studies (rats, mice, rabbits, and dogs). Furthermore, it should be noted that analysis of covariance is of questionable validity in analyzing body weight and related organ weight changes, since a primary assumption is the independence of treatment—that the relationship of the two variables is the same for all treatments (Ridgeman, 1975). Plainly, in toxicology this is not true.

In cases where the differences between the error mean squares are much greater, the ratio of F ratios will diverge in precision from the result of the efficiency of covariance adjustment. These cases are where either sample sizes are much larger or where the differences between means themselves are much larger. This latter case is one that does not occur in the designs under discussion in any manner that would leave analysis of covariance as a valid approach, because group means start out being very similar and cannot diverge markedly unless there is a treatment effect. As we have discussed earlier, a treatment effect invalidates a prime underpinning assumption of analysis of covariance.

Clinical Chemistry

A number of clinical chemistry parameters are commonly determined on the blood and urine collected from animals in chronic, subchronic, and, occasionally, acute toxicity studies. In the past (and still, in some places), the accepted practice has been to evaluate these data using univariate-parametric methods (primarily t-tests and/or ANOVA). However, this can be shown not to be the best approach on a number of grounds.

First, such biochemical parameters are rarely independent of each other. Our interest is rarely focused on just one of the parameters. Rather, there are batteries of the parameters associated with toxic actions at particular target organs. For example, increases in creatinine phosphokinase (CPK), γ-hydroxybutyrate dehydrogenase (γ-HBDH), and lactate dehydrogenase (LDH) occurring together are strongly indicative of myocardial damage. In such cases, we are not just interested in a significant increase in one of these, but in all three. Table 14.2 gives a brief overview of the association of various parameters with actions at particular target organs. A more detailed coverage of the interpretation of such clinical laboratory tests can be found in Gad and Chengelis (1992) and Loeb and Quimby (1989).

Similarly, the serum electrolytes (sodium, potassium, and calcium) interact with each other; a decrease in one is frequently tied, for instance, to an increase in one of the others. Furthermore, the nature of the data (in the case of some parameters), either because of the biological nature of the parameter or the way in which it is measured, is frequently either not normally distributed (particularly because of being markedly skewed) or not continuous in nature. This can be seen in some of the reference data for experimental animals in Mitruka and Rawnsley (1977) or Weil (1982) in, for example, creatinine, sodium, potassium, chloride, calcium, and blood urea nitrogen. It should be remembered that both normal distribution and continuous data are underlying assumptions in the parametric statistical techniques described in this chapter.

In recent acute, subchronic, and chronic studies we have been involved with, clinical chemistry statistical test methodologies were selected in accordance with the decision tree approach presented at the beginning of this volume. The methods this approach most frequently resulted in are outlined in Table 14.3. This may serve as a guide to the uninitiated. A more detailed discussion may be found in Martin et al. (1975) or Harris (1978).

TABLE 14.2

Association of Changes in Biochemical Parameters with Actions at Particular Target Organs

Parameter	Blood	Heart	Lung	Kidney	Liver	Bone	Intestine	Pancreas	Notes
Albumin				↓	↓				Produced by the liver. Very significant reductions indicate extensive liver damage.
ALP					↑	↑	↑		Elevations usually associated with cholestasis. Bone alkaline phosphatase tends to be higher in young animals.
Bilirubin (total)	↑				↑				Usually elevated due to cholestasis, either due to obstruction or hepatopathy.
BUN				↑	↓				Estimates blood filtering capacity of the kidneys. Does not become significantly elevated until the kidney function is reduced 60–75%.
Calcium				↓					Can be life threatening and result in acute death.
Cholinesterase				↓	↓				Found in plasma, brain, and RBC.
CPK		↑							Most often elevated due to skeletal muscle damage but can also be produced by cardiac muscle damage. Can be more sensitive than histopathology.
Creatinine				↑					Also estimates blood filtering capacity of kidney as BUN does.
GGT					↑				Elevated in cholestasis. This is a microsomal enzyme and levels often increase in response to microsomal enzyme induction.
Glucose								↑	Alterations other than those associated with stress are uncommon and reflect an effect on the pancreatic islets or anorexia.
HBDH		↑			↑				
LDH		↑	↑	↑	↑				Increase usually due to skeletal muscle, cardiac muscle, or liver damage. Not very specific.

Analyte			Comments
Protein (total)	↓	↓	Absolute alterations are usually associated with decreased production (liver) or increased loss (kidney). Can see increase in case of muscle "wasting" (catabolism).
SDH		↑↓	Liver enzyme which can be quite sensitive but is fairly unstable. Samples should be processed as soon as possible.
SGOT	↑	↑	Present in skeletal muscle and heart and most commonly associated with damage to these.
SGPT		↑	Elevations usually associated with hepatic damage or disease.

↑: increase in chemistry values; ↓: decrease in chemistry values.

ALP: alkaline phosphatase; BUN: blood urea nitrogen; CPK: creatinine phosphokinase; GGT: gamma glutamyl transferase; HBDH: hydroxybutyric dehydrogenase; LDH: lactic dehydrogenase; RBC: red blood cells; SDH: sorbitol dehydrogenase; SGOT: serum glutamic oxaloacetic transaminase (also called AST [aspartate amino transferase]); SGPT: serum glutamic pyruvic transaminase (also called ALT [alanine amino transferase]).

TABLE 14.3

Tests Often Used in Analysis of Clinical Chemistry Data

Clinical Chemistry Parameters	Statistical Test
Calcium	ANOVA, Bartlett's, and/or F-test, t-test
Glucose	ANOVA, Bartlett's, and/or F-test, t-test
Blood–urea–nitrogen	ANOVA, Bartlett's, and/or F-test, t-test
Creatinine	ANOVA, Bartlett's, and/or F-test, t-test
Cholinesterase	ANOVA, Bartlett's, and/or F-test, t-test
Total bilirubin	Kruskal–Wallis nonparametric, ANOVA
Total protein	ANOVA, Bartlett's, and/or F-test, t-test
Albumin	ANOVA, Bartlett's, and/or F-test, t-test
GGT	Kruskal–Wallis nonparametric, ANOVA
HBDH	ANOVA, Bartlett's, and/or F-test, t-test
AP	ANOVA, Bartlett's, and/or F-test, t-test
CPK	ANOVA, Bartlett's, and/or F-test, t-test
LDH	ANOVA, Bartlett's, and/or F-test, t-test
SGOT	ANOVA, Bartlett's, and/or F-test, t-test
SGPT	ANOVA, Bartlett's, and/or F-test, t-test
Hgb	ANOVA, Bartlett's, and/or F-test, t-test

Hematology

Much of what we said about clinical chemistry parameters is also true for the hematologic measurements made in toxicology studies. Which test to perform should be evaluated by use of a decision tree until one becomes confident as to the most appropriate methods. Keep in mind that sets of values and (in some cases) population distribution vary not only between species, but also between the commonly used strains of species and that "control" or "standard" values will "drift" over the course of only a few years.

Again, the majority of these parameters are interrelated and highly dependent on the method used to determine them. Red blood cell count (RBC), platelet counts, and mean corpuscular volume (MCV) may be determined using a device such as a Coulter counter to take direct measurements, and the resulting data are usually stable for parametric methods. The hematocrit, however, may actually be a value calculated from the RBC and MCV values and, if so, is dependent on them. If the hematocrit is measured directly, instead of being calculated from the RBC and MCV, it may be compared by parametric methods.

Hemoglobin is directly measured and is an independent and continuous variable. However, and probably because at any one time a number of forms and conformations (oxyhemoglobin, deoxyhemoglobin, methemoglobin, etc.) of hemoglobin are actually present, the distribution seen is not typically a normal one, but rather may be a multimodal one. Here a nonparametric technique such as the Wilcoxon or multiple rank sum is called for.

Consideration of the white blood cell (WBC) and differential counts leads to another problem. The total WBC is, typically, a normal population amenable to parametric analysis, but differential counts are normally determined by counting, manually, one or more sets of 100 cells each. The resulting relative percentages of neutrophils are then reported as either percentages or are multiplied by the total WBC count with the resulting "count" being reported as the "absolute" differential WBC. Such data, particularly in the case of eosinophils (where the distributions do not approach normality) should usually be analyzed by nonparametric methods. It is widely believed that "relative" (%) differential data should not be reported because they are likely to be misleading.

Last, it should always be kept in mind that it is rare for a change in any single hematologic parameter to be meaningful. Rather, because these parameters are so interrelated, patterns of changes in parameters should be expected if a real effect is present, and analysis and interpretation of results should focus on such patterns of changes. Classification analysis techniques often provide the basis for a useful approach to such problems.

Histopathologic Lesion Incidence

The last 20 years have seen increasing emphasis placed on histopathological examination of tissues collected from animals in subchronic and chronic toxicity studies. While it is not true that only those lesions which occur at a statistically significantly increased rate in treated/exposed animals are of concern (for there are the cases where a lesion may be of such a rare type that the occurrence of only one or a few such in treated animals "raises a flag"), it is true that, in most cases, a statistical evaluation is the only way to determine if what we see in treated animals is significantly worse than what has been seen in control animals. And although cancer is not our only concern, this category of lesions is that of greatest interest.

Typically, comparison of incidences of any one type of lesion between controls and treated animals is made using the multiple 2×2 chi-square test or Fisher's exact test with a modification of the numbers of animals as the denominators. Too often, experimenters exclude from consideration all those animals (in both groups) that died prior to the first animals being found with lesions at that site. The special case of carcinogenicity bioassays will be discussed in detail in the next chapter.

An option that should be kept in mind is that, frequently, a pathologist cannot only identify a lesion as present, but also grade those present as to severity. This represents a significant increase in the information content of the data that should not be given up by performing an analysis based only on the perceived quantal nature (present/absent) of the data. Quantal data, analyzed by chi-square or Fisher's exact tests, are a subset (the 2×2 case) of categorical or contingency table data. In this case it also becomes ranked (or "ordinal") data—the categories are naturally ordered (for example, no effect < mild lesion < moderate lesion < severe lesion). This gives a $2 \times R$

table if there are only one treatment and one control group, or an $N \times R$ ("multiway") table if there are three or more groups of animals.

The traditional method of analyzing multiple, cross-classified data has been to collapse the $N \times R$ contingency table over all but two of the variables and to follow this with the computation of some measure of association between these variables. For an N-dimensional table this results in $N(N-1)/2$ separate analyses. The result is crude, "giving away" information and even (by inappropriate pooling of data) yielding a faulty understanding of the meaning of data. Although computationally more laborious, a multiway ($N \times R$ table) analysis should be utilized.

_____

Reproduction

The reproductive implications of the toxic effects of chemicals are becoming increasingly important. Because of this, reproduction studies, together with other closely related types of studies (such as teratogenesis, dominant lethal, and mutagenesis studies, which are discussed later in this chapter), are now commonly made companions to chronic toxicity studies.

One point that must be kept in mind with all reproduction-related studies is the nature of the appropriate sampling unit. What is the appropriate N in such a study; the number of individual pups, the number of litters, pregnant females? Fortunately, it is now fairly well accepted that the first case (using the number of offspring as the N) is inappropriate (Weil, 1970). The real effects in such studies actually occur in the female exposed to the chemical, or mated to a male who was exposed. What happens to her, and to the development of the litter she is carrying, is biologically independent of what happens to every other female/litter in the study. This cannot be said for each offspring in each litter; for example, the death of one member of a litter can and will be related to what happens to every other member. Or the effect on all of the offspring might be similar for all of those from one female and different or lacking for those from another.

As defined by Oser and Oser (1956), there are four primary variables of interest in a reproduction study. First, there is the fertility index (FI), which may be defined as the percentage of attempted matings (i.e., each female housed with a male) that resulted in pregnancy, pregnancy being determined by a method such as the presence of implantation sites in the female. Second, there is the gestation index (GI), which is defined as the percentage of mated females, as evidenced by a vaginal plug being dropped or a positive vaginal smear, that delivers viable litters (i.e., litters with at least one live pup). Two related variables that may also be studied are the mean number of pups born per litter and the percentage of total pups per litter that are stillborn. Third, there is the viability index (VI), which is defined as the percentage of offspring born that survive at least 4 days after birth. Finally (in this four-variable system), there is the lactation index (LI), which is the percentage of animals per litter that survive 4 days and also weaning. In rats and mice, this is classically taken to be 21 days after birth. An additional variable that may reasonably be included in such a study is the mean weight gain per pup per litter.

Given that our N is at least 10 (we will further explore proper sample size under the topic of teratology), we may test each of these variables for significance using a method such as the Wilcoxon–Mann–Whitney U test, or the Kruskal–Wallis nonparametric ANOVA. If N is less than 10, we cannot expect the central limit theorem to be operative and should use the Wilcoxon sum of ranks (for two groups) or the Kruskal–Wallis nonparametric ANOVA (for three or more groups) to compare groups.

Developmental Toxicology

When the primary concern of a reproductive/developmental study is the occurrence of birth defects or deformations (terata, either structural or functional) in the offspring of exposed animals, the study is one of developmental toxicology (teratology). In the analysis of the data from such a study, we must consider several points.

First is sample size. Earlier in this book we reviewed this topic generally and presented a method to estimate sufficient sample size. The difficulties with applying these methods here revolve around two points: (1) selecting a sufficient level of sensitivity for detecting an effect and (2) factoring in how many animals will be removed from study (without contributing a datum) by either not becoming pregnant or not surviving to a sufficiently late stage of pregnancy. Experience generally dictates that one should attempt to have 20 pregnant animals per study group if a pilot study has provided some confidence that the pregnant test animals will survive the dose levels selected. Again, it is essential to recognize that the litter, not the fetus, is the basic independent unit for each variable.

A more fundamental consideration, alluded to in the section on reproduction, is that as we use more animals, the mean of means (each variable will be such in a mathematical sense) will approach normality in its distribution. This is one of the implications of the central limit theorem: even when the individual data are not normally distributed, their means will approach normality in their distribution. At a sample size of ten or greater, the approximation of normality is such that we may use a parametric test (such as a t-test or ANOVA) to evaluate results. At sample sizes less than ten, a nonparametric test (Wilcoxon rank sum or Kruskal–Wallis nonparametric ANOVA) is more appropriate. Other methodologies have been suggested (Kupper and Haseman, 1978; Nelson and Holson, 1978) but do not offer any prospect of widespread usage. One nonparametric method that is widely used is the Mann–Whitney U test, which was described earlier. Williams and Buschbom (1982) further discuss some of the available statistical options and their consequences, and Rai and Ryzin (1985) have recommended a dose-responsive model.

Dominant Lethal Assay

The dominant lethal study is essentially a reproduction study that seeks to study the endpoint of lethality to the fetuses after implantation and before delivery. The proper identification of the sampling unit (the pregnant female) and the design of an experiment so that a sufficiently large sample is available for analysis are the prime statistical considerations. The question of sampling unit has been adequately addressed in earlier sections. Sample size is of concern here because the hypothesis-testing techniques that are appropriate with small samples are of relatively low power, as the variability about the mean in such cases is relatively large. With sufficient sample size (e.g., from 30 to 50 pregnant females per dose level per week [Bateman, 1977]) variability about the mean and the nature of the distribution allow sensitive statistical techniques to be employed.

The variables that are typically recorded and included in analysis are (for each level/week): (a) the number of pregnant females, (b) live fetuses/pregnancy, (c) total implants/pregnancy, (d) early fetal deaths (early resorptions)/pregnancy, and (e) late fetal deaths/pregnancy.

A wide variety of techniques for analysis of these data have been (and are) used. Most common is the use of ANOVA after the data have been transformed by the arcsine transform (Mosteller and Youtz, 1961).

Beta binomial (Aeschbacher et al., 1977; Vuataz and Sotek, 1978) and Poisson distributions (Dean and Johnston, 1977) have also been attributed to these data, and transforms and appropriate tests have been proposed for use in each of these cases (in each case with the note that the transforms serve to "stabilize the variance" of the data). With sufficient sample size, as defined earlier in this section, the Mann–Whitney U test is recommended for use here. Smaller sample sizes necessitate the use of the Wilcoxon rank sum test.

Diet and Chamber Analysis

Earlier we presented the basic principles and methods for sampling. Sampling is important in many aspects of toxicology, and here we address its application to diet preparation and the analysis of atmospheres from inhalation chambers.

In feeding studies, we seek to deliver doses of a material to animals by mixing the material with their diet. Similarly, in an inhalation study we mix a material with the air the test animals breathe.

In both cases, we must then sample the medium (food or atmosphere) and analyze these samples to determine what levels or concentrations of material were actually present and to assure ourselves that the test material is homogeneously distributed. Having an accurate picture of these delivered concentrations, and how they varied over the course of time, is essential on a number of grounds:

The regulatory agencies and sound scientific practice require that analyzed diet and mean daily inhalation atmosphere levels be ±10% of the target level.

Excessive peak concentrations, because of the overloading of metabolic repair systems, could result in extreme acute effects that would lead to results in a chronic study that are not truly indicative of the chronic low-level effects of the compound, but rather of periods of metabolic and physiologic overload. Such results could be misinterpreted if true exposure or diet levels were not maintained at a relatively constant level.

Sampling strategies are not just a matter of numbers (for statistical aspects), but of geometry, so that the contents of a container or the entire atmosphere in a chamber is truly sampled; and of time, in accordance with the stability of the test compound. The samples must be both randomly collected and representative of the entire mass of what one is trying to characterize. In the special case of sampling and characterizing the physical properties of aerosols in an inhalation study, some special considerations and terminology apply. Because of the physiologic characteristics of the respiration of humans and of test animals, our concern is very largely limited to those particles or droplets that are of a respirable size. Unfortunately, "respirable size" is a complex characteristic based on aerodynamic diameter, density, and physiological characteristics. Unfortunately, while those particles with an aerodynamic diameter of less than ten microns are generally agreed to be respirable in humans (that is, they can be drawn down to the deep portions of the lungs), three microns in aerodynamic diameter is a more realistic value. The one favorable factor is that there are now available a selection of instruments that accurately (and relatively easily) collect and measure particles or droplets. These measurements result in concentrations in a defined volume of gas and can be expressed as either a number concentration or a mass concentration (the latter being more common). Such measurements generate categorical data—concentrations are measured in each of a series of aerodynamic size groups (such as > 100 microns, 100 to 25 microns, 25 to 10 microns, 10 to 3 microns, etc.). The appropriate descriptive statistics for this class of data are the geometric mean and its standard deviation. These aspects and the statistical interpretation of the data that are finally collected should be considered after sufficient interaction with the appropriate professionals. Typically, it then becomes a matter of the calculation of measures of central tendency and dispersion statistics, with the identification of those values that are beyond acceptable limits (Bliss, 1935).

Aerosol Statistics

An additional complication to characterizing atmospheres in inhalation chambers when the test material is delivered as an aerosol (either liquid or solid dispersed droplets or particles) is that the entire basis of descriptive statistics is different. Early in this book we introduced the geometric mean as the appropriate description of location for an aerosol population. A more rigorous and extensive discussion is necessary, however.

An aerosol atmosphere most commonly contains a wide range of particle sizes and cannot be defined adequately by an arithmetic average diameter. A knowledge of the size distribution and a mathematical expression to describe it are highly desirable, especially when it is necessary to estimate particulate characteristics that are not measured directly. Many mathematical relationships have been proposed to describe particle size distributions. Of these, the most widely used in aerosol work is the log-normal distribution, which is merely the normal distribution applied to the logarithms of the quantities actually measured. This discussion of the log-normal distribution, therefore, should be preceded by a review of the normal distribution.

If the quantity $x = \ln(D)$ is normally distributed, then the distribution of D (the particle diameter) is said to be log-normal. The log-normal distribution is particularly useful in particle size analysis because of the characteristics described in the following.

Consider a property of a particle that can be defined quantitatively by

$$Q_r(D) = X_r D^r$$

where Q_r is a constant (shape factor) for a given value of r. For a log-normal distribution, the relative number of particles having diameters whose logarithms fall in the interval $x \pm dx/2$ is given by

$$\mu_0 = \ln(\delta_{0g})$$

where μ is the true population mean, δ is the true population standard deviation, $f(x)$ is the normal probability distribution function

$$f(x)dx = \frac{e^{-\frac{(x-\mu)^2}{2\sigma^2}}}{\sigma\sqrt{2\pi}} dx$$

and δ_{0g} is the geometric mean diameter of the population.

The large amount of information available concerning the statistics of sampling from a normal distribution is directly applicable to particle size analysis when D is log-normally distributed. For a sample of N particles, the maximum likelihood statistics, m and s, are given by the following two equations (Mercer, 1973):

$$m = \ln(D_{0g}) = \sum_1^N \ln D_i / N_i$$

and

$$s = \ln(\sigma_g) = \sqrt{\frac{\sum_1^N (\ln D_i - \ln D_{0g})}{N-1}}$$

In practice, D_{0g}, which estimates the population count median diameter σ_{0g}, and σ_g, which estimates the population geometric standard deviation e^σ, are reported, rather than m and s. The mean logarithms of the D_r distributions are then estimated using the Hatch-Choate equation:

$$\ln(D_{rg}) = \ln(D_{0g}) + rs^2$$

or

$$D_{rg} = D_{0g} e^{r\sigma^2}$$

The diameter of the particle having the average amount of Q_r is calculated from

$$\ln(\bar{D}_r) = \ln D_{0g} + rs^2/2$$

or

$$\bar{D}_r = D_{0g} e^{rs^2/2}$$

For samples of N particles, the sampling distribution of the mean logarithm of diameters m is normal and has a mean of μ_0 and a standard deviation equal to $\sigma N^{1/2}$. For the values of N of interest in particle size work, the sampling distribution of the variance s^2 is also normal, having a mean of σ^2 and a standard deviation equal to $\sigma^2(2/N)^{1/2}$. The sampling distribution of $\ln(D_{rg})$ is normal, having a mean of μ_r and a standard deviation given by

$$\sigma(\ln D_{rg}) = (\sigma / \sqrt{N})\sqrt{1 + 2r^2\sigma^2}$$

The uncertainty in estimating μ_r (and hence D_{rg}) increases rapidly as $r\sigma$ increases.

The statistical relationships above are based on the assumptions that each particle diameter is measured independently and without error. In practice, however, it is common to sort the particles into a series of size intervals. The proper estimate of μ_0 and s then becomes quite complicated. The desired statistics can be approximated, however, by

$$\ln D_{0g} = \left(\sum_{i=1}^{k} n_i \cdot D_i \right) \Big/ \left(\sum_{i=1}^{k} n_i \right)$$

and

$$s = \ln \sigma_g = \sqrt{\left(\sum_{i=1}^{k} n_i \cdot (\ln D_i - \ln D_{0g})^2 \right) \Big/ \left(\sum_{i=1}^{k} n_i - 1 \right)}$$

where n_i is the number of particles in the ith size interval, D_i is an average diameter for that interval, and k is the total number of size intervals. Alternatively, the statistics can be approximated by plotting on logarithmic-probability paper the cumulative percent

$$P_j = 100 \left(\sum_{i=1}^{i} n_i \right) \Big/ \left(\sum_{i=1}^{k} n_i \right)_{(j<k)}$$

against the upper limit of the jth size interval for a number of values of j and drawing the straight line that the resulting points appear to estimate. D_{0g} is the diameter at which the line has the coordinate $P = 50\%$. Therefore,

$$\ln(\sigma_g) = \ln(D_{84}) - \ln(D_{0g}) = \ln(D_{84}/D_{0g})$$

and

$$\sigma_g = D_{84}/D_{0g}$$

where D_{84} is the diameter at which the line has the coordinate $P = 84\%$. The statistics can also be calculated from the cumulative distribution using the method of profit analysis described in an earlier chapter.

In the case

$$\text{Probit}(P_j/100) = a' + b \cdot \ln(D_j)$$

For each value of $P_i/100$, the corresponding probit can be calculated and a' and b are calculated by the method of least squares. The desired statistics are

$$m = \ln(D_{0g}) = (5 - a')/b$$

and

$$s = \ln \cdot g = 1/b$$

Because a large value of N is encountered in most particulate samples, the confidence limits on the various median diameters are given by

$$D_{rg}(\text{C.L.}) = D_{rg} \cdot \exp[\pm \tau_a \cdot \sigma[\ln D_{rg})]$$

The confidence limits on the population geometric standard deviation are found to be, approximately

$$\sigma_g(\text{C.L.}) = \sigma_g^a$$

where $a = [2(N - 1)]^{1/2}[\pm t_a + (2N - 3)^{1/2}]$, the positive value of t_a giving the lower confidence limit. Here t_a is used in place of z_a to relate it to the limits on the mean at the same level of confidence. A goodness-of-fit test can be carried out if the transformation $X = \ln(D)$ is employed.

Excellent reviews of the field of aerosol measurement are now available (such as Mercer, 1973; Stockham and Fochtman, 1979; and Willeke, 1980).

Mutagenesis

In the last 25 years a wide variety of tests (see Kilbey, et al., 1977, for an overview of those available) for mutagenicity have been developed and brought into use. These tests give us a quicker and cheaper (although not as conclusive) way of predicting whether a material of interest is a mutagen, and possibly a carcinogen, than do longer-term, whole-animal studies.

How to analyze the results of the multitude of tests (Ames, DNA repair, micronucleus, host-mediated, cell transformation, sister chromatid exchange, and *Drosophila* SLRL, to name a few) is a new and extremely important question. Some workers in the field hold that it is not possible (or necessary) to perform statistical analysis, that the tests can simply be judged to be positive or not positive on the basis of whether or not they achieve a particular increase in the incidence of mutations in the test organism. This is plainly not an acceptable response, when societal needs are not limited to yes/no answers but rather include at least relative quantitation of potencies (particularly in mutagenesis, where we have come to recognize the existence of a nonzero background level of activity from naturally occurring factors and agents). Such quantitations of potency are complicated by the fact that we are dealing with a nonlinear phenomenon; although low doses of most mutagens produce a linear response curve, in increasing the dose the curve will flatten out (and even turn into a declining curve) as the higher doses provoke an acute response.

Several concepts different from those we have previously discussed need to be examined, for our concern has now shifted from how a multicellular organism acts in response to one of a number of complex actions to how a mutational event is expressed, most frequently by a single cell. Given that we can handle much larger numbers of experimental units in systems that use smaller test organisms, we can seek to detect both weak and strong mutagens.

Conducting the appropriate statistical analysis, and utilizing the results of such an analysis properly, must start with an understanding of the biological system involved and, from this understanding, developing the correct model and hypothesis. We start such a process by considering each of five interacting factors (Grafe and Vollmar, 1977; Vollmar, 1977).

α, which is the probability of our committing a type I error (saying an agent is mutagenic when it is not, equivalent to our p in such earlier considered designs as the Fisher's exact test); false positive

β, which is the probability of our committing a type II error (saying an agent is not mutagenic when it is); false negative

Δ, our desired sensitivity in an assay system (such as being able to detect an increase of 10% in mutations in a population)

σ, the variability of the biological system and the effects of chance errors n, the single necessary sample size to achieve each of these (we can, by our actions, change only this portion of the equation) as n is proportional to:

$$\frac{\sigma}{\alpha, \beta, \text{ and } \Delta}$$

The implications of this are, therefore, that (a) the greater σ is, the larger n must be to achieve the desired levels of α, β, and Δ, (b) the smaller the desired levels of α, β, and/or Δ, if n is constant, the larger our σ is.

What is the background mutation level and the variability in our technique? As any good genetic or general toxicologist will acknowledge, matched concurrent control groups are essential. Fortunately, with these test systems large ns are readily attainable, although there are other complications to this problem, which we shall consider later. An example of the confusion that would otherwise result is illustrated in the intralaboratory comparisons on some of these methods done to date, such as that reviewed by Weil (1978).

New statistical tests based on these assumptions and upon the underlying population distributions have been proposed, along with the necessary computational background to allow one to alter one of the input variables (α, β, or Δ). A set that shows particular promise is that proposed by Katz (1978 and 1979) in his two articles. He described two separate test statistics: Φ for when we can accurately estimate the number of individuals in both the experimental and control groups, and θ for when we do not actually estimate the number of surviving individuals in each group, and we can assume that the test material is only mildly toxic in terms of killing the test organisms. Each of these two test statistics is also formulated on the basis of only a single exposure of the organisms to the test chemicals. Given this, we then may compute

$$\Phi = \frac{\alpha(M_E - 0.5) - K\beta(M_C + 0.5)}{\sqrt{K\alpha\beta(M_E + M_C)}}$$

where a and b are the number of groups of control (c) and experimental (e) organisms, respectively.

$K = N_E/N_C$.

N_C and N_E are the numbers of surviving microorganisms.

M_C and M_E are the numbers of mutations in experimental and control groups.

μ_r and μ_c are the true (but unknown) mutation rates (as μ_c gets smaller, Ns must increase).

We may compute the second case as

$$\theta = \frac{\alpha(M_E - 0.5) + (M_C + 0.5)}{\alpha\beta(M_E + M_C)}$$

with the same constituents.

In both cases, at a confidence level for α of 0.05, we accept that $\mu_c = \mu_e$ if the test statistic (either Φ or θ) is less than 1.64. If it is equal to or greater than 1.64, we may conclude that we have a mutagenic effect (at $\alpha = 0.05$).

In the second case (θ, where we do not have separate estimates of population sizes for the control and experimental groups) if K deviates widely from 1.0 (if the material is markedly toxic), we should use more containers of control organisms (tables for the proportions of each to use given different survival frequencies may be found in Katz, 1979). If different levels are desired, tables for Φ and θ may be found in Kastenbaum and Bowman (1970).

An outgrowth of this is that the mutation rate per surviving cells μ_c and μ_e can be determined. It must be remembered that if the control mutation rate is so high that a reduction in mutation rates can be achieved by the test compound, these test statistics must be adjusted to allow for a two-sided hypothesis (Ehrenberg, 1977). The α levels may likewise be adjusted in each case, or tested for, if we want to assure ourselves that a mutagenic effect exists at a certain level of confidence (note that this is different from disproving the null hypothesis).

It should be noted that there are numerous specific recommendations for statistical methods designed for individual mutagenicity techniques, such as that of Bernstein et al. (1982) for the Ames test. Exploring each of them is beyond the scope of this chapter, however.

Behavioral Toxicity

A brief review of the types of studies/experiments conducted in the area of behavioral toxicology, and a classification of these into groups, is in order. Although there are a small number of studies that do not fit into the following classifications, the great majority may be fitted into one of the following four groups. Many of these points were first covered by the author in an earlier article (Gad, 1982b).

1. Observational score-type studies (Irwin screens and functional observation batteries) are based on observing and grading the response of an animal to its normal environment or to a stimulus that is imprecisely controlled. This type of result is generated by one of two major sorts of studies. Open-field studies involve placing an animal in the center of a flat, open area and counting each occurrence of several types of activities (grooming, moving outside a designated central area, rearing, etc.) or timing until the first occurrence of each type of activity. The data generated are scalar of either a continuous or discontinuous nature, but frequently are not of a normal distribution. Tilson et al. (1980) presented some examples of this sort. Observational screen studies involve a combination of observing behavior and evoking a response to a simple stimulus, the resulting observation being graded as normal or as deviating from normal on a graded scale. Most of the data so generated are rank in nature, with some portions being quantal or interval. Irwin (1968) and Gad (1982a) have presented schemes for the conduct of such studies. Table 14.4 gives an example of the nature (and of one form of statistical analysis) of such data generated after exposure to one material.

2. The second type of study is one that generates rates of response as data. The studies are based on the number of responses to a discrete controlled stimulus or are free of direct connection to a stimulus. The three most frequently measured parameters are licking of a liquid (milk, sugar water, ethanol, or a psychoactive agent in water), gross locomotor activity (measured by a photocell or electromagnetic device), or lever pulling. Work presenting examples of such studies

TABLE 14.4

Irwin Screen Parameters Showing Significant Differences between Treated and Control Groups

Parameter	Control Sum of Ranks	N_c	18 Crown 6 Treated Sum of Ranks	N_T	Observed Difference in Treated Animals (Compared to Controls)
			Rats (18 Crown 6 Animals Given 40 mg/kg i.p.)		
Twitches	55.0	10	270.0	15	Involuntary muscle twitches
Visual placing	55.0	10	270.0	15	Less aware of visual stimuli
Grip strength	120.0	10	205.0	15	Considerable loss of strength, especially in hind limbs
Respiration	55.0	10	270.0	15	Increased rate of respiration
Tremors	55.0	10	270.0	15	Marked tremors

Note: All parameters above are significant at $p < 0.05$.

has been published by Annau (1972) and Norton (1973). The data generated are most often of a discontinuous or continuous scalar nature and are often complicated by underlying patterns of biological rhythm (to be discussed more fully later).

3. The third type of study generates a variety of data that are classified as error rate. These are studies based on animals learning a response to a stimulus or memorizing a simple task (such as running a maze or a Skinner box-type shock-avoidance system). These tests or trials are structured so that animals can pass or fail on each of a number of successive trials. The resulting data are quantal, although frequently expressed as a percentage.

4. The final major type of study is that which results in data that are measures of the time to an endpoint. Such studies are based on animals being exposed to or dosed with a toxicant and the time taken for an effect to be observed is measured. The endpoint is usually failure to continue to be able to perform a task and can, therefore, be death, incapacitation, or the learning of a response to a discrete stimulus. Burt (1972) and Johnson et al. (1972) present data of this form. The data are always of a censored nature; that is, the period of observation is always artificially limited as in measuring time-to-incapacitation in combustion toxicology data, where animals are exposed to the thermal decomposition gases to test materials for a period of 30 minutes. If incapacitation is not observed during these 30 minutes it is judged not to occur. The data generated by these studies are continuous, discontinuous, or rank in nature. They are discontinuous because the researcher may check or may be restricted to checking for the occurrence of the endpoint only at certain discrete points in time. On the other hand, they are rank if the periods to check for occurrence of the endpoint are far enough apart, in which case one may actually only know that the endpoint occurred during a broad period of time—but not where in that period.

TABLE 14.5

Overview of Statistical Testing in Behavioral Toxicology—Those Tests Commonly Used[a] as Opposed to Those Most Frequently Appropriate

Type of Observation	Most Commonly Used Procedures	Suggested Procedures
Observational scores	Either Student's t-test or one-way ANOVA	Kruskal–Wallis nonparametric ANOVA or Wilcoxon rank sum
Response rates	Either Student's t-test or one-way ANOVA	Kruskal–Wallis ANOVA or one-way ANOVA
Error rates	ANOVA followed by a *post hoc* test	Fisher's exact, or R × C chi-square, or Mann–Whitney U test
Times to endpoint	Either Student's t-test or one-way ANOVA	ANOVA then a *post hoc* test or Kruskal–Wallis ANOVA
Teratology and reproduction	ANOVA followed by a *post hoc* test	Fisher's exact test, Kruskal–Wallis ANOVA, or Mann–Whitney U test

[a] That these are the most commonly used procedures was established by an extensive literature review that is beyond the scope of this book. The reader need only, however, look at the example articles cited in the text of this chapter to verify this fact.

There is a special class of test that should also be considered at this point—the behavioral teratology or reproduction study. These studies are based on dosing or exposing either parental animals during selected periods in the mating and gestation process or pregnant females at selected periods during gestation. The resulting offspring are then tested for developmental defects of a neurological and behavioral nature. Analysis is complicated by a number of facts: (1) the parental animals are the actual targets for toxic effects, but observations are made on offspring; (2) the toxic effects in the parental generation may alter the performance of the mother in rearing its offspring, which in turn can lead to a confusion of prenatal and postnatal effects; (3) finally, different capabilities and behaviors develop at different times (discussed further below).

A researcher can, by varying the selection of the animal model (species, strain, sex), modify the nature of the data generated and the degree of dispersion of these data. In behavioral studies particularly, limiting the within-group variability of data is a significant problem and generally should be a highly desirable goal.

Most, if not all, behavioral toxicology studies depend on as least some instrumentation. Very frequently overlooked here (and, indeed, in most research) is that instrumentation, by its operating characteristics and limitations, goes a long way toward determining the nature of the data generated by it. An activity monitor measures motor activity in discrete segments. If it is a "jiggle cage" type monitor, these segments are restricted so that only a distinctly limited number of counts can be achieved in a given period of time and then only if they are of the appropriate magnitude. Likewise, technique can also readily determine the nature of the data. In measuring response to pain, for example, one could record it as a quantal measure (present or absent), a rank score (on a scale of 1 to 5 for from decreased to increased responsiveness, with 3 being "normal"), or as scalar data (by using an analgesia meter, which determines either how much pressure or heat is required to evoke a response).

Study design factors are probably the most widely recognized of the factors that influence the type of data resulting from a study. Number of animals used, frequency of measures, and length of period of observation are three obvious design factors that are readily under the control of the researcher and that directly help to determine the nature of the data.

Finally, it is appropriate to review each of the types of studies presently seen in behavioral toxicology, according to the classification presented at the beginning of this section, in terms of which statistical methods are used now and what procedures should be recommended for use. The recommendations, of course, should be viewed with a critical eye. They are intended with current experimental design and technique in mind and can only claim to be the best when one is limited to addressing the most common problems from a library of readily and commonly available and understood tests.

Table 14.5 summarizes this review and recommendation process.

References

Aeschbacher, H.U., Vautaz, L., Sotek, J., and Stalder, R. (1977) Use of the beta binomial distribution in dominant-lethal testing for "weak mutagenic activity," Part 1. *Mutat. Res.*, 44, 369–390.

Annau, Z. (1972) The comparative effects of hypoxia and carbon monoxide hypoxia on behavior, in *Behavioral Toxicology*, Weiss, B. and Laties, V.G., Eds., Plenum Press, New York, pp. 105–127.

Bateman, A.T. (1977) The dominant lethal assay in the male mouse, in *Handbook of Mutagenicity Test Procedures*, Kilbey, B.J., Legator, M., Nichols, W., and Ramel, C., Eds., Elsevier, New York, pp. 325–334.

Bernstein, L., Kaldor, J., McCann, J., and Pike, M.C. (1982) An empirical approach to the statistical analysis of mutagenesis data from the Salmonella test. *Mutation Res.*, 97, 267–281.

Bliss, C.I. (1935) Statistical relations in fertilizer inspection. Bulletin 674. Connecticut Agricultural Experiment Station, New Haven, CT.

Boyd, E.M. (1972) *Predictive Toxicometrics*. Williams & Wilkins, Baltimore.

Boyd, E.M. and Knight, L.M. (1963) Postmortem shifts in the weight and water levels of body organs. *Tox. Appl. Pharm.*, 5, 119–128.

Bruce, R.D. (1985) An up-and-down procedure for acute toxicity testing. *Fund. Appl. Toxicol.*, 5, 151–157.

Burt, G.S. (1972) Use of behavioral techniques in the assessment of environmental contaminants, in *Behavioral Toxicology*, Weiss, B. and Laties, V.G., Eds., Plenum Press, New York, pp. 241–263.

Dean, B.J. and Johnston, A. (1977) Dominant lethal assays in the male mice: evaluation of experimental design, statistical methods and the sensitivity of Charles River (CD1) mice. *Mutat. Res.*, 42, 269–278.

Debanne, S.M. and Haller, H.S. (1985) Evaluation of statistical methodologies for estimation of median effective dose. *Toxicol. Appl. Pharmacol.*, 79, 274–282.

DePass, L.R., Myers, R.C., Weaver, E.V., and Weil, C.S. (1984) An assessment of the importance of number of dosage levels, number of animals per dosage level, sex and method of LD_{50} and slope calculations in acute toxicity studies, in *Alternate Methods in Toxicology, Vol. 2: Acute Toxicity Testing: Alternate Approaches*, Goldberg, A.M., Ed., Mary Ann Liebert, Inc., New York.

Ehrenberg, L. (1977) Aspects of statistical inference in testing genetic toxicity, in *Handbook of Mutagenicity Test Procedures*, Kilbey, B.J., Legator, M., Nichols, W., and Ramel, C., Eds., Elsevier, New York, pp. 419–459.

Gad, S.C. (1982a) A neuromuscular screen for use in industrial toxicology. *J. Toxicol. Environ. Health*, 9, 691–704.

Gad, S.C. (1982b) Statistical analysis of behavioral toxicology data and studies. *Arch. Toxicol. Suppl.*, 5, 256–266.

Gad, S.C. and Chengelis, C.P. (1992) *Animal Models in Toxicology.* Marcel Dekker, New York.

Gad, S.C., Smith, A.C., Cramp, A.L., Gavigan, F.A., and Derelanko, M.J. (1984) Innovative designs and practices for acute systemic toxicity studies. *Drug Chem. Toxicol.*, 7, 423–434.

Grafe, A. and Vollmar, J. (1977) Small numbers in mutagenicity tests. *Arch. Toxicol.*, 38, 27–34.

Harris, E.K. (1978) Review of statistical methods of analysis of series of biochemical test results. *Ann. Biol. Clin.*, 36, 194–197.

Irwin, S. (1968) Comprehensive observational assessment, a systematic, quantitative procedure for assessing the behavioral and physiologic state of the mouse. *Psychopharmacologia*, 13, 222–257.

Jackson, B. (1962) Statistical analysis of body weight data. *Toxicol. Appl. Pharmacol.*, 4, 432–443.

Johnson, B.L., Anger, W.K., Setzer, J.V., and Xinytaras, C. (1972) The application of a computer controlled time discrimination performance to problems, in *Behavioral Toxicology*, Weiss, B., Laties, V.G., Eds., Plenum Press, New York, pp. 129–153.

Kastenbaum, M.A. and Bowman, K.O. (1970) Tables for determining the statistical significance of mutation frequencies. *Mutat. Res.*, 9, 527–549.

Katz, A.J. (1978) Design and analysis of experiments on mutagenicity. I. Minimal sample sizes. *Mutat. Res.*, 50, 301–307.

Katz, A.J. (1979) Design and analysis of experiments on mutagenicity. II. Assays involving microorganisms. *Mutat. Res.*, 64, 61–77.

Kilbey, B.J., Legator, M., Nicholas, W., and Ramel, C. (1977) *Handbook of Mutagenicity Test Procedures.* Elsevier, New York, pp. 425–433.

Kupper, L.L. and Haseman, J.K. (1978) The use of a correlated binomial model for the analysis of certain toxicological experiments. *Biometrics*, 34, 69–76.

Loeb, W.F. and Quimby, F.W. (1989) *The Clinical Chemistry of Laboratory Animals.* Pergamon Press, New York.

Martin, H.F., Gudzinowicz, B.J., and Fanger, H. (1975) *Normal Values in Clinical Chemistry.* Marcel Dekker, New York.

Mercer, T.T. (1973) *Aerosol Technology in Hazard Evaluation.* Academic Press, New York.

Mitruka, B.M. and Rawnsley, H.M. (1977) *Clinical Biochemical and Hematological Reference Values in Normal Experimental Animals.* Masson, New York.

Mosteller, F. and Youtz, C. (1961) Tables of the Freeman–Tukey transformations for the binomial and Poisson distributions. *Biometrika*, 48, 433–440.

Nelson, C.J. and Holson, J.F. (1978) Statistical analysis of teratologic data: problems and advancements. *J. Environ. Pathol. Toxicol.*, 2, 187–199.

Norton, S. (1973) Amphetamine as a model for hyperactivity in the rat. *Physiol. Behav.*, 11, 181–186.

Oser, B.L. and Oser, M. (1956) Nutritional studies in rats on diets containing high levels of partial ester emulsifiers. II. Reproduction and lactation. *J. Nutr.*, 60, 429.

Rai, K. and Ryzin, J.V. (1985) A dose-response model for teratological experiments involving quantal responses. *Biometrics*, 41, 1–9.

Ridgeman, W.J. (1975) *Experimentation in Biology.* Wiley, New York, pp. 214–215.

Stockham, J.D. and Fochtman, E.G. (1979) *Particle Size Analysis.* Ann Arbor Science, Ann Arbor, MI.

Tilson, H.A., Cabe, P.A., and Burne, T.A. (1980) Behavioral procedures for the assessment of neurotoxicity, in *Experimental and Clinical Neurotoxicology*, Spencer, P.S. and Schaumburg, N.H., Eds., Williams & Wilkins, Baltimore, pp. 758–766.

Vollmar, J. (1977) Statistical problems in mutagenicity tests. *Arch. Toxicol.*, 38, 13–25.

Vuataz, L. and Sotek, J. (1978) Use of the beta-binomial distribution in dominant-lethal testing for "weak mutagenic activity," Part 2. *Mutat. Res.*, 52, 211–230.

Weil, C.S. (1962) Applications of methods of statistical analysis to efficient repeated-dose toxicological tests. I. General considerations and problems involved. Sex differences in rat liver and kidney weights. *Toxicol. Appl. Pharmacol.*, 4, 561–571.

Weil, C.S. (1970) Selection of the valid number of sampling units and a consideration of their combination in toxicological studies involving reproduction, teratogenesis or carcinogenesis. *Food Chem. Toxicol.*, 8, 177–182.

Weil, C.S. (1972) Statistics vs. safety factors and scientific judgment in the evaluation of safety for man. *Toxicol. Appl. Pharmacol.*, 21, 459–472.

Weil, C.S. (1973) Experimental design and interpretation of data from prolonged toxicity studies, in *Proc. 5th Int. Congr. Pharmacol.*, Beacon Press, San Francisco, Vol. 2, pp. 4–12.

Weil, C.S. (1975) Toxicology experimental design and conduct as measured by inter-laboratory collaboration studies. *J. Assoc. Off. Anal. Chem.*, 58, 687–688.

Weil, C.S. (1978) A critique of the collaborative cytogenetics study to measure and minimize interlaboratory variation. *Mutat. Res.*, 50, 285–291.

Weil, C.S. (1982) Statistical analysis and normality of selected hematologic and clinical chemistry measurements used in toxicologic studies. *Arch. Toxicol.*, Suppl. 5, 237–253.

Weil, C.S., Carpenter, C.P., and Smith, H.I. (1953) Specifications for calculating the median effective dose. *Amer. Indust. Hyg. Assoc. Quart.*, 14, 200–206.

Weil, C.S. and Gad, S.C. (1980) Applications of methods of statistical analysis to efficient repeated-dose toxicologic tests. 2. Methods for analysis of body, liver and kidney weight data. *Toxicol. Appl. Pharmacol.*, 52, 214–226.

Willeke, K. (1980) *Generation of Aerosols and Facilities for Exposure Experiments.* Ann Arbor Science, Ann Arbor, MI.

Williams, R. and Buschbom, R.L. (1982) *Statistical Analysis of Litter Experiments in Teratology.* Battelle PNL–4425.

Zar, J.H. (1998) *Biostatistical Analysis.* Prentice Hall, Englewood Cliffs, NJ.

Zbinden, G. and Flury-Roversi, M. (1981) Significance of the LD_{50} test for the toxicological evaluation of chemical substances. *Arch. Toxicol.*, 47, 77–99.

15

Carcinogenesis

In the experimental evaluation of substances for carcinogenesis based on experimental results in a nonhuman species at some relatively high dose or exposure level, an attempt is made to predict the occurrence and level of tumorigenesis in humans at much lower levels. In this chapter we will examine the assumptions involved in this undertaking and review the aspects of design and interpretation of animal carcinogenicity studies.

Carcinogenicity Bioassays

To an increasing extent, science has come to understand what appears to be most of the mechanisms underlying chemical- and radiation-induced carcinogenesis. A review of these mechanisms is not germane to this chapter (readers are referred to Warshawsky, 2004, and Williams and Jatropoulus, 2001, for good reviews), but it is now clear that cancer as seen in humans is the result of a multifocal set of causes.

Mechanisms and Theories of Chemical Carcinogenesis

Genetic (all due to some mutagenic event)

Epigenetic (no mutagenic event)

Oncogene activation

Two-step (induction/promotion)

Multistep (combination of above)

The single most important statistical consideration in the design of bioassays in the past was based on the point of view that what was being observed and evaluated was a simple quantal response (cancer occurred or it did not), and that a sufficient number of animals needed to be used to have reasonable expectations of detecting such an effect. Although the single fact of whether or not the simple incidence of neoplastic tumors is increased due to an agent of concern is of interest, a much more complex model must now be considered. The time-to-tumor, patterns of tumor incidence, effects on survival rate, and age at first tumor all must now be included in a model.

Bioassay Design

As presented earlier in the section on experimental design, the first step that must be taken is to clearly state the objective of the study to be undertaken. Carcinogenicity bioassays have two possible objectives. The first objective is to detect possible carcinogens. Compounds are evaluated to determine if they can or cannot induce a statistically detectable increase in tumor rates over background levels, and only by happenstance is information generated that is useful in risk assessment. Most older studies have such detection as their objective. Current thought is that at least two species must be used for detection, although the necessity of a second species (the mouse) is increasingly questioned. The new transgenic mouse assays all address this single point.

The second objective for a bioassay is to provide a range of dose-response information (with tumor incidence being the response) so that a risk assessment may be performed. Unlike detection, which requires only one treatment group with adequate survival times (to allow expression of tumors), dose-response requires at least three treatment groups with adequate survival. We will shortly look at the selection of dose levels for this case. However, given that the species is known to be responsive, only one species of animal need be used for this objective.

To address either or both of these objectives, three major types of study designs have evolved. First is the classical skin painting study, usually performed in mice. A single, easily detected endpoint (the formation of skin tumors) is evaluated during the course of the study. Although dose-response can be evaluated in such a study (dose usually being varied by using different concentrations of test material in volatile solvent), most often detection is the objective of such a study. Although others have used different frequencies of application of test material to vary dose, there are data to suggest that this only serves to introduce an additional variable (Wilson and Holland, 1982). Traditionally, both test and control groups in such a test consist of 50 to 100 mice of one sex (males being preferred because of their very low spontaneous tumor rate). This design is also used in tumor initiation/promotion studies.

The second common type of design is the original National Cancer Institute (NCI) bioassay. The announced objective of these studies was detection of moderate to strong carcinogens, although the results have also been used in attempts at risk assessment. Both mice and rats were used in parallel studies. Each study used 50 males and 50 females at each of two dose levels (high and low) plus an equal-sized control group. The National Toxicology Program (NTP) has subsequently moved away from this design because of a recognition of its inherent limitations. More animals per group and more dose groups are now used.

TABLE 15.1

Sample Size Required to Obtain a Specified Sensitivity at $p < 0.05$ Treatment Group Incidence

Background Tumor Incidence	P^a	0.95	0.9	0.8	0.7	0.6	0.5	0.4	0.3	0.2	0.1
0.3	0.90	10	12	18	31	46	102	389			
	0.50	6	6	9	12	22	32	123			
0.2	0.90	8	10	12	18	30	42	88	320		
	0.50	5	5	6	9	12	19	28	101		
0.1	0.90	6	8	10	12	17	25	33	65	214	
	0.50	3	3	5	6	9	11	17	31	68	
0.05	0.90	5	6	8	10	13	18	25	35	76	464
	0.50	3	3	5	6	7	9	12	19	24	147
0.01	0.90	5	5	7	8	10	13	19	27	46	114
	0.50	3	3	5	5	6	8	10	13	25	56

[a] P = power for each comparison of treatment group with background tumor incidence.

Finally, there is the standard industrial toxicology design, which uses at least two species (usually rats and mice) in groups of no fewer that 100 males and females each. Each study has three dose groups and at least one control. Frequently, additional numbers of animals are included to allow for interim terminations and histopathological evaluations. In both this and the NCI design, a long list of organs and tissues are collected, processed, and examined microscopically. This design seeks to address both the detection and dose-response objectives with a moderate degree of success.

Selecting the number of animals to use for dose groups in a study requires consideration of both biological (expected survival rates, background tumor rates, etc.) and statistical factors. The prime statistical consideration is reflected in Table 15.1. It can be seen in this table that if, for example, we were studying a compound which caused liver tumors and were using mice (with a background or control incidence of 30%), we would have to use 389 animals per sex per group to be able to demonstrate that an incidence rate of 40% in treatment animals was significant compared to the controls at the $p \le 0.05$ level.

Perhaps the most difficult aspect of designing a good carcinogenicity study is the selection of the dose levels to be used. At the start, it is necessary to consider the first underlying assumption in the design and use of animal cancer bioassays — the need to test at the highest possible dose for the longest practical period.

The rationale behind this assumption is that although humans may be exposed at very low levels, detecting the resulting small increase (over background) in the incidence of tumors would require the use of an impractically large number of test animals per group. This point is illustrated in Table 15.2, where, for instance, while only 46 animals (per group) are needed to show a 10% increase over a zero background (that is, a rarely occurring tumor type), 770,000 animals (per group) would be needed to detect a 1/10 of a percent increase above a 5% background. As we increase dose, however, the incidence

TABLE 15.2

Average Number of Animals Needed to Detect a Significant Increase

Background Incidence (%)	Expected Increase in Incidence (%)					
	0.01	0.1	1	3	5	10
0	46,000,000	460,000	4,600	511	164	46
0.01	46,000,000	460,000	4,600	511	164	46
0.1	47,000,000	470,000	4,700	520	168	47
1	51,000,000	510,000	5,100	570	204	51
5	77,000,000	770,000	7,700	856	304	77
10	100,000,000	1,000,000	10,000	1,100	400	100
20	148,000,000	1,480,000	14,800	1,644	592	148
25	160,000,000	1,600,000	16,000	1,840	664	166

Note: Number of animals needed in each group — controls as well as treated — to detect a significant increase in the incidence of an event (tumors, anomalies, etc.) over the background incidence (control) at several expected incidence levels using the Fisher exact probability test ($p \leq 0.05$).

of tumors (the response) will also increase until it reaches the point where a modest increase (e.g., 10% over a reasonably small background level [say 1%]) could be detected using an acceptably small-sized group of test animals (in Table 15.2 we see that 51 animals would be needed for this example case). There are, however, at least two real limitations to the highest dose level. First, the test rodent population must have a sufficient survival rate after receiving a lifetime (or two years) of regular doses to allow for meaningful statistical analysis. Second, we really want the metabolism and mechanism of action of the chemical at the highest level tested to be the same as at the low levels where human exposure would occur. Unfortunately, toxicologists usually must select the high dose level based only on the information provided by a subchronic or range-finding study (usually 90 days in length), but selection of either too low or too high a dose will make the study invalid for detection of carcinogenicity and may seriously impair the use of the results for risk assessment.

There are several solutions to this problem. One of these has been the rather simplistic approach of the NTP Bioassay Program, which is to conduct a 3-month range-finding study with sufficient dose levels to establish a level that significantly (10%) decreases the rate of body weight gain. This dose is defined as the maximum tolerated dose (MTD) and is selected as the highest dose. Two other levels, generally one half MTD and one quarter MTD, are selected for testing as the intermediate and low dose levels. In many earlier NCI studies, only one other level was used.

The dose range-finding study is necessary in most cases, but the suppression of body weight gain is a scientifically questionable benchmark when dealing with establishment of safety factors. Physiologic, pharmacologic, or metabolic markers generally serve as better indicators of systemic response than body weight. A series of well-defined acute and subchronic studies designed to determine the "chronicity factor" and to study onset of pathology can be more predictive for dose setting than body weight suppression.

Also, the NTP's MTD may well be at a level where the metabolic mechanisms for handling a compound at real-life exposure levels have been saturated or overwhelmed, bringing into play entirely artifactual metabolic and physiologic mechanisms (Gehring and Blau, 1977). The regulatory response to questioning the appropriateness of the MTD as a high dose level (exemplified by Haseman, 1985) has been to acknowledge that occasionally an excessively high dose is selected, but to counter by saying that using lower doses would seriously decrease the sensitivity of detection.

Selection of levels for the intermediate and lower doses for a study is easy only in comparison to the selection of the high dose. If an objective of the study is to generate dose-response data, then the optimal placement of the doses below the high is such that they cover as much of the range of a response curve as possible and yet still have the lowest dose at a high enough level that one can detect and quantify a response. If the objective is detection, then having too great a distance between the highest and next highest dose creates a risk to the validity of the study. If the survival in the high dose is too low, yet the next highest dose does not show non-neoplastic results (that is, cause other-than-neoplastic adverse biological effects) such as to support it being a high enough dose to have detected a strong or moderate carcinogen, the entire study may have to be rejected as inadequate to address its objective. Portier and Hoel (1984) have proposed statistical guidelines (for setting dose levels below the high) based on response surfaces. In so doing they suggest that the lowest dose be no less than 10% of the highest.

Although it is universally agreed that the appropriate animal model for testing a chemical for carcinogenicity would be one whose metabolism, pharmacokinetics, and biological responses were most similar to humans, economic considerations have largely constrained practical choices to rats and mice. The use of both sexes of both species is preferred on the grounds that it provides for (in the face of a lack of understanding of which species would actually be most like humans for a particular agent) a greater likelihood of utilizing the more sensitive species.

Use of the mouse has been both advocated and defended on these grounds and because of the economic advantages and the species' historical utilization (Grasso and Crampton, 1972). Increasingly, however, it is believed that the use of the mouse is redundant and represents a diversion of resources while yielding little additional information. Wittenau and Estes (1983) cited a "unique contribution" for mouse data in 273 bioassays of only 13.6% of the cases (that is, 37 cases). Others have questioned the use of the mouse based on the belief that it gives artifactual liver carcinogenesis results, a point now generally conceded. One suggestion for the interpretation of mouse bioassays is that in those cases where there is only an increase in liver tumors in mice (or lung tumors in strain A mice) and no supporting mutagenicity findings (a situation characteristic of some classes of chemicals), the test compound should not be considered an overt carcinogen (Ward et al., 1979). ICH guidelines (ICH 1994, 1995, 1997a, and 1997b) in fact drop the requirement for a second species (mouse) bioassay when other data are available to supplement the rat bioassay results or allow for use of a transgenic mouse bioassay for satisfying the two species requirement.

The NCI/NTP (National Toxicology Program) currently recommends an F1 hybrid cross between two inbred strains, the C57BI/6 female and the C3H male, the results being commonly designated as the B6C3F1. This mouse was found to be very successful in a large-scale pesticide testing program in the mid-1960s. It is a hardy animal with good survival and disease resistance, easy to breed, and reported to have a relatively low spontaneous tumor incidence. Usually at least 80% of the control mice are still alive at a 24 months termination.

Unfortunately, while it was originally believed (Page, 1977) that the spontaneous liver tumor incidence in male B6C3F1 mice was 15.7%, it actually appears to be closer to 32.1% (Nutrition Foundation, 1983). The issue of spontaneous tumor rates and their impact on the design and interpretation of studies will be discussed more fully later. Thus, use of a cross of two increased mouse strains is also a point of controversy. Haseman and Hoel (1979) have presented data to support the idea that inbred strains have lower degrees of variability of biological functions and tumor rates, making them more sensitive detectors and quantitators. These authors also suggest that the use of cross from two such inbred strains allows one to more readily detect tumor incidence increases. On the other hand, it has been argued that such genetically homogeneous strains do not properly reflect the diversity of metabolic functions (particularly one that would serve to detoxify or act as defense mechanisms) that are present in the human population.

Study length and the frequency of treatment are design aspects that must also be considered. These are aspects where the objectives of detection and dose-response definition conflict.

For the greatest confidence in a "negative" detection result, an agent should be administered continuously for the majority of an animal's lifespan. The NTP considers that 2 years is a practical treatment period in rats and mice, although the animals currently used in such studies may survive an additional 6 to 12 months. Study lengths of 15 to 18 months are considered adequate for shorter-lived species, such as hamsters. An acceptable exposure/observation period for dogs is considered to be 5 to 7 years, an age equivalent to about 45 to 60 years in humans. For dietary treatments, continuous exposure is considered desirable and practical. With other routes, practical considerations may dictate interrupted treatments. For example, inhalation treatment for 6 to 8 hours per day on a 5 day/wk schedule is the usual practice. Regimens requiring special handling of animals, such as parenteral injections, are usually on a 5 day/wk basis. With some compounds intermittent exposures may be required because of toxicity. Various types of recovery can occur during exposure-free periods, which may either enhance or decrease chances of carcinogenicity. In view of the objective of assessing carcinogenicity as the initial step, intermittent exposure on a 3 to 5 day/wk basis is considered both practical and desirable for most compounds.

Following cessation of dosing or exposure, continued observation during a nontreatment period may be required before termination of the experiment. Such a period is considered desirable because (1) induced lesions may progress to more readily observable lesions, and (2) morphologically similar but noncarcinogenic proliferative lesions that are stress related may regress. Neoplastic or "neoplastic-like" lesions that persist long after removal of the stimulus are considered of serious consequences, from the hazard viewpoint. Many expert anatomical pathologists, however, feel able to diagnose and determine the biological nature of tumorous lesions existing at the time of treatment without the added benefit of a treatment-free period. Indeed, the proper interpretation of significance of neoplastic lesions (tumors) requires consideration of the accompanying occurrence of preneoplastic lesions. There should be a biologic as well as a statistical trend present. The occurrence of neoplasia without hyperplastic lesions at an increased incidence is not generally regarded as convincing evidence of tumorigenesis (Dinse, 1994).

In determining the length of an observation period, several factors must be considered: period of exposure, survival pattern of both treated and control animals, nature of lesions found in animals that have already died, tissue storage and retention of the chemical, and results of other studies that would suggest induction of late-occurring tumors. The usual length of a treatment-free observation period is 3 months in mice and hamsters and 6 months in rats. An alternative would be to terminate the experiment or an individual treatment group on the basis of survival (say at the point at which 50% of the group with the lowest survival have died).

The arguments against such prolonged treatment and maintenance on study revolve around the relationship between age and tumor incidence. As test animals (or humans) become older, the background ("naturally occurring") incidence of tumors increases (Dix and Cohen, 1980) and it becomes increasingly difficult to identify a treatment effect from the background effect. Salsburg (1980) has published an analysis of patterns of senile lesions in mice and rats, citing what he calls the principle of biological confounding. "If a particular lesion (e.g., pituitary tumor) is part of a larger syndrome induced by the treatment, it is impossible to determine whether the treatment has 'caused' that lesion."

This could lead to a situation where any real carcinogen would be nonidentifiable. If the usual pattern of old-age lesions for a given species or strain of animals include tumors, then almost every biologically active treatment can be expected to influence the incidence of tumors in a cluster of lesions at a sufficiently high dose.

Reconsidering our basic principles of experimental design, it is clear that we should try to design bioassays so that any carcinogenesis is a clear-cut, single event, unconfounded by the occurrence of significant numbers of lesions due to other causes (such as age). One answer to this problem is the use of interim termination groups. When an evaluation of tumor incidences in an interim sacrifice sample of animals indicates that background incidence is becoming a source of confounding data, termination plans for the study can be altered to minimize the loss of power. Several authors (such as Louis and Orva, 1985; Ciminera, 1985) have presented such adaptive sacrifice plans.

A number of other possible confounding factors can enter into a bioassay unless design precludes them. These include (a) cage and litter effects (Lagakos and Mosteller, 1981), which can be avoided by proper prestudy randomization of animals and rotation of cage locations; (b) vehicle (corn oil, for example, has been found to be a promoter for liver carcinogens); and (c) the use of the potential hazard route for man, e.g., dietary inclusion instead of gastric intubation.

In the last chapter of this book we will discuss several controversial statistical issues that apply to carcinogenesis. Other general aspects of the design of carcinogenicity bioassay may be found in Robens et al. (1994) and Neal and Gibson (1984).

Histopathologic Lesion Incidence

The basic experimental endpoint in a bioassay depends upon histopathologic examination of many tissues collected from animals in the studies. While it is not true that only those lesions that occur at a statistically significantly increased rate in treated/exposed animals are of concern (for there are the cases where a lesion may be of such a rare type that the occurrence of only one or a few such in treated animals "raises a flag"), it is true that, in most cases, a statistical evaluation is the only way to determine if what we are seeing in treated animals is significantly worse than what has been seen in control animals. And although cancer is not our only concern, it is the class of lesions that is of the greatest interest.

Typically, comparison of incidences of any one type of lesion between controls and treated animals are made using chi-square or Fisher's exact test with a modification of the numbers as the denominators. Too often, experimenters exclude from the consideration all those animals (in both groups) that died prior to the first animals being found with a tumor at that site.

Two major controversial questions are involved in such comparisons: (a) should they be based on one-tailed or two-tailed distribution? and (b) what are the effects and implications of multiple comparisons?

The one- or two-tailed controversy revolves around the question of which hypothesis we are properly testing in a study such as a chronic carcinogenicity study. Is the tumor incidence different between the control and treated groups? In such cases, it is a bidirectional hypothesis and, therefore, a two-tailed distribution we are testing against. Or are we asking if the tumor incidence is greater in the treated group than in the control group? In the latter case, it is a unidirectional hypothesis and we are contemplating only the right-hand tail of the distribution. The implications of the answer to this question are more than theoretical; significance is much greater (exactly double, in fact) in the one-tailed case than in the two-tailed. For example, a set of data analyzed by Fisher's exact test, which would have a two-tailed p-level of 0.098 and a one-tailed level of 0.049, would therefore be flagged as significantly different.

Feinstein (1979) provides an excellent discussion of the background in a nonmathematical way. Determination of the correct approach must rest on a clear definition by the researcher, beforehand, of the objective of his study and of the possible outcomes (if a bidirectional outcome is possible, are we justified in using a one-tailed test statistic?).

The multiple comparisons problem is a much more lively one. In chronic studies, we test lesion/tumor incidence in each of a number of tissues, for each sex and species, with each result being flagged if it exceeds the fiducial limit of $p \leq 0.05$.

The point we must ponder here is the meaning of "$p \leq 0.05$." This is the level of the probability of our making a type I error (incorrectly concluding we have an effect when, in fact, we do not). So we have accepted the fact that there is a 5% chance of our producing a false positive from this study. Our trade-off is a much lower change (typically 1%) of a type II error, that is, of our passing as safe a compound that is not safe. These two error levels are connected: to achieve a lower type II level inflates our type I level. The problem in this case is that if we make a large number of such comparisons, we are repeatedly taking the chance that we will "find" a false positive result. The set of lesions and/or tumor comparisons described above may number more than 70 tests for significance in a single study, which will result in a large inflation of our false positive level.

The extent of this inflated false-positive rate (and how to reduce its effects) has been discussed and estimated with a great degree of variability. Salsburg (1977) has estimated that the typical NCI-type cancer bioassay has a probability of type I error ranging between 20 and 50%. Fears and colleagues (Fears and Tarone, 1977; Fears et al., 1977), however, have estimated it as being between 6 and 24%. Without some form of correction factor, the "false positive" rate of a series of multiple tests can be calculated as being equal to 1 to $0.95N$ where N is the number of tests and the selected alpha level is 0.05. CDER sets the overall false-positive rate arising from the multiplicity of statistical testing to be 10% (FDA, 2001) and concludes that test results provide only equivocal evidence of a carcinogenicity positive finding unless all three test approaches yield consistent results.

Salsburg (1980) expressed the concern that such an exaggerated false-positive result may result in a good compound being banned. Although Haseman (1985) challenged this on the point that a much more mature decision process than this is used by the regulatory agencies, Salsburg has pointed out at least two cases where the decision to ban was based purely on such a single statistical significance.

What, then, is a proper use of such results? Or, conversely, how can we control for such an inflated error rate?

There are statistical methods available for dealing with this multiple comparisons problem. One such is the use of Bonferroni inequalities to correct for successive multiple comparisons (Bonferroni-Miller, 1966). These methods have the drawback that there is some accompanying loss of power expressed as an inability to identify true positives properly.

A second approach is to use the information in a more mature decision-making process. First, the historical control incidence rates (such as are given for the B6C3F1 mouse and the Fischer 344 rats in Fears et al., 1977) should be considered; some background incidences are so high that these tissues are "null and void" for making decisions. Second, we should look not just for a single significant incidence in a tissue but rather for a trend (Prentice et al., 1992; Smythe et al., 1986). For example, we might have the following percentages of a liver tumor incidence in the female rats of a study: (a) control — 3%, (b) 10 mg/kg — 6%, (c) 50 mg/kg — 17%, and (d) 250 mg/kg — 54%. In this study only the incidence at the 250 mg/kg level might be statistically significant. However, the trend through each of the levels is suggestive of a dose-response. Looking for such a trend is an essential step in a scientific assessment of the results, and one of the available trend analysis techniques, such as presented earlier in this chapter, should be utilized.

Another method for determining whether statistically significant incidences are merely random occurrences is to compare the results of the quantitative variables with two or more concurrently run control groups. Often the mean of one variable will differ from only one of these controls and be numerically within the range of this same variable of the two control means. If so, the statistical significance compared to that one control must be seriously questioned as to its being correlated with a biological significance.

Issues in Histopathological Analysis

There are four main situations where combining pathological morphological neoplastic alterations in a statistical analysis can be considered (McConnell et al., 1986). The first is when essentially the same lesion has been recorded under two or more different names, or even under the same name in different places. Here failure to combine these conditions in the analysis may severely limit the chances of detecting a true treatment effect. It should be noted, however, that grouping together conditions that are actually different might also result in the masking of a true treatment effect, particularly if the treatment has a very specific effect.

The second situation is when separately recorded lesions form successive steps on the pathway of the same process. The most important example of this phenomenon is for the incidence of related types of malignant tumor, benign tumor, and focal hyperplasia. It will normally be appropriate to carry out analyses of (1) incidence of malignant tumor, (2) incidence of benign or malignant tumor, and, where appropriate, (3) incidence of focal hyperplasia, benign, or malignant tumor. It will not normally be appropriate to carry out analyses of benign tumor incidence only or of the incidence of focal hyperplasia only.

The third situation for combining is when the same pathological condition appears in different organs as a result of the same underlying process. Examples of this are the multicentric tumors, such as myeloid leukemia, reticulum cell sarcoma, and lymphosarcoma, or certain nonneoplastic conditions, such as arteritis/periarteritis and amyloid degeneration. Here analysis will normally be carried out only if there's incidence at any site.

The final situation where an analysis of combined pathology findings is appropriate is for analysis of overall incidence of malignant tumor at any site, of benign or malignant tumor at any site, or of multiple tumor incidences. While analyses of tumor incidence at specific sites are normally more meaningful, as treatments often affect only a few specific sites, these additional analyses are usually required to guard against the possibility that treatment had some weak but general tumor-enhancing effect that would not be otherwise evident. In some situations, one might also envisage analyses of other combinations of specific tumors, such as tumors at related sites; e.g., endocrine organs if the compound demonstrates a hormonal effect, or of similar histopathologic type.

Taking Severity into Account

The argument that all data recorded by the toxicologic pathologist should be analyzed should be extended to grading of lesions. If the pathologist chooses to grade a condition for severity, the grade should be taken into account in the statistical analysis. There are two ways to analyze data when the grade is to be taken into account. In one, data are analyzed when the animal has a condition and the condition is at least grade 2, at least grade 3, etc. In the other approach, nonparametric methods utilizing ranked data are used. The latter approach is more powerful, as it uses all the information in one analysis. Unfortunately, those without some statistical training may not understand the output as easily.

Note that consistent grading is necessary for meaningful analyses based on grade. If a condition has been scored only as present/absent for some animals, but has been graded for others, it is not possible to use graded analyses unless the pathologist is willing to go back and grade the specific animals showing the condition.

Using Simple Methods that Avoid Complex Assumptions

Different statistical analytical methods can vary considerably in their complexity and in the number of assumptions they make. Statistical analysis should be used for the clarification of effects rather than being performed only to obtain a "p-value." Wherever possible, use statistical methods that are simple, robust, and make few assumptions. The use of statistical models has its place, however, more for effect estimation than for hypothesis testing and in studies of complex design rather than in those of simple design. There are three reasons for this. First, the toxicologic pathologist usually understands simpler methods, and hence can justify the outputs. Second, seldom are there adequate data in practice to validate fully the assumptions of any formal statistical model. Third, even if particular models are shown to be appropriate for use, the loss of power in using appropriate simpler methods is often very small.

Methods for the routine statistical analysis of tumor incidence do not use formal parametric statistical models (Armitage and Doll, 1961). For example, when evaluating the relationship of treatment to incidence of a well-defined pathological finding, and adjusting for other factors, in particular age at death, that might bias the comparison, methods involving "stratification" are recommended. These should be used in preference to a multiple regression approach or time-to-tumor models.

Analyses of variance (ANOVA) methods can be useful for estimating treatment effects when continuously distributed data are obtained. However, the appropriateness of ANOVA as a tool depends on the validity of the fundamental assumptions of normally distributed variables and equal variability in each group. If these assumptions are violated, or cannot be reasonably demonstrated or assumed, nonparametric methods are more appropriate for hypothesis testing. These may be based on the rank of observations, rather than their actual value , and do not depend on the assumptions common to the parametric methods.

Using All Data

In addition to evaluating treatment and effect relationships, there are situations where determination of the effect of other sources of systematic differences among individuals is warranted. Gender differences, differing times of sacrifice, and differing secondary treatments may be considered. A source of systematic difference can be considered a factor. It is often important to evaluate the effect of these factors so that a more powerful analysis of the treatment effects can be made. These factors can be evaluated for relationships within each level of the factor (e.g., factor: sex; level: male, female) and combined to see whether data can be pooled. Some scientists consider that conclusions for males and females should always be drawn separately, but there are strong statistical arguments for a joint analysis.

Combining, Pooling, and Stratification of Lesion Data

Table 15.3 is a hypothetical study of a toxic agent that induces tumors without shortening the lives of tumor-bearing animals. Hypothetical data for *t* number of animals with tumor number examined are shown in Table 15.3.

Table 15.3 shows that if the time of death is ignored and pooled data are evaluated, the incidence of tumors is the same in each group. This leads to the false conclusion that treatment had no effect. However, by evaluating

TABLE 15.3

Sample Data from a Hypothetical Bioassay

	Control	Exposed	Combined
Early deaths	1/20 (5%)	18/90 (20%)	19/110 (17%)
Late deaths	24/80 (30%)	7/10 (70%)	31/90 (34%)
Total	25/100 (25%)	25/100 (25%)	50/200 (25%)

the time to death on treatment an increased incidence in the exposed group can be seen. An appropriate statistical method would combine a measure of difference between the groups based on the early deaths and a measure of difference based on the late deaths, and conclude correctly that incidence, after adjustment for time of death, is greater in the exposed groups.

In this example, time of death is the stratifying variable with two strata: early deaths and late deaths. The essence of the methodology is to make comparisons only within strata so that one is always comparing like with like except in respect of treatment. Then combine the differences over strata. Stratification can be used to adjust for any variable, or combinations of variables.

Some studies are of factorial design, in which combinations of treatments are evaluated. The simplest such design is one in which four equal sized groups of animals receive no treatment, treatment A only, treatment B only, and treatments A and B. The basic assumption for analysis of this type of experiment is that the two treatment effects are independent. In this factorial experiment, one can use stratification to enable more powerful tests to be conducted of the possible individual treatment effects. Thus, to test for effects of treatment A, for example, one conducts comparisons in two strata, the first consisting of groups 1 and 2 (not given treatment B) and the second consisting of groups 3 and 4 (given treatment B). Combination of results from the two strata is based on twice as many animals and is therefore markedly more likely to detect possible effects of treatment A than is a simple comparison of groups 1 and 2. There is also the possibility of identifying interactions, such as synergism and antagonism, between the two treatments.

In some routine long-term screening studies, the study design involves five groups of 50 animals of each sex, three groups of which are treated with increasing doses of a compound, and two of which are untreated controls. Assuming that there is no systematic difference between the control groups, e.g., the second control group in a different room or from a different batch of animals, the main analyses will pool the control groups, resulting in a single group of 100 animals. Pooling the control groups should only occur following a preliminary analysis to show no difference in the incidence of effects in these control groups. If there is a difference between control groups, the "cause" of the difference needs to be evaluated and determined how this cause may have affected, or not, the experimental groups. Once this has been determined, the most appropriate method of analysis for the situation is used.

Trend Analysis, Low-Dose Extrapolation, and NOEL Estimation

While comparisons of individual treated groups with the control group are important, a more powerful test of a possible effect of treatment is a test for a dose-related trend. Trend tests use all data in a single analysis to evaluate the effects of treatment that result in a positive or negative dose-response relationship. In interpreting the results of trend tests, it should be noted that a significant trend does not necessarily imply a significant effect at lower doses. Conversely, a lack of a significant increase at lower doses does not necessarily indicate evidence of a threshold; i.e., a dose below which no increase occurs.

Testing for trend is a more sensitive method than simple pair-wise comparisons of treated and control groups for showing a possible treatment effect. Attempting to estimate the magnitude of effects at low doses, typically below the lowest positive dose tested in the study, is a much more complex procedure and is heavily dependent on the assumed functional form of the dose-response relationship.

Low dose extrapolation is typically conducted for tumors believed to be caused by a genotoxic effect. Given severe limits in understanding the biology of low-dose genotoxicity, and the inability to generate data in the very low response range of the dose-response curve, "no threshold" is ordinarily assumed. This assumption is based on a current understanding of initiation and a lack of evidence for a less conservative (i.e., protective) approach. However, some, but by no means all, scientists believe these tumors have no threshold. For other types of tumors, and for many nonneoplastic endpoints, a threshold cannot be estimated directly from data at a limited number of dose levels. A no-observed-effect level (NOEL) can be estimated by finding the highest dose level at which there is no significant increase in treatment-related effects. It should be noted that the NOEL addresses any treatment-related effect, whereas the no-observable-adverse-effect level (NOAEL) defines the effect as detrimental. The other statistic of great importance to understanding where on the dose curve the response occurs is the lowest-effect level (LOEL), or lowest-adverse-effect level (LOAEL). These levels are the lowest dose at which an effect, or adverse effect, occurred. The true threshold for the effect of interest under the conditions of the study is bracketed by the NOEL and LOEL (i.e., the threshold t is given by NOEL $< t <$ LOEL).

Need for Age Adjustment

If there are marked differences in survival among treated groups there is a need for an age adjustment, i.e., an adjustment for age at death or onset. This is illustrated in Table 15.4 where, because of the greater number of deaths occurring early in the treated group, the true effect of treatment disappears if no adjustment is made. Thus, a major purpose of age adjustment is to avoid temporal bias.

TABLE 15.4

The Effect of Age Adjustment on Mortality Data

	Control	Exposed
Early deaths	0/20	0/20
Middle deaths	1/10	9/10
Late deaths	20/20	20/20
Total	21/50	29/50

It is not always recognized that even where there are no survival differences, age adjustment can increase the power to detect group differences. This is illustrated by the example in Table 15.4.

In Table 15.4 treatment results in an earlier onset of a condition causing mortality, which eventually occurs in all animals. Failure to age adjust will result in a comparison of 29/50 with 21/50, which is not statistically significant. Age adjustment will essentially ignore early and late deaths, which contribute no comparative statistical information, and be based on the comparison of 9/10 with 1/10, which is statistically significant. By avoiding diluting data capable of detecting treatment effects with data that are of little or no value for this purpose, age adjustment sharpens the contrast rather than avoiding bias.

Taking the Context of Observation into Account

Age adjustment cannot be used unless the context of the endpoint is clear. There are three relevant contexts, with the first two relating to the situation where the condition is only observed at death, e.g., an internal tumor, and the third where it can be observed during life, e.g., a skin tumor. In the first context the condition is assumed to have caused the death of the animal, i.e., to be fatal. Here the incidence rate for a time interval and a group is calculated as the number of animals dying because of the lesion during the interval divided by the number of animals alive at the start of the interval. In the second context, the animal is assumed to have died of another cause; i.e., the internal tumor is incidental. In the case of incidental lesions the rate is calculated as the number of animals with the lesion dying during the interval divided by the total number of deaths during the interval. In the third context, where the lesion is visible, the rate is calculated as the number of animals with the condition during the interval divided by the number of animals without the condition at the start of the interval.

The method of Peto and colleagues (1980) takes the context of observation into account. Sometimes, the nature of the lesion does not always allow the toxicologic pathologist to decide whether a condition is fatal or incidental. However, in experiments where marked survival differences are seen, the toxicologic pathologist should attempt to decide upon the context of the lesion. Failure to do so may result in the inability to conclude reliably whether a treatment is beneficial or harmful.

For those reading carcinogenicity studies, the definition of what constitutes a fatal tumor sometimes conflicts with the needs of the statistician regarding hard measurable endpoints. Here we have a paradox: the statistician requires mutually inclusive and mutually exclusive groups of fatal vs. nonfatal tumors for the analysis, whereas the pathologist rarely is able to judge without question as to whether the lesion was fatal. To demonstrate the problem and pose a solution, we shall use examples to show the need for data for the statistician and the weakness of data generated by the toxicologic pathologist.

The effect of lack of definition of context, and the need to have context for statistical purposes, is well illustrated by the evaluation of N-nitrosodimethylamine (NDMA) for carcinogenicity. Here it was assumed that all pituitary tumors were fatal. This resulted in the false conclusion that NDMA was carcinogenic. In contrast, if it were assumed that these lesions were incidental, the false conclusion that NDMA was protective would result. By using the toxicologic pathologist's contextual opinion as to which tumors were, and which were not, likely to be fatal, the resulting analysis concluded correctly that NDMA had no carcinogenic effect in the pituitary. It is imperative that the toxicologic pathologist attempts to judge the context of the lesions in question, as he is the most competent to make such a judgment. Failure to put the lesions in context may result in either an erroneous conclusion or no conclusion at all.

Many toxicologic pathologists have entered the Peto variable, a quantal categorization as to whether the tumor in question was incidental or fatal, whenever he felt that a particular tumor was why the animal came to necropsy that day. For example, an animal that has been slowly deteriorating in a bioassay was found to have a 10-cm pituitary mass at necropsy. The histological evaluation of the mass revealed that is was a *pars distalis* adenoma. The clinical history of no neurological signs with this space-occupying lesion indicated that the tumor was slow growing. It may seem that the Peto variable in this case should be marked as "fatal;" however, such an action would be a mistake, as described in the following paragraph.

The Peto variable is collected for the purpose of determining the duration of the tumor's presence. A fatal classification represents a tumor that has been interpreted as "rapidly fatal." Tumors are classified by the pathologist as incidental, fatal, or mortality independent (observable). Incidental tumors are those tumors deemed not directly or indirectly responsible for the animal's death but observed at necropsy. Fatal tumors are tumors deemed to have killed the animal either directly or indirectly. Mortality-independent tumors are tumors that are detected at times other than at necropsy. The distinction between fatal and incidental tumors is important because it is essential to distinguish between a chemical that reduces survival by shortening the time to tumor onset or the time to death following tumor onset (a real carcinogenic effect) and one that reduces survival but for which tumors are observed earlier simply because they are dying of competing causes (noncarcinogenic effect). In the example given earlier, collection of a fatal rather than an incidental classification misrepresents the "real" findings for the Peto variable, as the tumor was present for a long time. In fact, the problem resides with the interpretation of what the Peto variable represents. Obviously, dichotomous data must result from a simple yes/no question, but often two questions are asked: was the tumor the cause for necropsy of the animal and/or was the tumor rapidly fatal?

The Peto variable represents the second question and not the first. For this reason it has been suggested that only tumors recognized as rapidly fatal receive the fatal Peto variable designation. To make sure that only rapidly fatal tumors are included as fatal, malignant lymphomas and leukemias would be marked as fatal, whereas all other tumors should be marked as incidental. It is our opinion that the conservative use of the fatal Peto variable is preferable to misclassification of an incidental datum to fatal to make sure that overly conservative assessments are not made in carcinogenesis studies that are quite conservative in their output without the introduction of this statistic. Regardless of violating the assumptions of the test, it is preferable to record tumor-related deaths for Peto's trend test evaluation than not to have any data regarding premature mortality in a bioassay.

Although it is normally good practice for the toxicologic pathologist to ascribe "factors contributing to unscheduled death" for each animal, it is not strictly necessary to determine the context of observation for all conditions at the outset. An alternative strategy is to analyze data under differing assumptions. Multiple analyses of data may be done using a different context for unscheduled deaths. Typically, analyses of decedents would assume that all cases are incidental and use the context of no cases fatal, all cases fatal, or all cases of the same defined severity occurring in decedents fatal.

If the conclusion for all scenarios is the same, or if the toxicologic pathologist states that one scenario is the most likely, it may not be necessary to know the context of observation for the condition in question for each individual animal. Using this alternative strategy might result in analysis cost reductions and time savings by allowing evaluations to focus on a limited number of lesions where the conclusion seems to hang on correct knowledge of the context of observation. Finally, it should be noted that although many nonneoplastic conditions observed at death are never causes of death, it is in principle as necessary to know the context of observation for nonneoplastic conditions as it is for tumors.

Experimental and Observational Units

Animals in a study are often both the "experimental unit" and the "observational unit," but this is not always so. For determining treatment effects by the methods of the next section, it is important that each experimental unit provide only one item of data for analysis, as the methods all assume that individual data items are statistically independent.

In many feeding studies, where the cage is assigned to a treatment, it is the cage rather than the animal that is the experimental unit. In contrast, histopathologic observations for a tissue may be based on multiple sections per animal. In this case the section is the observational unit, whereas the animal is the experimental unit. Similarly, in reproduction and teratology studies the dam is the experimental unit, whereas fetuses are the observational units. Multiple observations per experimental unit should be combined in some suitable way into an overall average for that unit before analysis. Often these multiple observations are not distributed normally, hence median and semi-quartile ranges are used commonly for describing multiple observation data for an experimental unit (Table 15.5).

TABLE 15.5

The Effect of Age Adjustment on Mortality Data

Animal Number	Treatment	Pup Number	Hepatic Pathology Score
E-9991	B	1	0
(0)		(0)	0
		2	0
		(0)	0
		3	0
		(0)	0
		4	0
		(0)	
E-9992	A	1	3
(2.5)		(2.5)	2
E-9993	D	—	—
E-9994	A	1	3
1.5		(3)	3
		2	1
		(1.5)	2
		3	1
		(1.5)	2
E-9995	C	1	1
(0.5)		(1)	1
		2	0
		(0)	0
		3	1
		(0.5)	0
E-9996	D	—	—
E-9997	B	1	0
(0)		(0)	0
		2	0
		(0)	0
		3	0
		(0)	0
		4	0
		(0)	0
		5	0
		(0)	0
E-9998		1	1
(1)		(0.5)	0
		2	0
		(1)	2
		3	1
		(1)	1
		4	2
		(1.5)	1

For these data, a series of calculations would be necessary before statistical analysis could be undertaken. As these observations are unlikely to be distributed normally, median and semi-quartile ranges would be used. To obtain a datum for each dam, first the median score per pup (in parentheses under pup number) is derived. Then the median score per pup per dam (in parentheses under animal number) would be used for describing the multiple observations for the experimental unit, the dam.

Missing Data

Missing data is a common problem in toxicologic pathology. Missing tissues and inadequate blood samples for analysis are not uncommon, even under GLP conditions. How missing data are handled in statistical evaluation is critical to obtaining a valid interpretation of results.

Animals with missing data can simply be removed from the analysis. There are, however, some situations where removal of the experimental unit can be inappropriate. Particularly of note is when a lesion assumed to have caused the death of the animal is analyzed. For example, an animal dies at week 83 of a bioassay for which the section was unavailable for microscopic examination. This animal cannot contribute to the group comparison at week 83. However, as it was alive in previous weeks it should contribute to the denominator of the calculations in all previous weeks.

Missing observations also occurs when histopathologic evaluation is only done when an abnormality is seen at postmortem. In such an experiment, hypothetical data may appear as in Table 15.6.

The following statistics could be derived for the data. Ignoring animals with no microscopic sections, one would compare $2/2 = 100\%$ with $14/15 = 93\%$ and conclude that treatment nonsignificantly decreased incidence. This is likely to be a false conclusion, and it would be better here to compare the percentage of animals that had a postmortem abnormality that turned out to be a tumor, i.e., $2/50 = 4\%$ with $14/50 = 28\%$. Unless some aspect of treatment made tumors much easier to detect at postmortem, one could then conclude that treatment did have an effect on tumor incidence.

Particular care has to be taken in studies where the procedures for histopathologic examination vary by group; otherwise, observer bias will affect

TABLE 15.6

Data Generated from a Hypothetical Experiment Where Histopathological Evaluation is Done Only in Animals Where Gross Lesions Were Seen

	Control Group	Treated Group
Number in group	50	50
Number with gross lesions	2	15
Histopathological evaluation	2	15
Classified neoplasm	2	14

the data generated. The protocol often requires a full microscopic examination of a given tissue list in decedents in all groups and in terminally killed controls and high-dose animals. In other animals (terminally killed low- and mid-dose animals), microscopic examination of a tissue is only conducted if the tissue is found to be abnormal at postmortem. Such a protocol is designed to save money but can lead to invalid comparisons among treatment groups. Suppose, for example, responses in terminally killed animals are 8/20 in the controls, 3/3 (with 17 unexamined) in the low-dose, and 5/6 (with 14 unexamined) in the mid-dose animals. Is one supposed to conclude that treatment at the low and mid doses increased response, based on a comparison of the proportions examined microscopically (40, 100, and 83%)? Or should one conclude that it decreased response, based on the proportion of animals in the group (40, 15, and 25%)? It could well be that treatment had no effect but that some small tumors were missed at postmortem. In this situation, a valid comparison can only be achieved by ignoring the low and mid dose groups when carrying out the comparison for the age stratum "terminal kill." This may seem wasteful of data, but actually represents the appropriate use of relevant data.

Use of Historical Control Data

In some situations, particularly where incidences are low, the results from a single study may suggest an effect of treatment on tumor incidence. Statistical analysis may fail to demonstrate a treatment effect. The possibility of comparing treated group results with those of historical control groups from other studies in the institution is then often raised. By doing this, a nonsignificant incidence of 2 out of 50 cases in a treated group may seem much more significant if no cases have been seen in, say, 1000 animals representing controls from 20 similar studies. Conversely, a significant incidence of 5 out of 50 cases in a treated group as compared with 0 out of 50 in the study controls may seem far less convincing if many other control groups had incidences around 5 out of 50.

While not understating the importance of looking at historical control data, it must be emphasized that there are a number of reasons why variation between studies may be greater than variation within study. Hence the utility of data from historical controls may be of lesser value than would appear on the surface. Differences in diet, duration of the study, intercurrent mortality, and the study pathologist may all contribute to variation between studies. Statistical techniques that ignore this variation and test treatment incidence against a pooled control incidence may give results that are seriously in error and are likely to overstate statistical significance.

Bioassay Interpretation

The interpretation of the results of even the best designed carcinogenesis bioassay is a complex statistical and biological problem. In addressing the statistical aspects, we shall have to review some biological points that have statistical implications as we proceed.

First, all such bioassays are evaluated by comparison of the observed results in treatment groups with those in one or more control groups. These control groups always include at least one group that is concurrent, but because of concern about variability in background tumor rates a historical control group is also considered in at least some manner.

The underlying problem in the use of concurrent controls alone is the belief that the selected populations of animals are subject both to an inordinate degree of variability in their spontaneous tumor incidence rates and that the strains maintained at separate breeding facilities are each subject to a slow but significant degree of genetic drift. The first problem raises concern that, by chance, the animals selected to be controls for any particular study will be either "too high" or "too low" in their tumor incidences, leading to either a false positive or false negative statistical test result when test animals are compared to these controls. The second problem leads to concern that, over the years, different laboratories will be using different standards (control groups) against which to compare the outcome of their tests, making any kind of relative comparison between compounds or laboratories impossible.

The last ten years have seen at least eight separate publications reporting five sets of background tumor incidences in test animals. These eight publications are summarized and compared in Table 15.7 and Table 15.8 for B6C3F1 mice and Fischer 344 rats, respectively.

It should be kept in mind in considering these separate columns of numbers that there are some overlaps in the populations being reported. For example, it is almost certain that some NCI/NTP study control groups were incorporated in several separate publications. At the same time, the related survival and growth data on control animals (broken out by type of treatment and vehicle) have also been published (Cameron et al., 1985) allowing for some assessment of comparability of control animal populations based on grounds other than just tumor incidences. It is interesting that in these NCI/ NTP bioassay program control populations, mean survival of B6C3F1 mice was greater than that of F344 rats (Seilkop, 1995).

Generally, historical control group data are used primarily as a check to ensure that the statistical evaluations used in comparing treatment groups to concurrent controls have a sound starting point (Chu et al., 1981).

Dempster et al. (1983) have, however, proposed a method for incorporating historical control data in the actual process of statistical analysis. A variable degree of pooling (combining) of historical with concurrent controls is performed based on the extent to which the historical data fit an assumed normal logistic (log transform) model.

Age (in either animals or humans) is clearly related to both "background" cancer incidence and chemically induced carcinogenesis. Indeed, one view of chemically induced carcinogenesis is that it serves in many (if not all) cases to accelerate the rate at which developing deficiencies in the body's defense system allow cancers to be expressed. As either a carcinogen becomes more potent or a larger dose is used, neoplasms successfully overcome or evade defense mechanisms and are expressed as tumors. In some cases, the

TABLE 15.7

Reported Background Tumor Incidences in B6C3F1 Mice

Organ/Tissue	Chu (1977) M	Chu (1977) F	Fears et al. (1977, p. 1977); Gart et al. (1979) M	Fears et al. (1977, p. 1977); Gart et al. (1979) F	Chu et al., (1981) M	Chu et al., (1981) F	Tarone et al. (1981) M (Ranges)	Tarone et al. (1981) F
Brain	0.1	0.1	<	0	<0.1	0.1		
Skin/subcutaneous	1.9	1.6	1	<1.0	<0.1	0.1		
Mammary gland	—	0.8	—	<1.0	—	1.3		
Circulatory system	2.4	1.7	<1.0	<1.0	2.9	2.4		
Lung/trachea	11.7	4.4	9.2	3.5	13.7	5.2	10.6–21.9	3.6–7.1
Heart	0.1	0.1	<1.0	0				
Liver	21.9	4	15.6	2.5	24.6	4.7	25.0–40.1	4.6–9.7
Pancreas	0.1	0.1	<1.0	<1.0	2.1	<0.1		
Stomach	0.3	0.3	1.1	<1.0	0.4	0.4		
Intestines	0.4	0.4	<1.0	<1.0	0.5	0.2		
Kidney	0.2	0.1	<1.0	<1.0	0.3	<0.1		
Urinary/bladder	0.1	0.1	0	1	<0.1	<0.1		
Preputial gland	—	—	—	—	—	—		
Testis	0.5	NA	<1.0	NA	0.4	NA		
Ovary	NA	0.7	NA	<1.0	NA	0.9		
Uterus	NA	1.2	NA	1.9	NA	1.6		
Pituitary	0.2	3.2	<1.0	3.5	0.3	3.6		
Adrenal	0.9	0.7	<1.0	<1.0	1.4	0.6		
Thyroid	1	1.3	1.1	<1.0	1	1.7		
Pancreatic islets	0.3	0.1	<1.0	<1.0	0.4	0.2		
Body cavities	0.1	0.3	<1.0	<1.0	0.4	0.3		
Leukemia/lymphoma	5.6	12.7	1.6	6.8	10.3	20.6	7.2–12.2	1.7–30.4
N	2355	2365	1132	1176	3543	3617	?	?

TABLE 15.8

Reported Background Tumor Incidences in Fischer 344 Rats

Organ/Tissue	Chu (1977) M	F	Fears et al. (1977, p. 1977); Gart et al. (1979) M	F	Goodman et al. (1979)[a] M	F	Chu et al. (1981) M	F	Tarone et al. (1981) (Ranges) M	F
Brain	0.9	0.6	1.3	<0	8.1	0.55	0.8	0.6	(Ranges)	
Skin/subcutaneous	6.6	3.2	5.7	2.5	6.4	3	7.8	3.2		
Mammary gland	1.4	17.9	0	18.8	1.54	8.5	1.5	20.9		
Circulatory system	0.4	0.5	<1.0	<1.0	3.8	0.27	0.7	0.4		
Lung/trachea	3.1	1.8	2.4	<1.0	2.9	2.0	3	1.9		
Heart	0.3	0.1	<1.0	<1.0	0.2	0.2	0.05			
Liver	1.8	3.1	1.2	1.3	1.74	3.9	2.2	1.9	0.7–3.4	0.5–2.9
Pancreas	0.2	—	<1.0	<0	0.16	0	0.2			
Stomach	0.3	0.2	<1.0	<1.0	0.32	0.2	0.3	0.2		
Intestines	0.3	0.5	<1.0	<1.0	0.31	0.36	0.6	0.3		
Kidney	0.4	0.2	<1.0	<1.0	0.38	0.16	0.5	0.2		
Urinary/bladder	0.1	0.2	<1.0	<1.0	0.1	0.22	0.1	0.3		
Preputial gland	1.4	1.2	—	—	1.4	1.2	2.4	1.8		
Testis	80.6	NA	76.2	NA	80.1	NA	2.3	NA		
Ovary	NA	0.3	NA	<1.0	NA	0.33	NA	0.4		
Uterus	NA	15.6	NA	16.8	NA	5.55	NA	17		
Pituitary	11.5	30.5	10.2	29.5	11.4	0.3	4.7	34.9	7.5–31.2	31.0–58.6
Adrenal	10.0	4.6	8.7	4.0	9.95	4.58	2.4	5.2		
Thyroid	7.1	6.5	5.1	5.6	7.16	6.65	8.2	6.8	3.6	4.7
Pancreatic islets	0.8	1.0	3.2	1.3	3.89	1.05	3.9	0.8		
Body cavities	1.1	0.3	<1.0	<1.0	2.51	0.38	2.6	0.4	2.8–9.0	1.0–1.9
Leukemia/lymphoma	11.7	9.1	6.5	5.4	12.3	9.9	9.9	13.4	9.1–23.6	7.5–15.4
N	1806	1765	846	840	1794	1754	b	b		

[a] Gives detailed breakdown of neoplastic and nonneoplastic lesions in aged animals.

[b] Range of averages, six different laboratories.

effect of a test chemical clearly results in the earlier appearance of tumors in a test animal population than in nontreated members of the same population. Unless a study is designed and conducted so that a reasonably accurate measurement of time-to-tumor can be made, one is left with only the incidence of tumors found at the end of the study and the variable incidence in animals that died on study and cannot rule out the possibility that, although the terminal incidences were comparable, the test chemical resulted in an earlier development or expression of these same tumors. This is one of the strengths of the traditional skin painting studies, which allow easy detection of skin tumors as soon as they appear and tracking of their progress.

If the target organ is not the skin, the only reasonably sensitive manner of evaluating time-to-tumor (unless the tumors are rapidly life threatening and there is an accordingly high early mortality rate leading to necropsy of spontaneous deaths in test animals) is to periodically, during the study, terminate, necropsy, and histopathologically evaluate random samples of test and control animals. The traditional NCI bioassay had no such interim or serial sacrifices (Chu et al., 1981), and therefore could not address such issues.

Such serial sacrifices are usually conducted on at least 20 animals per sex per group starting at 1 year into the study. Several statistical methods other than life table procedures are available for analysis of such data (Bratcher, 1977; Dinse, 1985).

A related issue is the age at which to terminate the animals. We have already stressed that as a study progresses, the rise in the background level of tumors makes it more and more difficult to clearly partition treatment-effect tumors from age-effect tumors. Swenberg (Swenberg, 1985; Solleveld et al., 1984) has made the point that the incidence of many tumor types has increased from 100 to 500% when control rat results from 2-year studies (rats 110 to 116 weeks of age) were compared to those from lifespan studies (140 to 146 weeks of age). If such an increase in age (25%) can result in such extreme increases in spontaneous tumors, what is the effect on interpretation of incidence rates seen in concurrent treatment groups? This is especially the case if, as Salsburg (1980) has suggested, any biologically active treatment will result in a shift in the patterns of neoplastic lesions occurring in aging animals. The current practice is to interpret tumor incidence on an independent site-by-site basis (on the assumption that what happens at each tissue site is independent of what happens elsewhere), and no allowance or factoring is made for the fact that what may be occurring in animals over their lifespan (as expressed by tumor incidence levels at an advanced age) is merely a shifting of patterns from one tumor site to another. In other words, commonly the "significantly" increased incidence of liver tumors is focused on, while the just as statistically significant decrease in kidney tumors compared to controls is ignored. Clearly, we should not be trying to analyze tumor data from animals that are advancing into senescence in the same manner that we do from those that lack these confounding factors. Where should a cut-off point be? This is a problem, but clearly Cameron's data (Cameron et al., 1985) suggest that the growth curves in NCI/NTP studies show consistent patterns of decline in body weights from these animals starting at the following ages (in weeks).

	Males	Females
B6C3F1 Mice	96	101
Fischer 344 Rats	91	106

The existence of similar data for tumor incidences (unfortunately not available from NCI/NTP studies) would certainly improve our confidence in selecting cutoff points for age, but the above ages merit consideration as termination points.

Having reviewed the preceding biological factors, we may now begin to directly address the statistical interpretation of carcinogenesis bioassays. Such interpretation, once believed to be a simple problem of calculating the statistical significance of increases of tumor incidences in treatment groups at each of a number of tissue sites, is now clearly a more complex task. Assuming dose level and route were appropriate, at least four separate questions must still be addressed in such an interpretation of incidence.

1. Are the data resulting from the bioassay sufficient to warrant analysis and interpretation? Factors that may invalidate a bioassay include inadequate survival in test or control groups, extreme (high or low) control group tumor incidence levels, excessive loss of tissues from autolysis, infection during the study, and the use of contaminated diet or water.

2. Are there increases in tumor incidences in test groups compared to those in control groups? If so, then we must proceed to an incidence comparison on some form of contingency table arrangement of the data. Such comparisons are traditionally performed using a series of Fisher's exact tests as presented in Chapter 6 (Bhapkar, 1958).

3. If there is a significant increase in tumor incidence, is there a trend (dose-response) in the data for these sites that concurs with what we know about biological responses to toxicants? That is, as dose increases, response should increase. A significant increase occurring only in a low dose group (with the incidence levels in the higher dose groups being comparable to controls), would be of very questionable biological significance.

4. If significant incidence and trend are present, is there supporting evidence of the material being a carcinogen? An example of this was cited earlier in the case of mouse liver tumors where the presence of positive mutagenicity findings would support a belief of biological significance and concern about real-life exposure to humans.

Two major controversial questions are involved in such comparisons: (a) should they be based on a one-tailed or a two-tailed distribution, and (b) what are the effects and implications of multiple comparisons? The one- or two-tailed controversy revolves around the question of which hypothesis we are properly testing in a study such as a chronic carcinogenicity study. We might be asking whether the tumor incidence differs between the control and treated groups. In such cases, it is a bidirectional hypothesis and, there-fore, we are testing against a two-tailed distribution. Or we might be asking whether the tumor incidence is greater in the treated group than in the control group. In the latter case, it is a unidirectional hypothesis and we are contemplating only the right-hand tail of the distribution. The implications of the question we ask are more of a theoretical interest; significance is much greater (exactly double, in fact, for Fisher's exact test) in the one-tailed case than in the two-tailed. For example, a set of data analyzed by Fisher's exact test that would have a two-tailed p-level of 0.098 and a one-tailed level of 0.049, would be flagged, therefore, as significantly different if the one-tailed test were employed. Feinstein (1979) provides an excellent nonmathematical discussion. Determination of the correct approach must rest on a clear def-inition by the researcher, beforehand, of the objective of his study and of the possible outcomes (if a bidirectional outcome is possible, are we justified in using a one-tailed test statistic?).

The multiple comparisons problem was presented earlier in this chapter. Another statistical method proposed by McKnight and Crowley (1984) for handling multiple comparisons provides a reasonably sensitive yet unbiased means of evaluating such data, if information from frequent interim termi-nations is present. Similarly, Meng and Dempster (1985) have proposed a Bayesian approach to such an analysis to solve the multiple comparisons prob-lem. In this, a logistically distributed (or log-transformed) model that accom-modates the incidences of all tumor types or sites observed in the current experiment, as well as their historical control incidences, is developed. Exchangeable normal expected values are assumed for certain linear terms in the model. Posterior means, standard deviations, and Bayesian p-values are computed for an overall treatment effect as well as for the effects on individual tumor types or tissue sites. Model assumptions are then evaluated using probability plots and the sensitivity of the parameter estimates to alternative expected values is analyzed.

The third and fourth questions presented earlier are parts of what is evolv-ing as a second set of approaches to the interpretation of bioassay results.

These new approaches use the information in a more mature decision-making process. First, the historical control incidence rates such as those given for the B6C3F1 mouse and the Fischer 344 rats in Table 15.3 and Table 15.4 should be considered; as we have seen, some background incidences are so high that these tissues are "null and void" for making decisions. Second, we should look not just for a single significant incidence in a tissue, but rather for a trend. For example, we might have the following percentages of a liver tumor incidence in the female rats of a study: (a) control — 3%, (b) 10 mg/kg — 6%, (c) 50 mg/kg — 17%, and (d) 250 mg/kg — 54%. In this study only the incidence at the 250 mg/kg level might be statistically significant. However, the trend through each of the levels is suggestive of a dose-response. Looking for such a trend is an essential step in a scientific assessment of the results, and one of the available trend analysis techniques, such as presented in Chapter 8, should be utilized. Another method for determining whether statistically significant incidences are merely random occurrences is to compare the results of the quantitative variables to two or more concurrently run control groups. Often the mean of one variable will differ from only one of these controls and be numerically within the range of this same variable of the two control means. If so, the statistical significance compared to the one control must be seriously questioned as to its being associated with a biological significance. Three different such stepwise interpretive procedures are common. These are the NCI method, the weight-of-evidence method, and the Peto method. The NCI approach is somewhat complex, involving each of the four steps outlined earlier in a process overviewed by Chu et al. (1981). The statistical aspects of this are outlined below.

NCI Bioassay Method

Survival analysis — by sex, species and organ. Exclude all animals dying prior to first incidence of tumor at that site. Do a life table analysis for survival at the same time.

Use Fisher's exact test to obtain one-tailed p at each site using the survival adjusted ratios obtained in 1 above.

Utilize the Bonferroni correction using r (where r = the number of dose levels, not k = the number of total comparisons). Multiply the computed p by r to maintain overall error rate. Significance is claimed only if p is less than a/r.

Perform tests for linear trend using Cochran-Armitage test (dose-response curve must be significantly different from zero, and positive).

(Note: in a 100 animal bioassay, you need 5 or more animals to have tumors to achieve a one-tailed $p \geq 0.05$. With the Bonferroni correction, 7 or more are needed.

NCI believes and practices the rare tumor incidence flag mechanism (Bonferroni-Miller, 1966; Tarone, 1975, 1990; Armitage, 1955.)

The nine possible interpretations of an analysis of tumor incidence and survival analysis, such as that presented in Chapter 9 (Cox, 1972; Byar, 1977; and Hammond et al., 1978) are summarized in Table 15.9.

Weight-of-Evidence

The weight-of-evidence approach consists primarily of the four steps of interpretation presented earlier, with emphasis on the last step (integration of related and supporting information into the evaluation process), as opposed to the NCI approach (which places emphasis on the two "statistical" steps).

TABLE 15.9

Interpretation of the Analysis of Tumor Incidence and Survival Analysis (Life Table)

Outcome Type	Tumor Association with Treatment	Mortality Association with Treatment	Interpretation
A	+	+	Unadjusted test may underestimate tumorigenicity of treatment.[a]
B	+	0	Unadjusted test gives valid picture of tumorigenicity of treatment.
C	+	–	Tumors found in treated groups may reflect longer survival of treated groups. Time-adjusted analysis is indicated.
D	–	+	Apparent negative findings in tumors may be due to the shorter survival in treated groups. Time-adjusted analysis and/or a retest at lower doses is indicated.
E	–	0	Unadjusted test gives a valid picture of the possible tumor-preventive capacity of the treatment.
F	–	–	Unadjusted test may underestimate the possible tumor-preventive capacity of the treatment.
G	0	+	High mortality in treated groups may lead to unadjusted test missing a possible tumorigen. Adjusted analysis and/or retest at lower doses is indicated.
H	0	0	Unadjusted test gives valid picture of lack of association with treatment.
I	0	–	Longer survival in treated groups may mask tumor-preventive capacity of treatment.

+ = yes, – = no, and 0 = no bearing on discussion.

[a] The unadjusted test referred to here is a contingency table type of analysis of incidence, such as a Fisher's exact test.

The weight-of-evidence approach poses difficulty in the regulatory and legal fields because it requires judgment and is not overly quantitative. However, it does represent a scientifically valid approach for distinguishing important differences in the potential of chemicals to induce cancer. The greatest weight of evidence should be given to chemicals that induce dose-related increases in malignant tumors at multiple sites, in both sexes, and in multiple species using appropriate routes of administration. At the other end of the spectrum, much less weight should be given to chemicals that induce only an increased incidence of a benign neoplasm whose incidence is normally quite variable in only the high dose group of one sex of a single species. One must also integrate a significant amount of additional information.

For example, the shape and extent of the dose-response curve should be known in relation to factors such as the chemical's pharmacokinetics, its overwhelming of host defenses, or saturation of metabolic systems. Is the chemical genotoxic? How does the site and dose-response for toxicity compare with those for carcinogenicity? This knowledge is highly relevant when attempting to understand the mechanisms involved in carcinogenesis for each specific chemical and can and should be incorporated into both hazard identification and risk assessment to improve their accuracy (Goddard, 1997). It is widely believed to be appropriate to test a chemical at the MTD in order to gain assurance that it has been adequately tested. However, if a chemical is not genotoxic, but induces frank cytotoxicity in the liver only at doses at which it also induces liver tumors, it should be considered differently than a chemical that is genotoxic and induces liver tumors over a large dose range, including noncytotoxic doses.

Peto Procedure

The Peto procedure is actually a collection of approaches arising from the central belief that it is possible to generate an additional vital set of data from a well-run bioassay and that we should utilize these same pieces of data in interpreting results.

The data in question constitute an evaluation of the likelihood that each individual tumor would (or would not) be life threatening. The approach calls for the pathologist on a study to not only identify a mass or tumor as neoplastic or not, but also to categorize each neoplasm in one of several possible classes as to the risk it presents to the survival of the host organism. Such classification is generally in one of at least five different categories:

Tumor did or would definitely cause death of animal.

Tumor probably did or could cause death of animal.

Cannot be determined.

Tumor probably didn't or wouldn't cause death of animal.

Tumor didn't or wouldn't cause death of animal.

Such data can then be employed in a more precise interpretation of the meaning of the bioassay. An entire separate, sensitive set of significance tests based on such data has been proposed by Peto et al. (1980).

The last point to be addressed under the topic of the carcinogenicity bioassay is the use of the resulting data for the conduct of carcinogenic potency comparisons. Such a potency comparison would be both valuable in a scientific sense and provide a basis for prioritization of regulatory actions.

Potency and dose-response of carcinogens for any single species of animals may be expressed in one of two manners — either as the incidence rate of tumors at the end of a set period of time or as the time lag from treatment to a specified incidence rate of tumors. This second manner has also been extended to determining time to death as a result of tumors produced by a carcinogen (Lijinsky et al. 1981).

Squire (1981) has proposed a ranking system for animal carcinogens based on data from NTP bioassays (that is, in the absence of time-to-tumor information). The major considerations are:

Number of species affected

Number of different types of neoplasms induced in one or more species

A negative correction for the spontaneous incidence in control groups of induced neoplasms

Cumulative dose or exposure per kilogram body weight in affected groups

The proportion of induced neoplasms which were malignant

The degree of support for genotoxicity (mutagenicity) data

Of course, our real interest in the potency of carcinogens is in humans, which means an interspecies comparison. Crouch and Wilson (1979), using the results of some 70 NCI/NTP bioassays where carcinogenicity was established in both rats and mice, reported that a comparison demonstrated empirically that good correlations exist between these two species for suitably defined carcinogenic potencies for various chemicals. Such a correlation would allow sufficient accuracy in extrapolating from animal data to human risk to support a logical scheme for the evaluation of such risks. In 1985, Bernstein et al. examined a larger NCI/NTP bioassay program database. They observed that there is a very high correlation between the maximum doses tested (max d) for rats and mice on a milligram per kilogram body weight per day basis. Calculating the carcinogenic potency (b, defined in their paper), they found it to be restricted to an approximately 30-fold range surrounding $\log(2)/\max d$, which has a biological as well as statistical basis. Since the max ds for the set of NCI/NTP test chemicals varied over many orders of magnitude, it necessarily follows statistically that the carcinogenic potencies will be highly correlated. This "artifact" of potency estimation does not imply that there is no basis for extrapolating animal results to man. They concluded that "it does suggest, however, that the interpretation of correlation studies of carcinogenic potency needs much further thought."

On an intermediate level, DuMouchel and Harris (1983) have suggested a class of Bayesian statistical methods for the interspecies extrapolation of potency functions that allows for the combining of data from different substances and species of animals, using the results as one constructs the model to estimate inter-experimental error between the different sources of data being combined.

Statistical Analysis

The actual statistical techniques used to evaluate the results of carcinogenicity bioassays basically utilize four sets of techniques, three of which have been presented earlier in this book. These methods are exact tests, trend tests, life tables (such as log rank techniques), and Peto analysis.

These are then integrated into the decision-making schemes discussed earlier in this chapter. FDA (2001) has provided a draft guidance on the analysis of such bioassays.

Exact Tests

The basic forms of these (the Fisher exact test and chi-square) have previously been presented, and the reader should review these. Carcinogenicity assays are, of course, conducted at doses that are at least near those that will compromise mortality. As a consequence, one generally encounters competing toxicity producing differential mortality during such a study. Also, often, particularly with certain agricultural chemicals, latency of spontaneous tumors in rodents may shorten as a confounded effect of treatment with toxicity. Because of such happenings, simple tests on proportions, such as chi-squared and Fisher–Irwin exact tests on contingency tables, may not produce optimal evaluation of the incidence data (Farrar and Crump, 1988). In many cases, however, statisticians still use some of these tests as methods of preliminary evaluation (Farrar and Crump, 1990). These are unadjusted methods without regard for the mortality patterns in a study. Failure to take into account mortality patterns in a study sometimes causes serious flaws in interpretation of the results. The numbers at risk are generally the numbers of animals histopathologically examined for specific tissues.

Some gross adjustments on the numbers at risk can be made by eliminating early deaths or sacrifices by justifying that those animals were not at risk to have developed the particular tumor in question. Unless there is dramatic change in tumor prevalence distribution over time, the gross adjusted method provides fairly reliable evidence of treatment effect, at least for nonpalpable tissue masses.

Trend Tests

Basic forms of the trend tests (such as that of Tarone) have previously been presented in this text. More recent forms, such as the Bailer–Portier poly-3 and poly-6 (Bailer and Portier, 1988) are likewise modifications of the Cochran-Armitage.

Group comparison tests for proportions notoriously lack power. Trend tests, because of their use of prior information (dose levels) are much more powerful. Also, it is generally believed that the nature of true carcinogenicity (or toxicity for that matter) manifests itself as dose-response. Because of the above facts, evaluation of trend takes precedence over group comparisons. In order to achieve optimal test statistics, many people use ordinal dose levels (0, 1, 2,..., etc.) instead of the true arithmetic dose levels to test for trend. However, such a decision should be made *a priori.* The following example demonstrates the weakness of homogeneity tests.

Example 15.1 — Trend vs. Heterogeneity

Number at Risk	Number with Tumor	Dose Level
50	2	0
50	4	1
50	6	2
50	7	3

Cochran-Armitage Test for Trend

	Calculated Chi-Square Subgroup	df	Alpha	Two-Tailed p
Trend	3.3446	1	0.05	0.0674
Departure	0.0694	2	0.05	0.9659
Homogeneity	3.4141	3	0.05	0.3321

One-Tailed Tests for Trend

Type	Probability
Uncorrected	0.0337[a]
Continuity corrected	0.0426[a]
Exact	0.0423[a]

Multiple Pairwise Group Comparisons by Fisher–Irwin Exact Test

Groups Compared	Alpha	One-Tail Prob.
1 vs. 2	0.05	0.33887
2 vs. 3	0.05	0.13433
1 vs. 4	0.05	0.07975

[a] Direction = +.

As is evident from this example, often group comparison tests will fail to identify significant treatment but trend tests will. The same arguments apply to survival-adjusted tests on proportions as well. In an experiment with more than one dose group ($K > 1$), the most convincing evidence for carcinogenicity is given by tumor incidence rates that increase with increasing dose. A test designed specifically to detect such dose-related trends is Tarone's (1975) trend test.

Letting $d = (O, d_1, d_2, ..., d_k)^T$ be the vector of dose levels in all $K + 1$ groups and letting

$$(O - E) = (O_1 - E_1, O_2 - E_2, \quad , O_k - E_k)^T \quad \text{and} \quad V = \begin{pmatrix} V_{11} & & V_{ik} \\ \cdot & \cdot & \cdot \\ V_{k1} & & V_{kk} \end{pmatrix}$$

contain elements as described in the previous section but for all $K + 1$ groups, the trend statistic is given by

$$X\frac{2}{T} = \frac{[d^T(O - E)]^2}{d^T V d}$$

The statistic will be large when there is evidence of a dose-related increase or decrease in the tumor incidence rates, and small when there is little difference in the tumor incidence between groups or when group differences are not dose related. Under the null hypothesis of no differences between groups, the statistic has approximately a chi-squared distribution with one degree of freedom.

Tarone's trend test is most powerful at detecting dose-related trends when tumor onset hazard functions are proportional to each other. For more power against other dose-related group differences, weighted versions of the statistic are also available; see Breslow (1984) or Crowley and Breslow (1984) for details.

These tests are based on the generalized logistic function (Cox, 1972). Specifically one can use Cochran-Armitage test (or its parallel, Mantel-Haenszel version) for monotonic trend as the heterogeneity test.

Life Table and Survival Analysis

These methods are essential when there is any significant degree of mortality in a bioassay. They seek to adjust for the differences in periods of risk individual animals undergo. Life table techniques can be used for those data where there are observable or palpable tumors. Specifically, one should use Kaplan-Meier product limit estimates from censored data graphically, Cox–Tarone binary regression (log-rank test), and Gehan–Breslow modification of Kruskal–Wallis tests (Thomas et al., 1977) on censored data.

The Kaplan-Meier estimates produce a step function for each group and are plotted over the lifetime of the animals. Planned, accidentally killed, and lost animals are censored. Moribund deaths are considered to be treatment related. A graphical representation of Kaplan-Meier estimates provide excellent interpretation of survival adjusted data except in the cases where the curves cross between two or more groups. When the curves cross and change direction, no meaningful interpretation of the data can be made by any statistical method because the proportional odds characteristic is totally lost over time. This would be a rare case where treatment initially produces more tumor or death and then, due to repair or other mechanisms, becomes beneficial.

Cox–Tarone binary regression (Tarone, 1975; Thomas et al., 1977): censored survival and tumor incidence data are expressed in a logistic model in dose over time. The log-rank test (Peto, 1974), based on the Weibull distribution, and the Mantel-Haenszel (Mantel and Haenszel, 1952) test are very similar to this test when there are no covariates or stratifying variables in the design. The logistic regression–based Cox–Tarone test is preferable because one can easily incorporate covariates and stratifying variables that one cannot in the IARC methods.

The Gehan–Breslow modification of Kruskal–Wallis test is a nonparametric test on censored observations. It assigns more weight to early incidences compared to the Cox–Tarone test.

Survival-adjusted tests on proportions: as mentioned earlier, in the case of survival-adjusted analyses, instead of having a single $2 \times k$ table, one has a series of such $2 \times k$ tables across the entire lifetime of the study. The numbers at risk for such analyses will depend on the type of tumor one is dealing with. These are shown below:

Palpable or lethal tumors: number at risk at time t = number of animals surviving at the end of time $t - 1$.

Incidental tumors: the number at risk at time t = number of animals that either died or were sacrificed whose particular tissue was examined histopathologically.

The methods of analyzing the incidences, once the appropriate numbers at risk are assigned for these tumors, are rather similar, either binary regression-based or by pooling evidence from individual tables (Gart et al., 1986).

Peto Analysis

The Peto method of analysis of bioassay tumor data is based on careful classification of tumors into five different categories, as defined by IARC.

1. Definitely incidental
2. Probably incidental

Comment: Combine 1 and 2.

3. Probably lethal
4. Definitely lethal

Comment: These categories may be combined into one (otherwise it requires a careful cause of death determination) (Bolden buck et al., 1977).

5. Mortality independent (such as mammary, skin, and other observable or superficial tumors)

Based on recent studies using real historical control data of CD mice and CD rats from Charles River Laboratory and simulation studies conducted by the FDA and in collaboration with NTP, a new statistical decision rule for tests for a positive trend in tumor incidence has been developed (Haseman, 1992, 1995). The new decision rule tests the positive trend in incidence rates in rare and common tumors at 0.025 and 0.005 levels of significance, respectively. The new decision rule achieves an overall false positive rate of around 10% in a standard two-species and two-sex study (Lin, 1995, 1997; Lin and Rahman, 1998a, b). The 10% overall false-positive rate is seen by CDER statisticians as appropriate in a new drug regulatory setting.

Regulatory statistical literature emphasizes methods of testing for positive trends in tumor rate (Lin, 1998, 2000; Lin and Ali, 1994; Chen and Gaylor, 1986; Dinse and Haseman, 1986; Dinse and Lagokos, 1983; Haseman et al., 1986). There are situations, however, in which pair-wise comparisons between control and individual-treated groups may be more appropriate than trend tests because trend tests assume that a carcinogenic effect is related to doses or systemic exposure weights, or ranks. The assumption may be true for simple direct-acting carcinogens in studies not complicated by excessive toxicity. However, there are many cases in which the response is to a drug metabolite, is mediated through a receptor (or enzyme) that may be saturated at low doses, is compounded by dose-related toxicity, or is complicated by other nonlinear effects. Under those situations, pair-wise comparisons may be appropriate and the decision rule described in Haseman et al. (1984) should be used in interpreting the results of the pair-wise tests. Sponsors should conduct both trend tests and pair-wise comparison tests and present the results of both types of tests.

Logistic Regression Method for Occult (Internal Organ) Tumors (Dinse, 1985)

Tumor prevalence is modeled as logistic function of dose and polynomial in age.

Comment: logistic tumor prevalence method is unbiased. Requires maximum likelihood estimation. Allows for covariates and stratifying variables. May be time consuming. May have convergence problem with sparse tables and clustering of tumors.

Methods to be Avoided

The following methods and practices should be avoided in evaluation of carcinogenicity:

Use of only the animals surviving after 1 year in the study.

Use of a two-strata approach: separate analyses for animals killed during the first year of the study and the ones thereafter.

Exclusion of all animals in the study that died during test and analyze only the animals that are sacrificed at the end of the study.

Exclusion of interim sacrifice animals from statistical analyses.

Evaluation of number of tumors of all sites as opposed to the number of animals with tumors for specific sites of specific organs.

Another issue is subjectivity in slide reading by most pathologists who do not want to read them in a coded fashion whereby they will not know the dose group an animal is coming from. This is not under statisticians' control but they should be aware of that in any case.

Often a chemical being tested is both toxic as well as potentially carcinogenic. When competing toxicity causes extreme differences in mortality or there is clustering effect in tumor prevalence in a very short interval of time, none of the adjusted methods works. One then must use biological intuition to evaluate the tumor data.

Use of historical control incidence data for statistical evaluation is controversial. There are too many sources of variation in these data. For example, different pathologists use different criteria for categorizing tumors (in fact, the same pathologist may change opinion over time); there is laboratory-to-laboratory variation; there may be genetic drift over time; location of suppliers may make a difference; and finally, these data are not part of the randomized concurrent control. Regulatory agencies and pathologists generally use these data for qualitative evaluation. My personal view is that is where they belong.

References

Armitage, P. (1955) Tests for linear trends in proportions and frequencies. *Biometrics*, 11, 375–386.

Armitage, P. and Doll, R. (1961) *Stochastic Models for Carcinogenesis from the Berkeley Symposium on Mathematical Statistics and Probability*. University of California Press, Berkeley, CA, pp. 19–38.

Bailer, A.J. and Portier, C.J. (1988) Effects of treatment-induced mortality and tumor-induced mortality on tests for carcinogenicity in small samples. *Biometrics*, 44(2), 417–431.

Bernstein, L., Gold, L.S., Ames, B.N., Pike, M.C., and Hoel, D.G. (1985) Some tautologous aspects of the comparison of carcinogenic potency in rats and mice. *Fund. App. Toxicol.*, 5, 79–86.

Bhapkar, V.P. (1958) On the analysis of contingency tables with a quantitative response. *Biometrics*, 24, 329–338, 1968.

Boldenbuck, D.H., Neuhaus, G., and Heimann, G. (1977) Analyzing carcinogenicity assays without cause of death information. *Drug Inform. J.*, 31, 489–507.

Bonferroni-Miller, R.G. (1966) *Simultaneous Statistical Interference*. McGraw Hill, New York, pp. 5–10.

Bratcher, T.L. (1977) Bayesian analysis of a dose–response experiment with serial sacrifices. *J. Env. Path. Toxicol.*, 1, 287–292.

Breslow, N. (1984) Comparison of survival curves, in *Cancer Clinical Trials: Methods and Practice*, Buyse, M.E., Staquet, M.J., and Sylvester, R.J., Eds., Oxford University Press, Oxford, pp. 381–406.

Byar, D.D. (1977) *Analysis of Survival Data in Heterogeneous Populations in Recent Developments in Statistics*, Barra, J.R., Brodeau, F., Romier, G., and Van Cutsem, B., Eds., North Holland, New York, pp. 51–68.

Cameron, T.P., Hickman, R.L., Korneich, M.R., and Tarone, R.E. (1985) History survival and growth patterns of B6C3F1 mice and F344 rats in the National Cancer Institute Carcinogenesis Testing Program. *Fund. App. Toxicol.*, 5, 526–538.

Chen, J.J. and Gaylor, D.W. (1986) The upper percentiles of the distribution of the logrank statistics for small numbers of tumors, *Commun. Stats.-Simulat. Computat.*, 15, 991–1002.

Chu, K. (1977) *Percent Spontaneous Primary Tumors in Untreated Species Used at NCI for Carcinogen Bioassays*; NCI Clearing House, Rockville, MD.

Chu, K.C., Cueto, C., and Ward, J.M. (1981) Factors in the evaluation of 200 National Cancer Institute Carcinogen Bioassays. *J. Toxicol. Environ. Health*, 8, 251–280.

Ciminera, J.L. (1985) Some Issues in the Design, Evaluation and Interpretation of Tumorigenicity Studies in Animals; Presented at the Symposium on Long-Term Animal Carcinogenicity Studies: A Statistical Perspective, March 4–6, 1985, Bethesda, MD.

Cox, D.R. (1972) Regression models and life-tables. *J. Roy. Stat. Soc.*, 34B, 187–220.

Crouch, E. and Wilson, R. (1979) Interspecies comparison of carcinogenic potency. *J. Tox. Environ. Health*, 5, 1095–1118.

Crowley, J. and Breslow, N. (1984) Statistical analysis of survival data. *Annu. Rev. Pub. Health*, 5, 385–411.

Dempster, A.P., Selwyn, M.R., and Weeks, B.J. (1983) Combining historical and randomized controls for assessing trends in proportions. *J. Am. Stat. Assoc.*, 78, 221–227.

Dinse, G.E. (1985) Estimating Tumor Prevalence, Lethality and Mortality, Presented at the Symposium on Long-Term Animal Carcinogenicity Studies: A Statistical Perspective, March 4–6, 1985, Bethesda, MD.

Dinse, G.E. (1994) A comparison of tumor incidence analyses applicable in single-sacrifice animal experiments. *Stat. Med.*, 13, 689–708.

Dinse, G.E. and Haseman, J.K. (1986) Logistic regression analysis of incidental-tumor data from animal carcinogenicity experiments. *Fund. Appl. Toxicol.*, 6, 751–770.

Dinse, G.E. and Lagokos, S.W. (1983) Regression analysis of tumor prevalence data, *J. Roy. Statis. Soc., Series C*, 32, 236–248.

Dix, D. and Cohen, P. (1980) On the role of aging in cancer incidence. *J. Theor. Biol.*, 83, 163–173.

DuMouchel, W.H. and Harris, J.E. (1983) Bayes methods for combining the results of cancer studies in humans and other species. *J. Am. Stat. Assoc.*, 78, 293–315.

Farrar, D.B. and Crump, K.S. (1988) Exact statistical tests for any carcinogenic effect in animal bioassays. *Fund. Appl. Toxicol.*, 11, 652–663.

Farrar, D.B. and Crump, K.S. (1990) Exact statistical tests for any carcinogenic effect in animal bioassays. II. Age-adjusted tests. *Fund. Appl. Toxicol.*, 15, 710–721.

FDA (2001) Statistical Aspects of the Design, Analysis, and Interpretation of Chronic Rodent Carcinogenicity Studies of Pharmaceuticals. May 2001.

Fears, T.R. and Tarone, R.E. (1977) Response to "Use of statistics when examining lifetime studies in rodents to detect carcinogenicity," *J. Tox. Environ. Health*, 3, 629–632.

Fears, T.R., Tarone, R.E., and Chu, K.C. (1977) False-positive and false-negative rates for carcinogenicity screens, *Canc. Res.*, 27, 1941–1945.

Feinstein, A.R. (1979) Clinical biostatistics XXII: biologic dependency, hypothesis testing, unilateral probabilities, and other issues in scientific direction vs. statistical duplexity. *Clin. Pharmacol. Therapeut.*, 17, 499–513.

Gart, J.J., Chu, K.C., and Tarone, R.E. (1979) Statistical issues in interpretation of chronic bioassay tests for carcinogenicity. *J. Nat. Canc. Inst.*, 62, 957–974.

Gart, J.J., Krewski, D., Lee, P.N. Tarone, R.E., and Wahrendorf, J. (1986) *The Design and Analysis of Long-Term Animal Experiments*, IARC Scientific Publications No. 79, Lyon, France.

Gehring, P.J. and Blau, G.E. (1977) Mechanisms of carcinogenicity: dose response. *J. Environ. Path. Toxicol.*, 1, 163–179.

Goddard, M.J. (1997) Evolving statistical issues in carcinogenicity studies. *Drug Inform. J.*, 31, 509–520.

Goodman, D.G., Ward, J.M., Squire, R.A., Chu, K.C., and Linhart, M.S. (1979) Neoplastic and nonneoplastic lesions in aging F344 rats. *Toxicol. Appl. Pharmacol.*, 48, 237–248.

Grasso, P. and Crampton, R.F. (1972) The value of the mouse in carcinogenicity testing. *Food Chem. Toxicol.*, 10, 418–426.

Hammond, E.C., Garfinkel, L., and Lew, E.A. (1978) Longevity, selective mortality, and competitive risks in relation to chemical carcinogenesis. *Environ. Res.*, 16, 153–173.

Haseman, J., Hajian, G., Crump, K., Selvyn, M., and Peace, K. (1990) Dual control groups in rodent carcinogenicity studies, in *Statistical Issues in Drug Research and Development*, Peace, K.E., Ed., Marcel Dekker, New York, pp. 351–361.

Haseman, J.K. (1985) False Positive Issues in Carcinogenicity Testing: An Examination of the FDA Color Studies. Presented at the Symposium on Long-Term Animal Carcinogenicity Studies: A Statistical Perspective, March 4–6, 1985, Bethesda, MD.

Haseman, J.K. (1985) Issues in carcinogenicity testing: dose selection. *Fund. App. Toxicol.*, 5, 66–78.

Haseman, J.K. (1992) Value of historical controls in the interpretation of rodent tumors data. *Drug Inform. J.*, 26, 191–200.

Haseman, J.K. (1995) Data analysis: statistical analysis and use of historical control data. *Reg. Toxicol. Pharmacol.*, 21, 52–59.

Haseman, J.K. and Hoel, D.G. (1979) Statistical design of toxicity assays: rodent genetic structure of test animal population. *J. Toxicol. Environ. Health*, 5, 89–101.

Haseman, J.K., Huff, J.E., and Boorman, G.A. (1984) Use of historical control data in carcinogenicity studies in rodents. *Toxicol. Pathol.*, 12, 126–135.

Haseman, J.K., Winbush, J.S., and O'Donnell, M.W. (1986) Use of dual control groups to estimate false positive rates in laboratory animal carcinogenicity studies. *Fund. Appl. Toxicol.*, 7, 573–584.

ICH. (1994) S1C Dose Selection for Carcinogenicity Studies of Pharmaceuticals. 27 October 1994.

ICH. (1995) S1A Guideline on the Need for Carcinogenicity Studies of Pharmaceuticals. 29 November 1995.

ICH. (1997a) S1B Testing for Carcinogenicity of Pharmaceuticals. 16 July 1997.

ICH. (1997b) S1C(R) Addendum to "Dose Selection for Carcinogenicity Studies of Pharmaceuticals" Addition of a Limit Dose and Related Notes. 17 July 1997.

Krewski, D., Smythe, T., Fung, K., and Burnett, R. (1991) Conditional and unconditional tests with historical controls. *Can. J. Stat.*, 19, 407–423.

Lagakos, S. and Mosteller, F. (1981) A case study of statistics in the regulatory process: the FD&C red no. 40 experiments. *J. Natl. Canc. Inst.*, 66, 197–212.

Lijinsky, W., Rueber, M.D., and Riggs, C.W. (1981) Dose response studies of carcinogenesis in rats by nitrosodiethylamine. *Canc. Res.*, 41, 4997–5003.

Lin, K.K. (1988) Peto prevalence method versus regression methods in analyzing incidental tumor data from animal carcinogenicity experiments: an emperical study, in *1988 American Statistical Association Annual Meeting Proceedings* (biopharmaceutical section), New Orleans, Louisiana.

Lin, K.K. (1995) A regulatory perspective on statistical methods for analyzing new drug carcinogenicity study data. *Bio/Pharm. Quart.*, 1(2), 18–20.

Lin, K.K. (1997) Control of Overall False Positive Rates in Animal Carcinogenicity Studies of Pharmaceuticals. Presented at 1997 FDA Forum on Regulatory Sciences, December 8–9, 1997, Bethesda, MD.

Lin, K.K. (2000) Carcinogenicity studies of pharmaceuticals, in *Encyclopedia of Biopharmaceutical Studies*, Chow, S.C., Ed., Marcel Dekker, New York.

Lin, K.K. and Ali, M.W. (1994) Statistical review and evaluation of animal tumorgenicity studies, in *Statistics in the Pharmaceutical Industry*, 2nd ed., Buncher, C.R. and Tsay, J.Y., Eds., Marcel Dekker, New York.

Lin, K.K. and Rahman, M.A. (1998a) Overall false positive rates in tests for linear trend in tumor incidence in animal carcinogenicity studies on new drugs, *J. Pharm. Stats.*, 8(1), 1–22.

Lin, K.K. and Rahman, M.A. (1998b) False positive rates in tests for trend and differences in tumor incidence in animal carcinogenicity studies of pharmaceuticals under ICH Guidance S1B unpubl. report, Division of Biometrics 2, Center for Drug Evalution and Research, Food and Drug Administration.

Louis, T.A. and Orva, J. (1985) Adaptive Sacrifice Plans for the Carcinogen Bioassay, Presented at The Symposium on Long-Term Animal Carcinogenicity Studies: A Statistical Perspective, March 4–6, 1985, Bethesda, MD.

Mantel, N. and Haenszel, W. (1952) Statistical aspects of the analysis of data from the retrospective studies of disease. *J. Nat. Canc. Inst.*, 22, 719–748.

McConnell, E.F., Solleveld, H.A., Swenberg, J.A., and Boorman, G.A. (1986) Guidelines for combining neoplasms for evaluation of rodent carcinogenesis studies. *J. Nat. Canc. Inst.*, 76, 283–289.

McKnight, B. and Crowley, J. (1984) Tests for differences in tumor incidence based on animal carcinogenesis experiments. *J. Am. Stat. Assoc.*, 79, 639–648.

Meng, C. and Dempster, A.P. (1985) A Bayesian Approach to the Multiplicity Problem for Significance Testing with Binomial Data, Presented at The Symposium on Long-Term Animal Carcinogenicity Studies: A Statistical Perspective, March 4–6, 1985, Bethesda, MD.

Neal, R.A. and Gibson, J.E. (1984) The uses of toxicology and epidemiology in identifying and assessing carcinogenic risks, in *Reducing the Carcinogenic Risks in Industry*, Deisler, P.F., Ed., Marcel Dekker, New York, pp. 39–65.

Nutrition Foundation. (1983) *The Relevance of Mouse Liver Hepatoma to Human Carcinogenic Risk*. Nutrition Foundation, Washington, DC, 1983.

Page, N.P. (1977) Concepts of a bioassay program in environmental carcinogenesis, in *Environmental Cancer*, Kraybill, H.F. and Mehlman, M.A., Eds., Hemisphere, New York, pp. 87–171.

Peto, R. (1974) Guidelines on the analysis of tumor rates and death rates in experimental animals. *Br. J. Canc.*, 29, 101–105.

Peto, R., Pike, M., Day, N., Gray, R., Lee, P., Parish, S., Peto, J., Richards, S., and Wahrendorf, J. (1980) Guidelines for simple, sensitive significance tests for carcinogenic effects in long-term animal experiments, in *IARC Monographs on the Evaluation of the Carcinogenic Risk of Chemicals to Humans, Supplement 2, Long-Term and Short-Term Screening Assays for Carcinogens: A Critical Appraisal, Lyon,* International Agency for Research in Cancer, Lyon, France pp. 311–346.

Portier, C.J. and Hoel, D.H. (1984) Design of animal carcinogenicity studies for goodness-of-fit of multistage models. *Fund. Appl. Toxicol.,* 4, 949–959.

Prentice, R.L., Smythe, R.T., Krewski, D., and Mason, M. (1992) On the use of historical control data to estimate dose response trends in quantal bioassay. *Biometrics,* 48, 459–478.

Robens, J.F., Calabrese, E.J., Piegorsch, W.D., Joiner, J.J., Schueler, R.L., and Hayes, A.W. (1994) Principles of testing for carcinogenesis, in *Principles and Methods of Toxicology,* Hayes, A.W., Ed., Raven Press, New York, pp. 697–738.

Salsburg, D. (1980) The effects of lifetime feeding studies on patterns of senile lesions in mice and rats. *Drug Chem. Toxicol.,* 3, 1–33.

Salsburg, D.S. (1977) Use of statistics when examining lifetime studies in rodents to detect carcinogenicity. *J. Toxicol. Environ. Health,* 3, 611–628.

Seilkop, S.K. (1995) The effect of body weight on tumor incidence and carcinogenicity testing in B6C3F1 mice and F344 rats. *Fund. Appl. Toxicol.,* 24, 247–259.

Smythe, R.T., Krewski, D., and Murdoch, D. (1986) The use of historical control information in modeling dose–response relationships in carcinogenesis. *Stat. Probab. Lett.,* 4, 87–93.

Solleveld, H.A., Haseman, J.K., and McConnel, E.E. (1984) Natural history of body weight gain, survival, and neoplasia in the F344 rat. *J. Nat. Canc. Inst.,* 72, 929–940.

Squire, R.A. (1981) Ranking animal carcinogens: a proposed regulatory approach. *Science,* 214, 877–880.

Swenberg, J.A. (1985): The interpretation and use of data from long-term carcinogenesis studies in animals. *CIIT Activities* 5(6), 1–6.

Tarone, R.E. (1975) Tests for trend in life table analysis. *Biometrika,* 62, 679–682.

Tarone, R.E. (1990) A modified Bonferroni method for discrete data. *Biometrics,* 46, 515–522.

Tarone, R.E., Chu, K.C., and Ward, J.M. (1981) Variability in the rates of some common naturally occurring tumors in Fischer 344 rats and (C57BL/6NXC3H/HEN)F1 (B6C3F1) mice. *J. Nat. Canc. Inst.,* 66, 1175–1181.

Thomas, D.G., Breslow, N., and Gart, J.J. (1977) Trend and homogeneity analyses of proportions and life table data. *Comput. Biomed. Res.,* 10, 373–381.

Ward, J.M., Griesemer, R.A., and Weisburger, E.K. (1979) The mouse liver tumor as an endpoint in carcinogenesis tests. *Tox. Appl. Pharmacol.,* 51, 389–397.

Warshawsky, D. (2004) *Molecular Carcinogenesis.* CRC Press, Boca Raton, FL.

Williams, G.M. and Jatropoulus, M.J. (2001) Principles of testing for carcinogenic activity, in *Principles and Methods of Toxicology,* Hawthorne, A.W., Ed., Taylor & Francis, Philadelphia.

Wilson, J.S. and Holland, L.M. (1982) The effect of application frequency on epidermal carcinogenesis assays. *Toxicology,* 24, 45–53.

Wittenau, M.S. and Estes, P. (1983) The redundancy of mouse carcinogenicity bioassays. *Fund. Appl. Toxicol.,* 3, 631–639.

16

Risk Assessment and Low-Dose Extrapolation

The broad realm of risk assessment (as it applies to toxicology) is based on experimental results in a nonhuman species at some relatively high dose or exposure level, from which an attempt is made to predict the level of impact in humans at much lower levels. In this chapter we will examine the assumptions involved in such undertakings, take a critical look at low dose extrapolation models and methods, present the framework on which risk assessment is based, and examine approaches such as the benchmark dose and threshold of toxicological concern (TTC).

The reader should first understand that, contrary to popular belief, risk assessment in toxicology is not limited to carcinogenesis. Rather, it may be applied to all the possible deferred toxicologic consequences of exposure to chemicals or agents that are of a truly severe and delayed (relative to point of exposure) nature. That is, those things (such as carcinogenesis, developmental and neurotoxicity, or reproductive impairment) that threaten life (either existing or prospective) at a time distant to the actual exposure to the chemical or agent. Because the consequences of these toxic events are extreme yet are distanced from the actual cause by time (unlike over exposure to an acutely lethal agent, such as carbon monoxide), society is willing to accept only a low level of risk while maintaining the benefits of use of the agent. Although the most familiar (and, to date, best developed) case is that of carcinogenesis, much of what is presented for risk assessment may also be applied to the other endpoints of concern.

Mutagens are not generally conceded to have thresholds, but the human health concern of those that are not established as carcinogens or teratogens is conjectural. Certainly materials identified as mutagens in an *in vivo* mammalian system (such as a dominant lethal study in mice or *in vivo* sister chromatid exchange in rabbits or rats) should be treated more conservatively than those whose mutagenicity has been established only in bacterial or biochemical test systems. The EPA (1984a,b) has proposed a regulatory process for the assessment of risks from mutagens that is not noticeably different from their approach to carcinogens and utilizes dose-response data from many test systems (even biochemical) as a starting point.

Developmental toxicants are now clearly established to have dose-response relationships that are subject to both time and quantity-of-dose thresholds

(Jusko, 1972). The exposure of the fertilized ovum must occur within a time window when the ovum is susceptible. Likewise, the dose must be sufficient to cause an effect and yet not so great as to cause a spontaneous abortion of the embryo or fetus. Rai and Van Ryzin (1985) recently proposed a dose-response model for developmental toxicity that included two approaches to low-dose extrapolation based on a one hit model. This model is also useful for mammalian dominant lethal data, which can be considered a subset of either developmental or reproductive toxicity data. Kimmel et al. (1990) have developed and proposed the benchmark dose approach for the same area, while Leroux et al. (1996) have put forward a biologically based model for dose-response in developmental toxicity.

Reproductive toxicants are a whole new wide world to worry about. Although Gross et al. (1970) suggested caution in evaluating data that might be improperly suggestive of a nontoxic threshold level, there is a clear consensus that there are thresholds below which reproductive agents are inactive. The difficulty is that such a wide variety of biochemical and physiological processes are involved in the successful operation of the reproductive process that we do not yet have adequate experimental methods to detect and quantitate all possible effects. Dixon and Nadolney (1985) and Roberts (1989) have presented brief overviews of the problems involved in going from our present state of knowledge in this area to the performance of meaningful risk assessments.

Although some of the fine points may vary, one should expect that the risk assessment process for any irreversible toxicant will follow the general form and steps presented in this chapter and such assessment should be undertaken with full knowledge of the involved uncertainties, weaknesses, and difficulties.

Low-Dose Extrapolation

Risk assessment, in the sense in which we will consider it in this book, and as it is performed by toxicologists, involves a number of separate steps, each of which involves some form of mathematical model to bridge gaps in biological knowledge. In a very crude sense, these steps can be categorized as answers to one of three problems:

Given knowledge of what happens at relatively high doses in one or more animal species, we must predict what would happen in the same species at much lower dose levels.

We must estimate what actual human exposures are. This means identifying groups or classes of exposed people and estimating "lifetime exposure."

Given the dose-response estimation made for animals (from step 1) and exposure estimates for classes of people (from step 2), we must finally couple the two by some model which translates "mice-to-men." That is, we must perform a species-to-species extrapolation.

The second and third steps will be separately addressed in the final section of this chapter. The first step, known as low-dose extrapolation, will be addressed here.

Threshold

The process of low-dose extrapolation, no matter which method is used, consists of three distinct steps. First, the actual dose-response data points available (which, for reasons discussed earlier, are invariably in the high dose and/or high response region) are identified, providing us with a starting point. Second, a mathematical method is selected and employed to extend the dose-response relationship from the region we know to the regions we are interested in or concerned about. Third, we make a basic assumption about the nature of the dose-response relationship in the extreme low dose and/or response region, then proceed to develop a model specific to the compound of interest.

This basic assumption about the extreme low-dose region is the question of threshold. For all biological phenomena except carcinogenesis and mutagenesis, it is a basic principle of biology that there is a dose level (threshold) below which no response is evolved. But there is an ongoing controversy as to applicability of the concept of a threshold for these two phenomena (Mantel, 1963). Regulatory bodies in the U.S. have based much of their risk assessment work on the belief (which cannot, of course, be either proved or disproved) that a single molecule of any agent found to be a carcinogen or mutagen at high dose levels will involve some increase in the incidence of that response. Epstein (1973), in quoting Umberto Saffiotti, presented one major argument against the concept of the use threshold.

Certain approaches to the problem of identifying a "safe threshold" for carcinogens are scientifically and economically unsound. Let's say one has in mind some proposals to test graded doses of one carcinogen down to extremely low levels, such as those to which a human population may be exposed through, say, residues in food. In order to detect possible low incidences of tumors, such a study would use large numbers of mice, of the order of magnitude of 100,000 mice per experiment. This approach seems to assume that such a study would reveal that there is a threshold dose below which the carcinogen is no longer effective, and, therefore, that a "safe dose" can be identified in this manner. Now, there is presently no scientific basis for assuming that such a threshold would appear. Chances are that such a "megamouse experiment" would actually confirm that no threshold can be determined. But let us assume that the results showed a lack of measurable tumor response below a certain dose level in the selected set of experimental conditions and for the single carcinogen under test. In order to base any generalization for safety extrapolations on such a hypothetical finding, one would have to confirm it and extend it to include other carcinogens and other experimental conditions such as variations in diet, in the vehicle used, in the age of the animals, their sex, etc. Each of these tests would then imply other "megamouse experiments." The task would be formidable: suffice it to say that an experiment on 100,000 mice would cost about $15 million if one did 20 such experiments, it would cost $300 million. All this to try and estimate the possible shape of a dose response curve which would still leave most of our problems in the evaluation of carcinogenesis hazards unsolved. This effort would also block the nation's resources for long-term bioassays for years to come and actually prevent the use of such resources for the detection of potent carcinogenic hazards from yet untested environmental chemicals. If 2 million mice are made available as resources, they can be used effectively to test 4,000 new compounds, each on 500 mice, thereby detecting among them those that are highly carcinogenic in the test conditions.

The argument against the existence of a threshold for carcinogens continues with the following points:

1. *In vitro* a single molecule of a chemical can achieve an alteration of the genetic elements of a cell, mutating it. Such mutations can be to neoplastic forms, therefore including the process that (in the end) would produce a cancer *in vivo*. Albert et al. (1979) have presented a variation on this theme with initiation promotion in mouse skin as a model.

2. Even if there is a biological threshold for any individual agent of concern due to various defense mechanisms, we cannot rule out the possibility of the presence of other agents in the environment that may either act as promoters for our agent of concern or saturate the existing defense mechanisms, effectively "humping over" the threshold. Mantel presented this argument in his 1963 review of the concept of threshold in carcinogenesis.

3. The presence of a threshold would preclude the possibility of linear dose-response. Mantel and Schneiderman (1975) presented this point of view, which is a variation on point one. Of course, the existence of a threshold would actually only mean that a linear (or any other dose-response relationship) would start at some point above zero, being discontinuous only in the extreme lowest dose range.

4. The presence of a background exposure of carcinogens and promoters, and of spontaneously occurring cancers in a population at risk as large and diverse as that of human beings, implies that even if there are thresholds for some or most individuals, there will still remain others who have been "jumped over" their individual thresholds by background events. Crump et al. (1976) termed this the existence of a "random threshold" such that there could be no exposure that was absolutely safe for absolutely everyone who might be exposed. Interestingly, Gross et al. (1970) presented a similar argument for "absolute safety" in addressing reproductive toxicity data for a food additive.

5. Existing methods of low-dose risk extrapolation implicitly account for the increase in time-to-tumor statistics insofar as they accord for the decrease in tumor incidence, invalidating the pro-threshold argument (to be reviewed later) that at very low doses time-to-tumor would become so long that it would exceed lifespan. Guess and Hoel (1977) proposed this argument against thresholds.

All five of these arguments against the existence of a threshold center on the belief that without proof of absolute safety, we must proceed in the most conservative manner possible. Calabrese (1983) summarized the essence of this approach as lack of belief in threshold together with six other principles.

1. Use of upper confidence limits on the estimated VSD (virtually safe dose) instead of on the VSDs themselves
2. Use of the most sensitive animal species
3. Use of the most sensitive sex of that species
4. Use of the most sensitive strain within a species
5. Expression of dosage given on a dietary concentration basis rather than on a bodyweight basis when extrapolating from animals to humans; this will result in about a 15-fold lower acceptable exposure for humans as compared to the mouse
6. The slope of unity by Mantel and Bryan (a model for low-dose extrapolation to be discussed later) is almost always less than the observed data, thereby resulting in lower acceptable exposure

The arguments for the existence of the threshold are as numerous and tend to be more mechanistic. These are summarized as follows:

1. Most (if not all) carcinogens and mutagens exhibit a dose-response relationship, resulting in an apparent or effective threshold for at least some agents. This is the classical toxicology argument, coming from the general case of all other toxic actions and finding no reason or data to support these actions being different (Klaassen and Doull, 1980). Theoretical analysis of the process of carcinogenesis as understood for radiation (the case in which we have the most human data) likewise suggests a lack of linearity at very low doses (Arley, 1961).
2. Toxicity, including carcinogenesis, is a dynamic process that includes absorption of an agent into the body, distribution to various tissues, reversible or irreversible reactions with cellular components, adaptation and repair by molecular and cellular components of the body, and ultimately clearance from the body by metabolism and/or excretion. Such pharmacokinetic processes are generally linear only within prescribed ranges. Gehring et al. (1977) have proposed that such pharmacokinetic processes provide a conceptual basis for understanding how metabolic thresholds may lead to a disproportionate increase in toxicity, including carcinogenesis, above certain dose levels. They have also conducted and presented work on vinyl chloride carcinogenesis and pharmacokinetic data to support this proposal.
3. Because an organism as large as either a mouse or a man has a tremendous number of cells, a large number of defense mechanisms of high efficiency, and low probability of a "hit" by a carcinogen being effective in initiating or promoting a neoplasm, there is a biological

threshold based on just stochastic or probabilistic grounds. The last aspect (low probability of a "meaningful" reaction) arises from consideration of the fact that the vast majority of molecules that a carcinogen comes in contact with (and reacts with) in a multicellular organism cannot then contribute to the development of a neoplasm—they are, in effect, a multitude of "dummies" acting to block the assassin's shot. Dinman (1972) originally proposed this concept with the following six elements:

a. A cell is estimated to be composed of approximately 1014 molecules and atoms with which a xenobiotic substance may interact.

b. A major factor influencing activity is molecular specificity, compared to the mere presence of an atom or molecule in a cell.

c. There are lower concentration limits for the occurrence of biologically significant intracellular molecular reactivity. Numerous examples of *in vitro* studies of specific inhibitors have demonstrated that a lower concentration limit for such inhibition is 10^{-8} M.

d. Binding or interaction with proteins or other molecules at sites where there is no resulting functional effect may not happen frequently.

e. All chemical components of the cells and cells themselves are in a dynamic flux. The major consideration is that the rate of loss exceeds the rate of normal replacement.

f. Cells of different types of tissues have the capacity to induce normal DNA repair mechanisms to repair genetic damage caused by environmental mutagens. In fact, the absence of DNA repair mechanisms in persons with xeroderma pigmentosum clearly demonstrates the lifesaving functional capacity of this process in normal individuals.

Based on these elements and the estimate that 104 molecules per cell is the lower limit for a material to be biologically active, Friedman (1973) has calculated threshold levels for a number of materials. The calculations were also based on the assumption that there are 6×10^{13} cells in a 70 kg "average man," therefore requiring a minimum effective dose of 8.6×10^{15} molecules per kilogram of cells. Some of these results are presented in Table 16.1.

Several such calculated values (vitamin A, estradiol, and diethylstilbestrol) were also shown to be conservative — below experimentally established effect levels.

As dose or exposure to a carcinogenic agent decreases, it is well established that the time it takes for tumors to be expressed gets longer and longer. At some point in the dose time-to-tumor curve, the time necessary for a tumor to develop will exceed the lifespan of the exposed member of a population. Doll (1971) has presented human (cigarette smoker) data to support this dose time-to-tumor hypothesis, along with an excellent compilation of human age-related tumor incidences at a large number of sites. Yanysheva and Antomonov (1976) have reported on the dose-time-effect relationship with respect to the carcinogen benzo(a)pyrene. They found that the number of animals with tumors decreases

TABLE 16.1

Estimated No-Effect Quantities of Some Potent Carcinogens and Toxic Agents

Agent	Molecular Weight	Calculated No-Effect Level (g/kg Body Wt.)	Molarity	Mol/kg Body Weight
Aflatoxim	312	5×10^{-9}	1.6×10^{-11}	9.6×10^{12}
1,2,5,6-diben-zanthracene (subcutaneous)	278	2.5×10^{-4}	9×10^{-7}	5.4×10^{17}
1,2,5,6-diben-zanthracene (subcutaneous)	278	2.5×10^{-5}	9×10^{-8}	5.4×10^{16}
3-Methylcholan-threne (subcutaneous)	268	2.5×10^{-5}	9.3×10^{-8}	5.6×10^{16}
2,4-Benz-a-pyrene (skin)	252	2.5×10^{-6}	1×10^{-8}	6×10^{16}
Aramite	335	1×10^{-1}	3×10^{-4}	1.8×10^{20}
Tetrachlorodi-benzodioxin	320	6×10^{-8}	1.9×10^{-10}	1.1×10^{14}
Botulinum toxin (mouse)	900,000	6.5×10^{-11}	7×10^{-17}	4.2×10^{7}

Source: From Friedman, L., in *Pharmacology and the Future of Man*, vol. 2, p. 33. Karger, Basel, Switzerland 1973. With permission.

TABLE 16.2

Calculated Time for Appearance of the First Lung Tumor Following Administration of Various Total Benzo(a)pyrene Doses in Ten Portions, Intratracheally

Benzo(a)pyrene Dose (mg)	Time of Tumor Occurrence (Months)
0.1	27.0
0.05	38.0
0.02	67.9
0.01	118.9
0.005	221.0
0.002	527.3

as exposure to benzo(a)pyrene decreases. Furthermore, the latency period varies inversely with the dose. Based on their results, Yanysheva and Antomonov (1976) developed a dose-time-effect relationship as shown in Table 16.2. Based on the information in Table 16.2, Yanysheva and Antomonov suggested a dose of benzo(a)pyrene that they believed would lead to a carcinogenic effect only after the normal lifespan of the exposed individual. Kraybill (1977) combined arguments in 3 (p. 285) and those for a time-to-tumor threshold by suggesting that "the fallacy of considering just singular insults in a biomedical assessment, the traditional approach, can thus be appreciated."

There are directly demonstrable physicochemical factors that cause some agents to be carcinogens above certain dose levels and, conversely, cause these particular materials not to be carcinogens (at least by the mechanism operative at the higher levels) at lower levels. Two examples of such mechanisms of threshold are xylitol and hexavalent chromium.

Elizabeth Miller used and presented the case of xylitol as part of her 1977 presidential address to the American Association for Cancer Research (Miller, 1978):

> The recent report on the development of tumors of the urinary bladder in male mice fed the sweetener xylitol as 10 to 20% of their diets in two-year tests and its possible implications for the use of xylitol in human foods provide an example of the problems to be resolved. These tumors apparently developed only in urinary bladders that contained stones (oxalates), a condition long known to predispose rodents to the development of bladder tumors. Neither bladder stones nor bladder tumors were reported in female mice fed the high levels of xylitol or in male mice fed 2% of the sweetener. Since xylitol is a normal intermediate in the metabolism of D-glucuronate, has not shown mutagenic activity and would not be expected to yield strong electrophilic reactants on metabolism, there seems to be little reason for concern of hazard to humans ingesting low levels of xylitol in foods. Yet, strict interpretation of the Delaney amendment to the Pure Food and Drug Act would prohibit the use of xylitol, since the act does not permit addition to food of any chemical that has caused tumors in either humans or animals.

Currently, of course, the Food Safety Protection Act would preclude such an action.

Hexavalent chromium is a somewhat more complex case, which has only now come to completion. This is presented as a case history.

Case History

Evidence Supportive of a Threshold for Hexavalent Chromium Carcinogenicity

In animal studies chromates produce cancers only at the site of contact (lung carcinomas from intratracheal implantation or instillation and muscle sarcomas from intramuscular injection). No tumors have been found at distant sites. It is believed that the explanation for this observation is that hexavalent chromium is rapidly converted to trivalent chromium and that only the hexavalent form is capable of causing cancer. Carefully performed epidemiology studies have also failed to identify any excess risk of cancer in organ systems other than the respiratory tract.

The rat intratracheal instillation study (detailed below) found lung tumors in 20 of 80 rats exposed to 1.25 mg/kg of sodium dichromate once a week. There were no tumors in 80 rats exposed to the same lifetime dose but at 0.25 mg/kg of sodium dichromate 5 days per week. One obvious explanation is that the animals were unable to detoxify the once a week dose and hence some tumors developed.

Several studies on the metabolism and/or detoxification of chromates have been published. These studies (Petrilli and De Flora, 1978a, b, 1980, 1982; De Flora, 1978; De Flora et al., 1982, 1984; Bennicelli et al., 1983) have demonstrated that:

Hexavalent chromium (Cr^{+6}) is converted to trivalent chromium (Cr^{+3}) by saliva and gastric juice.

Cr^{+6} is converted to Cr^{+3} in red blood cells.

Cr^{+6} is converted in cytoplasm to Cr^{+3} by endogenous glutathione and other reducing compounds.

Cr^{+6} is metabolized to Cr^{+3} in certain cells, probably by DT disphosphate. This reduction, very potent in liver cells, is weaker in lung tissue but appears to be quite potent in human lung macrophages.

Cr^{+6} is mutagenic while Cr^{+3} is not. Trivalent chromium is an essential mineral that is incorporated into glucose tolerance factor.

While some scientists have suggested Cr^{+3} is the ultimate carcinogen binding to DNA (perhaps following reduction from Cr^{+6} in the immediate vicinity of DNA), there is some evidence that the ultimate carcinogen may be the more reactive Cr^{+5} ion (Jannette, 1982). In any event, whether C^{+3} or Cr^{+5} is the ultimate carcinogen, reduction of Cr^{+6} to Cr^{+3} results in the practical elimination of carcinogenic risk as demonstrated by mutagenicity studies and the lack of tumors at distant sites following Cr^{+6} administration in animals.

	Five Times per Week	One Time per Week
$Na_2Cr_2O_7 \cdot 2H_2O$	0.25 mg/kg(40 ♂, 50 ♀)*	1.25 mg/kg(40 ♂, 40 ♀)
$Na_2Cr_2O_7 \cdot 2H_2O$	0.05 mg/kg(40 ♂, 45 ♀)**	0.25 mg/kg(40 ♂, 40 ♀)
$Na_2Cr_2O_7 \cdot 2H_2O$	0.01 mg/kg(40 ♂, 45 ♀)**	0.05 mg/kg(40 ♂, 40 ♀)
$CaCrO_4$	0.25 mg/kg(40 ♂, 50 ♀)*	0.25 mg/kg(40 ♂, 40 ♀)
Benxo(a)pyrene		5.00 mg/kg(–, 10 ♀)
Dimethylcarbamyl chloride		1.00 mg/kg(10 ♂, 10 ♀)
Physiological saline solution	1.00 ml/kg(40 ♂, 50 ♀)	1.00 ml/kg(40 ♂, 40 ♀)

Note: Untreated (40 ♂, 50 ♀)*.

* 10 additional rats for intermediate sacrifice after 12 months of treatment, to decide whether further treatment would be possible.
** 5 additional rats for assessment under the electron microscope as well as some rats indicated by   *   being so assessed.

A lifetime intratracheal instillation study was performed to specifically address the question of physicochemical defense system overload. Thirteen separate dose groups were included, as detailed below (Steinhoff et al., 1986).

For the groups receiving sodium dichromate and calcium chromate (each as solutions), what the above design achieved was to give the same total lifetime doses, but in one set of cases (the five times per week) the total dose was spread over five smaller increments (which were each hopefully within the range of possible detoxification by defense mechanisms) and in the other set of cases delivered these as weekly "bolus" doses, which overwhelmed defense mechanisms.

The survival in dosed test and negative control animals through 2 years of treatment was almost complete. The only increase in tumor incidence was in the lungs, as shown below.

		Lung Tumors	
		Benign	Malignant
Untreated		—	—
Phys. NaCl	5 ×/week	—	—
Phys. NaCl	1 ×/week	—	—
$Na_2Cr_2O_7 \cdot 2H_2O$	5 × 0.01 mg/kg/week	—	—
$Na_2Cr_2O_7 \cdot 2H_2O$	5 × 0.05 mg/kg/week	—	—
$Na_2Cr_2O_7 \cdot 2H_2O$	5 × 0.05 mg/kg/week	—	—
$Na_2Cr_2O_7 \cdot 2H_2O$	1 × 0.05 mg/kg/week	—	—
$Na_2Cr_2O_7 \cdot 2H_2O$	1 × 0.05 mg/kg/week	1 (1 ♂)	—
$Na_2Cr_2O_7 \cdot 2H_2O$	1 × 1.25 mg/kg/week	12 (6♂, 6♀)	8 (5♂, ♀)
$CaCrO_4$	5 × 0.25 mg/kg/week	5 (5♂)	1 (1♀).
$CaCrO_4$	1 × 1.25 mg/kg/week	11 (9♂, 2♀)	3 (2♂, 1♀)

TABLE 16.3

Evidence for Thresholds in Carcinogenesis (Stokinger, 1977)

Test Substance	Route	Species	Dose Levels Eliciting Tumors	Dose Levels Not Eliciting Tumors	Duration
Bis-chloromethyl ether	Inhalation	Rat	100 μg/M^3	10 μg/m^3 1 μg/m^3	6 months/ daily
1,4-Dioxane	Oral	Rat	1% in H$_2$0	0.1% in H$_2$0 0.01% in H$_2$0	2 years
	Inhalation	Rat	>1000 ppm	111 ppm	2 years/daily
Coal tar	Topical	Mouse	6400 mg 640 mg 64 mg	<0.64mg	64 weeks twice
β-Napthylamine	Inhalation Topical	Human	>5% in form	<0.5% in	22 years
Hexamethyl-phosphoramide	Inhalation	Rat	4000 ppm 400 ppm	50 ppm	8 months
Vinyl chloride	Inhalation	Rat	2500 ppm 200 ppm 50 ppm	<50 ppm >10 ppm	7 months
Vinylidene chloride	Inhalation	Human	>200 ppm	1950–1955, 160 ppm average; 30–170 ppm 1960, <50 ppm decreasing to 10 ppm	25 years

There are numerous other cases of data that support the concept of a threshold. As early as 1943 Bryan and Shimkin reported skin painting data for three carcinogenic hydrocarbons that at least suggested thresholds.

In 1977, Wolfe reported a similar case for radiation exposure and Stokinger presented evidence for the existence of thresholds for seven chemical carcinogens (Table 16.3).

The ultimate empirical answer to the question of whether or not there are thresholds for carcinogens and mutagens in multicellular organisms would be a massive study with sufficient animals to allow for establishing a dose-response curve across the broad dose range. Such a study is both logistically and economically unfeasible, but NCTR conducted a "megamouse" study that was designed to go part way toward this goal. In the ED$_{01}$ study 24,192 BALB/c female mice were fed 2-acetylaminofluorene (2-AAF), a carcinogen that results in tumors in two different organs (bladder and liver) by unrelated mechanisms. The spontaneous incidence rates for either of these two tumors in any one BALB/c female control was below 0.1% (Cairnes, 1980). The entire study was initially reported as a complete issue of a journal (Staffa and Mehlman, 1980).

The data resulting from this study (Littlefield et al., 1980) strongly suggest a threshold for bladder tumors, but also emphasize the importance of time-to-tumor in the interpretation of results.

Models

There are at least eight different models for extrapolating a line or curve across the entire range from a high-dose region to a low-dose region. In this section we will examine each of these models and compare them in terms of characteristics and outcome. Some of these models are such that they handle only quantal (also called dichotomous) data, while others will also accommodate time-to-tumor information.

In the models below, certain standard symbols are used. Most of these models express the probability of a response P as a function, f, of dosage D, so that $P = f(D)$ and the models differ only with respect to choice of function, f. The nonthreshold models assume that if proportion p of control animals respond to a dose that $f(D) = p$ only for D equal to zero, and that for any nonzero D, $f(D) \to p$ (that is, there is a response). Threshold models assume the existence of a D_0 such that for all $D < D_0$, $f(D) = p$ (that is, that there is some dose below which there is no response). If safety is defined as zero increase over control response, then a nonthreshold model would require that any nonzero dosage be associated with some finite risk.

One-Hit

The one-hit model is based on the assumption that cancer initiates from a single cell as a result of a random occurrence or "hit" that causes an irreversible alteration in the DNA of a susceptible cell type. It is also assumed that the likelihood of this hit is proportional to the level of carcinogen exposure. This suggests a direct linear dose-response such that if one is to diminish the risk from 10^{-2} to 10^{-8}, then the dose should be divided by 106.

Accordingly, the one-hit model is also called the linear model, although a number of the other models also behave in a linear manner at lower doses. Based on the concept that a single receptor molecule of some form responds after an animal has been exposed to some single unit of an agent, the probability of tumor induction by exposure to the agent is then

$$P(D) = 1 - \frac{1}{e^{\lambda D}}$$

where $D \geq 0$, $0 \leq P(D) \leq 1$, λ is an unknown rate constant (or slope) and D is the expected number of hits at dose level D. "Dose" is used in a very general sense. It may mean the total accumulated dose or the dosage rate in terms of body weight, surface area approximations, or concentration in the diet. Computing the one-hit model in terms of the exponential series gives

$$P(D) = \frac{\lambda D}{1!} + \frac{(\lambda D)^2}{2!} + \frac{(\lambda D)^3}{3!} +$$

which, for small values of λ, is well approximated by

$$P(D) \approx \lambda D$$

Although Hoel et al. (1975) have argued that this model is consistent with reasonable biological assumptions, there is now almost universal agreement that the model is excessively conservative. The concept of a hit is a metaphor for a variety of possible elementary biochemical events and the model must be considered phenomenologic rather than molecular. This model is essentially equivalent to assuming that the dose-response curve is linear in the low-dose region. Thus, the slope of the one-hit curve at dose D is $\lambda[1 - P(D)]$, and for dose levels at which $P(D) < 0.05$ varies by less than 5%, i.e., is essentially constant and equal to. The linear model is one of two models, the other being the probit model, specified by the Environmental Protection Agency (1976) in its interim guidelines for assessment of the health risk of suspected carcinogens. The assumption of low-dose linearity will generally lead to a very low, virtually safe dose (VSD), so low as to lead the Food and Drug Administration Advisory Committee (1971) to remark that assuming linearity "...would lead to few conflicts with the result of applying the Delaney clause." The one-hit model, having only one disposable parameter, l, will often fail to provide a satisfactory fit to dose-response data in the observable range. Other models described below, by introducing additional parameters, often lead to reasonable fits in the observable range.

An additional degree of conservatism is introduced by extrapolating back to zero from the upper confidence limit (UCL) for the net excess tumor rate (treated minus control rate). The linear model assumes that the tumor rate is proportional to dose, or that $P(D) = \lambda D$. The upper confidence limit for the slope is UCL divided by experimental dosage. Thus an estimate of an upper limit for the proportion of tumor-bearing animals, P_u, for a given dose D is

$$P(D) = \frac{(UCL)(D)}{D_e}$$

where D_e is the experimental dosage. Conversely, the dose D for a given P_u is

$$P(D) = \frac{P_u D_e}{UCL}$$

The linear model may serve as a conservative upper boundary for probit dose-response curves. This upper boundary on the proportion of tumor-bearing animals may not be as conservative as one might imagine. Crump et al. (1976), Peto (1978), and Guess et al. (1977) have shown that the curvilinear dose-response curve resulting from the multistage model is well approximated by the linear model at low dose levels. Gross et al. (1970) discussed the statistical aspects of a linear model for extrapolation.

Example 16.1 illustrates the linear extrapolation model.

Example 16.1

A compound is administered as 5% (50,000 ppm) in diet for 2 years to a group of 100 animals. Twenty-two of these test animals and 6 of 100 control animals are found to have developed liver tumors at the end of the study.

Thus upper confidence limit on the excess tumor rate is approximately

$$= (p_t - p_c) + Z \sqrt{\frac{p_t(1-p_t)}{n_y} + \frac{p_c(1-p_c)}{n_c}}$$

where p_t is the proportion of animals with tumors in n_t treated animals, p_c is the proportion of animals with tumors in n_c control animals, and Z is the normal deviate corresponding to the level of confidence desired.

The upper 99% confidence level for this example is thus

$$= (0.22 - 0.06) + 2.33 \sqrt{\frac{0.22(0.78)}{100} + \frac{0.06(0.94)}{100}}$$

$$= (0.16) + 2.33 \sqrt{0.001716 + 0.000564}$$

$$= (0.16) + 2.33(0.0477493)$$

$$= 0.271256$$

If it is then desired to estimate an upper limit of risk associated with exposure to 10 ppm of the material in diet this would be

$$= 0.271256 \times \frac{10}{50000} = 5.43 \times 10^{-5}$$

The Probit Model

This model assumes that the log tolerances have a normal distribution with mean μ and standard deviation σ. The proportion of individuals responding to dose D, say $P(D)$, is then simply

$$P(D) = \Phi[(\log D - \mu)/\sigma] = \Phi(\alpha + \beta \log D)$$

where $\Phi(x)$ is the standard normal integral from $-\infty$ to x, $\alpha = -\mu/\sigma$ and $\beta = 1/\sigma$.

This dose-response curve has $P(D)$ near zero if D is near zero and $P(D)$ increasing to unity as dose increases. A plot of a typical probit dose-response is given by an S-shaped (sigmoid) curve. The quantity above is referred to as the slope of the probit line, where

$$Y = \Phi^{-1}[P(D)] = \alpha + \beta \log D$$

and

$$Y + 5 \text{ is the probit of } P$$

This is the same model we presented earlier in this book for linearizing a special case of quantal response, the data for LD$_{50}$s. Despite its nonthreshold assumption it is a characteristic of the probit curve that as dose decreases, zero response is approached very rapidly, more rapidly than any power of dose. Other curves to be considered approach zero response more slowly than the probit.

An alternative derivation of the probit model that relates it to time-to-response has been given by Chand and Hoel (1974) using the Druckrey observation that median time-to-tumor, T, is related to dose, D, by the equation $Dt^n = C$, where n and C are constants unrelated to D (Druckrey, 1967). Combining this relation with an assumed log-normal distribution of response time then gives the $P(D)$ as probability of response to any given time, T_0, where α and β are simple functions of n, C, T_0, and the standard deviation of the distribution of response times.

The actual method, which has the probit model as its basis, is the Mantel-Bryan procedure. As originally proposed, this procedure used the probit model but with a preassigned slope of unity. The rationale for this slope was that all observed probit slopes at the time of the proposal exceeded that value, the procedure therefore being considered conservative. An additional conservative feature involves use of the upper 99% confidence limit of the proportion responding at a dose level, rather than the observed proportion. The procedure then extrapolates downward to a response level of 10^{-8}, using each separate dose level in the experiment, or combinations, taking as the VSD the highest of the values obtained. A conservative method of taking account of the response of the control group was also given. An improved version of the procedure, which included several sets of independent data and better methods of handling background response rates and responses at multiple doses, has since been published (Mantel et al., 1975).

A dosage D_0 is said to be virtually safe if $f(D_0) < p + (1-p)P_0$, where P_0 is some near-zero lifetime risk, such as 10^{-8}, the value proposed by Mantel and Bryan, or 10^{-6}, the value adopted by the FDA. The VSD is then calculated as $f^{-1}[p + (1 - p)P_0]$. The calculation thus requires choosing a model, f, determining the value of its disposable constants from observations in the observable range, and extrapolating down to the unobservable elevation in response, P_0, to determine the VSD. This is illustrated in Example 16.2.

Example 16.2

From the data in Example 16.1, we have already calculated an upper 99% confidence interval of 0.271, corresponding to a normal deviate of −0.61. If it is desired to determine the level corresponding to a tumor probability of less than one in a million (which has a normal deviate of −4.753), the extrapolation proceeds along the probit-log dosage line with a slope of 1 from the normal deviate of the upper 99% confidence limit on the observed result to the normal deviate for the selected probability or −0.61 − (−4.753) = 4.143 standard deviations.

TABLE 16.4

Mantel-Bryan Dosages for Various Sample Sizes with the Same
Proportion of Experimental Animals with Tumors

Sample Size	No. of Animals with Tumors[a]	Upper 99% Confidence Limit	Dosage (Fraction of Experimental Dosage)
50	2	0.158	1/5630
100	4	0.112	1/3430
200	8	0.085	1/2400
400	16	0.069	1/1860

[a] Predicted tumor probability $< 10^{-6}$.

The dose level corresponding to this risk is then

$$= \frac{50000 \text{ ppm}}{10^{4.143}} = 3.597 \text{ ppm}$$

One of the advantages of the Mantel-Bryan procedure is that it rewards a
larger experiment by reducing the upper confidence limit, which results in
a larger dose for a selected proportion of tumor-bearing animals. Table 16.4
shows some dosages for a series of sample sizes; all yield observed tumor
rates of 4%, with no tumors in the controls for a predicted tumor probability
of less than one in a million.

Some situations, such as cigarette smoking in man and diethylstilbestrol
in mice, have indicated slopes on the order of 1. Thus, one must be careful
to establish that the slope of the dose-response is sufficiently large before
applying the Mantel-Bryan procedure, indicating the desirability of multiple-
dose experiments.

According to Mantel and Schneiderman (1975), the Mantel-Bryan methodology has several advantages:

It does not need an experimental estimate of the slope

Statistical significance is not needed

It takes into account a nonzero spontaneous background tumor incidence

It considers multiple-dose studies

Any arbitrary acceptable risk can be calculated

It avoids categorizing a substance in absolute terms

It permits the investigator flexibility in study design

Mantel and Bryan (1961) provided an example of an actual study in which
the carcinogen 3-methylcholanthrene was given to mice as a single injection,
with 12 different dose levels used. Table 16.5 provides the methodology and
findings of the Mantel-Bryan procedure.

TABLE 16.5

Illustration of Methodology for Determining the "Safe" Dose from Results at Several Dose Levels (Mantel and Bryan, 1961)

Dose per Mouse (mg) (1)	Log Dose (2)	Result: No. of Tumors/No. of Mice (3)	Combined Result: No. of Tumors/No. of Mice (4)	Maximum p-Value 99% Assurance (5)	Corresponding Normal Deviate (6)	Calc. "Safe" (1/100 Million) Log Dose (2)–(6)–5.612
0.000244	6.388–10	0/79	0/158	0.0288	–1.899	2.675–10
0.000975	6.990–10	0/41	0/79	0.0566	–1.584	2.962–10
0.00195	7.291–10	0/19	0/38	0.1141	–1.205	2.884–10
0.0039	7.592–10	0/19	0/19	0.2152	–0.789	2.769–10
0.0078	7.893–10	3/17	3/17	0.430	–0.050	2.331–10
0.0156	8.194–10	6/18	6/18	0.729	+0.610	1.972–10
0.0312	8.495–10	13/20	13/20	0.871	+1.131	1.752–10
0.0625	8.796–10	17/21	17/21	0.958	+1.728	1.456–10
0.125	9.097–10	21/21	—	—	—	—
0.25	9.398–10	21/21	—	—	—	—
0.50	9.699–10	21/21	—	—	—	—
1.0	10.000–10	20/20	—	—	—	—

Some criticisms of the Mantel-Bryan procedure are:

The normal distribution may not offer as accurate a description in the tails of the distribution as it does in the central parts, especially if one proceeds out to 10^{-6} or 10^{-8}

The use of the arbitrarily low slope of unity for downward extrapolation has been criticized because of the lack of observational support

The argument does not incorporate any of the present understandings of the process of carcinogenesis

The model is insufficiently conservative, because the extrapolated probability approaches zero with decreasing dose more rapidly than any polynomial function of dose, and, in particular, more rapidly than a linear function of dose and hence may underestimate probability at low dose (Crump, 1979)

The model is excessively conservative, because it does not postulate a threshold or accommodate time to tumor data

Multistage

The multistage model (Armitage and Doll, 1961; Crump et al., 1976) represents a generalization of the one-hit model and assumes that the carcinogenic process is composed of an unknown number of stages that are needed for cancer expression. Inherent in this model is the additional assumption that the effect of the carcinogenic agent in question is additive to a carcinogenic effect produced by external stimuli at the same stages. Such an assumption generally leads one to expect a linear dose–response curve at low exposure levels.

This assumes that carcinogenesis occurs in a single cell as a point of origin and, according to the multistage model, is the result of several stages that can include somatic mutation. The transitional events are individually assumed to depend linearly on dose rate. This then leads in general to a model in which the probability of tumor approximates a low-order polynomial in dose rate. In the low-dose region, which would relate to environmental levels, one finds that the responses are well approximated by a linear function of dose rate. The characteristic in which the low dose probability is proportional to the kth power of dose, where k is the number of stages, was considered by Armitage and Doll (1961) to be quite inconsistent with observation. They derived a multistage model, which by assuming that the effect of the agent at some stages was additive to an effect induced by external stimuli at those stages, led to a lower power than k for D. Crump et al. (1976) discussed this model and, by assuming additivity at all stages, obtained as an expression for the required probability

$$P(D) = 1 - \exp\left\{-\sigma \sum_{i=0}^{x} \alpha_i D^i\right\}$$

where $\alpha = -\mu/\sigma$. Hartley and Sielken (1977) combined this model with time to response, obtaining a more general result. For $\alpha_1 > 0$ these models also imply low-dose linearity since

$$\lim_{D \to 0}[P'(D)] = \alpha_1 e^{-\alpha_0}$$

Armitage and Doll cited data relating lung cancer mortality to previous smoking habits as indicating to a linear dose-response curve, but errors in reporting the amount smoked would lead to such a curve even if the true curve were convex. This supports the view that the apparent low-dose linearity in many epidemiologic studies is an artifact of errors in the reporting of dose. Crump et al. (1976) stress the crucial nature of the additivity assumption, pointing out that it can make orders of magnitude differences in the estimated risk associated with the low dose exposure.

Crump (1979) describes a procedure for low-dose extrapolation in the presence of background, which, although based on the generalized model above, reduces (when upper confidence limits are used) to extrapolation using low-dose linearity. This is because the use of upper confidence limits on χ_1 on the model is equivalent to admitting the possibility of a positive value of χ_1, which at low doses dominates the expression. Once upper confidence limits on the VSD or risk at a given dose are used, there may be little practical difference, therefore, between use of the one-hit model and the generalization given by the Crump et al. equation above.

Hartley and Sielken (1977) have developed a procedure based on maximum likelihood for the Armitage-Doll model. Their program is very general and allows for the inclusion of the effect of the time to a tumor.

In practice, these two approaches result in fitting a polynomial model to the dose-response curve such that (where t is time):

$$\frac{p(D,t)}{1-P(D,t)} = gDh(t)$$

where $P(D,t)$ is the probability of the observance of a tumor in an animal by time t at a dosage D

$$p(D,t) = \frac{DP(D,t)}{Dt}$$

where $g(D)$ is a function of dose such that

$$g(\text{dose}) = (a_1 + b_1 \text{ dose})(a_2 + b_2 \text{ dose})\cdots(a_n + b_n \text{ dose})$$

$$= c_0 + c_1 \text{ dose} + c_2 \text{ dose}^2 + \cdots + c_n \text{ dose}^n$$

where a_i, b_i, $c_i \geq 0$ are parameters that vary from chemical to chemical and $h(t)$ is a function of time. The probability of a tumor by time t and dosage D is

$$P(D,t) = 1 - \exp[-g(D)H(t)]$$

where

$$H(t) = \int_0^t h(t)Dt$$

This function generally fits well in the experimental data range but has limited applicability to the estimation of potential risk at low doses. The limitations arise, first, because the model cannot reflect changes in kinetics, metabolism, and mechanisms at low doses and, second, because low-dose estimates are highly sensitive to a change of even a few observed tumors at the lowest experimental dose.

A logical statistical approach to account for the random variation in tumor frequencies is to express the results in terms of best estimates and measures of uncertainty.

Important biological mechanisms of activation and detoxification are not usually specifically considered. However, a steady-state kinetic model that incorporates the process of deactivation as well as other pharmacokinetic considerations has been offered by Cornfield (1977). He noted that whenever the detoxification response is irreversible, low exposure levels are predicted to be harmless. However, the presence of a reversible response suggests linearity at low-dose exposures. He additionally predicted that when multiple protective responses are sequentially operational, the dose-response relationship will look like a hockey stick "with the striking part flat or nearly flat and the handle rising steeply once the protective mechanisms are saturated." Despite its seemingly greater biological veracity, the Food Safety Council (1980) challenged the multistage model general assumption of low-dose linearity on the basis of (1) the general absence of support for dose-wise additivity seen in many studies in which additivity has been evaluated and (2) studies that showed the effects of one carcinogenic agent offset or prevented the carcinogenic effects of another.

Crump (1979) has noted that biostatistical models such as the multistage model assume that the quantity of carcinogen finding its way to the critical sites is proportional to the total body exposure, which is clearly not the case across the entire dose range covered by the model.

Criticisms of these models are summarized below:

These models do not consider the variation in susceptibility of the members of the population when deriving their dose–response relationships.

Low-dose linearity is not consistently found in experimental systems.

Low-dose linearity is assumed to occur by a mechanism of additivity to background levels, however, there is a lack of data supporting the additivity hypothesis.

They do not sufficiently recognize pharmacokinetic considerations including rates of absorption, tissue distribution, detoxification processes, repair, and excretion. (This would apply to the Mantel-Bryan model as well.)

Multi-Hit

This model is also called the k-hit or gamma multi-hit model. It is a generalization of the one-hit model.

If k hits of a receptor are required to induce cancer, the probability of a tumor as a function of exposure to a dose (D) is given by

$$P(D) = 1 - \sum_{t=0}^{k-1} \frac{(\lambda D)^i e^{-\lambda D}}{i!} \approx \frac{(\gamma D)^k}{k!}$$

For small values of D, the k-hit model may be approximated by

$$P(D) = \gamma D^k$$

or

$$\log P(D) = \log \gamma + k \log D$$

Thus k represents the slope of $\log P(D)$ vs. $\log D$. By the same reasoning, if at least k hits are required for a response, then

$$P(D) = P(X \geq k) = \int_0^{\lambda D} \frac{u^{k-r} e^{-u} du}{(k-1)!}$$

Because this equation contains an additional parameter, k, it will ordinarily provide a better description of dose-response data than the one-parameter curve. This can be further generalized by allowing k to be any positive number, not necessarily an integer. In this case the above formula can be described as that dose-response curve which assumes a gamma distribution of tolerances with shape parameter k. We note

$$\lim[P(D)/D^k] \text{ is constant}$$

$$D \to 0$$

Thus, in the low-dose region, the equation is linear for $k = 1$, concave for $k < 1$, and convex for $k > 1$. At higher doses the gamma and the log-normal distributions are hard to distinguish so that the model provides a blend of the probit model at high dose levels and the logit at low ones.

Procedures for estimating the parameters of the k-hit model by nonlinear maximum likelihood estimation have been developed by Rai and Van Ryzin (1979). This method has the advantage of permitting the data to determine the number of hits needed to describe the results without introducing more than two parameters. When only one dose level gives responses greater than zero and less than 100%, unique values of the two parameters can no longer be estimated. The background effect in this model is taken care of using Abbott's correction.

The multi-hit model is discussed in some detail in the Food Safety Council Report (1980). One derivation of this model follows from the assumption that k hits or molecular interactions are necessary to induce the formation of a tumor and the distribution of these molecular events over time follows a Poisson process. In practice the model appears to fit some data sets reasonably well and to give low-dose predictions that are similar to the other models. There are cases, however, in which the predicted values are inconsistent with the predictions of other models by many orders of magnitude. For instance, the virtually safe dose as predicted by the multi-hit model appears to be too high for nitrolotriacetic acid and far too low for vinyl chloride (Food Safety Council Report, 1980).

Pharmacokinetic Models

Pharmacokinetic models have often been used to predict the concentration of the parent compound and metabolites in the blood and at reactive sites, if identifiable. Cornfield (1977), Gehring and Blau (1977), and Anderson et al. (1980) have extended this concept to include rates for macromolecular events (e.g., DNA damage and repair) involved in the carcinogenic process. The addition of statistical distributions for the rate parameters and a stochastic component representing the probabilistic nature of molecular events and selection processes may represent a useful conceptual framework for describing the tumorigenic mechanism of many chemicals. Pharmacokinetic data are presently useful only in specific parts of the risk assessment process. A more complete understanding of the mechanism of chemically induced carcinogenesis would allow a more complete utilization of pharmacokinetic data. Pharmacokinetic comparisons between animals and humans are presently most useful for making species conversions and for understanding qualitative and quantitative species differences. The modeling of blood concentrations and metabolite concentrations identifies the existence of saturated pathways and adds to an understanding of the mechanism of toxicity in many cases.

Taking advantage of the similarity of the probit and pharmacokinetic models in the 5 to 95% range, Cornfield (1977) developed an approximate method of estimating its parameters, particularly the value of T, the saturation dose.

Risks at dosages below T are crucially dependent on K^*, the relative speed of the reverse, deactivation reaction and this cannot be well estimated from responses at dosages above T, so that low-dose assessment using this model may be more dependent on further pharmacokinetic experimentation than on further statistical developments.

This model considers an agent subjected to simultaneous activation and deactivation reactions, both reversible, with the probability of a response being proportional (linearly related) to the amount of active complex. Denoting total amount of substrate in the system by S and deactivating agent by T and the ratios of the rate constants governing the back and forward reactions by K for the activation step and K^*, for the deactivation step, the model is, for $D > T$

$$P(D) = \frac{D - S[P(D)] - y}{D - S[P(D)] - y + K}$$

Solving for y yields

$$y = K[P(D)]T / \{K[P(D)] + K^*[1 - P(D)]\}$$

and for $D < T$

$$P(D) \cong \frac{D}{S + K\left(1 + \frac{T}{K}\right)}$$

These equations follow from standard steady state mass action equations. Thus, at low dose levels, $D < T$, the dose-response curve is nearly linear, but for deactivating reactions in which the rate of the back reaction is small compared to that of the forward reaction, K^*, will be quite small and the slope will be near zero. In fact, in the limiting case in which $K^* = 0$ the dose-response curve has a threshold at $D = T$, but since the model is steady state and does not depend on the time course of the reaction, it cannot be considered to have established the existence of a threshold. For $K^* > 0$, the dose-response curve is shaped like a hockey stick with the striking part nearly flat and rising sharply once the administered dose exceeds the dose, T, that saturated the system. Because of the great sensitivity of the slope at low doses to the value of K^*/K, and insensitivity at high doses, responses at dose levels above $D = T$ probably cannot be used to predict those below T. This can be considered a limitation of the model, but it can equally well be considered a limitation of high-dose experimentation in the absence of detailed pharmacokinetic knowledge of metabolic pathways. The model can be generalized to cover a chain of simultaneous activating and deactivating reactions intervening between the introduction of D and the formation of activated complex, but this does not appear to change its qualitative characteristics. The kinetic constants, S, T, K, and K^*, are presumably subject to animal-to-animal variation. This variation is not formally incorporated in the model, so that the possibility of negative estimates of one or more of these constants cannot be excluded.

Weibull

Another generalization of the one-hit model is the Weibull model:

$$P(D) = 1 - \exp(\alpha - \beta D^m)$$

where m and β are parameters. Note that

$$\lim[P(D)/D^m] = \text{constant}$$

as the dose approaches zero.

Thus, in the low-dose region, this last equation is linear for $m = 1$, concave for $m < 1$, and convex for $m > 1$. With a typical set of data, the Weibull model tends to give an estimated risk at a low dose that lies between the estimates for the gamma multi-hit and the Armitage-Doll models. The Weibull distribution for time-to-tumors has been suggested by human cancers (Cook et al., 1969; Lee and O'Neill, 1971):

$$I = bD^m(t - w)^k$$

where I is the incidence rate of tumors at time t, b is a constant depending on experimental conditions, D is dosage, w the minimum time to the occurrence of an observable tumor, and m and k are parameters to be estimated. Also, Day (1967), Peto et al. (1972), and Peto and Lee (1973) have considered the Weibull distribution for time-to-tumor occurrence. Theoretical models of carcinogenesis also predict the Weibull distribution (Pike, 1966). Theoretical arguments and some experimental data suggest the Weibull distribution where tumor incidence is a polynomial in dose multiplied by a function of age. Hartley and Sielken (1977) adopted the form

$$(t) = \sum_{i=1}^{k} \xi_i t^i$$

where $\xi_i \geq 0$. They noted that this function regarded as a weighted average of Weibull hazard rates with positive weight coefficients, ξ_i. The conventional statistical procedure of weighted least-squares provides one method of fitting the Weibull model to a set of data. With a background response measured by the parameter p, the model, using Abbott's correction is:

$$P = p + (1 - p)(1 - \exp[-\beta D^m]) = 1 - \exp[-(\alpha + \beta D^m)]$$

where $\alpha + \beta D^m$.

With a nonlinear weighted least-squares regression program, one can estimate the three parameters (m, a, b) directly. With only a linear weighted least-squares regression program, one can use trial and error on m to find the values of the three parameters that produce a minimum error sum of squares. A program for one electronic calculator (the TI-59, discussed earlier) that conveniently handles up to nine data points is available from the Food Safety Council.

A nonlinear maximum likelihood method to obtain estimates of the parameters in the Weibull model can also be used. The use of the Weibull distribution for time-to-tumor leads to an extreme value distribution relating tumor response to dosage (Chand and Hoel, 1974). Hoel (1972) gives techniques for cases in which adjustments must be made for competing causes of death.

Logit

This model, like the probit model, leads to an S-shaped dose-response curve, symmetric about the 50% response point. Its equation (Berkson, 1944) is:

$$P(D) = 1/[1 + \exp\{-(\alpha + \beta \log D)\}]$$

It approaches zero response as D decreases more slowly than the probit curve, since

$$\lim 1/[P(D)/D^{\beta}] = K$$

as dose approaches zero where K is a constant.

The practical implication of this characteristic is that the logit model leads to lower VSD than the probit model, $1/25$th as much in calculations reported by Cornfield et al. (1978), even when both models are equally descriptive of the data in the observable range.

Albert and Altshuler (1973) have developed a related model for predicting tumor incidence and life shortening based on the work of Blum (1959) on skin tumor response and on Druckrey (1967) for a variety of chemical carcinogens in rodents. They had investigated cancer in mice exposed to radium. The basic relationship used was $D_i t^n = c$, where D is dosage, t is the median time to occurrence of tumors, n is a parameter greater than i, and c is a constant depending on the given experimental conditions. It is of interest to determine the time it takes for a small proportion of the population to develop tumors. With this formulation, as the dosage is increased, the time-to-tumor occurrence is shortened. Albert and Altshuler (1973) used the log-normal distribution to represent time-to-tumor occurrence, assuming the standard deviation to be independent of dosage.

The log-normal distribution of tumor times corresponds closely to the probit transformation as employed in the Mantel-Bryan procedure.

Log-Probit

The log-probit model assumes that the individual tolerances follow a log-normal distribution. Specific steps in the complex chain of events that lead to carcinogenesis are likely to have log-normal distributions. For example, it is reasonable to assume that the distribution of a population of kinetic rate constants for detoxification, metabolism, elimination, in addition to the distribution of immunosuppression surveillance capacity and DNA repair capacity, can be adequately approximated by normal or lognormal distributions.

Tolerance distribution models have been found to adequately model many types of biological dose–response data, but it is an overly simplistic expectation to represent the entire carcinogenic process by one tolerance distribution. A tolerance distribution model may give a good description of the observed data, but from a mechanistic point of view there is no reason to expect extrapolation to be valid. The probit model extrapolation has, however, fit well in some instances (Gehring et al., 1977).

The log-probit model has been used extensively in the bioassay of dichotomous responses (see Finney, 1952). A distinguishing feature of this model is that it assumes that each animal has its own threshold dose below which no response occurs and above which a tumor is produced by exposure to a chemical. An animal population has a range of thresholds encompassing the individual thresholds. The log-probit model assumes that the distribution of log dose thresholds is normal. This model states that there are relatively few extremely sensitive or extremely resistant animals in a population. For the log-probit model, the probability of a tumor induced by an exposure to a dose D of a chemical is given by

$$P(D) = \Phi(\alpha + \beta \log_{10} D)$$

where Φ denotes the standard cumulative Gaussian (normal) distribution. Chand and Hoel (1974) showed that the log-probit dose-response is obtained when the time-to-tumor distribution is log normal under certain conditions.

Miscellaneous

There are a large number of other proposed models for low-dose extrapolation, although these others have not gained any large following. Two examples of these are the extreme value and no-effect-level models.

Chand and Hoel (1974) showed that if the time-to-tumor distribution is a Weibull distribution, the dose-response model follows an extreme value model under certain conditions, with

$$P(D) = 1 - \exp[-\exp(\alpha + \beta \log D)]$$

Park and Snee (1983) made the observation that many biological responses vary linearly with the logarithm of dose, and that practical thresholds exist, and therefore the responses can be represented by the following model:

$$\text{Response} = B_1 \qquad \text{if dose} < D^*$$

$$\text{Response} = B_1 + B_2 \log(\text{dose}/D^*) \qquad \text{if dose} \geq D^*$$

This model incorporates a parameter D^* that represents a threshold below which no dose-related response occurs. In this model, B_1 is the constant response level at doses less than D^*, and B_2 is the slope of the log-dose response curve at doses $\geq D^*$. It has been empirically found that many quantitative toxicological endpoints can be adequately described by the no-effect-level model. This model may, therefore, be useful for establishing thresholds for endpoints related to the carcinogenic process in situations where information other than the simple presence or absence of a tumor is available. Both the model and predicted threshold are of value when carcinogenicity is a secondary event.

Critique and Comparison of Models

None of the models presented here (or any others) can be "proved" on the basis of biological arguments or available experimental data, but some are more attractive than others on these grounds. The multistage model appears to be the most general model according to the values of the parameters. Unfortunately, most of these models fit experimental data equally well for the observable response rates at experimental dosage levels, but they give quite different estimated responses when extrapolated to low dosage levels. There are now numerous sets of data that have been used to compare two or more of the models against each other. The comparisons presented below are examples from the literature.

Case 1

In 1971, the FDA Advisory Committee on Protocols for Safety Evaluation compared three models (probit, logit, and one-hit) using the data presented in Table 16.6.

TABLE 16.6

Experimentally Determined Incidences (%) of Animals with Tumor of Interest

Dose	Probit	Logit	One-Hit
2	69	70	75
1	50	50	50
0.5	31	30	29
0.25	16	16	16
0.125	7	8	8
0.0625	2	4	4

TABLE 16.7

Extrapolated Doses for Low Incidences of Tumors

Incidence of Animals with Tumors	Units of Dose		
	Probit	Logit	One-Hit
10^{-3}	1.5×10^{-2}	3.1×10^{-3}	1.4×10^{-3}
10^{-6}	1.4×10^{-3}	9.8×10^{-6}	1.4×10^{-6}
10^{-8}	4.1×10^{-4}	1.6×10^{-7}	1.4×10^{-8}

It should be clear that with any adequately designed and executed study, these three sets of results are indistinguishable. But in Table 16.7 the extrapolated doses needed to achieve certain incidences of response in a population are presented. They are seen to give values varying by as much as four orders of magnitude.

Case 2

Gaylor and Shapiro (1979) presented a comparison of the Mantel-Bryan and one-hit methods predicted upper confidence limits for a range of VSDs (from 10^{-2} to 10^{-6}). These calculations showed the Mantel-Bryan to be more conservative (i.e., giving a lower VSD than the one-hit method when the experimentally determined tumor rates were high but less so when they were low).

Case 3

In 1980 the Food Safety Council presented a comparison of the results of the data from 14 different experiments (presented in Table 16.8) to each of four different models (one-hit, Armitage-Doll multistage, Weibull, and the gamma multi-hit). The resulting estimates of background response, information about the parameter estimates for each model, a goodness-of-fit p-value for each of the models, and p-values for the improvement of Weibull and gamma multi-hit models are presented in Table 16.9. Note that the lower the goodness-of-fit p-value, the poorer the model handles the actual study data.

Table 16.9 shows that each of three models—the Armitage-Doll model, the Weibull model, and the multi-hit model—has a reasonable goodness-of-fit p-value (0.71, 0.80, and 0.93, respectively) in the experimental range. For example, assuming the Weibull model to be correct, the probability of seeing experimental data whose fit is not as good as the data obtained for this experiment is approximately 0.80, indicating a high degree of fit. However, in this case the one-hit model has a goodness-of-fit p-value < 0.001, indicating that this model is a poor fit to the data. Also in Table 16.9 the p-value for the improvement of fit for the Weibull model over the one-hit model is shown to be < 0.001. The Weibull model clearly provides a statistically significantly better fit to these data than does the one-hit model using conventional p- values of 0.01 or 0.05. The goodness-of-fit p-values for the one-hit, Weibull and gamma multi-hit model are based on usual chi-square tests while that for the Armitage-Doll model is

TABLE 16.8

Experimental Carcinogenicity Results for 13 Substances

Test No./ Substance	Species	Tumor or Lesion	Dose Units	Dose (Number of Responses/Number of Animals)						
A Aflatoxin B1	Rat	Liver tumor	ppb	0 (/18)	1 (2/22)	5 (1/22)	15 (4/2)	50 (20/25)	100 (28/28)	
B Bischloromethyl ether	Rat	Respiratory tumor	# of 6 h exposures at 100 ppb	10 (1/4)	20 (3/46)	40 (4/18)	60 (4/18)	80 (15/34)	100 (12/20)	
C Botulinum toxin	Mouse	Death	ng	0.027 (0/30)	0.030 (4/30)	0.034 (11/30)	0.037 (10/30)	0.040 (16/30)	0.045 (26/30)	0.050 (26/30)
D DDT	Mouse	Liver hepatoma	ppm	0 (4/11)	2 (4/105)	10 (11/124)	50 (/104)	250 (60/90)		
E Dieldrin	Mouse	Liver tumor	ppm	0 (17/156)	1.25 (11/60)	2.50 (25/58)	5.00 (44/60)			
F Dimethyl-nitrosoamine	Rat	Liver tumor	ppm	0 (0/29)	2 (0/18)	5 (4/62)	10 (2/5)	20 (15/23)		
G Ethylenethiourea	Rat	Thyroid carcinoma	ppm	0 (2/72)	5 (2/27)	25 (1/73)	125 (2/73)	250 (16/69)	500 (62/70)	
H Hexachlorobenzene	Rat	14th rib anomaly	mg/kg	0 (0/80)	10 (4/79)	20 (8/9)	40 (15/87)	60 (25/96)		
I NTA	Rat	Kidney tumor	% in diet	0 (0/127)	0.02 (0/48)	0.20 (0/48)	0.75 (1/9)	1.50 (2/9)	2.00 (12/48)	
J Sodium saccharin	Rat	Bladder tumors	% in diet	0.01 (0/25)	0.10 (0/27)	1.0 (0.27)	5.0 (1/25)	7.5 (7/29)		
K 2,3,7,8-Tetrachloro-dibenzo-p-dioxin	Rat	Intestinal anomaly	mg/kg	0 (0/24)	0.125 (0/38)	0.25 (1/33)	0.50 (3/3)	1.0 (3/10)		
L Rapeseed (span) oil	Rat	Cardiovascular lesion	% in diet	0 (1/10)	5 (1/10)	10 (4/10)	15 (4/10)	20 (5/10)		
M Vinyl chloride	Rat	Liver angiosarcoma	ppm	0 (1/59)	50 (1/59)	250 (4/59)	500 (7/59)	2500 (13/59)	6000 (13/60)	

TABLE 16.9

Results of Fitting Four Different Models to Data on 13 Substances

Cmpd.	One-Hit Model Estimated VSD[a]	One-Hit Model Goodness-of-fit (p-value)	Armitage-Doll Model Estimated VSD	Armitage-Doll Model Goodness-of-Fit (p-Value)	Weibull Model Estimated VSD	Weibull Model Goodness-of-Fit (p-Value)	Weibull Model Improvement over One-Hit (p-Value)	Gamma Multi-Hit Model Estimated VSD	Gamma Multi-Hit Model Goodness-of-Fit (p-Value)	Gamma Multi-Hit Model Improvement over One-Hit (p-Value)
A	3.4×10^{-5}	0.07	7.9×10^{-4}	0.49	4.0×10^{-2}	0.64	0.01	0.28	0.54	0.009
B	1.6×10^{-4}	0.32	4.0×10^{-4}	0.77	3.1×10^{-2}	0.81	0.04	3.7×10^{-2}	0.89	0.04
C	8.4×10^{-8}	<.001	4.2×10^{-3}	0.13	4.3×10^{-3}	0.22	<.001	1.3×10^{-2}	0.65	<.001
D	2.8×10^{-8}	0.16	6.4×10^{-4}	0.47	1.7×10^{-2}	0.22	0.1	4.9×10^{-2}	0.19	0.12
E	5.7×10^{-6}	0.07	2.2×10^{-5}	0.36	1.2×10^{-3}	0.44	0.03	6.7×10^{-3}	0.55	0.02
F	3.2×10^{-5}	0.04	1.9×10^{-2}	0.57	1.9×10^{-2}	0.63	0.003	7.7×10^{-2}	0.72	0.003
G	5.5×10^{-4}	<.001	4.5	0.71	6	0.8	<.001	33.5	0.93	<.001
H	2.1×10^{-4}	0.99	2.2×10^{-4}	0.94	2.6×10^{-4}	0.96	0.94	2.6×10^{-4}	0.96	0.95
I	2.0×10^{-5}	<.001	1.9×10^{-4}	0.09	0.52	0.48	<.001	0.8	0.48	<.001
J	4.3×10^{-5}	0.33	0.33	0.72	0.53	0.99	0.04	1.1	7.99	0.04
K	5.2×10^{-6}	0.53	1.6×10^{-3}	0.73	1.7×10^{-3}	0.85	0.12	3.8×10^{-3}	0.87	0.11
L	3.7×10^{-5}	0.78	5.7×10^{-3}	0.64	1.1×10^{-3}	0.62	0.67	3.8×10^{-3}	0.62	0.64
M	2.0×10^{-2}	0.03	2.1×10^{-9}	0.03	2.1×10^{-9}	0.56	0.002	3.9×10^{-10}	0.32	0.002

[a] Virtually safe dose at 10^{-6}.

based on the simulation procedure described in Crump et al. (1976). Likewise, the p-values for the improvement of fit for the Weibull and multi-hit models over the one-hit model are based on likelihood ratio procedures that are not generally available for the Armitage-Doll model.

Table 16.10 presents the estimates of VSD for 10^{-4} and 10^{-6} using each of the four models. These VSDs have been calculated for each model by taking $P(D_0) - P(0) = 10^{-4}$ or 10^{-6}, since $P(D_0) - P(0)$ represents the additional risk due to the added dose D_0.

Table 16.10 presents the VSD calculations for the four models at risk levels of 10^{-4} or 10^{-6}. Looking again at substance 11, ethylenethiourea, note that the range is from 5.5×10^{-2} to 63.0 ppm at a risk level of 10^{-4} and from 5.5×10^{-4} to 33.5 ppm at a risk level of 10^{-6}. Note that in this case the one-hit model, because of the imposed low-dose linearity, yields a much smaller VSD than the other three better-fitting models, which allow for low-dose nonlinearity but do not impose it. The nonlinearity in the observed data for ethylenethiourea is exhibited in the VSDs in Table 16.10 as well as in the estimates of the α_is for the Armitage-Doll model ($\alpha_1 = 0$) in Table 16.9 and the estimates of m (3.30) and k (8.23) for the Weibull and multi-hit models respectively.

Case 4

Similar attempts have been made to evaluate the validity of the one-hit model. Carlborg (1979b) has reported that the EPA has estimated, using this model, that current exposures to DDT, dieldrin, and aflatoxin are responsible for 153,000 excess liver cancers per year in the U.S. However, there are only about 7000 to 8000 liver cancers per year in the entire U.S. from all causes, including background. A similar type of assessment was made for carbon tetrachloride (Hartung, 1980), and arsenic (Downs, 1980), which led to the suggestion that the one-hit model grossly overestimated possible carcinogenic effects in humans. Carlborg's (1979a) attempt to compare the apparent occurrence of aflatoxin-induced liver cancer from several human epidemiological studies in Thailand, Kenya, Swaziland, and Mozambique, with the incidence of liver cancer predicted by animal exposure studies via several models (probit, logit, one-hit, Mantel-Bryan, Armitage-Doll multistage, Weibull, and gamma-hit) caused by environmentally relevant levels of aflatoxin, yielded similar results. While the need for such attempts at validation are critical, the actual conduct of such an undertaking is difficult because of the demands made upon the epidemiological studies to precisely quantify exposure levels and estimate background cancer incidences. Extrapolation of the data from the most sensitive rat study for the four models used (99% confidence limit) revealed that the Weibull model displayed the closest fit to the human response rates derived from the epidemiological studies. The other models predicted either greater (one-hit, Armitage-Doll, and Mantel-Bryan) or lesser (probit, logit, gamma-hit) human risks than were observed (Carlborg, 1979a).

Table 16.11 attempts to summarize the major characteristics of the eight models presented here in terms of their operating characteristics. The performance of each model in any one particular case is dependent on the nature

TABLE 16.10

Estimated Virtual Safe Dose (VSD) by 4 Models for 13 Substances

Cmpd.	Dose Unit	Estimated VSD at 10^{-4}				Estimated VSD at 10^{-6}			
		One-Hit	Armitage-Doll	Weibull	Multi-Hit	One-Hit	Armitage-Doll	Weibull	Multi-Hit
A	Ppb	3.4×10^{-3}	7.6×10^{-3}	0.4	1.2	3.4×10^{-5}	7.9×10^{-4}	4.0×10^{-2}	0.28
B	# of 6-hr exposures	1.6×10^{-2}	4.0×10^{-2}	0.47	0.48	1.6×10^{-4}	4.0×10^{-4}	3.1×10^{-2}	3.7×10^{w2}
C	ng	8.4×10^{-6}	9.1×10^{-3}	9.2×10^{-3}	1.7×10^{-2}	8.4×10^{-8}	4.2×10^{-3}	4.3×10^{-3}	1.3×10^{-2}
D	ppm	2.8×10^{-2}	6.4×10^{-2}	0.41	0.76	2.8×10^{-4}	6.4×10^{-4}	1.7×10^{-2}	4.9×10^{-2}
E	ppm	5.7×10^{-4}	2.2×10^{-3}	1.8×10^{-2}	5.1×10^{-2}	5.7×10^{-6}	2.2×10^{-5}	1.2×10^{-3}	6.7×10^{-3}
F	ppm	3.2×10^{-3}	0.19	0.19	0.41	3.2×10^{-5}	1.9×10^{-2}	1.9×10^{-2}	7.7×10^{-2}
G	ppm	5.5×10^{-2}	20.8	24.4	63	5.5×10^{-4}	4.5	6	33.5
H	mg/kg	2.1×10^{-2}	2.2×10^{-2}	2.4×10^{-2}	2.4×10^{-2}	2.1×10^{-4}	2.2×10^{-4}	2.6×10^{-4}	2.6×10^{-4}
I	% in diet	1.9×10^{-3}	1.9×10^{-2}	0.85	1	2.0×10^{-2}	1.9×10^{-4}	0.52	0.8
J	% in diet	4.3×10^{-3}	1.1	1.4	2	4.3×10^{-5}	0.33	0.53	1.1
K	mg/kg	5.2×10^{-4}	1.6×10^{-2}	1.7×10^{-2}	2.5×10^{-2}	5.2×10^{-6}	1.6×10^{-3}	3.8×10^{-3}	6.7×10^{-3}
L	% in diet	3.7×10^{-3}	5.7×10^{-3}	3.2×10^{-2}	6.7×10^{-2}	3.7×10^{-5}	5.7×10^{-5}	1.1×10^{-3}	3.8×10^{-3}
M	ppm	2	2	7.4×10^{-5}	3.0×10^{-2}	2.0×10^{-2}	2.0×10^{-2}	2.1×10^{-9}	3.9×10^{-10}

TABLE 16.11

Characteristics and Requirements for Use of Major Low-Dose Extrapolation Models

System	Low Dose Linearity	Extrapolates Low Dose Levels	Estimates Virtual Safe Dose	Mechanistic or Tolerance Distribution	Requires Metabolic Data	Accommodates Threshold	Takes Time-to-Tumor into Account	Estimate of Potential Risk of Low Dose
One-Hit (Linear)	X	X	X	M				Highest
Multistage (Armitage-Doll, 1961)	X	X	X	M				High
Weibull (Chand and Hoel, 1974)	X		X	T		X	X	High
Multi-Hit	X	X	X	M				Medium
Logit (Albert and Altshuler, 1973)	X	X	X	T		X	X	Medium
Probit (Mantel and Bryan, 1961)	X		X	T			X	Medium
Log-Probit (Gehring et al., 1977)		X	X	M	X	X	X	Low
Pharmacokinetic (Cornfield, 1977)		X	X	M	X	X	X	Lowest

of the observed dose-response curve. All fit true linear data well, but respond differently to concave or convex response curves. The actual choice of model must depend on what information is available and on the professional judgment of the investigator. The author believes that to attempt to use any purely mathematical model is wrong — that an understanding of the pharmacokinetics and mechanisms of toxicity across the dose range is an essential step in the risk assessment of carcinogens. Any mathematical model must utilize such data, and as there is now significant evidence that many of these actual response curves are multiphasic, only models that can accommodate such nonlinear response surfaces have a chance of being useful.

Risk Assessment

Having now investigated the methods that are available to extrapolate the risk of an irreversible event from a high dose range to a low dose range in the same animal species, we must now address the question of a means to extrapolate from an animal species to humans. This extrapolation has both qualitative aspects (to be discussed here) and quantitative aspects (called scaling, to be discussed later).

The qualitative aspects of species-to-species extrapolations are best addressed by a form of classification analysis (such as we reviewed in Chapter 9) tailored to the exact problem at hand. This approach identifies the physiological, metabolic, and other factors that may be involved in the risk-producing process in the model species (for example, the carcinogenesis process in test mice), establishes the similarities and differences between these factors and those in humans, and comes up with means to bridge the gaps between these two (or to identify the fact that there is no possible bridge).

Cross-Species Extrapolation

Tomatis (1979) has provided an excellent evaluation of the comparability of carcinogenicity findings between rodents and man, in general finding the former to be good predictors of the endpoint in the latter. However, in his 1984 Stokinger lecture, Weil pointed out that the model species should respond biologically to the material as similarly as possible to man; that the routes of exposure (actual and possible) should be the same; and that there are known wide variations in response to carcinogens. Deichmann (1975), for example, has reviewed studies demonstrating that 2-naphthylamine is a human and dog carcinogen, but not active in the mouse, rat, guinea pig, or rabbit.

Smith (1974) discussed interspecies variations of response to carcinogens, including N-2-fluorenyl-acetamide, which is potent for the dog, rabbit, hamster, and rat (believed to be due to formation of an active metabolite by N-hydroxylation), but not in the guinea pig or steppe lemming, which do not form this metabolic derivative.

TABLE 16.12

Classes of Factors to Be Considered in Species-to-Species Extrapolations
in Risk Assessment

I.	Sensitivity of Model Animal (Relative to Humans)
	A. Pharmacologic
	B. Receptor
	C. Lifespan
	D. Size
	E. Metabolic function
	F. Physiological
	G. Anatomic
	H. Nutritional requirements
	I. Reproductive and developmental processes
	J. Diet
	K. Critical reflex and behavioral responses (as emetic reflex)
	L. Behavioral
	M. Rate of cell division
	N. Other defense mechanisms
II.	Relative Population Differences
	A. Size
	B. Heterogeneity
	C. Selected "high class" nature of test population
III.	Differences between Test and Real World Environment
	A. Physical (temperature, humidity, etc.)
	B. Chemical
	C. Nutritional

Table 16.12 represents an overview of the classes of factors that should be considered in the first step of a species extrapolation. Examples of such actual differences that can be classified as one of these factors are almost endless.

The absorption of compounds from the gastrointestinal tract and from the lungs is comparable among vertebrate and mammalian species. There are, however, differences between herbivorous animals and omnivorous animals due to differences in stomach structure. The problem of distribution within the body probably relates less to species than to size and will be discussed later under scaling. Metabolism, xenobiotic metabolism of foreign compounds, metabolic activation, or toxification/detoxification mechanisms (by whatever name) are perhaps the critical factor, and this can differ widely from species to species. The increasing realization that the original compound administered is not necessarily the ultimate carcinogen makes the further study of these metabolic patterns critical.

In terms of excretory rates, the differences between the species are not very great; small animals tend to excrete compounds more rapidly than large ones in a rather systematic way. The various cellular and intercellular barriers seem to be surprisingly constant throughout the vertebrate phylum. In addition, it is beginning to be appreciated that the receptors, such as DNS, are comparable throughout the mammalian species.

There are lifespan (or temporal) differences that have not been considered adequately, either now or in the past. It takes time to develop a tumor, and at least some of that time may be taken up by the actual cell division process. Cell division rates appear to be significantly higher in smaller animals. Mouse and rat cells turn over faster than human cells—perhaps at twice the rate. On the other hand, the latent period for development of tumors is much shorter in small animals than in large ones (Hammond et al., 1978).

Another problem is that the lifespan of man is about 35 times that of the mouse or rat; thus there is a much longer time for a tumor to appear. These sorts of temporal considerations are of considerable importance.

Body size, irrespective of species, seems to be important in the rate of distribution of foreign compounds throughout the body. A simple example of this is that the cardiac output of the mouse is on the order of 1 ml/minute, and the mouse has a blood volume of about 1 ml. The mouse is turning its blood volume over every minute. In man, the cardiac output per minute is only 1/20 of its blood volume. So the mouse turns its blood over and distributes whatever is in the blood or collects excretory products over 20 times faster than man.

Another aspect of the size difference that should be considered is that the large animal has a much greater number of susceptible cells that may interact with potential carcinogenic agents, although there is also a proportionately increased number of "dummy" cells.

Rall (1977, 1979) and Borzelleca (1984) have published articles reviewing such factors and Calabrese (1983) and Gad and Chengelis (1988) have published excellent books on the subject.

Having delineated and quantified species differences (even if only having factored in comparative body weights and food consumption rates), we can now proceed to some form of quantitative extrapolation. This process is called scaling.

There are currently three major approaches to scaling in risk assessment. These are by fraction of diet, body weight, and body surface area (Calabrese, 1983; Schmidt-Nielsen, 1984).

The "fraction-of-diet" method is based on converting the results in the experimental animal model to man on a mg (of test substance) per kg (diet) per day basis. When the experimental model is the mouse, this leads to an extrapolation factor that is six-fold lower than on a body weight (mg/kg) basis (Association of Food and Drug Officials of the U.S., 1959). Fraction-of-diet factors are not considered accurate indices of actual dosages since the latter are influenced not only by voluntary food intake, as affected by palatability and caloric density of the diet and by single or multiple caging, but more particularly by the age of the animal. During the early stages of life, anatomic, physiologic, metabolic, and immunologic capabilities are not fully developed. Moreover, the potential for toxic effect in an animal is a function of the dose ingested — ultimately, of the number of active molecules reaching the target cell. Additionally, many agents of concern do not have ingestion as the major route of intake in man. Both the Environmental Protection Agency (EPA) and the Consumer Product Safety Commission (CPSC) frequently employ a fraction-of-diet scaling factor.

Human diets are generally assumed to be 600 to 700 g/day, while that in mice is 4 g/day and in rats 25 g/day (the equivalent of 50 g/kg/day).

There are several ways to perform a scaling operation on a body weight basis. The most common is to simply calculate a conversion factor (K) as

$$K = \frac{\text{weight of human (70 kg)}}{\text{weight of animal (0.4 kg for rat)}}$$

More exotic methods for doing this, such as that based on a form of linear regression, are reviewed by Calabrese (1983), who believes that the body weight method is preferable.

A difficulty with this approach is that the body weights of both animals and man change throughout life. An "ideal man" or "ideal rat" weight is therefore utilized.

Finally, there are the body surface area methods, which attempt to factor in differences in metabolic rates based on the principle that these change in proportion with body surface area (since as the ratio of body surface area to body weight increases, the more energy is required to maintain constant body temperature). There are several methods for doing this, each having a ratio of dose to the animal's body weight (in mg/kg) as a starting point, resulting in a conversion factor with mg/m² as the units.

The EPA version is generally calculated as:

$$(M_{\text{human}} / M_{\text{animal}})^{1/3} - \text{surface factor}$$

where M = mass in kilograms. Another form is calculated based on constants that have been developed for a multitude of species of animals by actual surface area measurements (Spector, 1956). The resulting formula for this is:

$$A = KW^{2/3}$$

where
A = surface area in cm²
K = constant, specific for each species
W = weight in grams.

A scaling factor is then simply calculated as a ratio of the surface area of man over that of the model species.

The "best" scaling factor is not generally agreed upon, although as will shortly be discussed, the FDA has recently (December 2002) settled on a body surface area approach called the HED (human equivalent dose) (FDA 2002). Although the majority opinion is that surface area is preferable where a metabolic activation or deactivation is known to be both critical to the risk producing process and present in both the model species and man, these assumptions may not always be valid. Table 16.13 presents a comparison of the weight and surface area extrapolation methods for eight species. Schneiderman et al. (1975) and Dixon (1976) have published comparisons of these methods,

TABLE 16.13

Extrapolation of a Dose of 100 mg/kg in the Mouse to Other Species

	Weight	Surface Area[a]	Extrapolated Dose (mg)		
Species	(g)	(cm²)	Body Weight (A)	Body Surface Area (B)	Ratio (A/B)
Mouse	20	46.4	2	2	1
Rat	400	516.7	40	22.3	1.79
Guinea pig	400	564.5	40	24.3	1.65
Rabbit	1500	1272	150	54.8	2.74
Dog	12,000	5766	1200	248.5	4.83
Cat	2000	1381	200	59.5	3.46
Monkey	4000	2975	400	128.2	3.12
Human	70,000	18,000	7000	775.8	9.02

[a] Surface area (except in case of man) calculated from the formula:

$$\text{surface area (cm}^2) = K(W^{2/3})$$

where K is a constant for each species and W is body weight (values of K and surface area of man taken from Spector, 1956).

but Schmidt-Nielsen (1984) should be considered the primary source on scaling in interspecies comparisons.

FDA HED Approach

In December of 2002, the therapeutic agent portions of FDA (that is, CDER and CBER) promulgated a document laying out in detail how to perform a risk assessment for a new therapeutic agent (one without previous clinical evaluation in humans) and derive a "safe" starting dose for FIM (first in man studies).

The approach was based on body surface area extrapolation between doses derived primarily from the work of Dr. Harold Boxenbaum (Boxenbaum and DiLea, 1995). Since the publication of the guidance document, other portions of FDA and other regulatory agencies have adopted similar approaches and the use of the listed data, factors, and considerations.

The approach consists of five steps and utilizes a number of acronyms, which need definitions before proceeding.

B Allometric exponent

BSA-CF Body surface area conversion factor: a factor that converts a dose (mg/kg) in an animal species to the equivalent dose in humans (also known as the human equivalent dose), based on differences in body surface area; a BSA-CF is the ratio of the body surface areas in the tested species to that of an average human

HED Human equivalent dose: a dose in humans anticipated to provide the same degree of effect as that observed in animals at a given dose. In this document, as in many communications from sponsors, the term HED is usually used to refer to the human equivalent dose of the NOAEL. When reference is made to the human equivalent of a dose other than the NOAEL, sponsors should explicitly and prominently note this usage

K A dimensionless factor that adjusts for differences in the surface area to weight ratio of species due to their different body shapes.

Km Factor for converting mg/kg dose to mg/m² dose

LOAEL Lowest-observable-adverse-effect level: the lowest dose tested in an animal species with adverse effects

MRSD Maximum recommended starting dose: the highest dose recommended as the initial dose in a clinical trial. In clinical trials of adult healthy volunteers, the MRSD is predicted to cause no adverse reactions. The units of the dose (e.g., mg/kg or mg/m²) may vary depending on practices employed in the area being investigated

MTD Maximum tolerated dose in toxicity studies: a dose that is significantly toxic

NOAEL No-observed-adverse-effect level: the highest dose tested in an animal species without adverse effects detected

NOEL No-observed-effect-level: the highest dose tested in an animal species with no detected effects

PAD Pharmacologically active dose: the lowest dose tested in an animal species with the intended pharmacologic activity

SF Safety factor: a number by which the HED is divided to introduce a margin of safety between the HED and the *maximum recommended starting dose*

W Body weight in kg

Step 1: No-Observed-Adverse-Effect (NOAEL) Determination

The first step in determining the MRSD is to review and evaluate the available animal data so that a NOAEL can be determined for each study. Several differing definitions of NOAEL exist, but for selecting a starting dose, the following is used here: the highest dose level that does not produce a significant increase in adverse effects. In this context, adverse effects that are statistically significant and adverse effects that may be clinically significant (even if they are not statistically significant) should be considered in the determination of the NOAEL. The NOAEL is a generally accepted benchmark for safety when derived from appropriate animal studies and can serve as the starting point for determining a reasonably safe starting dose of a new therapeutic in healthy (or asymptomatic) human volunteers.

The NOAEL is not the same as the NOEL, which refers to any effect, not just adverse ones, although in some cases the two might be identical. The definition of the NOAEL, in contrast to that of the NOEL, reflects the view that some effects observed in the animal may be acceptable pharmacodynamic actions of the therapeutic and may not raise a safety concern. The NOAEL should not be confused with LOAEL or MTD. Both of the latter concepts are based on findings of adverse effects and are not generally used as benchmarks for establishing safe starting doses in adult healthy volunteers. The term "level" refers to dose or dosage, generally expressed as mg/kg or mg/kg/day.

Initial IND submissions for first in human studies by definition lack human data or formal allometric comparison of pharmacokinetics. Measurements of systemic levels or exposure (i.e., AUC or Cmax) cannot be employed for setting a safe starting dose in humans, and it is critical to rely on dose and observed toxic response data from adequate and well-conducted toxicology studies. However, there are cases where data on bioavailability, metabolite profile, and plasma drug levels associated with toxicity may influence the choice of the NOAEL. One such case would be when saturation of drug absorption occurs at a dose that produces no toxicity. In this case, the lowest saturating dose, not the highest (non-toxic) dose, should be used for calculating the HED.

There are essentially three types of findings in nonclinical toxicology studies that can be used to determine the NOAEL: (1) overt toxicity (e.g., clinical signs, macro- and microscopic lesions); (2) surrogate markers of toxicity (e.g., serum liver enzyme levels); and (3) exaggerated pharmacodynamic effects. Although the nature and extent of adverse effects can vary greatly with different types of therapeutics and it is anticipated that in many instances experts will disagree on the characterization of effects as being adverse or not, the use of NOAEL as a benchmark for dose-setting in healthy volunteers should be acceptable to all responsible investigators. As a general rule, an adverse effect observed in nonclinical toxicology studies used to define a NOAEL for the purpose of dose-setting should be based on an effect that would be unacceptable if produced by the initial dose of a therapeutic in an initial (FIM) clinical trial conducted in adult healthy volunteers.

Step 2: Human Equivalent Dose (HED) Calculation

Conversion Based on Body Surface Area

After the NOAELs in the relevant animal studies have been determined, they are converted to HEDs. A decision should be made regarding the most appropriate method for extrapolating the animal dose to the equivalent human dose. Toxic endpoints for therapeutics administered systemically to animals, such as the MTD or NOAEL, are usually assumed to scale well between species when doses are normalized to body surface area (i.e., mg/m^2).

The basis for this assumption lies primarily with the work of Freireich et al. (1996) and Schein et al. (1970). These investigators reported that, for antineoplastic drugs, doses lethal to 10% of rodents ($LD_{10}s$) and MTDs in nonrodents both correlated with the human MTD when the doses were normalized to the same administration schedule and expressed as mg/m^2. Despite the subsequent analyses showing that the MTDs for this set of drugs scale best between species when doses are normalized to W 0.75 rather than W 0.67 (inherent in body surface area normalization), normalization to body surface area has remained a widespread practice for estimating an HED based on an animal dose.

An analysis of the impact of the allometric exponent on the conversion of an animal dose to the HED has been conducted. Based on this analysis and on the fact that correcting for body surface area increases clinical trial safety by resulting in a more conservative starting dose estimate, it was concluded that the approach of converting NOAEL doses to an HED based on body surface area correction factors (i.e., W 0.67) should be maintained for selecting starting doses for initial studies in adult healthy volunteers. Nonetheless, use of a different dose normalization approach, such as directly equating the human dose to the NOAEL in mg/kg, may be appropriate in some circumstances. Deviations from the surface area approach should be justified. The basis for justifying direct mg/kg conversion and examples in which other normalization methods are appropriate are described in the following subsection.

Although normalization to body surface area is an appropriate method for extrapolating doses between species, consistent factors for converting doses from mg/kg to mg/m^2 have not always been used. Given that body surface area normalization provides a reasonable approach for estimating an HED, the factors used for converting doses from each species should be standardized. Since surface area varies with W 0.67, the conversion factors are therefore dependent on the weight of the animals in the studies. However, analyses conducted to address the effect of body weight on the actual BSA-CF (body surface area conversion factor) demonstrated that a standard factor provides a reasonable estimate of the HED over a broad range of human and animal weights. The conversion factors and divisors shown in Table 16.14 are therefore recommended as the standard values to be used for interspecies dose conversions for NOAELs in CDER and CBER. These factors may also be applied when comparing safety margins for other toxicity endpoints (e.g., reproductive toxicity and carcinogenicity) when other data for comparison, i.e., AUCs, are unavailable or are otherwise inappropriate for comparison.

Basis for Using Mg/Kg Conversions

The factors in Table 16.14 for scaling animal NOAEL to HEDs are based on the assumption that doses scale 1:1 between species when normalized to body surface area. However, there are occasions for which scaling based on body weight (i.e., setting the HED [mg/kg] = NOAEL [mg/kg]) may be more appropriate. To consider mg/kg scaling for a therapeutic, the available data should show that

TABLE 16.14

Conversion of Animal Doses to Human Equivalent Doses (HED) based on Body Surface Area

Species	To Convert Animal Dose in mg/kg to Dose in mg/m², Multiply by km Below:	To Convert Animal Dose in mg/kg to HED[a] in mg/kg, Either	
		Divide Animal Dose by	Multiply Animal Dose by
Human	37	—	—
child (20 kg)[b]	25	—	—
Mouse	3	12.3	0.08
Hamster	5	7.4	0.13
Rat	6	6.2	0.16
Ferret	7	5.3	0.19
Guinea Pig	8	4.6	0.22
Rabbit	12	3.1	0.32
Dog	20	1.8	0.54
Primates:			
Monkeys[c]	12	3.1	0.32
Marmoset	6	6.2	0.16
Squirrel monkey	7	5.3	0.19
Baboon	20	1.8	0.54
Micro-pig	27	1.4	0.73
Mini-pig	35	1.1	0.95

[a] Assumes 60 kg human. For species not listed or for weights outside the standard ranges, human equivalent dose can be calculated from the formula: HED = animal dose in mg/kg x (animal weight in kg/human weight in kg) 0.33.

[b] This km is provided for reference only since healthy children will rarely be volunteers for phase 1 trials.

[c] For example, cynomolgus, rhesus, stumptail.

the NOAEL occurs at a similar mg/kg dose across species. The factors below should be satisfied before extrapolating to the HED on a mg/kg basis rather than using the mg/m² approach. Note that mg/kg scaling will give a 12–, 6–, and 2–fold higher HED than the default mg/m² approach for mice, rats, and dogs, respectively. If these factors cannot be met, the mg/m² scaling approach for determining the HED should be followed, as it will lead to a safer MRSD.

1. NOAELs occur at a similar mg/kg dose across test species (for the studies with a given dosing regimen relevant to the proposed initial clinical trial)

2. If only two NOAELs from toxicology studies in separate species are available, one of the following criteria should also be true

The therapeutic is administered orally and the dose is limited by local toxicities. Gastrointestinal (GI) compartment weights scales by W 0.94. GI volume determines the concentration of the therapeutic in the GI tract. It is thus reasonable that the toxicity of the therapeutic would scale by mg/kg

(W 1.0). The toxicity in humans (for a particular class) is dependent on an exposure parameter that is highly correlated across species with dose on a mg/kg basis. For example, complement activation by systemically administered antisense oligonucleotides in humans is believed to be dependent upon C_{max} (Geary et al., 1997). For some antisense drugs, the C_{max} correlates across nonclinical species with mg/kg dose and in such instances mg/kg scaling would be justified. Other pharmacologic and toxicologic endpoints also scale between species by mg/kg for the therapeutic. Examples of such endpoints include the MTD, lowest lethal dose, and the pharmacologically active dose.

Other Exceptions to mg/m² Scaling between Species
1. Therapeutics administered by alternative routes (e.g., topical, intranasal, subcutaneous, intramuscular) for which the dose is limited by local toxicities. Such therapeutics should be normalized to concentration (mg/area of application, for instance) or amount of drug (mg) at the application site.
2. Therapeutics administered into anatomical compartments that have little subsequent distribution outside of the compartment. Examples are intrathecal, intravesical, intraocular, intrapleural, and intraperitoneal administration. Such therapeutics should be normalized between species according to the compartmental volumes and concentrations of the therapeutic.
3. Biological products administered intravascularly with MW > 100,000 daltons. Such therapeutics should be normalized to mg/kg.

Step 3: Most Appropriate Species Selection

After the HEDs have been determined from the NOAELs from all toxicology studies relevant to the proposed human trial, the next step is to pick one HED for subsequent derivation of the MRSD. This HED should be chosen from the most appropriate species. In the absence of data on species relevance, a default position is that the most appropriate species for deriving the MRSD for a trial in adult healthy volunteers is the most sensitive species (i.e., the species in which the lowest HED can be identified).

Factors that could influence the choice of the most appropriate species rather than the default to the most sensitive species include: (1) differences in the absorption, distribution, metabolism, and elimination (ADME) of the therapeutic between the species; (2) class experience that may indicate a particular model is predictive of human toxicity; or (3) limited biological cross-species pharmacologic reactivity of the therapeutic. This latter point is especially important for biological therapeutics as many are human proteins that bind to human or nonhuman primate targets (see ICH guidance S6).

When determining the MRSD for the first dose of a new therapeutic in humans, absorption, distribution, and elimination parameters will not be known for humans. Comparative metabolism data, however, might be available based on *in vitro* studies. These data are particularly relevant when there are marked differences in both the *in vivo* metabolite profiles and HEDs in animals. Class experience implies that previous studies have demonstrated that a particular animal model is more appropriate for the assessment of safety for a particular class of therapeutics. For example, in the nonclinical safety assessment of the phosphorothioate antisense drugs, the monkey is considered the most appropriate species because monkeys experience the same dose-limiting toxicity as humans, (i.e., complement activation), whereas rodents do not. For this class of therapeutics, the MRSD would usually be based on the HED for the NOAEL in monkeys regardless of whether it was lower than that in rodents, unless unique dose-limiting toxicities were observed with the new antisense compound in the rodent species.

Similarities of biochemistry and physiology between the species and humans that are relevant to the limiting toxicities of the therapeutic should also be considered under class experience. If a species is the most sensitive but has differences in physiology compared to humans that sensitize it to the therapeutic, it may not be the most appropriate species for selecting the MRSD.

Step 4: Application of Safety Factor

Once the HED of the NOAEL in the most appropriate species has been determined, a safety factor is then applied in order to provide a margin of safety for protection of human subjects receiving the initial clinical dose. This safety factor allows for variability in extrapolating from animal toxicity studies to studies in humans resulting from: (1) uncertainties due to enhanced sensitivity to therapeutic activity in humans vs. animals, (2) difficulties in detecting certain toxicities in animals (e.g., headache, myalgias, mental disturbances), (3) differences in receptor densities or affinities, (4) unexpected toxicities, and (5) interspecies differences in absorption, distribution, metabolism, and excretion of the therapeutic. These differences may be accommodated by lowering the human starting dose from the HED of the selected species NOAEL.

In practice, the MRSD for the clinical trial is determined by dividing the HED derived from the animal NOAEL by the safety factor. The default safety factor used is 10. This is a historically accepted value, but, as described below, should be evaluated based on available information.

While a safety factor of 10 can generally be considered adequate for protection of human subjects participating in initial clinical trials, this safety factor may not be appropriate for all cases. The safety factor should be raised when there is reason for increased concern and lowered when concern is reduced due to available data that provide added assurance of safety. This can be visualized as a sliding scale, balancing findings that mitigate the concern for harm to healthy volunteers with those that suggest greater concern is warranted. The extent of the increase or decrease is largely a matter of judgment, using the available information. It is incumbent on the evaluator to clearly explain the reasoning behind the applied safety factor when it differs from the default value of 10, particularly if it is less than 10.

Considerations Increasing the Safety Factor

Of increasing concern in recent years is the realization that not all exposed populations are equally at risk. While experimental data in animals is primarily generated using healthy young adult animals, these are not necessarily representative of the entire populations that are exposed.

In particular, children, the elderly, and individuals with compromised health should at least be considered in exposure and risk assessments. Such populations may have greater degrees of susceptibility due to a range of factors (Guzelian et al., 1992; Perera 1977).

The following considerations indicate a safety concern that might warrant increasing the safety factor. In these circumstances, the MRSD would be calculated by dividing the HED by a safety factor that is greater than 10. If any of the following concerns are defined in review of the nonclinical safety database, an increase in the safety factor may be called for. If multiple concerns are identified, the safety factor should be increased accordingly.

1. Steep dose-response curve. A steep dose-response curve for significant toxicities in the most appropriate species or in multiple species may indicate a greater risk to the humans.

2. Severe toxicities. Qualitatively severe toxicities or damage to an organ system (e.g., central nervous system, CNS) indicates increased risk to humans.

3. Nonmonitorable toxicity. Nonmonitorable toxicities may include histopathologic changes in animals that are not readily monitored by clinical pathology markers.

4. Toxicities without prodromal indicators. If the onset of significant toxicities is not reliably associated with premonitory signs in animals, it may be difficult to know when toxic doses are approached in human trials.

5. Variable bioavailability. Widely divergent bioavailability in several species, with poor bioavailability in the test species used to derive the HED, suggest a greater possibility for underestimating the toxicity in humans.

6. Irreversible toxicity. Irreversible toxicities in animals suggest the possibility of permanent injury in human trial participants.

7. Unexplained mortality. Mortality that is not predicted by other parameters raises the level of concern.

8. Large variability in doses or AUC levels eliciting effect. When doses or exposure levels that produce a toxic effect differ greatly across species, the ability to predict a toxic level in humans is reduced and a greater safety factor may be called for.

9. Questionable study design or conduct. Poor study design or conduct casts doubt on the accuracy of the conclusions drawn from the data. For instance, few dose levels, wide dosing intervals, or large differences

in responses between animals within dosing groups may make it difficult to characterize the dose-response curve.

10. Novel therapeutic targets. Therapeutic targets that have not been previously clinically evaluated may increase the uncertainty of relying on the nonclinical data to support a safe starting dose in humans.

11. Animal models with limited utility. Some classes of therapeutic biologics may have very limited interspecies cross reactivity or pronounced immunogenicity or may work by mechanisms that are not known to be conserved between (nonhuman) animals and humans; in these cases, safety data from any animal studies may be very limited in scope and interpretability.

Considerations Decreasing the Safety Factor

Safety factors of less than 10 may be appropriate under some conditions. The toxicologic testing in these cases should be of the highest caliber in both conduct and design. Most of the time, candidate therapeutics for this approach would be members of a well-characterized class. Within the class, the therapeutics should be administered by the same route, schedule, and duration of administration; should have a similar metabolic profile and bioavailability; and should have similar toxicity profiles across all the species tested including humans. A smaller safety factor might also be used when toxicities produced by the therapeutic are easily monitored, reversible, predictable, and exhibit a moderate to shallow dose-response relationship with toxicities that are consistent across the tested species (both qualitatively and with respect to appropriately scaled dose and exposure).

An additional factor that could suggest a safety factor smaller than 10 would be a case where the NOAEL was determined based on toxicity studies of longer duration compared to the proposed clinical schedule in healthy volunteers. In this case, a greater margin of safety is often built into the NOAEL, as it was associated with a longer duration of exposure than that proposed in the clinical setting. This assumes that toxicities are cumulative, are not associated with acute peaks in therapeutic concentration (e.g., hypotension), and did not occur early in the repeat dose study.

Step 5: Consideration of the Pharmacologically Active Dose (PAD)

Once the MRSD has been determined, it may be of value to compare it to the PAD derived from pharmacodynamic models. If the PAD is from an *in vivo* study, an HED can be derived from a PAD estimate by using a body surface area conversion factor (BSA - CF). This HED value should be compared directly to the MRSD. If this pharmacologic HED is lower than the MRSD, it may be appropriate to decrease the clinical starting dose for pragmatic or scientific reasons. Additionally, for certain classes of drugs or biologics (e.g., vasodilators, anticoagulants, monoclonal antibodies, or growth factors), toxicity may arise from exaggerated pharmacologic effects. The PAD in these cases may be a more sensitive indicator of potential toxicity than the NOAEL and might therefore warrant lowering the MRSD.

Quantitation of Exposure

The remaining problem (or step) in performing a risk assessment is quantitating the exposure of the human population, both in terms of how many people are exposed by what routes (or means) and to what quantities of an agent they are exposed and adjusting for any special sensitivities/susceptibility factors of a group that are of specific concern (such as children or the elderly).

This process of identifying and quantitating exposure groups within the human population is beyond the scope of this text, except for some key points. Classification methods are again the key tool for identifying and properly delimiting human populations at risk. An investigator must first understand the process involved in making, shipping, using, and disposing of a material. The EPA recently proposed guidelines for such identification and exposure quantitation (EPA, 1984a). The exposure groups can be very large or relatively small subpopulations, each with a markedly different potential for exposure. For di-(2-ethyl-hexyl) phthalate (DEHP), for example, the following at-risk populations have been identified:

IV route:	3,000,000	receiving blood transfusions (50 mg/year)
	50,000	dialysis patients (4500 mg/year)
	10,000	hemophiliacs (760 mg/year)
Oral route:	10,800,000	children under 3 years of age (434 mg/year)
	220,000,000	adults (dietary contamination (1.1 mg/year)

Note: Not quantified were possible inhalation and dermal exposure.

All such estimates of exposure in humans (and of the number of humans exposed) are subject to a large degree of uncertainty. Additionally, the way in which total doses are divided in their delivery is well established as being significant (Wilson and Holland, 1982).

An alternative approach to achieving society's objective for the entire risk assessment procedures—protecting the human population from unacceptable levels of known risks—is the classical approach of using safety factors. In 1972, Weil summarized this approach as

> In summary, for the evaluation of safety for man, it is necessary to: (1) design and conduct appropriate toxicologic tests, (2) statistically compare the data from treated and control animals, (3) delineate the minimum effect and maximum no ill-effect levels (NIEL) for these animals, and (4) if the material is to be used, apply an appropriate safety factor, e.g., (a) 1/100 (NIEL) or 1/500 (NIEL) for some effects or (b) 1/500 (NIEL), if the effect was a significant increase in cancer in an appropriate test.

This approach has served society reasonably well over the years, once the experimental work has identified potential hazards and quantitated observable dose-response relationships. Until such time as the more elegant risk assessment procedures can install greater public confidence, the use of the safety factor approach should not be abandoned.

Benchmark Dose Approach

The entire process of risk assessment as discussed to this point is applicable not just to carcinogens, but also to the other classes of toxic agents that result in some form of irreversible harm. The only difference is that the concept of a threshold dose level below which no ill effects are evoked is accepted for most of these other classes (the exception being mutagens) (Olin et al., 1995).

In recent years, a new approach has been proposed by regulatory representatives for the assessment of risk of quantal endpoints in addition to carcinogenicity (Glowa, 1991). Initially the focus was on developmental toxicity (Auton, 1994; Barnes et al., 1995; National Research Council, 1994), but the proposals have now expanded to include reproductive and neurotoxicants (EPA, 1995). This approach has been labeled the "benchmark dose."

Common abbreviations used in such discussions include:

BMD   Benchmark dose

BMD/D   Benchmark dose/concentration

BMR   Benchmark response

ED10   Effective dose (10% response)

LD10   Lower bound on dose (ED10)

MLE   Maximum likelihood estimate

NOAEL   No-observed-adverse-effect level

LOAEL   Lowest-observed-adverse-effect level

RfD   Reference dose

Currently, human exposure guidelines for developmental toxicants are based on the NOAEL derived from laboratory studies. An NOAEL is defined as the highest experimental dose that fails to induce a significant increase in risk in comparison with the unexposed controls. A reference dose or reference concentration (RfD or RfC) is then obtained by dividing the NOAEL by a suitable uncertainty factor (UF) allowing for difference in susceptibility between animals and humans (Nair et al., 1995; Slob and Pieters, 1995). The resulting RfD is then used as a guideline for human exposure (Jarabek et al., 1990). Guidelines on the magnitude of the UF to be used in specific cases have been discussed by Barnes and Dourson (1988). Current U.S. Environmental Protection Agency uncertainty factors are given in Table 16.15.

The NOAEL, restricted in value to one of the experimental doses, fails to properly take sample size into account (smaller and less sensitive experiments lead to higher NOAELs than larger studies), and largely ignores the shape of the dose-response curve. The risk associated with doses at or above the NOAEL is not made explicit. However, Gaylor (1992) has shown, for a series of 120 developmental toxicity experiments, that the observed risk exceeding 1% is about $1/4$ of the cases. Leisenring and Ryan (1992) also found,

TABLE 16.15

U.S. EPA Guidelines for Uncertainty Factors

Guidelines	Factors
Average → sensitive human	≤10×
Animal → human	≤10×
LOAEL → NOAEL	≤10×
Database inadequacies	≤10×
Subchronic → chronic	≤10×
Modifying factors	0–10×

based on the statistical properties of the NOAELs, that the NOAEL may identify a dose level associated with unacceptably high risk with a reasonably high probability. Because of the limitations associated with the use of the NOAEL (Gaylor, 1983, 1989; Kimmel and Gaylor, 1988), the EPA (1991) is considering the use of the benchmark dose (BMD) method, proposed by Crump (1984), as the basis for deriving the RfD for developmental toxicity (Barnes et al., 1995).

The BMD is generally defined as the lower confidence limit (Crump, 1984, 1995; Crump et al., 1995) of the effective dose, d_α that induces α-percent increase in risk (ED$_\alpha$). Although the α-percent increase in risk may refer to the excessive risk $\pi(0) = \alpha$, or relative risk $[\pi(d_\alpha)-\pi(0)]/[1 - \pi(0)] = \alpha$, the latter takes into account the background risk in the absence of exposure and is more sensitive to high spontaneous risk. If the background risk is $\pi(0) - 0$, then the two measures of risk are equivalent. It can be shown that the relative risk also has additional mathematical properties that facilitate computation and interpretation. The ED$_\alpha$ may be defined as the solution to the equation

$$\frac{\pi(d_\alpha) - \pi(0)}{1 - \pi(0)} = \alpha$$

where $\pi(d)$ represents an appropriate dose-response model for a particular endpoint. Crump (1984) discussed the estimating of BMD based on dose-response model for a single endpoint. Allen et al. (1994a, b) estimated the BMDs using several dose-response models fitted to data from a large database. They found that the BMDs at 5% level are similar to NOAEL in magnitude on the average. Ryan (1992) and Krewski and Zhu (1995) used joint dose-response models to estimate the BMDs. A summary of such models is presented in Table 16.16.

Under the Weibull models for either the incidence of prenatal death or the incidence of fetal malformation, the ED$_\alpha$ is given by

$$d_\alpha = \left(\frac{-\log(1-\alpha)}{b} \right)^{1/\gamma}$$

TABLE 16.16

Models Used in the Study of BMD Approaches for Developmental Toxicity

Developmental Toxicity Models

RVR Model:
$P(d,s) = (1 - \exp\{-[\alpha + \beta(d - d_0)^w]\})^* \exp\{-s[\theta_1 + \theta_2(d - d_0)]\}$
NCTR Model:
$P(d,s) = 1 - \exp\{-[(\alpha + \theta_1 s) + (\beta + \theta_2 s)(d - d_0)^w]\}$
Log-Logistic Model:
$P(d,s) = \alpha + \theta_1 s + (1 - \alpha - \theta_1 s)/\{1 + \exp[\beta + \theta_2 s - \gamma \log(d - d_0)]\}$

where the subscript $k(k - 1,2)$ for the parameters (b_k, γ_k) is suppressed for simplicity of notation. Note that d is obtained by evaluating the equation at $(\hat{b}, \hat{\lambda})$. The variance of $\hat{d}_\alpha$ can be approximated by

$$\text{Var}(\hat{d}_\alpha) = \gamma^{-2} d_\alpha^2 C_1^T \Omega_1 C_1$$

using the δ-method, with the unknown parameters involved replaced by $(\hat{b}, \hat{\lambda})$. Here, Ω_1 is the covariance matrix for the estimates $(\hat{b}, \hat{\gamma})^T$ and

$$C_1 = (b^{-1}, \gamma^{-1} \log\{-b^{-1} \log(1 - \alpha)\})^T$$

The ED_α for overall toxicity, based on the trinomial model $\pi_3 = 1 - (1 - \pi_1)$ $(1 - \pi_2)$ is obtained as the solution to the equation

$$b_1 d_\alpha^\gamma + b_2^\gamma d_{2\alpha} = -\log(1 - \alpha)$$

The variance of $\hat{d}_\alpha$ based on the δ-method is given by

$$\text{Var}(\hat{d}_\alpha) = \left[b_{1\gamma_1} d_\alpha^{\gamma_1 - 1} + b_{2\gamma_2} d_\alpha^{\gamma_2 - 1} \right]^{-2} C_2^T \Omega_2 C_2$$

where Ω_2 is the covariance matrix of the estimates

$$(\hat{b}_1, \hat{\gamma}_1, \hat{b}_2, \hat{\lambda}_2)^T$$

and

$$C_2 = \left(d_\alpha^{\gamma_1}, b_1 d_\alpha^{\gamma_1} \log d_\alpha, d_\alpha^{\gamma_2}, b_2 d_\alpha^{\gamma_2} \log d_\alpha \right)^T$$

Since the variance estimates based on the d-method depend on the dose-response models and the estimates of the unknown parameters, alternative methods, such as those based on likelihood ratio (Chen and Kodell, 1989) for obtaining confidence limits of ED, may be used.

The ED_α for overall toxicity derived from the multivariate model $\pi_3 = 1 - (1 - \pi_1)(1 - \pi_2)$ is a more sensitive indicator of developmental toxicity than those for fetal malformation and prenatal death, in that the former is always below the minimum of the latter two (Ryan, 1992; Krewski and Zhu, 1995). In the absence of a strong dose-response relationship for one of the latter two endpoints, the ED_α for overall toxicity approximates their minimum. The estimates of the ED_α for overall toxicity-based multivariate dose-response models for the prenatal death rate r/m and fetal malformation rate y/s are generally expected to be more efficient than estimates based on the univariate models for the combined rate $(y + r)/m$ (Ryan, 1992). In general, risk assessment that is based on multivariate dose-response models is preferred on the ground that it can simultaneously account for each individual source of risk.

The generalized score tests for trend and dose-response modeling of multiple outcomes from developmental toxicity experiments have been discussed by others (Kavlock et al., 1995). Conditional on the number of implants per litter, joint analyses of several outcomes are numerically more stable and statistically more efficient than separate analysis of a single outcome. The extramultinomial variation induced by the litter effects may be characterized using a parametric covariance function, such as the extended Dirichlet-multinomial covariance. Alternatively, the Rao-Scott transformation, based on the concept of generalized design effects, may be used to allow for the approximation of the multinomial mean and covariance functions to the transformed data. Simple dose-response models—the Weibull model, for example—in conjunction with a power transformation for the dose can be used to describe the dose-response relationship in developmental toxicity data.

The method of generalized estimating equations, GEE, has been employed for model fitting (Fan and Chang, 1996). The GEEs are not only flexible in distributional assumptions, but also computationally simpler than the maximum likelihood estimation based, for example, on the Dirichlet-multinomial distribution. The GEE estimates of the model parameter are nearly as efficient as the maximum likelihood estimates, although estimates of the dispersion parameters that are based on the quadratic estimating equations are less efficient.

Generalized score functions after local orthogonalization can be used to construct a rich class of statistics for testing increasing trends in developmental toxicity data. These generalized score functions unify many of the specific statistics previously proposed in the literature. Further investigation on the behavior of these generalized score tests under various conditions would be useful.

Joint dose-response models can be directly applied to estimate the benchmark doses in risk assessment for developmental toxicity. The BMD, based on a multivariate dose-response model for multiple endpoints, has the advantage that it simultaneously takes into account different sources of risk. For example, the BMD, based on a multivariate dose-response model for multiple endpoints, has the advantage that it simultaneously takes into account different sources of risk. For example, the BMD for overall toxicity is a more sensitive measure of risk in that it is less than or equal to the minimum of the BMDs for fetal malformation or prenatal death (Starr, 1995a, b).

Threshold of Toxicological Concern (TTC)

The threshold of toxicological concern (TTC) is a principle that refers to the establishment of a human exposure threshold value for all chemicals, below which there would be no appreciable risk to human health (Kroes and Kozianowski, 2002; Kroes et al., 2003; Rulis, 1989). The concept that exposure thresholds, or safe levels of exposure, can be identified for individual chemicals in the diet (or pharmaceuticals) is already widely embodied in the practice of regulatory bodies in setting acceptable daily intakes (ADIs) for chemicals with known toxicological profiles. However, the TTC concept goes further than this in proposing that a *de minimus* value can be identified for many chemicals, including those of unknown toxicity, based on consideration of their chemical structures. The *de minimus* concept accepts that human exposure threshold levels exist for different types of chemicals/structures. Uncertainties are an inherent part of the risk characterization of chemicals, even where there is a full toxicological database. It is well understood that uncertainties normally exist in relation to the sensitivity of the test species studied compared to humans and the validity of the test methods to detect all adverse effects relevant to humans. The TTC can be used to assess the likelihood that a particular level of exposure to a chemical would be without toxic effects in the absence of chemical-specific toxicity data, based on the available toxicity data for a wide range of chemicals; in other words, knowledge from the "world of chemicals" is balanced against very low levels of intake of the chemical under evaluation. Reviews of available data on the "world of chemicals" are used to establish threshold levels of exposure, related to the chemical structure, that would be without significant risk. Exposure below the relevant threshold level would pose no appreciable risk to human health, despite the absence of toxicity data on the compound under consideration.

The establishment of more widely accepted TTC values would benefit consumers, industry, and regulators by avoiding unnecessary extensive toxicity testing and evaluations when human intakes are below such a threshold. The TTC approach would focus limited resources of time, animal use, cost, and expertise on the testing and evaluation of those substances with greater potential to pose risks to human health and would contribute to a reduction in the use of animals for safety testing. Application of the TTC principle would not only be used for priority setting for toxicity testing but could also be used to indicate analytical data needs and to set priorities for levels of "inherent concern." It is considered to be a preliminary step in safety assessment.

The concept of a TTC evolved from the review by Munro (1990) of the Threshold of Regulation in the U.S. The "TTC concept" formed the scientific basis of the U.S. Food and Drug Administration Threshold of Regulation for indirect food additives. The TTC principle has also been adopted by the Joint FAO/WHO Expert Committee on Food Additives (JEFCA) in its evaluations of flavoring substances. Since 1996 a decision tree incorporating different TTCs related to structural class has been used for the safety evaluation of over 1200 flavoring substances.

A threshold of regulation is used by the U.S. Food and Drug Administration to review components of food contact materials with low exposures and relates to a dietary concentration giving an intake of 1.5 μg per person per day (0.025 μg/kg bw/day). Below this level the FDA requires no specific toxicity testing and performs only an abbreviated safety assessment, mainly focused on intake assessment. Recently, EMEA likewise adopted such an approach for impurities in drug products, effectively setting 1.5 μg/day as a level below which mutagenic moieties would be considered to be too low to be a regulatory concern.

The thresholds used in the TTC approach are intakes, expressed in μg per person per day, below which a given compound of known structure is not expected to present a toxicological concern. Therefore, the TTC for a given compound has to be compared with an estimate of human exposure to this chemical to determine whether or not there is a safety concern and whether or not more detailed chemical-specific toxicity data are necessary. Thus, an appropriate human exposure estimate is a necessity for applying the TTC principle. The TTCs are calculated assuming a body weight of 60 kg, and this may need to be taken into account when the intake estimates for a compound are considered in relation to the relevant TTC (see below).

An Expert Group of the Threshold of Toxicological Concern Task Force of the European branch of the International Life Sciences Institute (ILSI Europe) has examined the TTC principle, which was based on general toxicity endpoints (including carcinogenicity), in relation to its applicability in food safety evaluation. The application of the TTC concept in food safety evaluation is not meant to replace other regulatory procedures but rather is a preliminary step in the risk assessment process to aid in the assessment of whether chemical-specific toxicity data are necessary.

The TTC principle has been evaluated for general toxicity endpoints (including carcinogenicity) as well as for specific endpoints, namely neurotoxicity and developmental neurotoxicity, developmental toxicity, and immunotoxicity. It was shown that the cumulative distribution of the no-observed-effect-levels (NOELs, equivalent in meaning to a no-observed-adverse-effect-level or NOAEL) for developmental toxicity did not differ greatly from the cumulative distribution of NOELs for chronic toxicity of Class III chemicals as described by Munro et al. (1996). The NOELs for immunotoxicity did not differ from the NOELs for other endpoints. In the case of neurotoxicants, the distribution was almost one order of magnitude lower than the distribution of NOELs for chronic toxicity of Class III chemicals. The distribution of the NOELs in the Class III chemicals differed considerably (by about three orders of magnitude higher) from the distribution of the 10^{-6} risk levels derived by linearized low-dose extrapolation for the carcinogens contained in the Gold database.

The present paper reports the considerations of the ILSI Europe Expert Group related to a number of further questions regarding the application of the TTC principle. Consideration was given to providing increased safety assurance by the identification of structural alerts for high potency carcinogens and to the question of whether neurotoxicants or teratogens should be considered as separate classes. In addition, further consideration was given to endocrine-disrupting chemicals and how food allergies, hypersensitivity reactions, and intolerances should be considered in relation to the application of the TTC principle. Finally, the Expert Group evaluated whether separate consideration of metabolism and accumulation was necessary in the application of a TTC.

Perception of Carcinogenic Risk Assessment

Both public and professional perception of the entire process leading to risk assessments of carcinogens is not good. The wide acceptance and large sales of Efron's *The Apocalyptics — Cancer and The Big Lie* (1984), which presents a broad-based critique on the entire process surrounding our understanding of environmental carcinogenesis, is an all too telling indicator of the public's increased loss of faith. The reverses of regulatory actions on benzene and formaldehyde, in part due to the faulty nature of the risk assessments presented to support these actions, have reinforced public doubts.

Such doubts are not limited to the laity. Gio Gori, formerly a deputy director of NCI, has presented an overview and general critique (Gori, 1982) of the entire process from a regulatory perspective. Hickey (1984) addressed the specific case of low-level radiation effects (where we have the most human data) from a statistician's point of view, pointing out that existing epidemiology data do not match the EPA's risk assessment. Certainly there is no consensus within toxicology, as the contents of this chapter should make clear. The consensus within the toxicology community is clearly that a more mechanistic and pharmacokinetic-based modeling process is called for. Park and Snee (1983) present an excellent outline of such an approach.

References

Albert, R.E. and Altshuler, B. (1973) Considerations relating to the formulation of limits for unavoidable population exposures to environmental carcinogens, in *Radionuclide Carcinogenesis*, Ballou, J.E. et al., Eds., pp. 233–253, AEC Symposium Series, CONF-72050, NTIS Springfield, IL.

Albert, R.E., Burns, F.J., and Altshuler, B. (1979) Reinterpretation of the linear non-threshold dose–response model in terms of the initiation–promotion mouse skin tumorigenesis, in *New Concepts in Safety Evaluation*, Mehlman, M.A., Shapiro, R.E., and Blumenthal, H., Eds., Hemisphere, New York, pp. 88–95.

Allen, B.C., Kavlock, R.J., Kimmell, C.A., and Faustman, E.M. (1994a) Dose–response assessment for developmental toxicity. II. Comparison of generic benchmark dose estimates with NOAELs. *Fundam. Appl. Toxicol.*, 23, 487–495.

Allen, B.C., Kavlock, R.J., Kimmell, C.A., and Faustman, E.M. (1994b) Dose–response assessment for developmental toxicity. III. Statistical models. *Fundam. Appl. Toxicol.*, 23, 496–509.

Anderson, M.W., Hoel, D.G., and Kaplan, N.L. (1980) A general scheme for the incorporation of pharmacokinetics in low-dose risk estimation for chemical carcinogenesis: example—vinyl chloride. *Toxicol. Appl. Pharmacol.*, 55, 154–161.

Arley, N. (1961): Theoretical analysis of carcinogenesis, in Proc. 4th Berkeley Symposium on Mathematical Statistics and Probability, Neyman, J., Ed., University of California Press, Berkeley, CA, pp. 1–18.

Armitage, P. and Doll, R. (1961) *Stochastic Models for Carcinogenesis from the Berkeley Symposium on Mathematical Statistics and Probability*, University of California Press, Berkeley, CA, pp. 19–38.

Association of Food and Drug Officials of the U.S. (1959) *Appraisal of the Safety of Chemicals in Foods, Drugs and Cosmetics*. Washington, DC.

Auton, T.R. (1994) Calculation of benchmark doses from teratology data. *Reg. Toxicol. Pharmacol.*, 19(2), 152–67.

Barnes, D.G., Daston, G.P., Evans, T.S., Jarabek, A.M., Kavlock, R.J., and Kimmel, C.A. (1995) Park C. Spitzer benchmark dose workshop: criteria for use of a benchmark dose to estimate a reference dose. *Reg. Toxicol. Pharmacol.*, 21(2), 296–306.

Barnes, D.G. and Dourson, M. (1988) Reference dose (RfD): description and use in health risk assessments. *Reg. Toxicol. Pharmacol*, 8, 471–486.

Bennicelli, C., Camoirano, A., Petruzzelli, S., Zanacchi, P., and De Flora, S. (1983) High sensitivity of *Salmonella* TA102 in detecting hexavalent chromium mutagenicity and its reversal by liver and lung preparations. *Mut. Res.*, 122, 1–5.

Berkson, J. (1944) Application of the logistic function to bio-assay. *J. Am. Stat. Assoc.*, 39, 357–365.

Blum, H.F. (1959) *Carcinogenesis by Ultraviolet Light*. Princeton University Press, Princeton, NJ.

Borzelleca, J.F. (1984) Extrapolation of animal data to man, in *Concepts in Toxicology*, vol. I, Tegeris, A.S., Ed., Karger, New York, pp. 294–304.

Boxenbaum, H. and DiLea, O. (1995) First time-in-human close selection: allometric thoughts and perspectives. *J. Clin. Pharm.*, 35, 957–966.

Bryan, W.R. and Shimkin, M.B. (1943) Quantitative analysis of dose–response data obtained with three carcinogenic hydrocarbons in strain C3H male mice. *J. Nat. Canc. Inst.*, 3, 503–531.

Cairnes, T. (1980) The ED_{01} study: introduction, objectives, and overview. *Environ. Path. Toxicol.*, 3(3), 1–7.

Calabrese, E.J. (1983) *Principles of Animal Extrapolation*. John Wiley, New York.

Carlborg, F.W. (1979a) Cancer, mathematical models and aflatoxin. *Food Chem. Toxicol.*, 17, 159–166.

Carlborg, F.W. (1979b) Comments on aspects of the EPA's water quality criteria. EPA methodology hearings, in *EPA Methodology Document*, Cincinnati, Ohio.

Chand, N. and Hoel, D.G. (1974) A comparison of models for determining safe levels of environmental agents, in *Reliability and Biometry Statistical Analysis of Lifelength*, Proschan, F. and Serfling, R.J., Eds., SIAM, Philadelphia.

Chen, J.J. and Kodell, R.L. (1989) Quantitative risk assessment for teratological effects. *J. Am. Stat. Assoc.*, 84, 966–971.

Cook, P.J., Doll, R., and Fellingham, S.A. (1969) A mathematical model for the age distribution of cancer in man. *JNCI*, 4, 93–112.

Cornfield, J. (1977) Carcinogenic risk assessment. *Science*, 198, 693–699.

Cornfield, J., Carlborg, F.W., and Van Ryzin, J. (1978) Setting Tolerance on the Basis of Mathematical Treatment of Dose–Response Data Extrapolated to Low Doses. Proc. 1st Int. Toxicology Congr., Academic Press, New York, pp. 143–164.

Crump, K. (1995) Calculation of benchmark doses from continuous data. *Risk Anal.*, 15, 79–85, 1995.

Crump, K., Allen, B., Faustman, E., Donison, M., Kimmel, C., and Zenich, H. (1995) *The Use of the Benchmark Dose Approach in Health Risk Assessment.* Risk Assessment Forum, EPA/630/R-94/007, February 1995.

Crump, K.S. (1984) A new method for determining allowable daily intakes. *Fundam. Appl. Toxicol.*, 4, 854–871.

Crump, K.S., Hoel, D.G., Langley, C.H., and Peto, R. (1976) Fundamental carcinogenic processes and their implications for low dose risk assessment. *Canc. Res.*, 36, 2973.

Crump, K.W. (1979) Dose response problems in carcinogenesis. *Biometrics*, 35, 157–167.

Day, T.D. (1967) Carcinogenic action of cigarette smoke condensate on mouse skin. *Br. J. Canc.*, 21, 56–81.

De Flora, S. (1978) Metabolic deactivation of mutagens in the *Salmonella*/microsome test. *Nature*, 271, 455–456.

De Flora, S., Bennicelli, C., Zanacchi, P., Camoriano, A., Petruzzelli, S., and Giuntini, C. (1984) Metabolic activation and deactivation of mutagens by preparations of human lung parenchyma and bronchial tree. *Mut. Res.*, 139, 9–14.

De Flora, S., Morelli, A., Zanacci, P., Bennicelli, C., and De Flora, A. (1982) Selective deactivation of ICR mutagens as related to their distinctive pulmonary carcinogenicity. *Carcinogenesis*, 3, 187–194.

Deichmann, W.B. (1975) Cummings memorial lecture—1975: the market basket: food for thought. *Am. Ind. Hyg. Assoc. J.*, 36, 411.

Dinman, B.D. (1972) "Non-concept" of "non-threshold" chemicals in the environment. *Science*, 175, 495–497.

Dixon, R.L. (1976) Problems in extrapolating toxicity data from laboratory animals to man. *Environ. Health Perspect.*, 13, 43–50.

Dixon, R.L. and Nadolney, C.H. (1985) Assessing risk of reproductive dysfunction associated with chemical exposure, in *Reproductive Toxicology*, Dixon, R.L., Ed., Raven Press, New York.

Doll, R. (1971) The age distribution of cancer: implications for models of carcinogenesis. *J. Roy. Stat. Soc.*, 134, 133–166.

Downs, T.D. (1980) Comments on the proposed guidelines and methodology for deriving water quality criteria for the protection of human health. EPA Methodology Hearings, in *EPA Methodology Document*, Cincinnati, Ohio.

Druckrey, H. (1967) Quantitative aspects in chemical carcinogenesis, in *Potential Carcinogenic Hazards from Drugs*, UICC Monogr. Ser., vol. 7., Truhaut, R., Ed., Springer-Verlag, Berlin, p. 60.

Efron, E. (1984) *The Apocalyptics.* Simon and Schuster, New York.

EPA. (1984a) Proposed guidelines for exposure assessment. *Federal Register*, 49(227), 46304–46312.

EPA. (1984b) Proposed guidelines for mutagenicity risk assessment. *Federal Register*, 49(227), 46314–46321.

EPA. (1991) Guidelines for developmental toxicity risk assessment. *Federal Register*, 56, 63797–63826.

EPA. (1995) Proposed guidelines for neurotoxicity risk assessment. *Federal Register*, 60, 52032–52056.

Epstein, S.S. (1973) The Delaney amendment. *Ecologist*, 3, 424–430.

Fan, A.M. and Chang, L.W. (1996) *Toxicology and Risk Assessment*. Marcel Dekker, New York.

FDA. (2002) Estimating the Safe Starting Dose in Clinical Trials for Therapeutics in Adult Healthy Volunteers. December 2002.

FDA Advisory Committee on Protocols for Safety Evaluation. (1971) Panel on carcinogenesis report on cancer testing in the safety evaluation of food additives and pesticides. *Toxicol. Appl. Pharmacol.*, 20, 419–438.

Finney, D.J. (1952) *Statistical Methods in Biological Assay.* Hafner, New York.

Food Safety Council. (1980) Quantitative risk assessment. *Food Chem. Toxicol.*, 18, 711–734.

Freireich, E.J., Gehan, E.A., Rall, D.P., Schmidt, L.H., and Skipper, H.E. (1966) Quantitative comparison of toxicity of anticancer agents in mouse, rat, hamster, dog, monkey and man. Cancer Chemother. Prep., 50, 219–244.

Friedman, L. (1973) Problems of evaluating the health significance of the chemicals present in foods, in *Pharmacology and the Future of Man*, vol. 2, Karger, Basel, Switzerland, pp. 30–41.

Gad, S.C. and Chengelis, C.P. (1988) *Animal Models in Toxicology,* Marcel Dekker, New York.

Gaylor, D.W. (1983) The use of safety factors for controlling risk. *J. Toxicol. Environ. Health*, 11, 329–336.

Gaylor, D.W. (1989) Quantitative risk analysis for quantal reproductive and developmental effects. *Environ. Health Perspect.*, 79, 243–246.

Gaylor, D.W. (1992) Incidence of developmental defects at the no observed adverse effects (NOAEL). *Regul. Toxicol. Pharmacol.*, 15, 151–160.

Gaylor, D.W. and Shapiro, R.E. (1979) Extrapolation and risk estimation for carcinogenesis, in *New Concepts in Safety Evaluation*, Mehlman, M.A., Shapiro, R.E., and Blumenthal, H., Eds., Hemisphere, New York, pp. 65–87.

Geary, R.S., Leeds, J.M., Henry, S.P., Monteith, D.K., and Levin, A.A. (1997) Antisense oligonucleotide inhibitors for the treatment of cancer: 1. QUESTION: pharmaco kinetic properties of phosphorothioate oligodeoxy nucleotides. *Anti Cancer Drug Des.*, 12, 383–393.

Gehring, P.J. and Blau, G.E. (1977) Mechanisms of carcinogenicity: dose response. *J. Environ. Pathol. Toxicol. Oncol.*, 1, 163–179.

Gehring, P.J., Watanabe, P.G., and Young, J.D. (1977) The relevance of dose-dependent pharmacokinetics in the assessment of carcinogenic hazards of chemicals, in *Origins of Human Cancer*, Hiatt, H., Watson, J.D., and Winsten, J.A., Eds., Cold Spring Harbor Laboratory, Cold Spring Harbor, NY, pp. 187–203.

Glowa, J. (1991) Dose-effect approaches to risk assessment. *Neusci. Biobehav. Rev.*, 15(1), 153–158.

Gori, G.B. (1982) Regulation of cancer-causing substances: utopia or reality. *Chem. Eng. News*, 56(23) 25–32.

Gross, M.A., Fitzhugh, O.G., and Mantel, N. (1970) Evaluation of safety for food additives: an illustration involving the influence of methylsalicylate on rat reproduction. *Biometrics*, 26, 181–194.

Guess, H.A., Crump, K.S., and Peto, R. (1977) Uncertainty estimates for low-dose-rate extrapolations of animal carcinogenicity data. *Canc. Res.*, 37, 3475–3483.

Guess, H.A. and Hoel, D.G. (1977) The effect of dose on cancer latency period. *J. Environ. Pathol. Toxicol. Oncol.*, 1, 279–286.

Guzelian, P.S., Henby, C.J., and Olin, S.S. (1992) *Similarities and Differences between Children and Adults: Implications for Risk Assessment.* ILSI Press, Washington, DC.

Hammond, E.C., Garfinkel, L., and Lew, E.A. (1978) Longevity, selective mortality, and competitive risks in relation to chemical carcinogenesis. *Env. Res.*, 16, 153–173.

Hartley, H.O. and Sielken, R.L. (1977) Estimation of "safe doses" in carcinogenic experiments. *Biometric*, 33, 1–30.

Hartung, R. (1980) Evaluation of the health effects section of the criteria document for ambient water for CCl4, *EPA Methodology Manual*,.

Hickey, R.J. (1984), Low-level radiation effects: extrapolation as "science." *Chem. Eng. News*, 58(42), 34–39.

Hoel, D.G. (1972) A representation of mortality data by competing risks. *Biometrics*, 28, 475–488.

Hoel, D.G., Gaylor, D.W., Kirschstein, R.L., Saffiotti, V., and Schneiderman, M.A. (1975) Estimation of risks of irreversible delayed toxicity. *J. Toxicol. Environ. Health*, 1, 133.

ICH (1997) Guidance for Industry: 56 Preclinical Safety Evaluation of Biotechnology-Derived Pharmaceuticals. *Federal Register*, 62, 222, 61515.

Jannette, K.W. (1982) Microsomal reduction of the carcinogen chromate produced chromium (V). *J. Am. Chem. Soc.*, 104, 874–875.

Jarabek, A.M., Menache, M.G., Overton, J.H., Jr., Dourson, M.L., and Miller, F.J. (1990) The U.S. Environmental Protection Agency's inhalation RfD methodology: risk assessment for air toxics. *Toxicol. Ind. Health*, 6, 279–301.

Jusko, W.J. (1972) Pharmacodynamic principles in chemical teratology: dose-effect relationships. *J. Pharm. Exp. Ther.*, 183, 469–480.

Kavlock, R.J., Allen, B.C., Faustman, E.M., and Mimmel, C.A. (1995) Dose–response assessments for developmental toxicity. IV. Benchmark doses for fetal weight changes. *Fundam. Appl. Toxicol.*, 26, 211–222.

Kimmel, C.A. and Gaylor, D.W. (1988) Issues in qualitative and quantitative risk analysis for developmental toxicity. *Risk Anal.*, 8, 15–20.

Kimmel, C.A., Ross, D.C., and Francis, E.S., Eds. (1990) Proceedings of the workshop on the qualitative and quantitative comparability of human and animal developmental neurotoxicity. *Neurotoxicol. Teratology*, 12(3):173–292.

Klaassen, C.D. and Doull, J. (1980) Evaluation of safety: toxicology evaluation, in *Casarett and Doull's toxicology*, Doull, J., Klaassen, C.D., and Amdur, M.O., Eds., Macmillan, New York, p. 26.

Kraybill, H.E. (1977) Newer approaches in assessment of environmental carcinogenesis, in *Mycotoxins in Human and Animal Health*, Rodricks, J.V., Ed. Pathotox Publishers, Park Forest South, IL, pp. 675–686.

Krewski, D. and Zhu, Y. (1995) A simple data transformation for estimating benchmark doses in developmental toxicity experiments. *Risk Anal.*, 15(1), 29–39.

Kroes, R. and Kozianowski, G. (2002) Threshold of toxicological concern (TTC) in food safety assessment. *Toxicol. Lett.*, 127, 43–46.

Kroes, R., Renwick, A.G., Cheeseman, M., Kliener, J., Mangelsdorf, I., Piersma, A., Schilter, B., Schlatter, J., van Schothorst, F., Vos, J.G., and Wurtzen, G. (2003) Structure-based thresholds of toxicological concern (TTC): guidance for application to substances present at low levels in the diet. *Food Chem. Toxicol*, 42, 65–83.

Lee, P.N. and O'Neill, J.A. (1971) The effect of both time and dose applied on tumor incidence rate in benzopyrene skin painting experiments. *Br. J. Canc.*, 25, 759–770.

Leisenring, W. and Ryan, L. (1992) statistical properties of the NOAEL. *Regul. Toxical. Pharmacol.*, 15(2), 161–171.

Leroux, B.G., Leisenring, W.M., Moolgavkar, S.H., and Faustman, E.M. (1996) A biologically based dose–response model for development. *Risk Anal.*, 16, 449–458.

Lijinksy, W., Rueber, M.D., and Riggs, C.W. (1981) Dose response studies of carcinogenesis in rats by nitrosodiethylamine. *Canc. Res.*, 41, 4997–5003.

Littlefield, N.A., Farmer, J.H., and Gaylor, D.W. (1980) Effects of dose and time in a long-dose carcinogenic study. *J. Environ. Pathol. Toxicol. Oncol.*, 3, 17–34.

Mantel, N. (1963) Part IV. The concept of threshold in carcinogenesis. *Clin. Pharmacol. Therapeutics*, 4, 104–109.

Mantel, N., Bohidar, N.R., Brown, C.C., Cimenera, J.L., and Tukey, J.W. (1975) An improved Mantel-Bryan procedure for "safety" testing of carcinogens. *Canc. Res.*, 35, 865–872.

Mantel, N. and Bryan, W.R. (1961) Safety testing of carcinogenic agents. *J. Nat. Canc. Inst.*, 27, 455–470.

Mantel, N. and Schneiderman, M.A. (1975) Estimating "safe" levels, a hazardous undertaking. *Canc. Res.*, 35, 1379–1386.

Miller, E.C. (1978) Some current perspectives on chemical carcinogens in humans and experimental animals: presidential address, *Canc. Res.*, 38, 1471.

Munro, I.C. (1990) Safety assessment procedures for indirect food additives: an overview. *Regul. Toxicol Pharmacol*, 12, 2–12.

Munro, I.C., Ford, R.A., Kennepohl, E., and Sprenger, J.G. (1996) Correlation of structural class with no-observed effect levels: a proposal for establishing a threshold of concern. *Food Chem. Toxicol.*, 34, 829–867.

Nair, R.S., Stevens, M.S., Martens, M.A., and Ekuta, J. (1995) Comparisons of BMD with NOAEL and LOAEL values derived from subchronic toxicity studies. *Arch. Toxicol. Suppl.*, 17, 44–54.

National Research Council. (1994) *Science and Judgment in Risk Assessment*. National Academy Press, Washington, DC.

Olin, G., Farland, W., Park, C., Rhomberg, L., Schenplein, R., Starr, T., and Wilson, J. (1995) *Low-Dose Extrapolation of Cancer Risks*. ILSI Press, Washington, DC.

Park, C.N. and Snee, R.D. (1983) Quantitative risk assessment. State-of-the-art for carcinogenesis. *Am. Stat.*, 37, 427–441.

Perera, F. (1977) Environment and cancer: who are susceptible? *Science*, 278, 1068–1073.

Peto, R. (1978) The carcinogenic effects of chronic exposure to very low levels of toxic substances. *Environ. Health Perspect.*, 22, 155–159.

Peto, R. and Lee, P.N. (1973) Weibull distributions for continuous carcinogenesis experiments. *Biometrics*, 29, 457–470.

Peto, R., Lee, P.N., and Paige, W.S. (1972) Statistical analysis of the bioassay of continuous carcinogens. *Roy. J. Canc.*, 26, 258–261.

Petrilli, F.L. and De Flora, S. (1978a) Metabolic deactivation of hexavalent chromium mutagenicity. *Mut. Res.*, 54, 139–147.

Petrilli, F.L. and De Flora, S. (1978b) Oxidation of inactive trivalent chromium to the mutagenic hexavalent form. *Mut. Res.*, 58, 167–173.

Petrilli, F.L. and De Flora, S. (1980) Mutagenicity of chromium compounds, in *Chromate Symposium 80. Focus of a* Standard, Industrial Health Foundation, Pittsburgh, PA, pp. 76–99.

Petrilli, F.L. and De Flora, S. (1982) Interpretations on chromium mutagenicity and carcinogenicity, in *Mutagens in Our Environment*, Sorsa, M. and Vainio, H., Eds., Alan R. Liss, New York, pp. 453–464.

Pike, M.C. (1966) A method of analysis of a certain class of experiments in carcinogenesis. *Biometrics*, 22, 142–161.

Portier, C.J. and Hoel, D.H. (1984) Design of animal carcinogenicity studies for goodness-of-fit of multistage models. *Fund. Appl. Toxicol.*, 4, 949–959.

Rai, K. and Van Ryzin, J. (1979) Risk assessment of toxic environmental substances based on a generalized multihit model, in *Energy and Health*, Breslow, N.E. Ed., SIAM Press, Philadelphia, pp. 99–177.

Rai, K. and Van Ryzin, J. (1985) A dose–response model for teratological experiments involving quantal responses. *Biometrics*, 41, 1–9.

Rall, D.P. (1977) Species differences in carcinogenicity testing, in *Origins of Human Cancer*, Hiatt, H.H., Watson, J.D., and Winsten, J.A., Eds., Cold Spring Harbor Laboratories, Cold Spring Harbor, NY, pp. 1283–1290.

Rall, D.P. (1979) Relevance of animal experiments to humans. *Environ. Health Perspect.*, 32, 297–300.

Roberts, C.N. (1989) *Risk Assessment—The Common Ground*. Life Science Research, Eye, Suffolk, U.K.

Rulis, A.M. (1989) Establishing a threshold of regulation. *Adv. Risk Anal.*, 7, 271–278.

Ryan, L. (1992) Quantitative risk assessment for developmental toxicity. *Biometrics*, 48, 163–174.

Schein, P.S., Davis, R.D., Carter, S., Newman, J., Schein, D.R., and Rall, D.P. (1970) The evaluation of anticancer drugs in dogs and monkeys for the prediction of qualitative toxicities in man. *Clin. Pharm. Therap*; 11, 3–40.

Schmidt-Nielsen, K. (1984) *Scaling: Why Is Animal Size So Important.* Cambridge University Press, New York.

Schneiderman, M.A., Mantel, N., and Brown, C.C. (1975) From mouse to man — or how to get from the laboratory to Park Avenue and 59th Street. *Ann. N.Y. Acad. Sci.*, 246, 237–248.

Slob, W. and Pieters, M.N. (1995) Probabilistic Approach to Assess Human RfDs and Human Health Risks from Toxicological Animal Studies. *Proc. Ann. Mtg. Society for Risk Analysis* and Japan Section of SRA, Abstract D8.04-A; 60.

Smith, R.L. (1974) The problem of species variations. *Ann. Nutr. Alim.*, 28, 335.

Spector, W.S. (1956) *Handbook of Biological Data.* W.B. Saunders, Philadelphia.

Staffa, J.A. and Mehlman, M.A. (1980) Innovations in cancer risk assessment (ED01 Study). *J. Environ. Pathol. Toxicol. Oncol.*, 3, 1–246.

Starr, T.B. (1995a) Concerns with the benchmark dose concept. *Toxicology Summer Forum*, Aspen, CO, July 14.

Starr, T.B. (1995b) The Benchmark Dose Concept: Questionable Utility for Risk Assessment? *Proc. Ann. Mtg. Society for Risk Analysis* and Japan Section of SRA, Abstract H2.03:86-7. Waikiki, Hawaii, December 3–6, 1995.

Steinhoff, D.G., Gad, S.C., Mohr, U., and Hatfield, G.K. (1986) Chronic intratracheal instillation study of sodium dichromate in rats. *Exp. Pathol.*, 30, 129–141.

Stokinger, H.E. (1977) Toxicology and drinking water contaminants. *J. Awwa*, 69(10), 399–402.

Tomatis, L. (1979) The predictive value of rodent carcinogenicity tests in the evaluation of human risks. *Ann. Rev. Pharmacol. Toxicol.*, 19, 511–530.

Weil, C.S. (1972) Statistics vs. safety factors and scientific judgment in the evaluation of safety for man. *Toxicol. Appl. Pharmacol.*, 21, 454–463.

Weil, C.S. (1984) Some questions and opinions: issues in toxicology and risk assessment. *Am. Ind. Hyg. Assoc. J.*, 45, 663–670.

Wilson, J.S. and Holland, L.M. (1982) The effect of application frequency on epidermal carcinogenesis assays. *Toxicology*, 24, 45–53.

Wolfe, B. (1977) Low-level radiation: predicting the effects. *Science*, 196, 1387–1389.

Yanysheva, N. Ya and Antomonov, G. Yu, (1976) Predicting the risk of tumor occurrence under the effect of small doses of carcinogens. *Environ. Health Perspect.*, 13, 95–99.

17

Epidemiology

Epidemiology is formally defined as the science of epidemics. In practice it is not limited only to the study of infectious diseases. *The Oxford English Dictionary* (Simpson and Weiner, 1989) now defines an epidemic in different terms as: "Epidemic, adjective. (1) Of a disease, prevalent among a people or community at a special time and produced by some special causes not generally present in the affected locality. (2) Widely prevalent; universal." The literal meaning is "upon the population." This includes much of "demography," which is given as "statistics of births, death, diseases, etc." "Epidemiology" as it has grown and matured has reverted to this literal meaning. It is about population studies of human health in the broadest sense.

Epidemiology looks at the association between adverse effects seen in humans and a selected potential "cause" of interest, such as use of or exposure to a medical device, pharmaceutical, or chemical contaminant. Those involved in the biologic safety of chemicals or devices must have a working knowledge of epidemiology, as it has been intimately involved in most of the major safety concerns (real and perceived) associated with drugs, devices, and chemicals during the last 20 years and provides an essential tool for risk assessment (McMahon and Pugh, 1970; Gordis, 1988; Kleinbaum et al., 1982; Lilienfeld et al., 1994).

In 1849, Snow collected statistics of cholera cases in central London and related the incidence to where their drinking water came from (Snow, 1936). Little was then known about infected water. We would now say that he was doing epidemiology. The whole process of how the disease is transmitted in polluted water would now be called the etiology of the disease. Once this became known, the pioneer studies had achieved their ends and were forgotten over the years.

Similarly, it was an empirical finding that dairymaids working with cows did not get smallpox. They contracted a mild disease, cowpox, which made them immune. All those vaccinated because of this discovery were inoculated with cowpox.

The empirical stage of our knowledge is certainly not always short lived. The evidence that "(cigarette) smoking may damage your health" is only epidemiological 30 years after the first results were widely publicized. It is strong evidence because it is made up of many different strands, which separately are breakable. The etiology of bronchitis, emphysema, and lung cancer, as related to cigarette smoking, is still complex and largely unknown.

In other situations, the combined evidence is less strong. "On the basis of epidemiological findings we are now being advised, as ordinary citizens, to avoid tobacco and alcohol; to limit our consumptions of sugar, milk, dairy products, eggs, and fatty meats; to rid ourselves of obesity, and to engage in violent — although some suggest moderate — physical exercise...." (Burch, 1978).

The classic book by Bradford Hill (1971) consists largely of elementary statistics and, for its time, advanced scientific method. He provided many examples of misuses of statistics comparable with those in a well-known popular exposition.

Small-sample statistics and "exact" inference have helped in epidemiology; for example, when the data being interpreted consist of very many small groups. Usually, however, there are disturbing factors that make rigorous inference impossible. Changes in the underlying assumptions, and in the models, make far more difference than that between a moderately good and an optimal method of data analysis or inference.

Experimental design of the kind that was so brilliantly successful in agriculture is generally impracticable for human subjects; even methods used in clinical trials have little place. But there have been exceptional situations where experiments are practically and ethically allowable. For example, there were the pioneer studies on coal miners begun before 1950. This was in an isolated valley in the south Wales mining area in Great Britain. It was thought that a disabling disease, pulmonary massive fibrosis, was brought about by a tubercular infection. The coal workers in this population were therefore provided with intensive medical examinations to detect and treat tuberculosis at a very early stage. In a control valley the corresponding routine medical checks were given. In both populations chest x-rays and various lung physiological variables were measured for many years in the same individuals. This was therefore a longitudinal study. Follow-up studies are still being published. Much more information has been obtained than the original plan provided for (Cochrane et al., 1979).

This is a good instance where the statistician has more to contribute. The blood pressure of an individual increases with age and also with his or her weight and body fat. For much of adult life, evidence from cross-sectional studies is that the increase with age is accounted for by the increase in arm girth. These are studies in which the population is measured at many or all ages, but at the same time. In a longitudinal study the same individuals are measured over the years. If they are born within a few years of one another, this population is called a cohort. The blood pressure of an individual fluctuates even over a few hours or less. The distribution of blood pressures in people of the same age varies itself with age; the variance increases, and the distribution is of log-normal type. But among old people the distribution is truncated: very high blood pressure reduces the chance of survival.

The example of blood pressures (or, more recently, of heart valve defect rate in obese individuals) is one where epidemiologists need to know what normal values are. This can be a major problem when lung variables are being investigated, such as forced expiratory volume (FEV) and forced vital capacity (FVC). These are measured and assessed by performance tests; for FEV the subject blows as hard as he can into a bag, after one or two practice blows. Sometimes he improves with practice; sometimes he tires. There have been many arguments and analyses on the question: is his best performance the best measure? (Or should it be the mean of the best three, or the mean of all except the first two practice blows?) Fortunately it seems to make little difference which is chosen. However, the actual distribution of the performances ought to yield extra information. The object is to assess all kinds of external or environmental factors on the lungs; e.g., industrial dusts, atmospheric pollution, cigarette smoking, the weather, the climate, and the season. We also want to detect early signs of chronic bronchitis and asthma. Again, we need to know what happens in normal individuals. Age, weight, and height are important, and interesting empirical laws have been found relating these to the lung variables.

Epidemiology is thus sometimes simply defined as the study of patterns of health in groups of people (Paddle, 1988). Behind this deceptively simple definition is an incredibly complex and diverse science, rich in concepts and methodology. The group of people of interest can be very small, historically consisting of as few as two people (Goudie et al., 1985). At the opposite extreme, studies of the geographic distribution of diseases using national mortality and cancer incidence rates have provided clues about the etiology of several diseases such as cardiovascular disease and stomach cancer. The patterns of health studied are also wide ranging and may include the distribution, course, and spread of disease. The term "disease" also has a loose definition in the context of epidemiology and might include ill-defined conditions such as Ginger Jake and toxic shock syndromes or consist of an indirect measure of impairment such as biochemical and hematological parameters or lung function measurements.

Epidemiology and toxicology differ in many other ways but principally in that epidemiology is essentially an observational science, in contrast to the generally experimental nature of toxicology. The opportunistic nature of epidemiology has been commented upon by several authors (e.g., Paddle, 1988; Utidjian, 1987). The epidemiologist often has to make do with historical data that have been collected for reasons that have nothing to do with epidemiology. Nevertheless, the availability of personnel records such as lists of new employees and terminations, payrolls and work rosters, and exposure monitoring data collected for compliance purposes has enabled many epidemiological studies to be conducted in the occupational setting. Thus, the epidemiologist has no control over who is exposed to an agent or device, the levels at which they are exposed to the agent of interest, or to what other agents they may be exposed. The epidemiologist has great difficulty in ascertaining what exposure has taken place and certainly has no control over lifestyle variables such as diet, exercise, and smoking.

Despite the lack of precise data, the epidemiologist has one major advantage over the toxicologist: an epidemiology study documents the actual health experiences of human beings subjected to real-life exposures in an occupational, environmental, or clinical use setting. Indeed, Smith (1988) has recently expressed the view that the uncertainty in epidemiology studies resulting from exposure estimation may be no greater than or less than the uncertainty associated with extrapolation of results from animals to man. Regulatory bodies such as the U.S. Environmental Protection Agency (EPA) are starting to change their attitudes toward epidemiology and to recognize that it has a role to play in the process of risk assessment (Greenland, 1987). However, there is a continuing need for epidemiologists to introduce more rigor into the conduct of their studies and to introduce standards equivalent to the GLPs under which animal experiments are performed.

Measurement of Exposure

Wegman and Eisen (1988) made the point that epidemiologists place much greater emphasis on the measure of response than on the measure of exposure. They claim that this is because most epidemiologists have been trained as physicians and are consequently more oriented toward measuring health outcomes. It is certainly true that a modern textbook of epidemiology such as Rothman's (1986) provides minimal guidance about what the epidemiologist should do with exposure assessments. However, this is probably as much a reflection of the historical paucity of quantitative exposure information as a reflection on the background of epidemiologists. Nevertheless, it is surprising how many epidemiological studies do not contain even a basic qualitative assessment of exposure. The contrast between epidemiology and toxicology is never more marked than in the area of estimation of dose-response. Not only can the toxicologists carefully control the conditions of the exposure to the agent of interest, but they can also generally be sure that his animals have not come into contact with any other toxic agents. The medical epidemiologist conducting a study of patients exposed to a hepatotoxin would certainly have to control for alcohol intake and possibly for exposure to other hepatotoxins in the clinical, work, and home environments. Nevertheless, it can be argued that epidemiologic studies more accurately measure the effect on human health of real-life exposures. The epidemiologist must therefore frequently develop sampling strategies that generate exposure data suitable for both compliance and epidemiological purposes. For patients, of course, the situation is usually different. Quantitating exposures is generally straightforward.

If an exposure matrix has been constructed with quantitative estimates of the exposure in each device use and time period, then it is a simple matter to estimate cumulative exposure. It is a more difficult process when, as is commonly the cause occupationally, only a qualitative measure of exposure is available, e.g., high, medium, or low. Even when exposure measurements are available, it may not be sensible to make an assumption that an exposure that occurred 20 years ago is equivalent to the same exposure yesterday. The use of average exposures may also be questionable, and peak exposures may be more relevant in the case of outcomes such as asthma and chronic bronchitis.

Epidemiological Study Designs

What follows is a brief introduction to the most important types of studies conducted by epidemiologists, with an attempt to briefly describe the principles of the major types of epidemiological studies in order to assist the toxicologist in understanding the reporting of epidemiological studies and the assumptions made by epidemiologists.

Cohort Studies

Historical Cohort Study

When the need arises to study the health status of a group of individuals, there is often a large body of historical data that can be utilized. If sufficient information exists on individuals exposed in the past to a potential hazardous device or material, then it may be possible to undertake a retrospective cohort study. The historical data will have been collected for reasons that have nothing to do with epidemiology. Nevertheless, the availability of medical records and morbidity and mortality indices has enabled many epidemiological studies to be conducted (in particular, mortality studies).

The principles of a historical cohort study can also be applied to follow a cohort of patients prospectively. It should be emphasized that many historical data studies have a prospective element insofar as they are updated after a further period of follow-up. There is no reason why reproductive performance, incidence of serious bacterial infections, or almost any measure of the health status of an individual should not be studied retrospectively if sufficient information is available.

Mortality and cancer incidence studies are unique among retrospective cohort studies in that they can be conducted using national cancer and mortality reports even if there has been no medical surveillance of the patient population of interest. A historical cohort study also has the advantages of being cheaper and providing estimates of the potential hazard much earlier than a prospective study. However, historical cohort studies are beset by a variety of problems. Principal among these is the problem of determining which individuals have been exposed to the exact agent of interest and, if so, to what degree. In addition, it may be difficult to decide what an appropriate comparison group is. It should also be borne in mind that in epidemiology, unlike animal experimentation, random allocation is not possible and there is no control over the factors that may distort the effects of the exposure of interest, such as smoking, alcohol use, and standard of living.

Cohort Definition and Follow-Up Period

A variety of sources of information are used to identify patients exposed to a particular potential hazard, to construct a medical and occupational history, and to complete the collection of information necessary for follow-up. It is essential that the cohort be well defined and that criteria for eligibility be strictly followed. This requires that a clear statement be made about membership of the cohort so that it is easy to decide whether a patient is a member or not. It is also important that the follow-up period be carefully defined. For instance, it is readily apparent that the follow-up period should not start before exposure has occurred. Furthermore, it is uncommon for the health effect of interest to manifest itself immediately after the initiation of agent exposure, and allowance for an appropriate biological induction (or latency) period may need to be made when interpreting the data.

Comparison Subjects

The usual comparison group for many studies is the appropriate portion of the national population. However, it is known that there are marked regional differences in the mortality rates for many causes of death. Regional mortality rates exist in most industrialized countries, but have to be used with caution because they are based on small numbers of deaths and estimated population sizes. In some situations the local rates for certain causes may be highly influenced by the mortality of the patients being studied. Furthermore, it is not always easy to decide what the most appropriate regional rate for comparison purposes is, as many employees may reside in a different region from that in which the site of exposure is situated.

An alternative or additional approach is to establish a cohort of unexposed but otherwise comparable individuals for comparison purposes. For example, some studies of breast implant recipients have been restricted to female healthcare professionals and use others from this same group who have not received implants as a control group. However, patients with very low exposures to the device type of interest can often provide similar information. A good discussion of the issues is found in the proceedings of a conference entirely devoted to the subject (MRC, 1984).

Analysis and Interpretation

In a cohort study the first stage in the analysis consists of calculating the number of deaths expected during the follow-up period. In order to calculate the expected deaths for the cohort, the survival experience of the cohort is broken down into individual years of survival known as "person-years." Each person-year is characterized by the age of the cohort member and the time period when survival occurred and the sex of the cohort member. The person-years are then multiplied by age, sex, and time-period-specific mortality rates to obtain the expected number of deaths. The ratio between observed and expected deaths is expressed as a standardized mortality ratio (SMR) as follows:

$$\text{SMS} = 100 \times \frac{\text{observed deaths}}{\text{expected deaths}}$$

Thus, an SMR of 125 represents an excess mortality of 25%. An SMR can be calculated for different causes of death and for subdivision of the person-years by factors such as level of exposure and time since first exposure.

Interpretation of cohort studies is not always straightforward, and there are a number of selection effects and biases that must be considered (Rothman, 1986). Occupational cohort studies routinely report that the mortality of active workers is less than that of the population as a whole. It is not an unexpected finding, since workers usually have to undergo some sort of selection process to become or remain workers. Nevertheless, this selection effect, known as the "healthy worker" effect, can lead to considerable arguments over the interpretation of study results, particularly if the cancer mortality is as expected, but the all-cause mortality is much lower than expected. Similar possibilities must be considered for device recipient cohorts, who are likely to receive more medical services than the general population.

Proportional Mortality Study

There are often situations where one has no accurate data on the comparison of a cohort but does possess a set of death records (or cancer registrations). Under these circumstances a proportional mortality study may sometimes be substituted for a cohort study. In such a mortality study the proportions of deaths from a specific cause among the study deaths is compared with the proportion of deaths from that cause in a comparison population. The results of a proportional mortality study are expressed in an analogous way to those of the cohort study with follow-up. Corresponding to the observed deaths from a particular cause, it is possible to calculate an expected number of deaths based on mortality rates for that cause and all causes of death in a comparison group and the total number of deaths in the study. The ratio between observed and expected deaths from a certain cause is expressed as a proportional mortality ratio (PMR) as follows:

$$SMS = 100 \times \frac{\text{observed deaths}}{\text{expected deaths}}$$

Thus, a PMR of 125 for a particular cause of death represents a 25% increase in the proportion of deaths due to that cause. A proportional mortality study has the advantage of avoiding the expensive and time-consuming establishment and tracing of a cohort, but the disadvantage of little or no exposure information.

Prospective Cohort Study

Prospective cohort studies are no different in principle from historical cohort studies in terms of scientific logic, the major differences being timing and methodology. The study starts with a group of apparently healthy individuals whose health and exposure is studied over a period of time. As it is possible to define in advance the information that is to be collected, prospective studies are theoretically more reliable than retrospective studies. However, long periods of observation may be required to obtain results.

Prospective cohort studies or longitudinal studies of continually changing health parameters such as lung function, incidence of inflammatory joint disease, blood biochemistry, and hematological measurements pose different problems from those encountered in mortality and cancer incidence studies. The relationships between changes in the parameters of interest and device exposure measurements have to be estimated and, if necessary, a comparison made of changes in the parameters between groups. These relationships may be extremely complicated, compounded by factors such as aging, and difficult to estimate, as there may be relatively few measurement points. Furthermore, large errors of measurement in the variables may be present because of factors such as within-laboratory variation and temporal variation within individuals. Missing observations and withdrawals may also cause problems, particularly if they are dependent on the level and change of the parameter of interest. These problems may make it difficult to interpret and judge the validity of statistical conclusions. Nevertheless, prospective cohort studies provide the best means of measuring changes in health parameters and relating them to exposure.

Case-Control Study

In a case-control study (also known as a case-referent study) two groups of individuals are selected for study, of which one has the disease whose causation is to be studied (the cases) and the other does not (the control). In the context of the chemical industry, the aim of a case-control study is to evaluate the relevance of past exposure to the development of a disease. This is done by obtaining an indirect estimate of the rate of occurrence of the disease in an exposed and unexposed group by comparing the frequency of exposure among cases and controls.

Principal Features of Case-Control Studies

Case-control and cohort studies complement each other as types of epidemiological study (Schlesselman, 1982). In a case-control study the groups are defined on the basis of the presence or absence of a given disease and, hence, only one disease can be studied at a time. The case-control study compensates for this by providing information on a wide range of exposures or other causes (background health problems of patients, for example) that may play a role in the development of the disease. In contrast, a cohort study generally focuses on a single exposure but can be analyzed for multiple disease outcomes. A case-control study is a better way of studying rare diseases because a very large cohort would be required to demonstrate an excess of a rare disease. In contrast, a case-control study is an inefficient way of assessing the effect of an uncommon exposure, when it might be possible to conduct a cohort study of all those exposed.

The complementary strengths and weaknesses of case-control and cohort studies can be used to one's advantage. Increasingly, mortality studies are being reported that utilize "nested" case-control studies to investigate the association between the exposures of interest and a cause of death for which an excess has been discovered. However, case-control studies have traditionally been held in low regard, largely because they are often badly conducted and interpreted. There is also a tendency to overinterpret the data and misuse statistical procedures. In addition, there is still considerable debate among leading epidemiologists themselves as to how controls should be selected, e.g., Poole (1986) and Schlesselman and Stadel (1987).

Analysis and Interpretation

In a case-control study it is possible to compare the frequencies of exposures in the cases and controls. However, what one is really interested in is a comparison of the frequencies of the disease in the exposed and the unexposed. The latter comparison is usually expressed as a relative risk (RR), which is defined as:

$$RR = \frac{\text{rate of disease in exposed group}}{\text{rate of disease in unexposed group}}$$

It is clearly not possible to calculate the RR directly in a case-control study, since exposed and unexposed groups have not been followed in order to determine the rates of occurrence of the disease in the two groups. Nevertheless, it is possible to calculate another statistic, the odds ratio (OR), which, if certain assumptions hold, is a good estimate of the RR. For cases and controls the exposure odds are simply the odds of being exposed, and the OR is defined as

$$OR = \frac{\text{cases with exposure}}{\text{controls with exposure}} \bigg/ \frac{\text{cases without exposure}}{\text{controls without exposure}}$$

An OR of 1 indicates that the rate of disease is unaffected by the treatment being studied. An OR greater than 1 indicates an increase in the rate of disease in exposed workers.

Matching

Matching is the selection of a comparison group that is, within stated limits, identical with the study group with respect to one or more factors, such as age, years of device use or treatment, smoking history, etc., that may distort the effect of the exposure of interest. The matching may be done on an individual or group basis. Although matching may be used in all types of studies, including follow-up and cross-sectional studies, it is more widely used in case-control studies. It is common to see case-control studies in which each case is matched to as many as three or four controls.

Nested Case-Control Study

In a cohort study the assessment of exposure for all cohort members may be extremely time consuming and demanding of resources. If an excess of death of incidence has been discovered for a small number of conditions, it may be much more efficient to conduct a case-control study to investigate the effect of exposure. Thus, instead of all members being studied, only the cases and a sample of non-cases would be compared with regard to treatment history. Thus, there is no need to investigate the exposure histories of all those who are neither cases nor controls. However, the nesting is only effective if there are a reasonable number of cases and sufficient variation in the treatment of the cohort members.

Other Study Designs

Descriptive Studies

There are large numbers of records in existence that document the health of various groups of people. Mortality statistics are available for many countries and even for certain devices and treatment types (e.g., Pell et al. 1978; Paddle, 1981). Similarly, there is a wide range of routine morbidity statistics; in particular, those based on cancer registrations (Waterhouse et al., 1982). These health statistics can be used to study differences between geographic regions (e.g., maps of cancer mortality and incidence presented at a recent symposium; Boyle et al., 1989), device use, and time periods. Investigations based on existing records of the distribution of disease and of possible causes are known as descriptive studies. It is sometimes possible to identify hazards associated with the development of rare conditions from observation of clustering in occupational groups, treatment groups, or geographical areas.

Cross-Sectional Studies

Cross-sectional studies measure the cause (treatment) and the effect (disease) at the same point in time. They compare the rates of diseases or symptoms of a treated group with an untreated group. Strictly speaking, the treatment information is ascertained simultaneously with the disease information. In practice, such studies are usually more meaningful from an etiological or causal point of view if the treatment assessment reflects treatment and past medical history. Current information is often all that is available but may still be meaningful because of the correlation between current treatment and relevant past treatments.

Cross-sectional studies are widely used to study the health of groups of individuals who are exposed to possible hazards but do not undergo complete regular surveillance (asthmatics and diabetics, for example). They are particularly suited to the study of subclinical parameters such as blood biochemistry and hematological values. Cross-sectional studies are also relatively straightforward to conduct in comparison with prospective cohort studies and are generally simpler to interpret.

Longitudinal and Cross-Sectional Data and Growth

Data on growth are needed for many problems. They provide a clear example of the difference between cross-sectional and longitudinal data. At the beginning of puberty the growth rate of an offspring slows down, and after this there is a spurt. This then slowly flattens out. But the time of onset of puberty varies. Its mean is earlier in girls, but for both girls and boys in defined populations it is a distribution. The consequence is that a cross-sectional curve of mean height against age, for populations of boys and girls separately, shows almost no sign of puberty: the structure shown in the individual curves is lost (Tanner et al., 1966).

In a related example, it is generally believed that the onset of puberty became earlier in many countries during periods when the general standard of living rose. This would show up in cross-sectional mean heights for different years, for, say, 14-year-olds, although from such data alone, they could simply be becoming taller at all ages. In fact, evidence from countries that have or had compulsory military service indicates that both may be true. The poorer children are shorter as teenagers, they go on growing for longer, but they do not catch up completely.

Form of Age-Incidence Curves in General

We have been moving from small to large populations. Data on age-incidence curves for many diseases does not need to be obtained from an epidemiological study, but from annual mortality statistics. These are, of course, cross-sectional; we would expect cohort data to be more realistically related to a model. Despite this, these positive power laws, which can be valid over several decades of age, are found everywhere. Burch's collection contains over 200 examples. Mostly these are given in the cumulative form

$$G(t) = \theta[1 - \exp\{-k(t - \lambda)^r\}]^n$$

Here θ is the proportion of the population that is susceptible, $G(t)$ the proportion that gets the disease at age t or earlier, and λ is assumed to be a constant latent period between the end of an initiation or promotion period and the event actually observed. In this theory, n and r are integers and are related to numbers of clones of cells and numbers of mutations, which finally lead to this event. Often either n or $r = 1$.

Whether Burch's underlying theory is right or not, a great variety of data are described economically in this way; they can be used as input in testing almost anything relevant. In epidemiology it is well nigh impossible to ignore age distributions in the populations under study.

There are several related models in which observed powers are related to numbers of mutations. It seems certain that a summation of random time intervals is involved. An alternative interpretation for those positive power laws is that they are simply parts of the increasing part of probability curves. Then there is no restriction to integral powers. However, it seems possible to fit the same data to a function like

$$b(t - v)^k$$

with different values of v, so that k depends on v. On both interpretations the curves have maxima, and the best evidence comes when a decrease at high ages is actually seen, but of course at high ages, say over 60, such a survival population can be very different from that for the age range of, say 50 to 60 (Cook et al., 1969; Defares et al., 1973; and Doll, 1971).

An Example Where Age Must Be Allowed for: Side Effects from Adding Fluoride to Drinking Water

Obviously, whether fluoridation is effective in preventing dental caries depends on longitudinal population studies of the teeth of children drinking altered water over several years. There has also been a heated controversy over possible side effects. In the U.S., data were available from 1950 to 1970 for 20 large cities, of which 10 had fluoridated water (F+) and 10 did not (F–). The crude death rates from all malignant neoplasms showed a greater increase, starting from near equality, in the F+ cities. The two groups of cities, however, had different distributions of sex, age, and race, which *changed* differently during these 20 years. Allowing for these, the excess cancer rate increased during these 20 years by 8.8 per 100,000 in F+ and by 7.7 in F–, but in proportional terms the excess cancer rate increased by 1% in F+ and 4% in F–. Then it was found that 215 too many cancer deaths had been entered for Boston, F+, because those for Suffolk County, which contained Boston, had been transcribed instead (Oldham and Newell, 1977).

After this had been put right, the relative increase was 1% less for F+ than for F–. On the whole, this is in line with other evidence. This example shows, however, how obvious and trivial pitfalls are interspersed with ones that are far from obvious.

Intervention Studies

Not all epidemiology is observational, and experimental studies have a role to play in evaluating the efficiency of an intervention program to prevent disease (e.g., fluoridation of water). An intervention study at one extreme may closely resemble a clinical trial with individuals randomly selected to receive some form of intervention (e.g., use of latex gloves). However, in some instances it may be a whole community that is selected to form the intervention group. The selection may or may not be random. The toxicologist might argue that even if selection were random, such a study of two communities, each consisting of many individuals, was in a sense a study of only two subjects. However, he should ask himself first whether the "three rats to a cage" design of many subacute toxicity studies really generates three independent responses per cage.

Finally, the reliability of the collected data in epidemiology studies is, as with toxicology studies, reflective of the effort and care put into collecting the data. In particular, it should be kept in mind that data arising from medical assessments are much preferred to that generated by patient self-reports.

Scope of Epidemiology in the 1990s

On any live problem in this field, active researchers disagree. Such controversies are usually well aired. It is less obvious when a major question is whether the field belongs to epidemiology at all. In relation to typical risk problems, for example on the long-term effects of ionizing radiation or of taking "soft" drugs, one may have to decide whether to study human populations only or also to experiment on animals on a large scale. It is probably the case that in all countries with good welfare services and medical records, studies on human beings are less costly than those on animals. "Record linkage," e.g., for records for an individual obtained at different times and places, is in its early stages and involves ethical problems of secrecy.

The tendency is now more toward social and "preventive medicine," that is, observing human subjects before they become ill and early detection of causes of health problems (FDA, 1981); for the age group 15 to 25, road accidents now provide the largest single cause of death in most of the countries listed. Interpreting accident data seems to belong far more to epidemiology than to, say, traffic engineering. Epidemiological statisticians studying such data seem less likely to be misled by inappropriate indices. An example is the number of fatal accidents to passengers traveling by air per million of miles traveled. But most such accidents take place at or near takeoff or landing; hence the increase in risk for a long flight compared with a short one is very little. On the other hand, such an index is probably reasonable for railway passengers, and possibly so for car drivers. For them we have to distinguish between accidents where another vehicle is or is not involved. Where it is involved, there are more accidents in urban areas where traffic is dense than in rural areas or on motorways. This leads in turn to another misuse of such statistics, in that in towns, motorists have to drive more slowly — namely, to a doubtful conclusion that it is not more risky to drive faster. Naturally, effects of speed have to be compared when other things are comparable; data for motorways and streets should not be pooled.

This nonmedical example shows up pitfalls that epidemiological statisticians are always being confronted with: consider the population at risk, consider third variables correlated with both of two other variables, ask whether the underlying model is reasonable with any set of comparative statistics (Checkoway et al., 1989).

In conclusion, the main object in this account has been to give instructive examples of what epidemiological statisticians have been and are doing. As such, it is far from complete; the geographical and reference coverage is even less so. However, there are many references in the books and articles available.

Conclusion

There are fundamental differences between human and animal studies. Some of these are summarized in Table 17.1. But, in the end it is, of course, effects in humans that we are most concerned about. This gives well-conducted and conclusive studies significantly greater weight than preclinical studies in assessing human risk (Brown and Paddle, 1988; Glocklin, 1987; Freireich et al., 1966; Davidson et al., 1986; Calabrese, 1986).

TABLE 17.1

Differences between Animal and Human Studies

Parameter	Animal Study	Human Study
Ethics	Provided that governmental animal cruelty/rights acts are not contravened, it is perfectly acceptable to knowingly expose the animal to carcinogens, mutagens, teratogens, etc.	It is unethical to knowingly and deliberately expose humans to carcinogens, mutagens, teratogens, etc.
Conduct	Good laboratory practice (strict adherence to GLP)	Protocol for study (protocol may change during study)
Subject observation	Monitored case histories (record of animal health throughout study)	Exhaustive followup (sometimes subjects are untraceable/disappear)
Dose	Regulated exposure (defined dose at defined intervals)	Defined exposed group (it may only be known whether there was a potential for exposure but not at what level)
Length of exposure	Depending on the suspected effect of the chemical (generally lifetime for carcinogens, throughout organogenesis for teratogens, generations for reprotoxins)	Various (depending on whether chemical is an occupational, marketplace, or environmental hazard)
Pattern of exposure	Single chemicals at around the maximal tolerated dose, dose levels constant	Mixed exposure at varied levels (usually "pulse" exposure)
Comparison groups	Randomized uniformity (control group known to have no exposure, otherwise identical with exposed group)	Valid unexposed group (it can be assumed that the only different variable is exposure)
Genetic homogeneity	Generally "inbred" strain used; hence, high degree of genetic homogeneity	High degree of heterogeneity
Death	Standardized necropsy (every animal subject to pathological examination)	High degree of heterogeneity
Relevance	Extremely relevant to the species in which data were generated (trans-species relevance unknown)	Extremely relevant to man

References

Boyle, P., Muir, C.S., and Grundmann, E. (1989) *Cancer Mapping*. Springer-Verlag, Berlin.

Brown, L.P. and Paddle, G.M. (1988) Risk assessment: animal or human model? *Pharm. Med.*, 3, 361–374.

Burch, P.R.J. (1978) Smoking and lung cancer: the problem of infering cause. *J. Roy. Statist. Soc. A*, 141, 437–458.

Calabrese, E.J. (1986) Animal extrapolation and the challenge of human heterogeneity. *J. Pharm. Sci.*, 75, 1041–1046.

Checkoway, H., Pearce, N., and Crawford-Brown, D.J. (1989) *Research Methods in Occupational Epidemiology.* Oxford University Press, New York.

Cochrane, A.L., Haley, T.J.L., Moore, F., and Hole, D. (1979) The mortality of men in the Rhondda Fach, 1950–1970. *Br. J. Ind. Med.*, 36, 15–22.

Cook, P.J., Doll, R., and Fellingham, S.A. (1969) A mathematical model for the age distribution of cancer in man. *Int. J. Canc.* 4, 93–112.

Davidson, I.W.F., Parker, J.C., and Beliles, R.P. (1986) Biological basis for extrapolation across mammalian species. *Regul. Toxicol. Pharmacol.*, 6, 211–237.

Defares, J.G., Sneddon, I.N., and Wise, M.E. (1973) *An Introduction to the Mathematics of Medicine and Biology*, 2nd ed. North Holland, Amsterdam, pp. 589–601.

Doll, R. (1971) The age distribution of cancer: implication for models of carcinogenesis. *J. Roy. Stat. Soc. Stat. Soc.*, 134, 133–166.

FDA. (1981) Protection of human subjects/informed consent/standards for institutional review boards for clinical investigations. *Federal Register*, 53 FR45678.

Freireich, E.J., Gehan, E.A., Rall, D.P., Schmidt, L.H., and Skipper, H.E. (1966) Quantitative comparison of toxicity of anticancer agents in the mouse, rat, hamster, dog, monkey and man. *Canc. Chemother. Rep.*, 80, 219–244.

Glocklin, V.C. (1987) Current FDA perspective on animal selection and extrapolation, in *Human Risk Assessment: The Role of Animal Selection and Extrapolation*, Roloff, M.V., Ed., Taylor and Francis, London, pp. 15–22.

Gordis, L., Ed. (1988) *Epidemiology and Health Risk Assessment.* Oxford University Press, New York.

Goudie, R.B., Jack, A.S., and Goudie, B.M. (1985) Genetic and developmental aspects of pathological pigmentation patterns. *Curr. Top. Pathol.*, 74, 132–138.

Greenland, S., Ed. (1987) *Evolution of Epidemiologic Ideas: Annotated Readings on Concepts and Methods.* Epidemiology Resources, Chapel Hill, NC.

Hill, A.B. (1971) *Principles of Medical Statistics*, 9th ed., Oxford University Press, London.

Kleinbaum, D.G., Kupper, L.L., and Morgenstern, H. (1982) *Epidemiologic Research.* Van Nostrand Reinhold, New York.

Lilienfeld, D.E., Stolley, P.D., and Abraham, M. (1994) *Foundations of Epidemiology.* Oxford University Press, New York.

McMahon, B. and Pugh, T.F. (1970) *Epidemiology, Principles and Methods.* Little, Brown, Boston.

MRC. (1984) Expected Numbers in Cohort Studies. *Medical Research Council Environmental Epidemiology Unit*, Scientific Report No. 6, Southampton, U.K.

Oldham, P.D. and Newell, D.J. (1977) Fluoridation of water supplies and cancer — a possible association? *Appl. Statist.*, 26, 125.

Paddle, G.M. (1981) A strategy for the identification of carcinogens in a large, complex chemical company, in *Quantification of Occupational Cancer*, Banbury Report 9, Petro R., and Schneiderman, M. Eds., Cold Spring Harbor Laboratory, Cold Spring Harbor, NY, pp. 177–186.

Paddle, G.M. (1988) Epidemiology, in *Experimental Toxicology: The Basic Principles*, Anderson, D. and Conning, D.M., Eds., Royal Society of Chemistry, London, pp. 436–456.

Pell, S., O'Berg, M., and Karrh, B. (1978) Cancer epidemiologic surveillance in the Du Pont company. *J. Occup. Med.*, 20, 725–740.

Poole, C. (1986) Exposure opportunity in case-control studies. *Am. J. Epidemiol.*, 123, 352–358.

Rothman, K.J. (1986) *Modern Epidemiology.* Little, Brown, Boston.

Schlesselman, J.J. (1982) *Case-Control Studies: Design, Conduct, Analysis.* Oxford University Press, New York.

Schlesselman, J.J. and Stadel, B.V. (1987) Exposure opportunity in epidemiologic studies. *Am. J. Epidemiol.*, 125, 174–178.

Simpson, J.A. and Weiner, E.S.C. (Eds.) (1989) *The Oxford English Dictionary,* 2nd ed. Oxford University Press, New York.

Smith, A.H. (1988) Epidemiologic input to environmental risk assessment. *Arch. Environ. Health*, 43, 124–127.

Snow, J. (1855) *On the Mode of Communication of Cholera.* John Churchill, London.

Snow, J. (1936) *Snow on Cholera.* Commonwealth Fund, New York, pp. 1–175.

Tanner, J., Whitehouse, R., and Takaishi, M. (1966) Standards from birth to maturity for height, weight, height velocity, and weight velocity: British children 1965, parts I and II. *Arch. Dis. Child*, 41, 454–471, 613–635.

Utidjian, H. (1987) The interaction between epidemiology and animal studies in industrial toxicology, in *Perspectives in Basic and Applied Toxicology,* Ballantyne, B., Ed., John Wright, Bristol, U.K., pp. 309–329.

Waterhouse, J.A.H., Muir, C.J., Shanmugaratnam, K., and Powell, J., Eds., (1982) *Cancer Incidence in Five Continents*, vol. IV. International Agency for Research on Cancer, Lyon, France IARC Scientific Publication No. 42.

Wegman, D.H. and Eisen, E.A. (1988) Epidemiology, in *Occupational Health, Recognizing and Preventing Work-Related Disease*, Levy, B.S. and Wegman, D.H., Eds., Little, Brown, Boston, pp. 55–73.

18

Structure Activity Relationships

Structure activity relationship (SAR) methods have become a legitimate and useful part of toxicology since the mid-1970s. These methods are various forms of mathematical or statistical models that seek to predict the adverse biological effects of chemicals based on their structure. The prediction may be of either a qualitative (carcinogen/noncarcinogen) or quantitative (LD_{50}) nature, with the second group usually being denoted as QSAR (quantitative structure activity relationship) models. It should be obvious at the onset that the basic techniques utilized to construct such models are those that were discussed earlier in Chapter 8 (modeling and extrapolation) and Chapter 10 (methods for the reduction of dimensionality).

The concept that the biological activity of a compound is a direct function of its chemical structure is now at least a century old (Crum-Brown and Fraser, 1869). During most of this century, the development and use of SARs were the domain of pharmacology and medicinal chemistry. These two fields are responsible for the beginnings of all the basic approaches in SAR work, usually with the effort being called drug design. An introductory medicinal chemistry text (such as Foye et al., 1995) is strongly recommended as a starting point for SAR. Additionally, *Burger's Medicinal Chemistry* (Wolff, 1995), with its excellent overview of drug structures and activities, should enhance at least the initial stages of identifying the potential biological actions of *de novo* compounds using a pattern recognition approach.

Having already classified SAR methods into qualitative and quantitative, it should also be pointed out that both of these can be approached on two different levels. The first is on a local level, where prediction of activity (or lack of activity) is limited to other members of a congeneric series or structure near neighbors. The accuracy of predictions via this approach is generally greater but is of value only if one has sufficient information on some of the structures within a series of interest.

The second approach is prediction of activity over a wide range, generally based on the presence or absence of particular structural features (functional groups).

For toxicology, SARs have a small but important and increasing number of uses at present. These can all be generalized as identifying potentially toxic effects, or restated as three main uses:

1. For the selection and design of toxicity tests to address endpoints of possible concern.

2. If a comprehensive or large testing program is to be conducted, SAR predictions can be used to prioritize the tests, so that outlined questions (the answers to which might preclude the need to do further testing) may be addressed first.

3. As an alternative to testing at all. Although in general it is not believed that the state-of-the-art for SAR methods allows such usage, in certain special cases (such as selecting which of several alternative candidate compounds to develop further and then test) this use may be valid and valuable.

Basic Assumptions

Starting with the initial assumption that there is a relationship between structure and biological activity, we can proceed to more readily testable assumptions.

First, that the dose of chemical is subject to a number of modifying factors (such as membrane selectivities and selective metabolic actions) that are each related in some manner to chemical structure. Indeed, absorption, metabolism, pharmacologic activity, and excretion are each subject to not just structurally determined actions, but also to (in many cases) stereo-specific differential handlings.

Given these assumptions, actual elucidation of SARs requires the following:

1. Knowledge of the biological activities of existing structures.

2. Knowledge of structural features that serve to predict activity (also called molecular parameters of interest).

3. One or more models that relate 2 to 1 with some degree of reliability.

There are now extensive sources of information as to both toxic properties of chemicals and, indeed, biological activities. These include books, journals, and manual and computerized databases. The reader is directed to Wexler et al.'s 2000 book as a guide to accessing the different sources of toxicology information, but is cautioned to remember that there is also extensive applicable information in the realms of medicinal chemistry and pharmacology, as exemplified by *Burger's* (Wolff, 1995).

Molecular Parameters of Interest

Which structural and physicochemical properties of a chemical are important in predicting its toxicologic activity are both open to considerable debate (Kaufman et al., 1983; Tamura, 1983; Tute, 1983). Table 18.1 presents a partial

TABLE 18.1

Molecular Parameters of Interest

ELECTRONIC EFFECTS
Ionization constants
Sigma substituent constant
Distribution constant
Resonance effect
Field effect
Molecular orbital indices[a]
Atomic/electron net charge
Nucleophilic superdelocalizability
Electrophilic superdelocalizability
Free radical superdelocalizability
Energy of the lowest empty molecular orbital
Energy of the highest occupied molecular orbital
Frontier atom–atom polarizability
Intermolecular coulombic interaction energy
Electric field created at point [A] by a set of charges on a molecule

HYPROPHOBIC PARAMETERS
Partition coefficients
Pi substituent constants
R_m value in liquid–liquid chromatography
Elution time in high-pressure liquid chromatography (HPLC)
Solubility
Solvent partition coefficients
pKa

STERIC EFFECTS
Intramolecular steric effects
Steric substituent constant
Hyperconjugation correction
Molar volume
Molar refractivity, MR substituent constants
Molecular weight
Van der Waals radii
Interatomic distances

SUBSTRUCTURAL EFFECTS
Three-dimensional geometry
Fragment and molecular properties (see Kubinyi, 1995 for substituent effects)
Chain lengths

[a] Calculated or theoretical parameters.

list of such parameters. The reader is referred to a biologically oriented physical chemistry text (such as Chang, 1981) both for explanations of these parameters and for references to sources from which specific values may be obtained.

There are now several systems available to study the three-dimensional structural aspects of molecules and their interactions. Various molecular modeling sets, molecular design and analysis packages, and molecular graphics software packages are available for personal computers and larger systems. Use of such forms of graphic structural examination as a tool or method in SAR analysis has been discussed by Cohen et al. (1974) and Gund et al. (1980). Such methods are generally called topological methods.

SAR Modeling Methods

A detailed review of even the major methodologies available for SAR/QSAR modeling in toxicology is beyond the scope of this book. Although we will briefly discuss the major approaches, the reader is directed to one of the several very readable introductory articles (Chang, 1981) or books (Olson and Christoffersen, 1979; Topliss, 1983; or Goldberg, 1983) for somewhat detailed presentations.

To begin with, it should be made clear that all the actual techniques involved in the performance of SAR analysis have already been presented in this text. It is only their actual application to data that sets such analysis apart from the forms of modeling we have previously looked at.

All the current major SAR methods used in toxicology can be classified based on what kinds of compound-related or structural data they use and what method is used to correlate this structural data with the existing biological data.

The more classical approaches use physicochemical data (such as molecular weight, free energies, etc.) as a starting point. The major approaches to this are by manual pattern recognition methods, cluster analysis, or regression analysis. It is this last, in the form of Hansch or linear-free energy relationships (LFER), that actually launched all SAR work (other than that on limited congeneric cases) into the realm of being a useful approach. Indeed, still foremost among the QSAR methods is the model proposed by Hansch and his co-workers (Hansch, 1971). It was the major contribution of this group to propose the incorporation of earlier observations of the importance of the relative lipophilicity to biologic activity into the formal LFER approach to provide a general QSAR model for biological effects. As a suitable measure of lipophilicity, the partition coefficient ($\log P$) between 1-octanol and water was proposed, and it was demonstrated that this was an approximately additive and constitutive property and that it was therefore calculable, in principle, from molecular structure. Using a probabilistic model for the Hansch equation, which can be expressed as:

$$\log(1/C) = -k\pi^2 + k'\pi + p\sigma + k''$$

or

$$\log(1/C) = -k(\log P)^2 + k'(\log P) + p\sigma + k''$$

where C is the dose that elicits a constant biological response (e.g., ED_{50}, LD_{50}), p is the substituent lipophilicity, $\log P$ is the partition coefficient, σ is the substituent electronic effect of Hammet, and k, k', p, and k'' are the regression coefficients derived from the statistical curve-fitting. The reciprocal of the concentration reflects the fact that higher potency is associated with lower dose, and the negative sign for the π_0 or $\log P_0$.

The statistical method used to determine the coefficients above is multiple linear regression. A number of statistics are derived in conjunction with such a calculation, which allows the statistical significance of the resulting correlation to be assessed. The most important of these are s, the standard deviation, r^2, the coefficient of determination or percentage of data variance accounted for by the model (r, the correlation coefficient is also commonly cited), and F, a statistic for assessing the overall significance of the derived equation, values, and confidence intervals (usually 95%) for the individual regression coefficients in the equation. These most be low to assure true "independence" or orthogonality of the variables, a necessary condition for meaningful results.

In a like manner, there are a number of approaches for using structural and substructural data and correlating these to biological activities. Such approaches are generally classified as regression analysis methods, pattern recognition methods, and miscellaneous others (such as factor analysis, principal components, and probabilistic analysis).

The regression analysis methods that use structural data have been, as we will see when we survey the state-of-the-art in toxicology, the most productive and useful. "Keys," or fragments of structure, are assigned weights as predictors of an activity, usually in some form of the Free–Wilson model (Free and Wilson, 1964), which was developed at virtually the same time as the Hansch. According to this method, the molecules of a chemical series are structurally decomposed into a common moiety (or core) that may be substituted in multiple positions. A series of linear equations of the form

$$BA_i = \sum_j a_j XX_{ij} + \mu$$

are constructed where BA is the biological activity, X_j is the jth substituent with a value of 1 if present and 0 if not, a_j is the contribution of the jth substituent to BA, and μ is the overall average activity. All activity contributions at each position of substitution must sum to zero. The series of linear equations thus generated is solved by the method of least squares for the a_j and μ. There must be several more equations than unknowns and each substituent should appear more than once at a position in different combinations with substituents at other positions. The favorable aspects of this model are:

1. Any set of quantitative biological data may be employed as the dependent variable
2. No independently determined substituent constants are required
3. The molecules comprising a sample of interest may be structurally dismembered in any desired or convenient manner
4. Multiple sites of variable substitution are readily handled by the model

There are also several limitations: a substantial number of compounds with varying substituent combinations is required for a meaningful analysis; the derived substituent contributions give no reasonable basis for extrapolating predictions from the substituent matrix analyzed; and the model will break down if nonlinear dependence on substituent properties is important or if there are interactions between the substituents.

Pattern recognition methods comprise yet another approach to examining structural features and/or chemical properties for underlying patterns that are associated with differing biological effects. Accurate classification of untested molecules is again the primary goal. This is carried out in two stages. First, a set of compounds, designated the training set, is chosen for which the correct classification is known. A set of molecular or property descriptors (features) is generated for each compound. A suitable classification algorithm is then applied to find some combination and weight of the descriptors that allow perfect classification. Many different statistical and geometric techniques for this purpose have been used and were presented in earlier chapters. The derived classification function is then applied in the second step to compounds not included in the training set to test predictability. In published works these have generally been other compounds of known classification. Performance is judged by the percentage of correct predictions. Stability of the classification function is usually tested by repeating the procedure several times with slightly altered, but randomly varied, sets or samples.

The main difficulty with these methods is in "decoding" the QSAR in order to identify particular structural fragments responsible for the expression of a particular activity. Even if identified as "responsible" for activity, far harder questions for the model to answer are whether the structural fragment so identified is "sufficient" for activity, whether it is always "necessary" for activity, and to what extent its expression is modified by its molecular environment.

TABLE 18.2

Existing SAR Models for Toxicology Endpoints

Endpoint	Prediction		Reference
	Quantitative	Qualitative	
Mutagenicity	X		Asher and Zervos (1977)
		X	Niculescu-Duvaz et al. (1981)
Carcinogenicity	X		Enslein et al. (1983)
	X		Franke (1973)
	X		Asher and Zervos (1977)
	X	X	Niculescu-Duvas et al. (1981)
	X		Enslein and Craig (1982)
Sensitization	X		Dupuis and Benezra (1982)
LD$_{50}$		X	Enslein et al. (1983)
Developmental toxicity	X		Enslein et al. (1983)
Biological oxygen demand (BOD)		X	Enslein et al. (1984)
Reproductive toxicity		X	Bernstein (1984)

Most pattern recognition methods use as weighting factors either the presence or absence of a particular fragment or feature (coded 1 or 0) or the frequency of occurrence of a feature. They may be made more sophisticated by coding the spatial relationship between features.

Enslein et al. (1984) have published a good brief description of the problems involved in applying these methods in toxicology.

Applications in Toxicology

SAR methods have been developed to predict a number of toxicological endpoints (mutagenesis, carcinogenesis, dermal sensitization, lethality [LD$_{50}$ values], biological oxygen demands, and teratogenicity) with varying degrees of accuracy, and models for the prediction of other endpoints are under development. Some of these existing models are presented by category of use in Table 18.2. Additionally, both EPA and FDA have models for mutagenicity/carcinogenicity that they utilize to "flag" possible problem compounds.

It should be expected that qualitative models are more "accurate" than quantitative ones, and that the more possible mechanisms associated with an endpoint, the less accurate (or more difficult) a prediction.

Reproductive

Structure–activity relationships have not been well studied in reproductive toxicology. Data are available that suggest structure–activity relationships for certain classes of chemicals (e.g., glycol ethers, phthalate esters, heavy metals). Yet, for other agents, nothing in their structure would have identified them as male reproductive toxicants (e.g., chlordecone). Bernstein (1984) reviewed the literature and has offered a set of classifications relating structure to reported male reproductive activity. Although limited in scope and in need of rigorous validation, such schemes do provide hypotheses that can be tested. Enslein et al. (1983) have proposed a commercial computer model for this and developmental endpoints. Comparison of the chemical or physical properties of an agent with those of known male reproductive toxicants may provide some indication of a potential for reproductive toxicity. Such information may be helpful in setting priorities for testing of agents or for the evaluation of potential toxicity when only minimal data are available.

Eye Irritation

Quantitative structure–activity relationship (QSAR) analysis, widely used to predict various physiological and biochemical activities of novel chemicals, also has been used to predict eye irritancy of structurally related chemicals. Using QSAR, Sugai et al. (1990) examined the eye irritancy (opacity and conjunctivitis) of 131 structurally heterogeneous substances. The accuracy was 86.3% for classifying irritancy of the chemicals. Overall accuracy rates as high as 91% have been reported. Although this approach may provide useful information on structurally related chemicals, its utility is limited for formulated products.

Lethality

Analysis of the structure–activity relationships within a class of chemicals can yield valuable information and may reduce the number of bioassays conducted. QSAR analysis is particularly useful during the discovery stage for selection of chemicals for further development. QSAR also can be used for prioritization of chemicals for various actions related to health and safety and environmental assessment. The elements generally needed for QSAR include a verified bioassays database for the endpoint to be predicted; a set of chemical–physical parameters that described the chemical structures so that the endpoint can be modeled in terms of these parameters, statistical techniques, i.e., principally multivariate regression and discriminant analysis for weighing these parameters in a near-optimum fashion for the explanation of the endpoint; and computer technology to make it all practical. Using QSAR, Enslein et al. (1983) has analyzed 2066 chemicals of various chemical structures and found that the oral rat LD_{50} of almost 50% of the compounds examined was predicted within a factor of 2, and 95% within a factor of 8. Obviously, there are limitations for the QSAR approach to predict a complex toxic response in whole animals. These include limited databases on which to base a QSAR model, the temptation to extrapolate beyond the confines of the model, and the noise inherent in the bioassays on which the models are based. The results from QSAR have to be used with caution and, at this stage, QSAR is useful during the discovery stage and for prioritizing chemicals.

Carcinogenicity

Given the $1 to $2 million cost and the 3 to 5 years required to test a single chemical in a lifetime rodent carcinogenicity bioassay, initial decisions on whether to continue the development of a chemical, submit a premanufacturing notice (PMN), or require additional testing may be based largely on structure–activity relationships (SARs) and limited short-term assays. A test agent's structure, solubility stability, pH sensitivity, electrophilicity, and chemical reactivity can represent important information for hazard identification. Historically, certain key molecular structures have provided regulators with some of the most readily available information on which to assess hazard potential. For example, 8 of the first 14 occupational carcinogens were regulated together by the Occupational Safety and Health Administration (OSHA) as belonging to the aromatic amine chemical class. The EPA Office of Toxic Substances relies on structure–activity relationships to meet deadlines for responding to PMN for new chemical manufacture under the Toxic Substances Control Act (TSCA). Structural alerts such as *n*-nitroso or aromatic amine groups, amino azo dye structures, and phenanthrene nuclei are clues to prioritizing agents for additional evaluation as potential carcinogens. The limited database of known developmental toxicants limits structure–activity relationships to only a few chemical classes, including chemicals with structures related to those of valproic acid and retinoic acid.

SARs are useful in assessing the relative toxicity of chemically related compounds. The EPA's 1994 reassessment of the risk of 2,3,7,8-tetrachlorodibenzo-*p*-dioxin and related chlorinated and brominated dibenzo-*p*-dioxins, dibenzofurans, and planar biphenyls might have relied too heavily on toxicity equivalence factors (TEFs) based on induction of the Ah receptor (EPA, 1994). The estimated toxicity of environmental mixtures containing those chemicals is a product of the concentration of each chemical times its TEF value. However, it is difficult to predict activity across chemical classes and especially across multiple toxic endpoints by using a single biological response. Many complex chemical–physical interactions are not easily understood and may be oversimplified by researchers. Several computerized SAR methods gave disappointing results in the National Toxicology Program's (NTP) 44-chemical rodent carcinogenicity prediction challenge (Ashby and Tennant, 1994).

References

Ashby, J. and Tennant, R.W. (1994) Prediction of rodent carcinogenicity of 44 chemicals: results. *Mutagenesis*, 9, 7–15.

Asher, I.M. and Zervos, C. (1977) *Structural Correlates of Carcinogenesis and Mutagenesis*, Office of Science, FDA, Washington, DC, 1977.

Bernstein, M.E. (1984) Agents affecting the male reproductive system — effects of structure on activity. *Drug Metab. Rev.*, 15, 941–996.

Chang, R. (1981) *Physical Chemistry with Application to Biological Systems.* MacMillan, New York.

Cohen, J.L., Lee, W., and Lien, E.J. (1974) Dependence on toxicity on molecular structure: group theory analysis. *J. Pharm Sci.*, 63, 1068–1072.

Crum-Brown, A. and Fraser, T.R. (1869) On the connection between chemical constitution and physiological action. Part II. - On the physiological action of the ammonium bases derived from *Atropia* and *Conia*. *Trans. Roy. Soc. Edinburgh*, 25, 693–739.

Dupuis, G. and Benezra, C. (1982) *Allergic Contact Dermatitis to Simple Chemicals—A Molecular Approach.* Marcel Dekker, New York.

Enslein, K. (1984) Estimation of toxicological endpoints by structure–activity relationships. *Pharmacol. Rev.*, 36, 131–134.

Enslein, K. and Craig, P.N. (1982) Carcinogenesis: a predictive structure activity model. *J. Toxicol. Environ. Health*, 10, 521–530.

Enslein, K., Lander, T.R., Tomb, M.E., and Landis, W.G. (1983) Mutagenicity (Ames): a structure–activity model. *Terato. Carcino. Mutagen.*, 3, 503–514.

Enslein, K., Tomb, M.E., and Lander, T.R. (1984) Structure–activity models of biological oxygen demand, in *QSAR in Environmental Toxicology*, Kaiser, K.L.E., Ed., Reidel, Dordrecht.

EPA. (1994) *Health Assessment Document of 2,3,7,8 Tetrachlorodibenzo-p-dioxin (TCDD) and Related Compounds.* External Review Draft. EPA/600/BP-921001b, Washington, DC.

Foye, W.O., Lemke, T.L., and Williams, D.A. (1995) *Principles of Medicinal Chemistry*, Lippincott, Williams & Wilkins, Philadelphia.

Franke, R. (1973) Structure–activity relationships in polycyclic aromatic hydrocarbons: induction of microsomal aryl hydrocarbon hydroxylase and its possible importance in chemical carcinogenesis. *Chem. Biol. Inter.*, 6, 1–17.

Free, S.M. and Wilson, J.W. (1964) A mathematical contribution to structure–activity studies. *J. Med. Chem.*, 7, 395–399.

Goldberg, L. (1983) *Structure–Activity Correlations as a Predictive Tool in Toxicology.* Hemisphere, New York.

Gund, P., Andosf, J.D., Rhodes, J.B., and Smith, G.M. (1980) Three-dimensional molecular modeling and drug design. *Science*, 208, 1425–1431.

Hansch, C. (1971) *Drug Design*, vol. I, ch. 2, Ariens, E.J., Ed., Academic Press, New York.

Kaufman, J.J., Lewchenko, V., Hariharan, P.C., and Kuski, W.S. (1983) Theoretical predictions of toxicity, in *Product Safety Education*, Goldberg, A.M., Ed., Mary Ann Liebert, New York, pp. 333–355.

Kubinyi, H. (1995) The quantitative analysis of structure–activity relationships, in *Burger's Medicinal Chemistry*, vol. I, Wolff, M.E., Ed., John Wiley, New York, pp. 495–572.

Niculescu-Duvaz, I., Craescu, T., Tugulea, M., Croisy, A., and Jacquignon, P.C. (1981) A quantitative structure–activity analysis of the mutagenic and carcinogenic action of 43 structurally related heterocyclic compounds. *Carcinogenesis*, 2, 269–275.

Olson, E.C. and Cristoffersen, R.E. (1979) *Computer Assisted Drug Design.* ACS, Washington, DC.

Sugai, S., Murata, K., Kitagahi, T., and Tomita, I. (1990) Studies on eye irritation caused by chemicals in rabbits, I: a quantitative structure–activity relationship approach to primary eye irritations of chemicals in rabbits. *J. Toxicol. Soc.*, 15, 245–262.

Tamura, R.M. (1983) A model for toxicologic prediction, in *Product Safety Evaluation*, Goldberg, A.M., Ed., Mary Ann Liebert, New York, pp. 317–330.

Topliss, J.G. (1983) *Quantitative Structure–Activity Relationships of Drugs*. Academic Press, New York.

Tute, M.S. (1983) Mathematical modeling, in *Animals and Alternatives in Toxicity Testing*, Balls, M., Riddell, R.J., and Worden, A.N., Eds., Academic Press, New York, pp. 137–166.

Wexler, P., Hakkinen, P.J., Kennedy, G., and Stoss, F. (2000) *Information Sources in Toxicology*. Academic Press, New York.

Wolff, M.E. (1995) *Burger's Medicinal Chemistry*. John Wiley, New York.

19

Good Laboratory Practices

Good Laboratory Practices (GLPs) present a standard documentation and procedures compliance monitoring program that assures the quality and integrity of nonclinical test data submitted to the Food and Drug Administration (FDA) and the Environmental Protection Agency (EPA). These were originally developed by the U.S. FDA to develop minimum research standards for laboratories to protect the quality and integrity of studies as a result of procedures conducted in some studies that were not conducted according to conventional laboratory procedures. As a result, FDA promulgated the GLP Regulations, 21 CFR Part 58, on December 22, 1978 (43 FR 59986). The regulations became effective in June 1979; sections have since been amended. The EPA adopted GLPs in August 1989. Other countries have also adopted GLPs. While there may be some slight variations, they all have the same central focus. Most countries have regular inspections and data audits to monitor laboratory compliance with the GLP requirements.

Sponsors of FDA-regulated products are required by the FFDCA and Public Health Service Act to submit evidence of their product's safety in research and/or marketing applications (FDA). These products include food and color additives, animal drugs, human drugs and biological products, human medical devices, diagnostic products, and electronic products. These data are then used to answer questions regarding the toxicity profile of the article; the observed no-adverse-effect dose level in the test system; the risks associated with clinical studies involving humans or animals; the potential teratogenic, carcinogenic, or other adverse effects of the article; and the level of use that can be approved.

Sponsors of EPA-regulated products are required by the Federal Insecticide, Fungicide, and Rodenticide Act (FIFRA) and the Toxic Substances Control Act (TSCA) to submit evidence that assures the quality and integrity of test data submitted to the agency (EPA). These data are used by the agency to regulate pesticides and industrial chemicals.

The importance of nonclinical laboratory studies demand that they be conducted according to scientifically sound protocols and with meticulous attention to quality. GLPs provide that guidance. GLP regulations cover a large part of the aspects of nonclinical research. They include:

- Inspection of a testing facility
- Personnel
- Testing facility management
- Study director
- QA unit
- Animal care facilities
- Facilities for handling test and control articles (EPA also has "reference substances")
- Lab operation areas
- Specimen and data storage
- Equipment design, maintenance, and calibration
- Standard operating procedures
- Animal care
- Characterization and handling of articles (EPA also includes this for reference substances)
- Protocol
- Reporting results
- Record retention
- Disqualification of testing facilities

Details of the GLPs and GLP inspections may be found on government-specific Web sites. Many are listed in the references below. Briefly, the U.S. FDA GLPs define the scope of the regulation and have a detailed listing of definitions. If the FDA is to consider a nonclinical laboratory study in support of an application for a research or marketing permit, the testing facility, records, and specimens must be available for inspection by an authorized employee of the FDA.

Every person who is responsible for any part of any GLP study must have the appropriate education, training, and experience, or combination thereof, to enable that person to perform the assigned functions. Each testing facility shall maintain up-to-date records of training, experience, and job description for everyone involved in the conduct of a nonclinical laboratory study. There shall be adequate personnel to conduct the study according to the protocol and to take appropriate precautions to avoid contamination of the test and control articles and the test systems.

The study director is responsible for the aspects of the study, but management also has responsibilities. For every nonclinical laboratory study, the management needs to assign a study director and assure that there is a quality assurance unit. Management must also assure that test and control articles or mixtures have been appropriately tested for identity, strength, purity, stability, and uniformity, as applicable. They need to assure that personnel understand the procedures they are to perform and are available in addition to the availability of resources, facilities, equipment, materials, and methodologies. The study director has overall responsibility for the technical conduct of the study, as well as for the interpretation, analysis, documentation, and reporting of results, and represents the single point of study control. This point has been interpreted in many ways, but the intent is that the study director is responsible for all aspects of the study, including procedures that may take place at another facility. Examples may include analysis of analytical samples for confirmation of concentration or homogeneity of a test article mixture, or for analysis of biological samples (e.g., pharmacokinetic or histology).

The quality assurance unit is responsible for monitoring each study to assure management that the facilities, equipment, personnel, methods, practices, records, and controls are in conformance with the regulations in this part. The quality assurance unit shall be entirely separate from and independent of the personnel engaged in the direction and conduct of that study.

The GLPs specify certain requirements for the facilities. In particular, there need to be dedicated areas for separation of species or test systems, isolation of individual projects, quarantine of animals, routine or specialized housing of animals, safe sanitary storage of waste before removal from the testing facility, feed, bedding, supplies, and equipment. Storage areas for feed and bedding shall be separated from areas housing the test systems and shall be protected against infestation or contamination. There also need to be separate areas for receipt and storage of the test and control articles, mixing of the test and control articles with a carrier, and storage of the test and control article mixtures. In addition, space needs to be provided for a limited-access archives.

Requirements for equipment also exist. The equipment must be tested, inspected, maintained, and calibrated on a regular basis, according to written standard operating procedures (SOPs). These procedures need to be documented. Written records also need to be maintained for nonroutine repairs performed on equipment as a result of failure and malfunction.

The testing facility shall have SOPs in writing setting forth nonclinical laboratory study methods that management is satisfied are adequate to insure the quality and integrity of the data generated in the course of a study. All deviations in a study from SOPs shall be authorized by the study director and shall be documented in the raw data. Significant changes in established SOPs shall be properly authorized in writing by management. The facility needs to maintain a historical file of SOPs and all revisions. These SOPs and any appropriate laboratory manuals must be immediately available when laboratory procedures are being performed.

There are many requirements for animal care. Information can be found in the GLPs but primarily in the National Academy of Sciences' (1996) *Guide for the Care and Use of Laboratory Animals.* Besides proper husbandry practices, animals need to be individually identified under most circumstances.

All of the reagents and solutions used in the laboratory areas need to be labeled to indicate identity, titer or concentration, storage requirements, and expiration date. These must be discarded if the reagents or solutions are deteriorated or outdated. The GLPs have strict guidelines for the test and control articles (U.S. EPA discusses test substances, control substances, and reference substances).

According to the FDA GLPs, a test article is "any food additive, color additive, drug, biological product, electronic product, medical device for human use, or any other article subject to regulation under the act or under sections 351 and 354-360F of the Public Health Service Act." A control article means "any food additive, color additive, drug, biological product, electronic product, medical device for human use, or any article other than a test article, feed, or water that is administered to the test system in the course of a nonclinical laboratory study for the purpose of establishing a basis for comparison with the test article."

The EPA GLPs for both the TOSCA EPA, 2004a and the FIFRA EPA, 2004b have basically the same definitions for control substances and reference substances. A control substance means "any chemical substance or mixture, or any other material other than a test substance, feed, or water, that is administered to the test system in the course of a study for the purpose of establishing a basis for comparison with the test substance for chemical or biological measurements." A reference substance means "any chemical substance or mixture, or analytical standard, or material other than a test substance, feed, or water, that is administered to or used in analyzing the test system in the course of a study for the purposes of establishing a basis for comparison with the test substance for known chemical or biological measurements." Test substances are defined differently. The TOSCA GLPs state that a test substance means "a substance or mixture administered or added to a test system in a study, which substance or mixture is used to develop data to meet the requirements of a TSCA section 4(a) test rule and/or is developed under a TSCA section 4 testing consent agreement or section 5 rule or order to the extent the agreement, rule or order references this part." The GLPs for FIFRA state that a test substance means a substance or mixture administered or added to a test system in a study, which substance or mixture: (1) is the subject of an application for a research or marketing permit supported by the study, or is the contemplated subject of such an application; or (2) is an ingredient, impurity, degradation product, metabolite, or radioactive isotope of a substance described by paragraph (1) of this definition, or some other substance related to a substance described by that paragraph, which is used in the study to assist in characterizing the toxicity, metabolism, or other characteristics of a substance described by that paragraph.

The FDA GLPs state that the identity, strength, purity, and composition, or any other characteristics that appropriately define the test or control article shall be determined for each batch and shall be documented. In addition, the methods of synthesis of the test and control articles need to be documented. The stability of each test or control article needs to be determined by the testing facility or by the sponsor either before study initiation, or concomitantly. Specific requirements for each storage container for a test or control article shall be labeled by name, chemical abstract number or code number, batch number, expiration date, if any, and, where appropriate, storage conditions. Retention samples are needed for any study longer than four weeks in duration.

Test and control article handling procedures are addressed. Procedures need to be established for a system for the handling of the test and control articles to ensure that proper handling and storage are utilized, the materials are allocated so that there is no possibility of contamination, deterioration, or damage. At all times, identification of the contents must be maintained throughout the distribution process and all actions are documented. Identification of the contents of a container helps to ensure that the material will be used properly. It is also prudent to label temporary or transport containers to avoid mix-ups.

Documentation and the same type of procedures are important for mixtures of articles with carriers. In addition, for each test or control article that is mixed with a carrier, appropriate analytical methods shall be conducted to determine the uniformity and stability of the mixture and the concentration of the test or control article in the mixture. In GLP studies, these assays should incorporate validated methods. SOPs need to define general ranges for standard parameters used for analytical acceptability.

The protocol for nonclinical studies must address specific concerns. In addition, it must be approved and written so that it clearly indicates the objectives and all methods for the conduct of the study. The nonclinical GLP studies have to be conducted in accordance with the protocol. This includes documentation of all aspects of the study. Over time, personnel leave laboratories; therefore, the only way to reproduce a study is to have original documentation that is adequate and legible. Data need to be signed and dated by the person making the observations. Any change in entries shall be made so as not to obscure the original entry, shall indicate the reason for such change, and shall be dated and signed or identified at the time of the change. With computerized systems that incorporate automated data collection, the individual responsible for direct data input shall be identified at the time of data input. Any change in automated data entries shall be dated and made so as not to obscure the original entry, the reason for the change needs to be indicated, and the responsible individual needs to be identified.

A final report for each nonclinical laboratory study shall be prepared. Details are provided in the GLPs. The final report summarizes most of the experimental details of the study. The final report needs to include the name and address of the facility(s); objectives and procedures as stated in the approved protocol, including any changes in the original protocol; test and control article information; and information on the preparations used to administer the material, including dosage, dosage regimen, route of administration. Also needed are duration of the study, a description of the methods, and test system used. Any circumstances that may have affected the quality or integrity of the data must be addressed. In addition, the locations where all specimens, raw data, and the final report are to be stored as needed. The quality assurance unit prepares and signs a statement of GLP compliance, and the final report is signed and dated by the study director. If there is any question regarding the GLP compliance to a study, the person(s) responsible should declare this information in the report, as ignoring any issues may lead to legal concerns. The study director addresses any corrections, as in the case of changes to protocols, as an amendment.

Currently, data recording is generally done either electronically or onto prepared forms, both of which seek to insure completion of the captured data and minimize validation in the manner of its collections. Table 19.1 presents some rules for form design that apply to both paper or electronic formats.

The final report and any amendments, all raw data, documentation, protocols and any amendments, and specimens (with the exception of specimens subject to degradation) generated as a result of a nonclinical laboratory study shall be retained in an archive. The archive facility needs to be set up for orderly storage and expedient retrieval. Conditions of storage shall minimize the deterioration of the documents or specimens. The archives do not necessarily have to be an in-house facility; the laboratory may contract with commercial archives to store materials in a GLP fashion. Any off-site data storage locations need to be indexed and documented so that this information is easily obtainable. In either case, the FDA requires that documentation records, raw data, and specimens pertaining to a nonclinical laboratory study shall be retained in the archive(s) for a period of at least 5 years following the date on which the results of the nonclinical laboratory study are submitted in support of an application for a research or marketing permit. With the exception of investigational new drug applications (IND) or applications for investigational device exemptions (IDE), if an application is approved for a research or marketing permit, for which the results of the nonclinical laboratory study were submitted, data need to be held for a period of at least 2 years following the date of approval. Other situations, e.g., where the nonclinical laboratory study does not result in the submission of the study in support of an application for a research or marketing permit, a period of at least 2 years following the date on which the study ends. Appropriate wet specimens, samples of test or control articles, and materials that may be subjected to degradation even under proper storage conditions, shall be retained only as long as the quality of the preparation affords evaluation.

TABLE 19.1

Rules for Form Design and Preparation

Forms should be used when some form of repetitive data must be collected. They may be either paper or electronic.

If only a few (two or three) pieces of data are to be collected, they should be entered into a notebook and not onto a form. This assumes that the few pieces are not a daily event, with the aggregate total of weeks/months/years ending up as lots of data to be pooled for analysis.

Forms should be self-contained, but should not try to repeat the content of SOPs or method descriptions.

Column headings on forms should always specify the units of measurement and other details of entries to be made. The form should be arranged so that sequential entries proceed down a page, not across. Each column should be clearly labeled with a heading that identifies what is to be entered in the column. Any fixed part of entries (such as °C) should be in the column header.

Columns should be arranged from left to right so that there is logical sequential order to the contents of an entry as it is made. An example would be date/time/animal number/body weight/name of the recorder. The last item for each entry should be the name or unique initials of the individual who made the data entry.

Standard conditions that apply to all the data elements to be recorded on a form or the columns of the form should be listed as footnotes at the bottom of the form.

Entries of data on the form should not use more digits than are appropriate for the precision of the data being recorded.

Each form should be clearly titled to indicate its purpose and use. If multiple types of forms are being used, each should have a unique title or number.

Before designing the form, carefully consider the purpose for which it is intended. What data will be collected, how often, with what instrument, and by whom? Each of these considerations should be reflected in some manner on the form.

Those things that are common/standard for all entries on the form should be stated as such once. These could include such things as instrument used, scale of measurement (°C, F, or K), or the location where the recording is made.

All GLP laboratories are subject to periodic audits by both internal and external quality assurance personnel. Figure 19.1 presents a standard checklist for such an audit.

Other documentation targeted for storage in archives include the master schedule sheet, copies of protocols, and records of quality assurance inspections, summaries of training and experience, job descriptions, records, and reports of the maintenance and calibration and inspection of equipment. In any case, ensure that the protocol or SOPs address the proper archiving of appropriate materials.

The FDA may find that it needs to disqualify testing facilities if the facility has not complied with the requirements of the GLP regulations. All studies completed after the date of disqualification can be excluded from consideration. It is prudent to ensure that all laboratories used are GLP compliant, or studies and entire projects may be compromised.

Client	Revision:	Title: **CONTRACT LABORATORY AUDIT CHECKLIST** **(GLP)**	Page 1 of 5

Audit#:_____   Date: _____   Auditor:_____

1.   Study Title:_____

2.   Laboratory:_____

3.   Address:_____

4.   Date of Audit:_____

5.   Auditor:_____

6.   Date of Last FDA Inspection of Laboratory:_____

7.   Date of Last CarboMedics Audit of Laboratory:_____

8.   Facility Manager:_____

9.   Study Director:_____

10.   Quality Assurance Unit:_____

		Unaccept	Needs Imp.	Accept	Excellent
11.	**Protocol:**				
a.	Title and Purpose of Study				
b.	Identification of Test and Control Articles				
c.	Name of Sponsor and Name and Address of Testing Facility				
d.	Description of Animal Model				
e.	Rationale for Animal Model				
f.	Procedure for Identification of Test System				
g.	Description of Experimental Design				
h.	Description of Animal Diet				
i.	Administration of Test or Control Article				
j.	Type and Frequency of Tests, Analyses, and Measurements				
k.	Records to be Maintained				
l.	Date of Approval and Dated Signature of Study Director				
m.	Statistical Methods to be Used				
n.	Changes (with Reasons) Approved and Maintained with Protocol				
12.	Master Schedule Sheet (Test system, Nature of Study, Date Study was Initiated, Current Status, Identity of Sponsor, and Name of Study Director)				

FIGURE 19.1

Contract laboratory audit checklist. (From Singer, D.C. and Upton, R., *Guidelines for Quality Auditing*; ASQC Quality Press, Milwaukee, WI, 1993. With permission.).

Client	Revision:	Title: CONTRACT LABORATORY AUDIT CHECKLIST (GLP)				Page 2 of 5

		Unaccept	Needs Imp.	Accept	Excellent
13.	Current Summary of Training and Experience and Job Description for Each Individual				
14.	Personnel Qualifications				
15.	**Quality Assurance (QA) Unit:**				
	a. Independent of Personnel Engaged in Study				
	b. Written Procedure for Operation of QA Unit				
	c. Maintains copy of Master Schedule Sheet				
	d. Maintains Copy of All Protocols				
	e. Inspections at Intervals Adequate to Assure Integrity				
	f. Written Reports of Periodic Inspections				
	g. Significant Problems Reported to Study Director and Management				
	h. Written Status Reports on Each Study				
	i. Reviews Final Study Report				
	j. All QA Unit Records are Kept in One Location				
16.	**Written Procedures:**				
	a. Animal Care				
	b. Animal Care Facilities				
	c. Animal Transfer and Identification				
	d. Characterization of Test and Control Articles				
	e. Handling of Test and Control Articles				
	f. Methods of Synthesis, Fabrication, or Derivation of Test and Control Articles				
	g. Determination of Stability of Test and Control Articles				
	h. Determination of Stability of Carrier Mixtures				
	i. Test System Observations				
	j. Laboratory Testing				
	k. Handling of Moribund or Dead				
	l. Necropsy or Postmortem Examination of Animals				
	m. Collection and Identification of Specimens				
	n. Histopathology				
	o. Inspection, Cleaning, Maintenance, Testing, Calibration, and Standardization of Equipment				
	p. Data Handling and Storage				
17.	Testing Facilities of Suitable Size and Construction				
18.	Spaces for Cleaning, Sterilizing, and Maintaining Equipment and Supplies				
19.	**Equipment:**				
	a. Adequate Equipment Including Environmental Control Equipment				
	b. Equipment Cleanliness				
	c. Adherence to Cleaning, Maintenance, Calibration, and Standardization Schedules				

FIGURE 19.1
Continued.

Client	Revision:	Title: **CONTRACT LABORATORY AUDIT CHECKLIST** **(GLP)**			Page 3 of 5

			Unaccept	Needs Imp.	Accept	Excellent
	d.	Records of All Inspection, Maintenance, Testing, Calibration, and Standardization Operations				
	e.	Records Include Defects, How and When Defects were Found, and Remedial Action				
20.		Labeling of Reagents and Solutions (Identity, Titer or Concentration, Storage Requirements, and Expiration Date)				
21.		**Test and Control Articles:**				
	a.	Records of Identity, Strength, Purity, and Composition of Each Batch				
	b.	Stability Determined				
	c.	Records of Stability Testing				
	d.	Labeling of Storage Containers				
	e.	Storage				
	f.	Retention of Reserve Samples				
	g.	Handling				
	h.	Testing of Carrier Mixtures				
	i.	Records of Stability Testing of Carrier Mixtures				
	j.	Labeling of Carrier Mixtures				
22.		**Animal Facilities:**				
	a.	Sufficient Number of Animal Rooms and Areas:				
		(1) Separation of Species and Test Systems				
		(2) Isolation of Individual Projects				
		(3) Isolation of Newly Received Animals				
		(4) Routine and Specialized Housing of Animals				
		(5) Isolation of Studies Using Biohazardous Materials				
		(6) Separate Areas, as appropriate, for Diagnosis, Treatment, and Control of Animal Diseases				
	b.	Facilities for Collection and Disposal of Animal Waste and Refuse				
	c.	Storage Areas for Feed, Bedding, Supplies, and Equipment				
	d.	Areas for Handling Test and Control Articles				
	e.	Space for Aseptic Surgery, Intensive Care, Necropsy, Histology, Radiography, and Handling of Biohazardous Materials				
23.		**Animal Care:**				
	a.	Isolation of Newly Received Animals				
	b.	Animals Free of Any Disease or Condition that Could Interfere with Study				
	c.	Records or Diagnosis and Treatment of Animal Disease				
	d.	Animal Identification				
	e	Separation of Different Species				
	f.	Cleaning of Cages and Equipment				
	g.	Records of Periodic Analyses of Feed and Water				
	h.	Bedding Does Not Interfere with Study Purpose or Conduct				
	i.	Records of Use of Pest Control Materials				

FIGURE 19.1
Continued.

		CONTRACT LABORATORY AUDIT CHECKLIST (GLP)			4 of 5
			Unaccept	Needs Imp.	Accept	Excellent
	j.	Pest Control Materials Do Not Interfere with Study				
24.		Identification of Specimens (Test System, Study, Nature, and Date of Collection)				
25.		Records Available to Pathologists when Examining Specimens Histopathologically				
26.		Records of All Deviations from Written Procedures, Including Authorization				
27.		All Records Specified in Protocol are Maintained				
28.		Data Entries (Manual and Computer)				
29.		Availability of Laboratory Manuals and Written Procedures				
30.		Study Conducted in Accordance with Protocol				
31.		Test Systems Monitored in Conformity with Protocol				
32.		Personnel Report Adverse Health or Medical Condition				
33.		Final Study Reports Include (as a Minimum) Name and Address of Facility Performing Study, Start and Completion Dates of Study, Objectives and Procedures Stated in the Protocol, Changes to Protocol, Statistical Methods for Data Analysis, Test and Control Articles Used, Stability of Test and Control Articles, Methods Used, Test System Used, Dosage and its Administration, All Circumstances That Could Have Affected the Data, Names of Key members of Study Team, Operations Performed on the Data, Summary and Analysis of Data, Conclusions Drawn, Signed and Dated Reports of Key Members of Study Team, Data and Specimen Storage Locations, Statement Prepared and Signed by QA Unit, Dated Signature of Study Director, and Corrections and Additions (in the Form of Amendments) to Final Study Reports				
34.		**Data Handling and Storage:**				
a.		Retention of All Raw Data, Documentation, Protocols, Required Specimens, and Final Study Reports				
b.		Archives Orderly and Minimize Deterioration of Documents and Specimens				
c.		An Individual is Responsible for Archives				
d.		Index of Material in Archives				
e.		Historical File of all Obsolete Documents				
f.		Retention Period of at Least 2 Years from Date of Approval by FDA of a Research or Marketing Permit or from Study Termination Date for Studies that are not included in an FDA Submission, Except at Least 5 Years from Date of Submittal to FDA if in Support of an IND or IDE				

FIGURE 19.1

Continued.

Client	Revision:	Title: **CONTRACT LABORATORY AUDIT CHECKLIST** (GLP)	Page 5 of 5

35.   Comments:_____

_____

_____

_____

_____

_____

36.   Auditor's Signature:_____ 37. Date:_____

cc: _____

_____

_____

References: Singer, D.C.; Upton, Ronald P.; *Guidelines for Quality Auditing***; ASQC**
Quality Press, 1993

Robert E. Spinock Consultants; Sample Audit Checklist, 1988

Audit Coordinator Approval:

_____   _____
Audit Coordinator   Date

FIGURE 19.1
Continued.

References

EPA (2004a) Good Laboratory Practice Standards: FIFRA. Code of Federal Regulation, Title 40, Part 792.

EPA (2004b) Good Laboratory Practice Standards: TOSCA. Code of Federal Regulation, Title 40, Part 160.

FDA (1978) Good Laboratory Practice For Nonclinical Laboratory Studies. Federal Register, 43 FR 59986.

National Academy of Sciences (1996) Guide for the Care and Use of Laboratory Animals. National Academy Press, Washington, D.C.

Singer, D.C. and Upton, R. (1993) Guidelines for Quality Auditing. ASQC Quality Press, Milwaukee, WI.

Further Reading

EPA (2005) Good Laboratory Practice Standards (GLPS). http://www.epa.gov/compliance/monitoring/programs/fifra/glp.html

FDA (2001) Bioresearch Monitoring Good Laboratory Practices: GLP References and Guidance. http://www.fda.gov/ora/compliance_ref/bimo/glp/default.htm

FDA (2004) Good Laboratory Practice For Nonclinical Laboratory Studies. Code of Federal Regulations, Title 21, Part 58.

Gad, S.C. (2003) *The Selection and Use of Contract Research Organizations*. Taylor & Francis, New York.

Gad, S.C. and Taulbee, S.M. (1996) *Handbook of Data Recording Maintenance and Management for the BioMedical Sciences*. CRC Press, New York.

Gad, S.C., Ed. (2001) Regulatory Toxicology, 2nd ed., Taylor & Francis, New York

20

Areas of Controversy in Statistics as Used for Toxicology

It should now be clear to the reader that the use of statistics in toxicology is not a "cut-and-dried" matter. There are a number of areas that are (and have been) the subject of honest controversy, and it should be expected that others will arise as the two fields advance.

This volume has already presented many of the problem areas. In Chapter 7, it was seen that there is no consensus as to which *post hoc* tests should be used after ANOVA because of some of the real-life characteristics of toxicologic data. In this same chapter, the arguments for and against the different methods of hypothesis testing for differences in organ weights were presented. In Chapter 8, the different experimental designs and computational methods for determining LD_{50} values were laid out. As shown in Chapter 12, meta-analysis is still not standardized to any extent and some still question whether it should be used as a quantitative tool at all. In Chapter 15, the matters of risk assessment and threshold were addressed.

There remain, however, three areas to be addressed that are somewhat particular to toxicology. These are the effects of censoring on data, the direction of hypothesis testing, and the use of unbalanced designs.

Censoring

Censoring is practiced when not all the possible data arising from an experiment are available for or used in analysis. Although some would make the distinction that censored data are different from missing data in that the values for the former can be accurately estimated and those for the latter cannot, here the term is used to mean all data not included in analysis, for whatever reasons. There are four major reasons for data being censored in toxicology studies, and the degree of accuracy for which the value of such censored values can be estimated varies depending on the reason for censoring.

The most common reason for censoring in toxicology is death — not all the animals that start a study end it. In these cases we have no basis to accurately estimate the observations that would have been made had the animal (or animals) lived. Censoring by death is an example of "left censoring" — unplanned, without recourse, and generally during a period when the information lost would be of interest (Gad and Smith, 1984). Right censoring, on the other hand, generally is planned, there is recourse to get the information if needed, and the information potentially lost is of minimal, if any, interest.

Data may be censored by having samples lost to measurement at intermittent periods. Such losses are the result of, for example, clotting of blood samples prior to analysis, loss of tissues during necropsy, or breakdown of instruments at critical times. Usually the values of the last observations can be estimated with some accuracy. Most such cases can be remedied by resampling (collecting more blood, for example).

When we judge an extreme value to be an outlier and reject it, we are censoring it. If the value is cleanly discarded, we are *de facto* saying we cannot accurately estimate its value. If, however, we use a procedure such as Winsoring and replace it with a less extreme value, we are in fact estimating the most probable, true value of the observations.

Finally, some observations may be censored because their values are beyond the range at which the instruments we use can accurately measure. An example is in measuring rabbit methemoglobin with an instrument designed for humans. Extreme low values are not accurately measured and are reported as negative percentage values. In this case, we can accurately estimate a censored value as being "less than" or "greater than" a known value.

What are the consequences of censoring? The answer depends on the nature and extent of the censoring process. If only a few of a large number of values are lost and the pattern of loss is randomly distributed among all groups on study, little if any harm is done. If the extent of data loss is too severe — say, because the majority of the animals in a group die — the entire experiment may have to be discarded. An intermediate case would be low, but nonrandom, censoring. This is not an uncommon case in toxicology, where censoring because of death tends to be concentrated in high-dose groups. In these cases, the experiment is not lost but rather truncated — some effects cannot be addressed with reference to the treatment used in highly censored groups. And, as we will discuss a little later, it may unbalance a design.

An additional common effect of censoring that should be kept in mind is its effect on the normality of the sample. If all values above a certain level (say of serum electrolytes) are censored because animals having such values die before the measurements are made, we are left with a truncated normal distribution. A special case of this was discussed by Gad and Smith (1984), in that the time-to-incapacitation values in combustion toxicology are censored because they are only measured to 30 minutes, and not beyond. Such truncated populations cannot be treated as normal for purposes of statistical analysis.

There is an entire family of methods that have been developed to address censored data sets. Bishop et al. (1971) present an excellent overview of some of these.

Direction of Hypothesis Testing

In which direction (or directions) we are testing a hypothesis can be restated as asking whether we are to use a one-tailed or a two-tailed test. This is of consequence because one-tailed tests are always more sensitive (more likely to find an effect) than are two-tailed tests.

Generally, such a selection must be made prior to the start of an experiment, based on a clear statement of the question being asked (that is, the objective of the study). If we are asking if a chemical increases the incidence of cancer, then our question is one-tailed — we are not interested in the detection of any significant decrease in the incidence of cancer. Most toxicology studies, however, are of a "shotgun" nature. They are designed to detect and identify any and all effects. This is a two-tailed question — can a chemical either increase or decrease the incidence of cancer?

Feinstein (1979) provides a clear discussion of questions of direction of effect as they relate to biostatistics.------

Unbalanced Designs

One of the principles of experimental design presented in Chapter 3 was that of balance. This held that group sizes should be at least approximately equal. And, as we have reviewed the different methods presented in this book, we have noted that a number of them have impaired performance if the sizes of the groups are not equivalent.

Yet it is not uncommon to lose data because of censoring in toxicology studies. And if such censoring is related to a compound or treatment effect, it is very likely that the most affected groups will not be equivalent in size to our control group at the end of an experiment.

At the same time, it should be clear that it is easier to statistically detect large effects than small effects. In the vast majority of cases, larger effects occur in high dose groups (not infrequently to the extent that no statistical analysis is necessary), while it is in the lower dose groups that the guidance provided by statistical analysis is most needed.

These reasons argue for the use of unbalanced designs in toxicology. That is, those treatments where it is expected that more statistical power will be needed or that are expected to suffer from an increased level of censoring due to death (where death itself is not the variable of interest) should be administered to test groups that are larger than other test groups. Farmer et al. (1977) have reviewed a number of options for deciding on the degree of imbalance with which to start a study.

Use of Computerized Statistical Packages

The increased power and ease-of-use of statistical program sets used for toxicology is hardly controversial, but can lead to two troublesome areas. First, even with large data sets, the packages and computers available make it possible for most personal computers to calculate statistics on them quickly and easily. This can lead to "fishing expeditions" where various tests are run until the desired result is obtained. In order to conserve objectivity, selection of a statistical plan designed to answer the specific question(s) we are looking at should occur prior to viewing and analysis of the data.

Secondly, we must recognize that, for many toxicology laboratories, the approach to statistical analysis of data is to use one of the packages that automatically selects and utilizes statistical tests. It is critically important in these cases to understand the limitations and proper uses of the statistical tests that are automatically employed.

References

Bishop, Y., Fujii, K., Arnold, E., and Epstein, S.S. (1971) Censored distribution techniques in analysis of toxicological data. *Experientia*, 27, 1056–1059.

Farmer, J.H., Uhler, R.J., and Haley, T.J. (1977) An unbalanced experimental design for dose response studies. *J. Environ. Pathol. Toxicol. Oncol.*, 1, 293–299.

Feinstein, A.R. (1979) Clinical biostatistics XXXII: biological dependency, "Hypothesis testing, unilateral probabilities, and other issues in scientific direction vs. statistical duplexity." *Clin. Pharmacol. Therapeut.*, 17, 499–513.

Gad, S.C. and Smith, A.C. (1984) Influence of heating rates on the toxicity of evolved combustion products: results and a system for research. *J. Fire Science*, 1, 465–479.

Appendix 1

Tables

TABLE A

Logarithms (Base 10)

N	0	1	2	3	4	5	6	7	8	9
10	0000	0043	0086	0128	0170	0212	0253	0294	0334	0374
11	0414	0453	0492	0531	0569	0607	0645	0682	0719	0755
12	0792	0828	0864	0899	0934	0969	1004	1038	1072	1106
13	1139	1173	1206	1239	1271	1303	1335	1367	1399	1430
14	1461	1492	1523	1553	1584	1614	1644	1673	1703	1732
15	1761	1790	1818	1847	1875	1903	1931	1959	1987	2014
16	2041	2068	2095	2122	2148	2175	2201	2227	2253	2279
17	2304	2330	2355	2380	2405	2430	2455	2480	2504	2529
18	2553	2577	2601	2625	2648	2672	2695	2718	2742	2765
19	2788	2810	2833	2856	2878	2900	2923	2945	2967	2989
20	3010	3032	3054	3075	3096	3118	3139	3160	3181	3201
21	3222	3243	3263	3284	3304	3324	3345	3365	3385	3404
22	3424	3444	3464	3843	3502	3522	3541	3560	3579	3598
23	3617	3636	3655	3674	3692	3711	3729	3747	3766	3784
24	3802	3820	3838	3856	3874	3892	3909	3927	3945	3962
25	3979	3997	4014	4031	4048	4065	4082	4099	4116	4133
26	4150	4166	4183	4200	4216	4232	4249	4265	4281	4298
27	4314	4330	4346	4362	4378	4393	4409	4425	4440	4456
28	4472	4487	4502	4518	4533	4548	4564	4579	4594	4609
29	4624	4639	4654	4669	4683	4698	4713	4728	4742	4757
30	4771	4786	4800	4814	4829	4843	4857	4871	4886	4900
31	4914	4928	4942	4955	4969	4983	4997	5011	5024	5038
32	5051	5065	5079	5092	5105	5119	5132	5145	5159	5172
33	5185	5198	5211	5224	5237	5250	5263	5276	5289	5302
34	5315	5328	5340	5353	5366	5378	5391	5403	5416	5428
35	5441	5453	5465	5478	5490	5502	5514	5527	5539	5551
36	5563	5575	5587	5599	5611	5623	5635	5647	5658	5670
37	5682	5694	5705	5717	5729	5740	5752	5763	5775	5786
38	5798	5809	5821	5832	5843	5855	5866	5877	5888	5899
39	5911	5922	5933	5944	5955	5966	5977	5988	5999	6010
40	6021	6031	6042	6053	6064	6075	6085	6096	6107	6117
41	6128	6138	6149	6160	6170	6180	6191	6201	6212	6222
42	6232	6243	6253	6263	6274	6284	6294	6304	6314	6325
43	6335	6345	6355	6365	6375	6385	6395	6405	6415	6425
44	6345	6444	6454	6464	6474	6484	6493	6503	6513	6522
45	6532	6542	6551	6561	6571	6580	6590	6599	6609	6618
46	6628	6637	6646	6656	6665	6675	6684	6693	6702	6712

(*Continued*)

TABLE A

(Continued)

N	0	1	2	3	4	5	6	7	8	9
47	6721	6730	6739	6749	6758	6767	6776	6785	6794	6803
48	6812	6821	6830	6839	6848	6857	6866	6875	6884	6893
49	6902	6911	6920	6928	6937	6946	6955	6964	6972	6981
50	6990	6998	7007	7016	7024	7033	7042	7050	7059	7067
51	7076	7084	7093	7101	7110	7118	7126	7135	7143	7152
52	7160	7168	7177	7185	7193	7202	7210	7218	7226	7235
53	7243	7251	7259	7267	7275	7284	7292	7300	7308	7316
54	7324	7332	7340	7348	7356	7364	7372	7380	7388	7396
55	7404	7412	7419	7427	7435	7443	7451	7459	7466	7474
56	7482	7490	7497	7505	7513	7520	7528	7536	7543	7551
57	7559	7566	7574	7582	7589	7597	7604	7612	7619	7627
58	7634	7642	7649	7657	7664	7672	7679	7686	7694	7701
59	7709	7716	7723	7731	7738	7745	7752	7760	7767	7774
60	7782	7789	7796	7803	7810	7818	7825	7832	7839	7846
61	7853	7860	7868	7875	7882	7889	7896	7903	7910	7917
62	7924	7931	7938	7945	7952	7959	7966	7973	7980	7987
63	7993	8000	8007	8014	8021	8028	8035	8041	8048	8055
64	8062	8068	8075	8082	8089	8096	8102	8109	8116	8122
65	8129	8136	8142	8149	8156	8162	8169	8176	8182	8189
66	8195	8202	8209	8215	8222	8228	8235	8241	8248	8254
67	8261	8267	8274	8280	8287	8293	8299	8306	8312	8319
68	8325	8331	8338	8344	8351	8357	8363	8370	8376	8382
69	8388	8395	8401	8407	8414	8420	8426	8432	8439	8445
70	8451	8457	8463	8470	8476	8482	8488	8494	8500	8506
71	8513	8519	8525	8531	8537	8543	8549	8555	8561	8567
72	8573	8579	8585	8591	8597	8603	8609	8615	8621	8627
73	8633	8639	8645	8651	8657	8663	8669	8675	8681	8686
74	8692	8698	8704	8710	8716	8722	8727	8733	8739	8745
75	8751	8756	8762	8768	8774	8779	8785	8791	8797	8802
76	8808	8814	8820	8825	8831	8837	8842	8848	8854	8859
77	8865	8871	8876	8882	8887	8893	8899	8904	8910	8915
78	8921	8927	8932	8938	8943	8949	8954	8960	8965	8971
79	8976	8982	8987	8993	8998	9004	9009	9015	9020	9025
80	9031	9036	9042	9047	9053	9058	9063	9069	9074	9079
81	9085	9090	9096	9101	9106	9112	9117	9122	9128	9133
82	9138	9143	9149	9154	9159	9165	9170	9175	9180	9186
83	9191	9196	9201	9206	9212	9217	9222	9227	9232	9238
84	9243	9248	9253	9258	9263	9269	9274	9279	9284	9289
85	9294	9299	9304	9309	9315	9320	9325	9330	9335	9340
86	9345	9350	9355	9360	9365	9370	9375	9380	9385	9390
87	9395	9400	9405	9410	9415	9420	9425	9430	9435	9440
88	9445	9450	9455	9460	9465	9469	9474	9479	9484	9489
89	9494	9499	9504	9509	9513	9518	9523	9528	9533	9538
90	9542	9547	9552	9557	9562	9566	9571	9576	9581	9686
91	9590	9595	9600	9605	9609	9614	9619	9624	9628	9633
92	9638	9643	9647	9652	9657	9661	9666	9671	9675	9680
93	9685	9689	9694	9699	9703	9708	9713	9717	9722	9727
94	9731	9736	9741	9745	9750	9754	9759	9763	9768	9773
95	9777	9782	9786	9791	9795	9800	9805	9809	9814	9818
96	9823	9827	9832	9836	9841	9845	9850	9854	9859	9863
97	9868	9872	9877	9881	9886	9890	9894	9899	9903	9908
98	9912	9917	9921	9926	9930	9934	9939	9943	9948	9952
99	9956	9961	9965	9969	9974	9978	9983	9987	9991	9996

TABLE B

Probit Transform Values

%	0.0	0.1	0.2	0.3	0.4	0.5	0.6	0.7	0.8	0.9
0	...	1.9098	2.1218	2.2522	2.3479	2.4242	2.4879	2.5427	2.5911	2.6344
1	2.6737	2.7096	2.7429	2.7738	2.8027	2.8299	2.8556	2.8799	2.9031	2.9251
2	2.9463	2.9665	2.9859	3.0046	3.0226	3.0400	3.0569	3.0732	3.0890	3.1043
3	3.1192	3.1337	3.1478	3.1616	3.1750	3.1881	3.2009	3.2134	3.2256	3.2376
4	3.2493	3.2608	3.2721	3.2831	3.2940	3.3046	3.3151	3.3253	3.3354	3.3454
5	3.3551	3.3648	3.3742	3.3836	3.3928	3.4018	3.4107	3.4195	3.4282	3.4368
6	3.4452	3.4536	3.4618	3.4699	3.4780	3.4859	3.4937	3.5015	3.5091	3.5167
7	3.5242	3.5316	3.5389	3.5462	3.5534	3.5605	3.5675	3.5745	3.5813	3.5882
8	3.5949	3.6016	3.6083	3.6148	3.6213	3.6278	3.6342	3.6405	3.6468	3.6531
9	3.6592	3.6654	3.6715	3.6775	3.6835	3.6894	3.6953	3.7102	3.7070	3.7127
10	3.7184	3.7241	3.7298	3.7354	3.7409	3.7464	3.7519	3.7574	3.7628	3.7681
11	3.7735	3.7788	3.7840	3.7893	3.7945	3.7996	3.8048	3.8099	3.8150	3.8200
12	3.8250	3.8300	3.8350	3.8399	3.8448	3.8497	3.8545	3.8593	3.8641	3.8689
13	3.8736	3.8783	3.8830	3.8877	3.8923	3.8969	3.9015	3.9161	3.9107	3.9152
14	3.9197	3.9242	3.9286	3.9331	3.9375	3.9419	3.9463	3.9506	3.9550	3.9593
15	3.9636	3.9678	3.9721	3.9763	3.9806	3.9848	3.9890	3.9931	3.9973	4.0014
16	4.0055	4.0096	4.0137	4.0178	4.0218	4.0259	4.0299	4.0339	4.0379	4.0419
17	4.0458	4.0498	4.0537	4.0576	4.0615	4.0654	4.0693	4.0731	4.0770	4.0808
18	4.0846	4.0884	4.0922	4.0960	4.0998	4.1035	4.1073	4.1110	4.1147	4.1184
19	4.1221	4.1258	4.1295	4.1331	4.1367	4.1404	4.1440	4.1476	4.1512	4.1548
20	4.1584	4.1619	4.1655	4.1690	4.1726	4.1761	4.1796	4.1831	4.1866	4.1901
21	4.1936	4.1970	4.2005	4.2039	4.2074	4.2108	4.2142	4.2176	4.2210	4.2244
22	4.2278	4.2312	4.2345	4.2379	4.2412	4.2446	4.2479	4.2512	4.2546	4.2579
23	4.2612	4.2644	4.2677	4.2710	4.2743	4.2775	4.2808	4.2840	4.2872	4.2905
24	4.2937	4.2969	4.3001	4.3033	4.3065	4.3097	4.3129	4.3160	4.3192	4.3224
25	4.3255	4.3287	4.3318	4.3349	4.3380	4.3412	4.3443	4.3474	4.3505	4.3536
26	4.3567	4.3597	4.3628	4.3659	4.3689	4.3720	4.3750	4.3781	4.3811	4.3842
27	4.3872	4.3902	4.3932	4.3962	4.3992	4.4022	4.4052	4.4082	4.4112	4.4142
28	4.4172	4.4201	4.4231	4.4260	4.4290	4.4319	4.4349	4.4378	4.4408	4.4437
29	4.4466	4.4495	4.4524	4.4554	4.4583	4.4612	4.4641	4.4670	4.4698	4.4727
30	4.4756	4.4785	4.4813	4.4842	4.4871	4.4899	4.4923	4.4956	4.4985	4.5013
31	4.5041	4.5070	4.5098	4.5126	4.5155	4.5183	4.5211	4.5239	4.5267	4.5295
32	4.5323	4.5351	4.5379	4.5407	4.5435	4.5462	4.5490	4.5518	4.5546	4.5573
33	4.5601	4.5628	4.5656	4.5684	4.5711	4.5739	4.5766	4.5793	4.5821	4.5848
34	4.5875	4.5903	4.5930	4.5957	4.5984	4.6011	4.6039	4.6066	4.6093	4.6120
35	4.6147	4.6174	4.6201	4.6228	4.6255	4.6281	4.6308	4.6335	4.6362	4.6389
36	3.6415	4.6442	4.6469	4.6495	4.6522	4.6549	4.6575	4.6602	4.6628	4.6655
37	4.6681	4.6708	4.6734	4.6761	4.6787	4.6814	4.6840	4.6866	4.6893	4.6919
38	4.6945	4.6971	4.6998	4.7024	4.7050	4.7076	4.7102	4.7129	4.7155	4.7181
39	4.7207	4.7233	4.7259	4.7285	4.7311	4.7337	4.7363	4.7389	4.7614	4.7441
40	4.7467	4.7492	4.7518	4.7544	4.7570	4.7596	4.7622	4.7647	4.7673	4.7699
41	4.7724	4.7750	4.7776	4.7802	4.7827	4.7853	4.7879	4.7904	4.7930	4.7955
42	4.7981	4.8007	4.8032	4.8058	4.8083	4.8109	4.8134	4.8160	4.8185	4.8211
43	4.8236	4.8262	4.8287	4.8313	4.8338	4.8363	4.8389	4.8414	4.8440	4.8465
44	4.8490	4.8516	4.8541	4.8566	4.8592	4.8617	4.8642	4.8668	4.8693	4.8718
45	4.8743	4.8769	4.8794	4.8819	4.8844	4.8870	4.8895	4.8920	4.8945	4.8970
46	4.8996	4.9021	4.9046	4.9071	4.9096	4.9122	4.9147	4.9172	4.9197	4.9222
47	4.9247	4.9272	4.9298	4.9323	4.9348	4.9373	4.9398	4.9423	4.9488	4.9473
48	4.9498	4.9524	4.9549	4.9574	4.9599	4.9624	4.9649	4.9674	4.9699	4.9724

(Continued)

TABLE B

(Continued)

%	0.0	0.1	0.2	0.3	0.4	0.5	0.6	0.7	0.8	0.9
49	4.9749	4.9774	4.9799	4.9825	4.9850	4.9875	4.9900	4.9925	4.9950	4.9975
50	5.0000	5.0025	5.0050	5.0075	5.0100	5.0125	5.0150	5.0175	5.0201	5.0226
51	5.0251	5.0276	5.0301	5.0326	5.0351	5.0376	5.0401	5.0426	5.0451	5.0476
52	5.0502	5.0527	5.0552	5.0577	5.0602	5.0627	5.0652	5.0677	5.0702	5.0728
53	5.0753	5.0778	5.0803	5.0828	5.0853	5.0878	5.0904	5.0929	5.0954	5.0979
54	5.1004	5.1030	5.1055	5.1080	5.1105	5.1130	5.1156	5.1181	5.1206	5.1231
55	5.1257	5.1282	5.1307	5.1332	5.1358	5.1383	5.1408	5.1434	5.1459	5.1484
56	5.1510	5.1535	5.1560	5.1586	5.1611	5.1637	5.1662	5.1689	5.1713	5.1738
57	5.1764	5.1789	5.1815	5.1840	5.1866	5.1891	5.1917	5.1942	5.1968	5.1993
58	5.2019	5.2045	5.2070	5.2096	5.2121	5.2147	5.2173	5.2198	5.2224	5.2250
59	5.2275	5.2301	5.2327	5.2353	5.2378	5.2404	5.2430	5.2456	5.2482	5.2508
60	5.2533	5.2559	5.2585	5.2611	5.2637	5.2666	5.2689	5.2715	5.2741	5.2767
61	5.2793	5.2819	5.2845	5.2871	5.2898	5.2924	5.2950	5.2976	5.3002	5.3029
62	5.3055	5.3081	5.3107	5.3134	5.3160	5.3186	5.3213	5.3239	5.3266	5.3292
63	5.3319	5.3345	5.3372	5.3398	5.3425	5.3451	5.3478	5.3505	5.3531	5.3558
64	5.3585	5.3611	5.3638	5.3665	5.3692	5.3719	5.3745	5.3772	5.3799	5.3826
65	5.3853	5.3880	5.3907	5.3934	5.3961	5.3989	5.4016	5.4043	5.4070	5.4097
66	5.4125	5.4152	5.4179	5.4207	5.4234	5.4261	5.4289	5.4316	5.4344	5.4372
67	5.4399	5.4427	5.4454	5.4482	5.4510	5.4538	5.4565	5.4693	5.4621	5.4649
68	5.4677	5.4705	5.4733	5.4761	5.4789	5.4817	5.4845	5.4874	5.4902	5.4930
69	5.4959	5.4987	5.5015	5.5044	5.5072	5.5101	5.5129	5.5158	5.5187	5.5215
70	5.5244	5.5273	5.5302	5.5330	5.5359	5.5388	5.5417	5.5446	5.5476	5.5505
71	5.5534	5.5563	5.5592	5.5622	5.5651	5.5681	5.5710	5.5740	5.5769	5.5799
72	5.5828	5.5858	5.5888	5.5918	5.5948	5.5978	5.6008	5.6038	5.6068	5.6098
73	5.6128	5.6158	5.6189	5.6219	5.6250	5.6280	5.6311	5.6341	5.6372	5.6403
74	5.6433	5.6464	5.6495	5.6426	5.6557	5.6588	5.6620	5.6651	5.6682	5.6713
75	5.6745	5.6776	5.6808	5.6840	5.6871	5.6903	5.6935	5.6967	5.6999	5.7031
76	5.7063	5.7095	5.7128	5.7160	5.7192	5.7225	5.7257	5.7290	5.7323	5.7356
77	5.7388	5.7421	5.7454	5.7488	5.7521	5.7554	5.7588	5.7621	5.7655	5.7688
78	5.7722	5.7756	5.7790	5.7824	5.7858	5.7892	5.7926	5.7961	5.7995	5.8030
79	5.8064	5.8099	5.8134	5.8169	5.8204	5.8239	5.8274	5.8310	5.8345	5.8381
80	5.8416	5.8452	5.8488	5.8524	5.8560	5.8596	5.8633	5.8669	5.8705	5.8742
81	5.8779	5.8816	5.8853	5.8890	5.8927	5.8965	5.9002	5.9040	5.9078	5.9116
82	5.9154	5.9192	5.9230	5.9269	5.9307	5.9346	5.9385	5.9424	5.9463	5.9502
83	5.9542	5.9581	5.9621	5.9661	5.9701	5.9741	5.9782	5.9822	5.9863	5.9904
84	5.9945	5.9986	6.0027	6.0069	6.0110	6.0152	6.0194	6.0237	6.0279	6.0322
85	6.0364	6.0407	6.0450	6.0494	6.0537	6.0581	6.0625	6.0669	6.0714	6.0758
86	6.0803	6.0848	6.9893	6.0939	6.0985	6.1031	6.1077	6.1123	6.1170	6.1217
87	6.1264	6.1311	6.1359	6.1407	6.1455	6.1503	6.1552	6.1601	6.1650	6.1700
88	6.1750	6.1800	6.1850	6.1901	6.1952	6.2004	6.2055	6.2107	6.2160	6.2212
89	6.2265	6.2319	6.2372	6.2426	6.2481	6.2536	6.2591	6.2646	6.2702	6.2759
90	6.2816	6.2873	6.2930	6.2988	6.3047	6.3106	6.3165	6.3225	6.3285	6.3346
91	6.3408	6.3469	6.3532	6.3595	6.3658	6.3722	6.3787	6.3852	6.3917	6.3984
92	6.4051	6.4118	6.4187	6.4255	6.4325	6.4395	6.4466	6.4538	6.4611	6.4684
93	6.4758	6.4833	6.4909	6.4985	6.5063	6.5141	6.5220	6.5301	6.5382	6.5464
94	6.5548	6.5632	6.5718	6.5805	6.5893	6.5982	6.6072	6.6164	6.6258	6.6352
95	6.6449	6.6546	6.6646	6.6747	6.6849	6.6954	6.7060	6.7169	6.7279	6.7392
96	6.7507	6.7624	6.7744	6.7866	6.7991	6.8119	6.8250	6.8384	6.8522	6.8663
97	6.8808	6.8957	6.9110	6.9268	6.9431	6.9600	6.9774	6.9954	7.0141	7.0335
98	7.0537	7.0749	7.0969	6.1201	7.1444	7.1701	7.1973	7.2262	7.2571	7.2904
99	7.3263	7.3656	7.4087	7.4571	7.5120	7.5758	7.6520	7.7478	7.8782	8.0902

TABLE C

Chi-Square (χ^2)*

df	.99	.98	.95	.90	.80	.70	.50	.30	.20	.10	.05	.02	.01	.001
1	.000157	.000628	.00393	.0158	.0642	.148	.455	1.074	1.642	2.706	3.841	5.412	6.635	10.827
2	.0201	.0404	.103	.211	.446	.713	1.386	2.408	3.219	4.605	5.991	7.824	9.210	13.815
3	.115	.185	.352	.584	1.005	1.424	2.366	3.665	4.642	6.251	7.815	9.837	11.345	16.268
4	.297	.429	.711	1.064	1.649	2.195	3.357	4.878	5.989	7.779	9.488	11.668	13.277	18.465
5	.554	.752	1.145	1.610	2.343	3.000	4.351	6.064	7.289	9.236	11.070	13.388	15.086	20.517
6	.872	1.134	1.635	2.204	3.070	3.828	5.348	7.231	8.558	10.645	12.592	15.033	16.812	22.457
7	1.239	1.546	2.167	2.833	3.822	4.671	6.346	8.383	9.803	12.017	14.067	16.622	18.475	24.322
8	1.646	2.032	2.733	3.490	4.594	5.527	7.344	9.524	11.030	13.362	15.507	18.168	20.090	26.125
9	2.088	2.523	3.325	4.168	5.380	6.393	8.343	10.656	12.242	14.684	16.919	19.679	21.666	27.877
10	2.558	3.059	3.940	4.865	6.179	7.267	9.342	11.781	13.422	15.987	18.307	21.161	23.209	29.588
11	3.053	3.609	4.575	5.578	6.989	8.148	10.341	12.899	14.631	17.275	19.675	22.618	24.725	31.264
12	3.571	4.178	5.226	6.304	7.807	9.034	11.310	14.011	15.812	18.549	21.026	24.054	26.217	32.909
13	4.107	4.765	4.892	7.042	8.634	9.926	12.340	15.119	16.985	19.812	22.362	25.472	27.688	34.528
14	4.660	5.368	6.571	7.790	9.467	10.821	13.338	16.222	18.151	21.064	23.685	26.873	29.141	36.123
15	5.229	5.985	7.261	8.547	10.307	11.721	14.339	17.332	19.311	22.307	24.996	28.259	30.578	37.697
16	5.812	6.614	7.962	9.312	11.152	12.624	15.338	18.418	20.465	23.542	26.296	29.633	32.000	39.252
17	6.408	7.255	8.672	10.085	12.002	13.531	16.338	19.511	21.615	24.769	27.587	30.995	33.409	40.790
18	7.015	7.906	9.390	10.865	12.857	14.440	17.338	20.601	22.760	25.989	28.869	32.346	34.805	42.312
19	7.633	8.567	10.117	11.651	13.716	15.352	18.338	21.689	23.900	27.204	30.144	33.687	36.191	43.280
20	8.260	9.237	10.851	12.443	14.578	16.266	19.337	22.775	25.038	28.412	31.410	35.020	37.566	45.315
21	8.897	9.915	11.591	13.240	15.445	17.182	20.337	23.858	26.171	29.615	32.671	36.343	38.932	46.797
22	9.542	10.600	12.338	14.041	16.314	18.101	21.337	24.939	27.301	30.813	33.924	37.659	40.289	48.268
23	10.196	11.293	13.091	14.848	17.187	19.021	22.337	26.018	28.429	32.007	35.172	38.968	41.638	49.728
24	10.856	11.992	13.848	15.659	18.062	19.943	23.337	27.096	29.553	33.196	36.415	40.270	42.980	51.179
25	11.524	12.697	14.611	16.473	18.940	20.867	24.337	28.172	30.675	34.382	37.652	41.566	44.314	52.620
26	12.198	13.409	15.379	17.292	19.820	21.792	25.336	29.246	31.795	35.563	38.885	42.856	45.642	54.052
27	12.879	14.125	16.151	18.114	20.703	22.719	26.336	30.319	32.912	36.741	40.113	44.140	46.963	55.476
28	13.565	14.847	16.928	18.939	21.588	23.647	27.336	31.391	34.027	37.916	41.337	45.419	48.278	56.893
29	14.256	15.574	17.708	19.768	22.475	24.577	28.336	32.461	35.139	39.087	42.557	46.693	49.588	58.302
30	14.953	16.306	18.493	20.599	23.364	25.508	29.336	33.530	36.250	40.256	43.773	47.962	50.892	59.703

* One-tailed distribution.

Statistics and Experimental Design

TABLE D

H Values*

	(Critical Value) a			
n_1	n_2	n_3	Sample Sizes	H
2	1	1	2.7000	0.500
2	2	1	3.6000	0.200
2	2	2	4.5714	0.067
			3.7143	0.200
3	1	1	3.2000	0.300
3	2	1	4.2857	0.100
			3.8571	0.133
3	2	2	5.3572	0.029
			4.7143	0.048
			4.5000	0.067
			4.4643	0.105
3	3	1	5.1429	0.043
			4.5714	0.100
			4.0000	0.129
3	3	2	6.2500	0.011
			5.3611	0.032
			5.1389	0.061
			4.5556	0.100
			4.2500	0.121
3	3	3	7.2000	0.004
			6.4889	0.011
			5.6889	0.029
			5.6000	0.050
			5.0667	0.086
			4.6222	0.100
4	1	1	3.5714	0.200
4	2	1	4.8214	0.057
			4.5000	0.076
			4.0179	0.114
4	2	2	6.0000	0.014
			5.333	0.033
			5.1250	0.052
			4.4583	0.100
			4.1667	0.105
4	3	1	5.8333	0.021
			5.2083	0.050
			5.0000	0.057
			4.0556	0.093
			3.8889	0.129
			4.7000	0.101
4	4	1	6.6667	0.010
			6.1667	0.022
			4.9667	0.048
			4.8667	0.054
			4.1667	0.082
			4.0667	0.102
4	4	2	7.0364	0.006
			6.8727	0.011
			5.4545	0.046

TABLE D

(*Continued*)

	(Critical Value) a			
n_1	n_2	n_3	Sample Sizes H	
			5.2364	0.052
			4.5545	0.098
			4.4455	0.103
4	4	3	7.1439	0.010
			7.1364	0.011
			5.5985	0.049
			5.5758	0.051
			4.5455	0.099
			4.4773	0.102
4	4	4	7.6538	0.008
			7.5385	0.011
			5.6923	0.049
			5.6538	0.054
			4.6539	0.097
			4.5001	0.104
5	1	1	3.8571	0.143
5	2	1	5.2500	0.036
			5.0000	0.048
			4.4500	0.071
			4.2000	0.095
			4.0500	0.119
5	2	2	6.5333	0.008
			6.1333	0.013
			5.1600	0.034
			5.0400	0.056
			4.3733	0.090
			4.2933	0.122
5	3	1	6.4000	0.012
			4.9600	0.048
4	3	2	6.4444	0.008
			6.3000	0.011
			5.4444	0.046
			5.4000	0.051
			4.5111	0.098
			4.4444	0.102
4	3	3	6.7455	0.010
			6.7091	0.013
			5.7909	0.046
			5.7273	0.050
			4.7091	0.092
5	3	3	5.6485	0.049
			5.5152	0.051
			4.5333	0.097
			4.4121	0.109
5	4	1	6.9545	0.008
			6.8400	0.011
			4.9855	0.044
			4.8600	0.056

(*Continued*)

TABLE D

(*Continued*)

(Critical Value) a				
n_1	n_2	n_3	Sample Sizes H	
			3.9873	0.098
			3.9600	0.102
5	4	2	7.2045	0.009
			7.1182	0.010
			5.2727	0.049
			5.2682	0.050
			4.5409	0.098
			5.5182	0.101
5	4	3	7.4449	0.010
			7.3949	0.011
			5.6564	0.049
			5.6308	0.050
			4.5487	0.099
			4.5231	0.103
5	4	4	7.7604	0.009
			7.7440	0.011
			5.6571	0.049
			5.6176	0.050
			4.6187	0.100
			4.5527	0.102
5	5	1	7.3091	0.009
			4.8711	0.052
			4.0178	0.095
			3.8400	0.123
5	3	2	6.9091	0.009
			6.8218	0.010
			5.2509	0.049
			5.1055	0.052
			4.6509	0.091
			4.4945	0.101
5	3	3	7.0788	0.009
			6.9818	0.011
5	5	1	6.8364	0.011
			5.1273	0.046
			4.9091	0.053
			4.1091	0.086
			4.0364	0.105
5	5	2	7.3385	0.010
			7.2692	0.010
			5.3385	0.047
			5.2462	0.051
			4.6231	0.097
			4.5077	0.100
5	5	3	7.5780	0.010
			7.5429	0.010
			5.7055	0.046
			5.6264	0.051
			5.5451	0.100
			4.5363	0.102

TABLE D

(Continued)

	(Critical Value) a			
n_1	n_2	n_3	Sample Sizes H	
5	5	4	7.8229	0.010
			7.7914	0.010
			5.6657	0.049
			5.6429	0.050
			4.5229	0.099
			4.5200	0.101
5	5	5	8.0000	0.009
			7.9800	0.010
			5.7800	0.049
			5.6600	0.051
			4.5600	0.100
			4.5000	0.102

* Test statistics for Kruskal–Wallis nonparametric ANOVA.

TABLE E

Mann–Whitney U Values

n_1	p	$n_2=2$	3	4	5	6	7	8	9	10	11	12	13	14	15	16	17	18	19	20
2	.001																			
	.005	0											0	0	0	0	0	0	0	0
	.01	0		0	0	0		0	0	0	0	0	1	1	1	1	1	1	2	2
	.025	0		0	0	0	0	1	1	1	1	2	2	2	2	2	3	3	3	3
	.05	0		0	1	1	1	2	2	2	2	3	3	4	4	4	4	5	5	5
	.10	0	0	1	2	2	2	3	3	4	4	5	5	5	6	6	7	7	8	8
3	.001	0				0	0	0	0	0	0	0	0	0	0	0	1	1	1	1
	.005	0	0	0	0	0	0	0	1	1	1	2	2	2	3	3	3	3	4	4
	.01	0	0	0	1	0	1	1	2	2	2	3	3	3	4	4	5	5	5	6
	.025	0	0	0	1	2	2	3	3	4	4	5	5	6	6	7	7	8	8	9
	.05	0	1	1	2	3	3	4	5	5	6	6	7	8	8	9	10	10	11	12
	.10	1	2	2	3	4	5	6	6	7	8	9	10	11	11	12	13	14	15	16
4	.001	0				0	0	0	0	1	1	1	2	2	2	3	3	4	4	4
	.005	0	0	0	0	1	1	2	2	3	3	4	4	5	6	6	7	7	8	9
	.01	0	0	0	1	2	2	3	4	4	5	6	6	7	9	8	9	10	10	11
	.025	0	0	1	2	3	4	5	5	6	7	8	9	10	11	12	12	13	14	15
	.05	0	1	2	3	4	5	6	7	8	9	10	11	12	13	15	16	17	18	19
	.10	1	2	4	5	6	7	8	10	11	12	13	14	16	17	18	19	21	22	23
5	.001	0	0	0	0	0	0	1	2	2	3	3	4	4	5	6	6	7	8	8
	.005	0	0	0	1	2	2	3	4	5	6	7	8	8	9	10	11	12	13	14
	.01	0	0	1	2	3	4	5	6	7	8	9	10	11	12	13	14	15	16	17
	.025	0	1	2	3	4	6	7	8	9	10	12	13	14	15	16	18	19	20	21
	.05	1	2	3	5	6	7	9	10	12	13	14	16	17	19	20	21	23	24	26
	.10	2	3	5	6	8	9	11	13	14	16	18	19	21	23	24	26	28	29	31
6	.001	0	0	0	0	0	0	2	3	4	5	5	6	7	8	9	10	11	12	13

6	.005	0	0	1	2	3	4	5	6	7	8	10	11	12	13	14	16	17	18	19
	.01	0	0	2	3	4	5	7	8	9	10	12	13	14	16	17	19	20	21	23
	.025	0	2	3	4	6	7	9	11	12	14	15	17	18	20	22	23	25	26	28
	.05	1	3	4	6	8	9	11	13	15	17	18	20	22	24	26	27	29	31	33
	.10	2	4	6	8	10	12	14	16	18	20	22	24	26	28	30	32	35	37	39
7	.001	0	0	0	0	1	2	3	4	6	7	8	9	10	11	12	14	15	16	17
	.005	0	0	1	2	4	5	7	8	10	11	13	14	16	17	19	20	22	23	25
	.01	0	1	2	4	5	7	8	10	12	13	15	17	18	20	22	24	25	27	29
	.025	0	2	4	6	7	9	11	13	15	17	19	21	23	25	27	29	31	33	35
	.05	1	3	5	7	9	12	14	16	18	20	22	25	27	29	31	34	36	38	40
	.10	2	5	7	9	12	14	17	19	22	24	27	29	32	34	37	39	42	44	47
8	.001	0	0	0	1	2	3	5	6	7	9	10	12	13	15	16	18	19	21	22
	.005	0	0	2	3	5	7	8	10	12	14	16	18	19	21	23	25	27	29	31
	.01	0	1	3	5	7	8	10	12	14	16	18	21	23	25	27	29	31	33	35
	.025	1	3	5	7	9	11	14	16	18	20	23	25	27	30	32	35	37	39	42
	.05	2	4	6	9	11	14	16	19	21	24	27	29	32	34	37	40	42	45	48
	.10	3	6	8	11	14	17	20	23	25	28	31	34	37	40	43	46	49	52	55
9	.001	0	0	0	2	3	4	6	8	9	11	13	15	16	18	20	22	24	26	27
	.005	0	1	2	4	6	8	10	12	14	17	19	21	23	25	28	30	32	34	37
	.01	0	2	4	6	8	10	12	15	17	19	22	24	27	29	32	34	37	39	41
	.025	1	3	5	8	11	13	16	18	21	24	27	29	32	35	38	40	43	46	49
	.05	2	5	7	10	13	16	19	22	25	28	31	34	37	40	43	46	49	52	55
	.10	3	6	10	13	16	19	23	26	29	32	36	39	42	46	49	53	56	59	63
10	.001	0	0	1	2	4	6	7	9	11	13	15	18	20	22	24	26	28	30	33
	.005	0	1	3	5	7	10	12	14	17	19	22	25	27	30	32	35	38	40	43
	.01	0	2	4	7	9	12	14	17	20	23	25	28	31	34	37	39	42	45	48

(Continued)

TABLE E
(Continued)

n_1	p	$n_2=2$	3	4	5	6	7	8	9	10	11	12	13	14	15	16	17	18	19	20
	.025	1	4	6	9	12	15	18	21	24	27	30	34	37	40	43	46	49	53	56
	.05	2	5	8	12	15	18	21	25	28	32	35	38	42	45	49	52	56	59	63
	.10	4	7	11	14	18	22	25	29	33	37	40	44	48	52	55	59	63	67	71
11	.001	0	0	1	3	5	7	9	11	13	16	18	21	23	25	28	30	33	35	38
	.005	0	1	3	6	8	11	14	17	19	22	25	28	31	34	37	40	43	46	49
	.01	0	2	5	8	10	13	16	19	23	26	29	32	35	38	42	45	48	51	54
	.025	1	4	7	10	14	17	20	24	27	31	34	38	41	45	48	52	56	59	63
	.05	2	6	9	13	17	20	24	28	32	35	39	43	47	51	55	58	62	66	70
	.10	4	8	12	16	20	24	28	32	37	41	45	49	53	58	62	66	70	74	79
12	.001	0	0	1	3	5	8	10	13	15	18	21	24	26	29	32	35	38	41	43
	.005	0	2	4	7	10	13	16	19	22	25	28	32	35	38	41	45	48	52	55
	.01	0	3	6	9	12	15	18	22	25	29	32	36	39	43	46	50	54	57	61
	.025	2	5	8	12	15	19	23	27	30	34	38	42	46	50	54	58	62	66	70
	.05	3	6	10	14	18	22	27	31	35	39	43	48	52	56	61	65	69	73	78
	.10	5	9	13	18	22	27	31	36	40	45	50	54	59	64	68	73	78	82	87
13	.001	0	0	2	4	6	9	12	15	18	21	24	27	30	33	36	39	43	46	49
	.005	0	2	4	8	11	14	18	21	25	28	32	35	39	43	46	50	54	58	61
	.01	1	3	6	10	13	17	21	24	28	32	36	40	44	48	52	56	60	64	68
	.025	2	5	9	13	17	21	25	29	34	38	42	46	51	55	60	64	68	73	77
	.05	3	7	11	16	20	25	29	34	38	43	48	52	57	62	66	71	76	81	85
	.10	5	10	14	19	24	29	34	39	44	49	54	59	64	69	75	80	85	90	95
14	.001	0	0	2	4	7	10	13	16	20	23	26	30	33	37	40	44	47	51	55
	.005	0	2	5	8	12	16	19	23	27	31	35	39	43	47	51	55	59	64	68
	.01	1	3	7	11	14	18	23	27	31	35	39	44	48	52	57	61	66	70	74

	.025	2	6	10	14	18	23	27	32	37	41	46	51	56	60	65	70	75	79	84
	.05	4	8	12	17	22	27	32	37	42	47	52	57	62	67	72	78	83	88	93
	.10	5	11	16	21	26	32	37	42	48	53	59	64	70	75	81	86	92	98	103
	.001	0	0	2	5	8	11	15	18	22	25	29	33	37	41	44	48	52	56	60
	.005	0	3	6	9	12	17	21	25	30	34	38	43	47	52	56	61	65	70	74
15	.01	1	4	8	12	16	20	25	29	34	38	43	48	52	57	62	67	71	76	81
	.025	2	6	11	15	20	25	30	35	40	45	50	55	60	65	71	76	81	86	91
	.05	4	8	13	19	24	29	34	40	45	51	56	62	67	73	78	84	89	95	101
	.10	6	11	17	23	28	34	40	46	52	58	64	69	75	81	87	93	99	105	111
	.001	0	0	3	6	9	12	16	20	24	28	32	36	40	44	49	53	57	61	66
	.005	0	3	6	10	14	19	23	28	32	37	42	46	51	56	61	66	71	75	80
16	.01	1	4	8	13	17	22	27	32	37	42	47	52	57	62	67	72	77	83	88
	.025	2	7	12	16	22	27	32	38	43	48	54	60	65	71	76	82	87	93	99
	.05	4	9	15	20	26	31	37	43	49	55	61	66	72	78	84	90	96	102	108
	.10	6	12	18	24	30	37	43	49	55	62	68	75	81	87	94	100	107	113	120
	.001	0	1	3	6	10	14	18	22	26	30	35	39	44	48	53	58	62	67	71
	.005	0	3	7	11	16	20	25	30	35	40	45	50	55	61	66	71	76	82	87
17	.01	1	5	9	14	19	24	29	34	39	45	50	56	61	67	72	78	83	89	94
	.025	3	7	12	18	23	29	35	40	46	52	58	64	70	76	82	88	94	100	106
	.05	4	10	16	21	27	34	40	46	52	58	65	71	78	84	90	97	103	110	116
	.10	7	13	19	26	32	39	46	53	59	66	73	80	86	93	100	114	121	128	128
	.001	0	1	4	7	11	15	19	24	28	33	38	43	47	52	57	62	67	72	77
	.005	0	3	7	12	17	22	27	32	38	43	48	54	59	65	71	76	82	88	93
18	.01	1	5	10	15	20	25	31	37	42	48	54	60	66	71	77	83	89	95	101
	.025	3	8	13	19	25	31	37	43	49	56	62	68	75	81	87	94	100	107	113
	.05	5	10	17	23	29	36	42	49	56	62	69	76	83	89	96	103	110	117	124
	.10	7	14	21	28	35	42	49	56	63	70	78	85	92	99	107	114	121	129	136

(Continued)

TABLE E
(Continued)

n_1	p	$n_2=2$	3	4	5	6	7	8	9	10	11	12	13	14	15	16	17	18	19	20
19	.001	0	1	4	8	12	16	21	26	30	35	41	46	51	56	61	67	72	78	83
	.005	1	4	8	13	18	23	29	34	40	46	52	58	64	70	75	82	88	94	100
	.01	2	5	10	16	21	27	33	39	45	51	57	64	70	76	83	89	95	102	108
	.025	3	8	14	20	26	33	39	46	53	59	66	73	79	86	93	100	107	114	120
	.05	5	11	18	24	31	38	45	52	59	66	73	81	88	95	102	110	117	124	131
	.10	8	15	22	29	37	44	52	59	67	74	82	90	98	105	113	121	129	136	144
20	.001	0	1	4	8	13	17	2	27	33	38	43	49	55	60	66	71	77	83	89
	.005	1	4	9	14	19	25	31	37	43	49	55	61	68	74	80	87	93	100	106
	.01	2	6	11	17	23	29	35	41	48	54	61	68	74	81	88	94	101	108	115
	.025	3	9	15	21	28	35	42	49	56	63	70	77	84	91	99	106	113	120	128
	.05	5	12	19	26	33	40	48	55	63	70	78	85	93	101	108	116	124	131	139
	.10	8	16	23	31	39	47	55	63	71	79	87	95	103	111	120	128	136	144	152

TABLE F

t-Test Critical Values*

df	.1	p ≤ .05	.01	.001
1	6.314	12.706	63.657	636.619
2	2.920	4.303	9.925	31.598
3	2.353	3.182	5.841	21.941
4	2.132	2.776	4.604	8.610
5	2.015	2.571	4.032	6.859
6	1.943	2.447	3.707	5.959
7	1.895	2.365	3.499	5.405
8	1.860	2.306	3.355	5.041
9	1.833	2.262	3.250	4.781
10	1.812	2.228	3.169	4.587
11	1.796	2.201	3.106	4.437
12	1.782	2.179	3.055	4.318
13	1.771	2.160	3.012	4.221
14	1.761	2.145	2.977	4.140
15	1.753	2.131	2.947	4.073
16	1.746	2.120	2.921	4.015
17	1.740	2.110	2.898	3.965
18	1.734	2.101	2.878	3.922
19	1.729	2.093	2.861	3.883
20	1.725	2.086	2.845	3.850
21	1.721	2.080	2.831	3.819
22	1.717	2.074	2.819	3.792
23	1.714	2.069	2.807	3.767
24	1.711	2.064	2.797	3.745
25	1.708	2.060	2.787	3.725
26	1.706	2.056	2.779	3.707
27	1.703	2.052	2.771	3.690
28	1.701	2.048	2.763	3.674
29	1.699	2.045	2.756	3.659
30	1.697	2.042	2.750	3.646
40	1.684	2.021	2.704	3.551
60	1.671	2.000	2.660	3.460
120	1.658	1.980	2.617	3.373
	1.645	1.960	2.576	3.294

* Two-tailed *t* distribution values.

TABLE G

F Distribution Values*

(1) p < 0.05

Sample size (N) for greater mean square (MS_{wg})

	1	2	3	4	5	6	7	8	9	10	12	15	20	24	30	40	60	120	∞
1	161.4	199.5	215.7	224.6	230.2	234.0	236.8	238.9	240.5	241.9	243.9	245.9	248.0	249.1	250.1	251.1	252.2	253.3	254.3
2	18.51	19.00	19.16	19.25	19.30	19.33	19.35	19.37	19.38	19.40	19.41	19.43	19.45	19.45	19.46	19.47	19.48	19.49	19.50
3	10.13	9.55	9.28	9.12	9.01	8.94	8.89	8.85	8.81	8.74	8.74	8.70	8.66	8.64	8.62	8.62	8.57	8.55	8.53
4	7.71	6.94	6.59	6.39	6.26	6.16	6.09	6.04	6.00	5.96	5.91	5.86	5.80	5.77	5.75	5.72	5.69	5.66	5.63
5	6.61	5.79	5.41	5.19	5.05	4.95	4.88	4.82	4.77	4.74	4.68	4.62	4.56	4.53	4.50	4.46	4.43	4.40	4.36
6	5.99	5.14	4.76	4.53	4.39	4.28	4.21	4.15	4.10	4.06	4.00	3.94	3.87	3.84	3.81	3.77	3.74	3.70	3.67
7	5.59	4.74	4.35	4.12	3.97	3.87	3.79	3.73	3.68	3.64	3.57	3.51	3.44	3.41	3.38	3.34	3.30	3.27	3.23
8	5.32	4.46	4.07	3.84	3.69	3.58	3.50	3.44	3.39	3.35	3.28	3.22	3.15	3.12	3.08	3.04	3.01	2.97	2.93
9	5.12	4.26	3.86	3.63	3.48	3.37	3.29	3.23	3.18	3.14	3.07	3.01	2.94	2.90	2.86	2.83	2.79	2.75	1.71
10	4.96	4.10	3.71	3.48	3.33	3.22	3.14	3.07	3.02	2.98	2.91	2.85	2.77	2.74	2.70	2.66	2.62	2.58	2.54
11	4.84	3.98	3.59	3.36	3.20	3.09	3.01	2.95	2.90	2.85	2.79	2.72	2.65	2.61	2.57	2.53	2.49	2.45	2.40
12	4.75	3.89	3.49	3.26	3.11	3.00	2.91	2.85	2.80	2.75	2.69	2.62	2.54	2.51	2.47	2.43	2.38	2.34	2.30
13	4.67	3.81	3.41	3.18	3.03	2.92	2.83	2.77	2.71	2.67	2.60	2.53	2.46	2.42	2.38	2.34	2.30	2.25	2.21
14	4.60	3.74	3.34	3.11	2.96	2.85	2.76	2.70	2.65	2.60	2.53	2.46	2.39	2.35	2.31	2.27	2.22	2.18	2.13
15	4.54	3.68	3.29	3.06	2.90	2.79	2.71	2.64	2.59	2.54	2.48	2.40	2.33	2.29	2.25	2.20	2.16	2.11	2.07
16	4.49	3.63	3.24	3.01	2.85	2.74	2.66	2.59	2.54	2.49	2.42	2.35	2.28	2.24	2.19	2.15	2.11	2.06	2.01
17	4.45	3.59	3.20	2.96	2.81	2.70	2.61	2.55	2.49	2.45	2.38	2.31	2.23	2.19	2.13	2.10	2.06	2.01	1.96
18	4.41	3.55	3.16	2.93	2.77	2.66	2.58	2.51	2.46	2.41	2.36	2.27	2.19	2.15	2.11	2.06	2.02	1.97	1.92
19	4.38	3.52	3.13	2.90	2.74	2.63	2.54	2.48	2.42	2.38	2.31	2.23	2.16	2.11	2.07	2.03	1.98	1.93	1.88
20	4.35	3.49	3.10	2.87	2.71	2.60	2.51	2.45	2.39	2.35	2.28	2.20	2.12	2.08	2.04	1.99	1.95	1.90	1.84
21	4.32	3.47	3.07	2.84	2.68	2.57	2.49	2.42	2.37	2.32	2.25	2.18	2.10	2.05	2.01	1.96	1.92	1.87	1.81
22	4.30	3.44	3.05	2.82	2.66	2.55	2.46	2.40	2.34	2.30	2.23	2.15	2.07	2.03	1.98	1.94	1.89	1.84	1.78
23	4.28	3.42	3.03	2.80	2.64	2.53	2.44	2.37	2.32	2.27	2.20	2.13	2.05	2.01	1.96	1.91	1.86	1.81	1.76

24	4.26	3.40	3.01	2.78	2.62	2.51	2.42	2.36	2.30	2.25	2.18	2.11	2.03	1.98	1.94	1.89	1.84	1.79	1.73
25	4.24	3.39	2.99	2.76	2.60	2.49	2.40	2.34	2.28	2.24	2.16	2.09	2.01	1.96	1.92	1.87	1.82	1.77	1.71
26	4.23	3.37	2.98	2.74	2.59	2.47	2.39	2.32	2.27	2.22	2.15	2.07	1.99	1.95	1.90	1.85	1.80	1.75	1.69
27	4.21	3.35	2.96	2.73	2.57	2.46	2.37	2.31	2.25	2.20	2.13	2.06	1.97	1.93	1.88	1.84	1.79	1.73	1.67
28	4.20	3.34	2.95	2.71	2.56	2.45	2.36	2.29	2.24	2.19	2.12	2.04	1.96	1.91	1.87	1.82	1.77	1.71	1.65
29	4.18	3.33	2.93	2.70	2.55	2.43	2.35	2.28	2.22	2.18	2.10	2.03	1.94	1.90	1.85	1.81	1.75	1.70	1.64
30	4.17	3.32	2.92	2.69	2.53	2.42	2.33	2.27	2.21	2.16	2.09	2.01	1.93	1.89	1.84	1.79	1.74	1.68	1.62
40	4.08	3.23	2.84	2.61	2.45	2.34	2.25	2.18	2.12	2.08	2.00	1.92	1.84	1.79	1.74	1.69	1.64	1.58	1.51
60	4.00	3.15	2.76	2.53	2.37	2.25	2.17	2.10	2.04	1.99	1.92	1.84	1.75	1.70	1.65	1.59	1.53	1.47	1.39
120	3.92	3.07	2.68	2.45	2.29	2.17	2.09	2.02	1.96	1.91	1.83	1.75	1.66	1.61	1.55	1.50	1.43	1.35	1.25
∞	3.83	3.00	2.60	2.37	2.21	2.10	2.01	1.94	1.88	1.83	1.75	1.67	1.57	1.52	1.46	1.39	1.32	1.22	1.00

* For F test, ANOVA, ANCOVA

(2) $p < 0.01$

Sample size (N) for greater mean square (MS_{wg})

	1	2	3	4	5	6	7	8	9	10	12	15	20	24	30	40	60	120	∞
1	4052	4999.50	5403	5625	5764	5859	5928	5982	6022	6056	6106	6157	6209	6235	6261	6287	6313	6339	6366
2	98.50	99.00	99.17	99.25	99.30	99.33	99.36	99.37	99.39	99.40	99.42	99.43	99.45	99.46	99.47	99.47	99.48	99.49	99.50
3	34.12	30.82	29.46	28.71	28.24	27.91	27.67	27.49	27.35	27.23	27.05	26.87	26.69	26.60	26.50	26.41	26.32	26.22	26.13
4	21.20	18.00	16.69	15.98	15.52	15.21	14.98	14.80	14.66	14.55	14.37	14.20	14.02	13.93	13.84	13.75	13.65	13.56	13.46
5	16.26	13.27	12.06	11.39	10.97	10.67	10.46	10.29	10.16	10.05	9.89	9.72	9.55	9.47	9.38	9.29	9.20	9.11	9.02
6	13.75	10.92	9.78	9.15	8.75	8.47	8.26	8.10	7.98	7.87	7.72	7.56	7.40	7.31	7.23	7.14	7.06	6.97	6.88
7	12.25	9.55	8.45	7.85	7.46	7.19	6.99	6.84	6.72	6.62	6.47	6.31	6.16	6.07	5.99	5.91	5.82	5.74	5.65
8	11.26	8.65	7.59	7.01	6.63	6.37	6.18	6.03	5.91	5.81	5.67	5.52	5.36	5.28	5.20	5.12	5.03	4.95	4.86
9	10.56	8.02	6.99	6.42	6.06	5.80	5.61	5.47	5.35	5.26	5.11	4.96	4.81	4.73	4.65	4.57	4.48	4.40	4.31
10	10.04	7.56	6.55	5.99	5.64	5.39	5.20	5.06	4.94	4.85	4.71	4.56	4.41	4.33	4.25	4.17	4.08	4.00	3.91
11	9.65	7.21	6.22	5.67	5.32	5.07	4.89	4.74	4.63	4.54	4.40	4.25	4.10	4.02	3.94	3.86	3.78	3.69	3.60

(Continued)

TABLE G
(Continued)

(2) p < 0.01

Sample size (N) for greater mean square (MS_{wg})

	1	2	3	4	5	6	7	8	9	10	12	15	20	24	30	40	60	120	∞
12	9.33	6.93	5.95	5.41	5.06	4.82	4.64	4.50	4.39	4.30	4.16	4.01	3.86	3.78	3.70	3.62	3.54	3.45	3.36
13	9.07	6.70	5.74	5.21	4.86	4.62	4.44	4.30	4.19	4.10	3.96	3.82	3.66	3.59	3.51	3.43	3.34	3.25	3.17
14	8.86	6.51	5.56	5.04	4.69	4.46	4.28	4.14	4.03	3.94	3.80	3.66	3.51	3.43	3.35	3.27	3.18	3.09	3.00
15	8.68	6.36	5.42	4.89	4.56	4.32	4.14	4.00	3.89	3.80	3.67	3.52	3.37	3.29	3.21	3.13	3.05	2.96	2.87
16	8.53	6.23	5.29	4.77	4.44	4.20	4.03	3.89	3.78	3.69	3.55	3.41	3.26	3.18	3.10	3.02	2.93	2.84	2.75
17	8.40	6.11	5.18	4.67	4.34	4.10	3.93	3.79	3.68	3.59	3.46	3.31	3.16	3.08	3.00	2.92	2.83	2.75	2.65
18	8.29	6.01	5.09	4.58	4.25	4.01	3.84	3.71	3.60	3.51	3.37	3.23	3.08	3.00	2.92	2.84	2.75	2.66	2.57
19	8.18	5.93	5.01	4.50	4.17	3.94	3.77	3.63	3.52	3.43	3.30	3.15	3.00	2.92	2.84	2.76	2.67	2.58	2.49
20	8.10	5.85	4.94	4.43	4.10	3.87	3.70	3.56	3.46	3.37	3.23	3.09	2.94	2.86	2.78	2.69	2.61	2.52	2.42
21	8.02	5.78	4.87	4.37	4.04	3.81	3.64	3.51	3.40	3.31	3.17	3.03	2.88	2.80	2.72	2.64	2.55	2.46	2.36
22	7.95	5.72	4.82	4.31	3.99	3.76	3.59	3.45	3.35	3.26	3.12	2.98	2.83	2.75	2.67	2.58	2.50	2.40	2.31
23	7.88	5.66	4.76	4.26	3.94	3.71	3.54	3.41	3.30	3.21	3.07	2.93	2.78	2.70	2.62	2.54	2.45	2.35	2.26
24	7.82	5.61	4.72	4.22	3.90	3.67	3.50	3.36	3.26	3.17	3.03	2.89	2.74	2.66	2.58	2.49	2.40	2.31	2.21
25	7.77	5.57	4.68	4.18	3.85	3.63	3.46	3.32	3.22	3.13	2.99	2.85	2.70	2.62	2.54	2.45	2.36	2.27	2.17
26	7.72	5.53	4.64	4.14	3.82	3.59	3.42	3.29	3.18	3.09	2.96	2.81	2.66	2.58	2.50	2.42	2.33	2.23	2.13
27	7.68	5.49	4.60	4.11	3.78	3.56	3.39	3.26	3.15	3.06	2.93	2.78	2.63	2.55	2.47	2.38	2.29	2.20	2.10
28	7.64	5.45	4.57	4.07	3.75	3.53	3.36	3.23	3.12	3.03	2.90	2.75	2.60	2.52	2.44	2.35	2.26	2.17	2.06
29	7.60	5.42	4.54	4.04	3.73	3.50	3.33	3.20	3.09	3.00	2.87	2.73	2.57	2.49	2.41	2.33	2.23	2.14	2.03
30	7.56	5.39	4.51	4.02	3.70	3.47	3.30	3.17	3.07	2.98	2.84	2.70	2.55	2.47	2.39	2.30	2.21	2.11	2.01
40	7.31	5.18	4.31	3.83	3.51	3.29	3.12	2.99	2.89	2.80	2.66	2.52	2.37	2.29	2.20	2.11	2.02	1.92	1.80
60	7.08	4.98	4.13	3.65	3.34	3.12	2.95	2.82	2.72	2.63	2.50	2.35	2.20	2.12	2.03	1.94	1.84	1.73	1.60
120	6.85	4.79	3.95	3.48	3.17	2.96	2.79	2.66	2.58	2.47	2.34	2.19	2.03	1.95	1.86	1.76	1.66	1.53	1.38
∞	6.63	4.61	3.78	3.32	3.02	2.80	2.64	2.51	2.41	2.32	2.18	2.04	1.88	1.79	1.70	1.59	1.47	1.32	1.00

(3) p < 0.001

Sample size (N) for greater mean square (MS_{wg})

	1	2	3	4	5	6	7	8	9	10	12	15	20	24	30	40	60	120	8
1	4053*	5000*	5404*	5625*	5764*	5859*	5921*	5981*	6023*	6056*	6107*	6153*	6209*	6235*	6261*	6287*	6313*	6340*	6366*
2	998.5	999.0	999.2	999.2	999.3	999.3	999.4	999.4	999.4	999.4	999.4	999.4	999.4	999.5	999.5	999.5	999.5	999.5	999.5
3	167.0	148.5	141.1	137.1	134.6	132.8	131.6	130.6	129.9	129.2	128.3	127.4	126.4	125.9	125.4	125.0	124.5	124.0	123.5
4	74.14	61.25	56.18	53.44	51.71	50.53	49.66	49.00	48.47	48.05	47.41	46.76	46.10	45.77	45.43	45.09	44.75	44.40	44.05
5	47.18	37.12	33.20	31.09	29.75	28.84	28.16	27.64	27.24	26.92	26.42	25.91	25.39	25.14	24.87	24.60	24.33	24.06	23.79
6	35.51	27.00	23.70	21.92	20.81	20.03	19.46	19.03	18.69	18.41	17.99	17.56	17.12	16.89	16.67	16.44	16.21	15.99	15.75
7	29.25	21.69	18.77	17.19	16.21	15.52	15.02	14.63	14.33	14.08	13.71	13.32	12.93	12.73	12.53	12.33	12.12	11.91	11.70
8	25.42	18.49	15.83	14.39	13.49	12.86	12.40	12.04	11.77	11.54	11.19	10.84	10.48	10.30	10.11	9.92	9.73	9.53	9.33
9	22.86	16.39	13.90	12.56	11.71	11.13	10.70	10.37	10.11	9.89	9.57	9.24	8.90	8.72	8.55	8.37	8.19	8.00	7.81
10	21.04	14.91	12.55	11.28	10.48	9.92	9.52	9.20	8.96	8.75	8.45	8.13	7.80	7.64	7.47	7.30	7.12	6.94	6.76
11	19.69	13.81	11.56	10.35	9.58	9.05	8.66	8.35	8.12	7.92	7.63	7.32	7.01	6.85	6.68	6.52	6.35	6.17	6.00
12	18.65	12.97	10.80	9.63	8.89	8.38	8.00	7.71	7.48	7.29	7.00	6.71	6.40	6.25	6.09	5.93	5.76	5.59	5.42
13	17.81	12.31	10.21	9.07	8.35	7.86	7.49	7.21	6.98	6.80	6.52	6.23	5.93	5.78	5.63	5.47	5.30	5.14	4.97
14	17.14	11.78	9.73	8.62	7.92	7.43	7.08	6.80	6.58	6.40	6.13	5.85	5.56	5.41	5.25	5.10	4.94	4.77	4.60
15	16.59	11.34	9.34	8.25	7.57	7.09	6.74	6.47	6.26	6.08	5.81	5.54	5.25	5.10	4.95	4.80	4.64	4.47	4.31
16	16.12	10.97	9.00	7.94	7.27	6.81	6.46	6.19	5.98	5.81	5.55	5.27	4.99	4.85	4.70	4.54	4.39	4.23	4.06
17	15.72	10.66	8.73	7.68	7.02	6.56	6.22	5.96	5.75	5.58	5.32	5.05	4.78	4.63	4.48	4.33	4.18	4.02	3.85
18	15.38	10.39	8.49	7.46	6.81	6.35	6.02	5.76	5.56	5.39	5.13	4.87	4.59	4.45	4.30	4.15	4.00	3.84	3.67
19	15.08	10.16	8.28	7.26	6.62	6.18	5.85	5.59	5.39	5.22	4.97	4.70	4.43	4.29	4.14	3.99	3.84	3.68	3.51
20	14.82	9.95	8.10	7.10	6.46	6.02	5.69	5.44	5.24	5.08	4.82	4.56	4.29	4.15	4.00	3.86	3.70	3.54	3.38
21	14.59	9.77	7.94	6.95	6.32	5.88	5.56	5.31	5.11	4.95	4.70	4.44	4.17	4.03	3.88	3.74	3.58	3.42	3.26
22	14.38	9.61	7.80	6.81	6.19	5.76	5.44	5.19	4.99	4.83	4.58	4.33	4.06	3.92	3.78	3.63	3.48	3.32	3.05
23	14.19	9.47	7.67	6.69	6.08	5.65	5.35	5.09	4.89	4.73	4.48	4.23	3.96	3.82	3.68	3.53	3.38	3.22	3.05
24	14.03	9.34	7.55	6.59	5.98	5.55	5.23	4.99	4.80	4.64	4.39	4.14	3.87	3.74	3.59	3.45	3.29	3.14	2.97

TABLE H

Z Scores for Normal Distributions

	PROPORTIONAL PARTS									
z	0.00	0.01	0.02	0.03	0.04	0.05	0.06	0.07	0.08	0.09
0.0	0.5000	0.4960	0.4920	0.4880	0.4840	0.4801	0.4761	0.4721	0.4681	0.4641
0.1	0.4602	0.4562	0.4522	0.4483	0.4443	0.4404	0.4364	0.4325	0.4286	0.4247
0.2	0.4207	0.4168	0.4129	0.4090	0.4052	0.4013	0.3974	0.3936	0.3897	0.3859
0.3	0.3821	0.3783	0.3745	0.3707	0.3669	0.3632	0.3594	0.3557	0.3520	0.3483
0.4	0.3446	0.3409	0.3372	0.3336	0.3300	0.3264	0.3228	0.3192	0.3156	0.3121
0.5	0.3085	0.3050	0.3015	0.2981	0.2946	0.2912	0.2877	0.2843	0.2810	0.2776
0.6	0.2743	0.2709	0.2676	0.2643	0.2611	0.2578	0.2546	0.2514	0.2483	0.2451
0.7	0.2420	0.2389	0.2358	0.2327	0.2296	0.2266	0.2236	0.2206	0.2177	0.2148
0.8	0.2119	0.2090	0.2061	0.2033	0.2005	0.1977	0.1949	0.1922	0.1894	0.1867
0.9	0.1841	0.1814	0.1788	0.1762	0.1736	0.1711	0.1685	0.1660	0.1635	0.1611
1.0	0.1587	0.1562	0.1539	0.1515	0.1492	0.1469	0.1446	0.1423	0.1401	0.1379
1.1	0.1357	0.1335	0.1314	0.1292	0.1271	0.1251	0.1230	0.1210	0.1190	0.1170
1.2	0.1151	0.1131	0.1112	0.1093	0.1075	0.1056	0.1038	0.1020	0.1003	0.0985
1.3	0.0968	0.0958	0.0934	0.0918	0.0901	0.0885	0.0869	0.0853	0.0838	0.0823
1.4	0.0808	0.0793	0.0778	0.0764	0.0749	0.0735	0.0721	0.0708	0.0694	0.0681
1.5	0.0668	0.0655	0.0643	0.0630	0.0618	0.0606	0.0594	0.0582	0.0571	0.0559
1.6	0.0548	0.0537	0.0526	0.0516	0.0505	0.0495	0.0485	0.0475	0.0465	0.0455
1.7	0.0446	0.0436	0.0427	0.0418	0.0409	0.0401	0.0392	0.0384	0.0375	0.0367
1.8	0.0359	0.0351	0.0344	0.0336	0.0329	0.0322	0.0314	0.0307	0.0301	0.0294
1.9	0.0287	0.0281	0.0274	0.0268	0.0262	0.0256	0.0250	0.0244	0.0239	0.0233
2.0	0.0228	0.0222	0.0217	0.0212	0.0207	0.0202	0.0197	0.0192	0.0188	0.0183
2.1	0.0179	0.0174	0.0170	0.0166	0.0162	0.0158	0.0154	0.0150	0.0146	0.0143
2.2	0.0139	0.0136	0.0132	0.0129	0.0125	0.0122	0.0119	0.0116	0.0113	0.0110
2.3	0.0107	0.0104	0.0102	0.0099	0.0096	0.0094	0.0091	0.0089	0.0087	0.0084
2.4	0.0082	0.0080	0.0078	0.0075	0.0073	0.0071	0.0069	0.0068	0.0066	0.0064
2.5	0.0062	0.0060	0.0059	0.0057	0.0055	0.0054	0.0052	0.0051	0.0049	0.0048
2.6	0.0047	0.0045	0.0044	0.0043	0.0041	0.0040	0.0039	0.0038	0.0037	0.0036
2.7	0.0035	0.0034	0.0033	0.0032	0.0031	0.0030	0.0029	0.0028	0.0027	0.0026
2.8	0.0026	0.0025	0.0024	0.0023	0.0023	0.0022	0.0021	0.0021	0.0020	0.0019
2.9	0.0019	0.0018	0.0018	0.0017	0.0016	0.0016	0.0015	0.0015	0.0014	0.0014
3.0	0.0013	0.0013	0.0013	0.0012	0.0012	0.0011	0.0011	0.0011	0.0010	0.0010
3.1	0.0010	0.0009	0.0009	0.0009	0.0008	0.0008	0.0008	0.0008	0.0007	0.0007
3.2	0.0007	0.0007	0.0006	0.0006	0.0006	0.0006	0.0006	0.0005	0.0005	0.0005
3.3	0.0005	0.0005	0.0005	0.0004	0.0004	0.0004	0.0004	0.0004	0.0004	0.0004
3.4	0.0003	0.0003	0.0003	0.0003	0.0003	0.0003	0.0003	0.0003	0.0003	0.0002
3.5	0.0002	0.0002	0.0002	0.0002	0.0002	0.0002	0.0002	0.0002	0.0002	0.0002
3.6	0.0002	0.0002	0.0001	0.0001	0.0001	0.0001	0.0001	0.0001	0.0001	0.0001
3.7	0.0001	0.0001	0.0001	0.0001	0.0001	0.0001	0.0001	0.0001	0.0001	0.0001
3.8	0.0001	0.0001	0.0001	0.0001	0.0001	0.0001	0.0001	0.0001	0.0001	0.0001
3.9	0.0000	0.0000	0.0000	0.0000	0.0000	0.0000	0.0000	0.0000	0.0000	0.0000

TABLE I

Table for Calculation of Median-Effective Dose by Moving Averages

n = 2, K = 3

r-values	f	σ f
0,0,1,2	1.00000	0.50000
0,0,2,2	0.50000	0.00000
0,1,1,2	0.50000	0.70711
0,1,2,2	0.00000	0.50000
1,0,1,2	1.00000	1.00000
1,0,2,2	0.00000	1.00000
1,1,1,2	0.00000	1.73205
0,0,2,1	1.00000	1.00000
0,1,1,1	1.00000	1.73205
0,1,2,1	0.00000	1.00000
0,0,3,3	1.00000	0.47140

n = 3, K = 3

r-values	f	σ f
0,0,4,3	0.66667	0.22222
0,1,2,3	1.00000	0.60858
0,0,2,3	0.83333	0.33333
0,0,3,3	0.50000	0.00000
0,1,1,3	0.83333	0.47140
0,1,2,3	0.50000	0.47140

n = 5, K = 3

r-values	f	f
2,2,2,4	0.7500	0.95607
2,2,3,4	0.25000	0.98821
0,0,5,3	0.83333	0.34021
0,1,4,3	0.83333	0.58134

n = 4, K = 3

r-values	f	σ f
2,0,3,4	0.50000	0.57735
2,0,4,4	0.00000	0.57735
2,1,1,4	1.00000	0.70711
2,1,2,4	0.50000	0.81650
2,1,3,4	0.00000	.91287
2,2,2,4	0.00000	1.00000
3,0,2,4	1.00000	1.15470
3,0,3,4	0.00000	1.41421
3,1,1,4	1.00000	1.41421
3,1,2,4	0.00000	1.82574
1,0,3,5	0.87500	0.30778
1,0,4,5	0.62500	0.26700
1,0,5,5	0.37500	0.15625
0,1,3,3	0.66667	0.52116
0,1,4,3	0.33333	0.35136
0,2,2,3	0.66667	0.58794
0,2,3,3	0.33333	0.52116

n = 6, K = 3

r-values	f	σ f
1,4,4,6	0.00000	0.36878
2,0,3,6	1.00000	0.33541
2,0,4,6	0.75000	0.32596
2,0,5,6	0.50000	0.29580

n = 5, K = 3

r-values	f	σ f
0,1,2,5	.90000	0.31623
0,1,3,5	0.7000	0.31623
0,1,4,5	0.50000	0.28284
0,1,5,5	0.30000	0.20000
0,2,2,5	0.70000	0.34641
0,2,3,5	0.50000	0.34641
0,2,4,5	0.30000	0.31623
0,2,5,5	0.10000	0.24495
0,3,3,5	0.30000	0.34641
0,3,4,5	0.10000	0.31623
1,1,2,5	0.87500	0.39652
1,1,3,5	0.62500	0.40625
1,1,4,5	0.37500	0.38654
1,1,5,5	0.12500	0.33219

n = 6, K = 3

r-values	f	σ f
0,3,6,5	0.00000	0.26833
0,4,4,5	0.20000	0.36000
0,4,5,5	0.00000	0.32249
1,0,4,5	1.00000	0.40311

(Continued)

TABLE I
(Continued)

n = 6, K = 3

r-values	f	σ f
0,1,5,3	0.50000	0.39087
0,2,3,3	0.83333	0.67013
0,2,4,3	0.50000	0.56519
0,2,5,3	0.16667	0.41388
0,3,3,3	0.50000	0.61237
0,3,4,3	0.16667	0.53142
1,0,5,3	0.75000	0.47598
1,1,4,3	0.75000	0.85239
1,1,5,3	0.25000	0.64348
1,2,3,3	0.75000	0.98821
1,2,4,3	0.25000	0.88829
1,3,3,3	0.25000	0.95607

n = 6, K = 3

r-values	f	σ f
0,0,3,6	1.00000	.022361

n = 6, K = 3

r-values	f	σ f
0,0,4,6	0.83333	0.21082
0,0,5,6	0.6667	0.21082
0,0,6,6	0.50000	0.00000
0,1,2,6	1.00000	0.26874
0,1,3,6	0.83333	0.27889
0,1,4,6	0.66667	0.26874
0,1,5,6	0.50000	0.23570

n = 6, K = 3

r-values	f	σ f
2,0,6,6	0.25000	0.23717
2,1,2,6	1.00000	0.40311
2,1,3,6	0.75000	0.42573
2,1,4,6	0.50000	0.43301
2,1,5,6	0.25000	0.42573
2,1,6,6	0.00000	0.29580
2,2,2,6	0.75000	0.45415
2,2,3,6	0.50000	0.48734
2,2,4,6	0.25000	0.50621
2,2,5,6	0.00000	0.43301
2,3,3,6	0.25000	0.53033
2,3,4,6	0.00000	0.48734
3,0,3,6	1.00000	0.44721
3,0,4,6	0.66667	0.43885
3,0,5,6	0.33333	0.44721
3,0,6,6	0.00000	0.44721

n = 6, K = 3

r-values	f	σ f
3,1,2,6	1.00000	0.53748
3,1,3,6	0.66667	0.57090
3,1,4,6	0.33333	0.61464
3,1,5,6	0.00000	0.64979
3,2,2,6	0.66667	0.60858
3,2,3,6	0.33333	0.68313
3,2,4,6	0.00000	0.74536

n = 6, K = 3

r-values	f	σ f
1,0,5,5	0.75000	0.31869
1,0,6,5	0.50000	0.17678
1,1,3,5	1.00000	0.48734
1,1,4,5	0.75000	0.44896
1,1,5,1	0.50000	0.39528
1,1,6,5	0.25000	0.31869
1,2,2,5	1.00000	0.51235
1,2,3,5	0.75000	0.50156
1,2,4,5	0.50000	0.48088
1,2,5,5	0.25000	0.44896
1,2,6,5	0.00000	0.40311
1,3,3,5	0.50000	0.50621
1,3,4,5	0.25000	0.50156
1,3,5,5	0.00000	0.48734
1,4,4,5	0.00000	0.51235
2,0,4,5	1.00000	0.53748

n = 6, K = 3

r-values	f	σ f
2,0,5,5	0.66667	0.42455
2,0,6,5	0.33333	0.30225
2,1,3,5	1.00000	0.64979
2,1,4,5	0.66667	0.59835
2,1,5,5	0.33333	0.55998
2,1,6,5	0.00000	0.53748
2,2,2,5	1.00000	0.68313

n = 6, K = 3		
r-values	f	σf
0,1,6,6	0.33333	0.16667
0,2,2,6	0.83333	0.29814
0,2,3,6	0.66667	0.30732
0,2,4,6	0.50000	0.29814
0,2,5,6	0.33333	0.26874
0,2,6,6	0.16667	0.21082
0,3,3,6	0.50000	0.31623
0,3,4,6	0.33333	0.30732
0,3,5,6	0.16667	0.27889
0,3,6,6	0.00000	0.22361
0,4,4,6	0.16667	0.29814
0,4,5,6	0.00000	0.266874
1,0,3,6	1.00000	0.26833
1,0,4,6	0.80000	0.25612
1,0,5,6	0.60000	0.21541
1,0,6,6	0.40000	0.12000
1,1,2,6	1.00000	0.32249
1,1,3,6	0.80000	0.33704
1,1,4,6	0.60000	0.33226
1,1,5,6	0.40000	0.30724
1,1,6,6	0.20000	0.25612
1,2,2,6	0.80000	0.36000
1,2,3,6	0.60000	0.37736
1,2,4,6	0.4000	0.37736
1,2,5,6	0.20000	0.36000
1,2,6,6	0.00000	0.26833
1,3,3,6	0.40000	0.39799
1,3,4,6	0.20000	0.40200

n = 6, K = 3		
r-values	f	σf
3,3,3,6	0.00000	0.77460
4,0,3,6	1.00000	0.67082
4,0,4,6	0.50000	0.70711
4,0,5,6	0.00000	0.80622
4,1,2,6	1.00000	0.80622
4,1,3,6	0.50000	0.89443
4,1,4,6	0.00000	1.02470
4,2,2,6	0.50000	0.94888
4,2,3,6	0.00000	1.11803
5,0,3,6	1.00000	1.34164
5,0,4,6	0.00000	1.61245
5,1,2,6	1.00000	1.61245
5,1,3,6	0.00000	1.94936
5,2,2,6	0.00000	2.04939
0,0,4,5	1.00000	0.26833
0,0,5,5	0.80000	0.25612
0,0,6,5	0.60000	0.12000
0,1,3,5	1.00000	0.34641
0,1,4,5	0.80000	0.36000
0,1,5,5	0.60000	0.30724
0,1,6,5	0.40000	0.21541
0,2,2,5	1.00000	0.36878
0,2,3,5	0.80000	0.40200
0,2,4,5	0.60000	0.37736
0,2,5,5	0.40000	0.33226
0,2,6,5	0.20000	0.25612
0,3,3,5	0.60000	0.39799
0,3,4,5	0.40000	0.37736

n = 6, K = 3		
r-values	f	σf
2,2,3,5	0.66667	0.66852
2,2,4,5	0.33333	0.66852
2,2,5,5	0.00000	0.68313
2,3,3,5	0.33333	0.70097
2,3,4,5	0.00000	0.74536
3,0,4,5	1.00000	0.80622
3,0,5,5	0.50000	0.65192
3,0,6,5	0.00000	0.67082
3,1,3,5	1.00000	0.92195
3,1,4,5	0.50000	0.90830
3,1,5,5	0.00000	0.92195
3,2,2,5	1.00000	1.02470
3,2,3,5	0.50000	1.01242
3,2,4,5	0.00000	1.11803
3,3,3,5	0.00000	1.16190
4,0,4,5	1.00000	1.61245
4,0,5,5	0.00000	1.61245
4,1,3,5	1.00000	1.94936
4,1,4,5	0.00000	2.04939
4,2,2,5	1.00000	1.94936
4,2,3,5	0.00000	2.23607
0,0,5,4	1.00000	0.29580
0,0,6,4	0.75000	0.23717
0,1,4,4	1.00000	0.43301
0,1,5,4	.075000	0.42573
0,1,6,4	0.50000	0.29580
0,2,3,4	1.00000	0.48734
0,2,4,4	0.75000	0.50621

(Continued)

TABLE I
(Continued)

	n = 6, K = 3			n = 6, K = 3			n = 6, K = 3	
r-values	f	σ f	r-values	f	σ f	r-values	f	σ f
1,3,5,6	0.00000	0.34641	0,3,5,5	0.20000	0.33704	0,2,5,4	0.50000	0.43301
0,2,6,4	0.25000	0.32596	2,3,4,4	0.00000	1.11803	1,3,5,3	0.00000	0.92195
0,3,3,4	0.75000	0.53033	3,0,5,4	1.00000	1.67332	1,4,4,3	0.00000	1.02470
0,3,4,4	0.50000	0.48734	3,0,6,4	0.00000	1.34164	2,0,6,3	1.00000	1.34164
0,3,5,4	0.25000	0.42573	3,1,4,4	1.00000	2.09762	2,1,5,3	1.00000	1.94936
0,3,6,4	0.00000	0.33541	3,1,5,4	0.00000	1.94936	2,1,6,3	0.00000	1.67332
0,4,4,4	0.25000	0.45415	3,2,3,4	1.00000	2.28035	2,2,4,3	1.00000	2.23607
0,4,5,4	0.00000	0.40300	3,2,4,4	0.00000	2.23607	2,2,5,3	0.00000	2.09762

	n = 6, K = 3			n = 6, K = 3			n = 6, K = 3	
r-values	f	σ f	r-values	f	σ f	r-values	f	σ f
1,0,5,4	1.00000	0.53748	3,3,3,4	0.00000	2.32379	2,3,3,3	1.00000	2.32379
1,0,6,4	0.66667	0.30225	0,0,6,3	1.00000	0.44721	2,3,4,3	0.00000	2.28035
1,1,4,4	1.00000	0.68313	0,1,5,3	1.00000	0.64979	0,1,6,2	1.00000	0.80622
1,1,5,4	0.66667	0.55998	0,1,6,3	0.66667	0.44721	0,2,5,2	1.00000	1.02470
1,1,6,4	0.33333	0.42455	0,2,4,3	1.00000	0.74536	0,2,6,2	0.50000	0.70711
1,2,3,4	1.00000	0.74536	0,2,5,3	0.66667	0.61464	0,3,4,2	1.00000	1.11803
1,2,4,4	0.66667	0.66852	0,2,6,3	0.33333	0.43885	0,3,5,2	0.50000	0.89443
1,2,5,4	0.33333	0.59835	0,3,3,3	1.00000	0.77460	0,3,6,2	0.00000	0.67082
1,2,6,4	0.00000	0.53748	0,3,4,3	0.66667	0.68313	0,4,4,2	0.50000	0.94868
1,3,3,4	0.66667	0.70097	0,3,5,3	0.33333	0.57090	0,4,5,2	0.00000	0.80622
1,3,4,4	0.33333	0.66852	0,3,6,3	0.00000	0.44721	1,1,6,2	1.00000	1.61245
1,3,5,4	0.00000	0.64979	0,4,4,3	0.33333	0.60858	1,2,5,2	1.00000	2.04939
1,4,4,4	0.00000	0.68313	0,4,5,3	0.00000	0.53748	1,2,6,2	0.00000	1.61245
2,1,6,4	0.00000	0.80622	1,2,5,3	0.50000	0.90830	0,3,5,1	1.00000	1.94936
2,2,3,4	1.00000	1.11803	1,2,6,3	0.00000	0.80622	0,3,6,1	0.00000	1.34164
2,2,4,4	0.50000	1.00000	1,3,3,3	1.00000	1.16190	0,4,4,1	1.00000	2.04939
2,2,5,4	0.00000	1.02470	1,3,4,3	0.50000	1.01242	0,4,5,1	0.00000	1.61245
2,3,3,4	0.50000	1.04881						

r-values	n = 10, K = 3 f	σ f	r-values	n = 10, K = 3 f	σ f	r-values	n = 10, K = 3 f	σ f
0,0,5,10	1.0	0.16667	1,1,5,10	0.88889	0.21631	2,2,7,10	0.50000	0.26021
0,0,6,10	0.9	0.16330	1,1,6,10	0.77778	0.21419	2,2,8,10	0.37500	0.24694
1,1,3,6	0.80000	0.33704	0,1,3,5	1.00000	0.34641	4,1,3,5	1.00000	1.94936
1,1,4,6	0.60000	0.33226	0,1,4,5	0.80000	0.36000	4,1,4,5	0.00000	2.04939
1,1,5,6	0.40000	0.30724	0,1,5,5	0.60000	0.30724	4,2,2,5	1.00000	2.04939
1,1,6,6	0.20000	0.25612	0,1,6,5	0.40000	0.21541	4,2,3,5	0.00000	2.23607
1,2,2,6	0.80000	0.36000	0,2,2,5	1.00000	0.36878	0,0,5,4	1.00000	0.29580
1,2,3,6	0.60000	0.37736	0,2,3,5	0.80000	0.40200	0,0,6,4	0.75000	0.23717
1,2,4,6	0.40000	0.37736	0,2,4,5	0.60000	0.37736	0,1,4,4	1.00000	0.43301
1,2,5,6	0.20000	0.36000	0,2,5,5	0.40000	0.33226	0,1,5,4	0.75000	0.42573
1,2,6,6	0.00000	0.26833	0,2,6,5	0.20000	0.25612	0,1,6,4	0.50000	0.29580
1,3,3,6	0.40000	0.39799	0,3,3,5	0.60000	0.39799	0,2,3,4	1.00000	0.48734
1,3,4,6	0.20000	0.40200	0,3,4,5	0.40000	0.37736	0,2,4,4	0.75000	0.50621
1,3,5,6	0.00000	0.34641	0,3,5,5	0.20000	0.33704	0,2,5,4	0.50000	0.43301
0,2,6,4	0.25000	0.32596	2,3,4,4	0.00000	1.11803	1,3,5,3	0.00000	0.92195
0,3,3,4	0.75000	0.53033	3,0,5,4	1.00000	1.67332	1,4,4,3	0.00000	1.02470
0,3,4,4	0.50000	0.48734	3,0,6,4	0.00000	1.34164	2,0,6,3	1.00000	1.34164
0,3,5,4	0.25000	0.42573	3,1,4,4	1.00000	2.09762	2,1,5,3	1.00000	1.94936
0,3,6,4	0.00000	0.33541	3,1,5,4	0.00000	1.94936	2,1,6,3	0.00000	1.67332
0,4,4,4	0.25000	0.45415	3,2,3,4	1.00000	2.28035	2,2,4,3	1.00000	2.23607
0,4,5,4	0.00000	0.40311	3,2,4,4	0.00000	2.23607	2,2,5,3	0.00000	2.09762

(Continued)

TABLE I
(Continued)

r-values	n = 6, K = 3 f	σ f	r-values	n = 6, K = 3 f	σ f	r-values	n = 6, K = 3 f	σ f
1,0,5,4	1.00000	0.53748	3,3,3,4	0.00000	2.32379	2,3,3,3	1.00000	2.32379
1,0,6,4	0.66667	0.30225	0,0,6,3	1.00000	0.44721	2,3,4,3	0.00000	2.28035
1,1,4,4	1.00000	0.68313	0,1,5,3	1.00000	0.64979	0,1,6,2	1.00000	0.80622
1,1,5,4	0.66667	0.55998	0,1,6,3	0.66667	0.44721	0,2,5,2	1.00000	1.02470
1,1,6,4	0.33333	0.42455	0,2,4,3	1.00000	0.74536	0,2,6,2	0.50000	0.70711
1,2,3,4	1.00000	0.74536	0,2,5,3	0.66667	0.61464	0,3,4,2	1.00000	1.11803
1,2,4,4	0.66667	0.66852	0,2,6,3	0.33333	0.43885	0,3,5,2	0.50000	0.89443
1,2,5,4	0.33333	0.59835	0,3,3,3	1.00000	0.77460	0,3,6,2	0.00000	0.67082
1,2,6,4	0.00000	0.53748	0,3,4,3	0.66667	0.68313	0,4,4,2	0.50000	0.94868
1,3,3,4	0.66667	0.70097	0,3,5,3	0.33333	0.57090	0,4,5,2	0.00000	0.80622
1,3,4,4	0.33333	0.66852	0,3,6,3	0.00000	0.44721	1,1,6,2	1.00000	1.61245
1,3,5,4	0.00000	0.64979	0,4,4,3	0.33333	0.60858	1,2,5,2	1.00000	2.04939
1,4,4,4	0.00000	0.68313	0,4,5,3	0.00000	0.53748	1,2,6,2	0.00000	1.61245
2,1,6,4	0.00000	0.80622	1,2,5,3	0.50000	0.90830	0,3,5,1	1.00000	1.94936
2,2,3,4	1.00000	1.11803	1,2,6,3	0.00000	0.80622	0,3,6,1	0.00000	1.34164
2,2,4,4	0.50000	1.00000	1,3,3,3	1.00000	1.16190	0,4,4,1	1.00000	2.04939
2,2,5,4	0.00000	1.02470	1,3,4,3	0.50000	1.01242	0,4,5,1	0.00000	1.61245
2,3,3,4	0.50000	1.04881						

r-values	n = 10, K = 3 f	σ f	r-values	n = 10, K = 3 f	σ f	r-values	n = 10, K = 3 f	σ f
0,0,5,10	1.0	0.16667	1,1,5,10	0.88889	0.21631	2,2,7,10	0.50000	0.26021
0,0,6,10	0.9	0.16330	1,1,6,10	0.77778	0.21419	2,2,8,10	0.37500	0.24694

Tables 415

n = 10, K = 3

r-values	f	σ f
0,0,7,10	0.8	0.15275
0,0,8,10	0.7	0.13333
0,0,9,10	0.6	0.10000
0,0,10,10	0.5	0.00000
0,1,4,10	1.0	0.19149
0,1,5,10	0.9	0.19436
0,1,6,10	0.8	0.19149
0,1,7,10	0.7	0.18257
0,1,8,10	0.6	0.16667
0,1,9,10	0.5	0.14142
0,1,10,10	0.4	0.10000
0,2,3,10	1.0	0.20276
0,2,4,10	0.9	0.21082
0,2,5,10	0.8	0.21344
0,2,6,10	0.7	0.21082
0,2,7,10	0.6	0.20276
0,2,8,10	0.5	0.18856
0,29,10	0.4	0.16667
0,2,10,10	0.3	0.13333
0,3,3,10	0.9	0.21602

n = 10, K = 3

r-values	f	σ f
0,3,4,10	0.8	0.22361
0,3,5,10	0.7	0.22608
0,3,6,10	0.6	0.22361
0,3,7,10	0.5	0.21602
0,3,8,10	0.4	0.20276

n = 10, K = 3

r-values	f	σ f
1,1,7,10	0.66667	0.20621
1,1,8,10	0.55556	0.19166
1,1,9,10	0.44444	0.16882
1,1,10,10	0.33333	0.13354
1,2,3,10	1.00000	0.22529
1,2,4,10	0.88889	0.23457
1,2,5,10	0.77778	0.23843
1,2,6,10	0.66667	0.23715
1,2,7,10	0.55556	0.23064
1,2,8,10	0.44444	0.21842
1,2,9,10	0.33333	0.19945
1,2,10,10	0.22222	0.17151
1,3,3,10	0.88889	0.24034
1,3,4,10	0.77778	0.24968
1,3,5,10	0.66667	0.25391
1,3,6,10	0.55556	0.25331
1,3,7,10	0.44444	0.24784
1,3,8,10	0.33333	0.23715
1,3,9,10	0.22222	0.22050
1,3,10,10	0.1111 1	0.19637

n = 10, K = 3

r-values	f	σ f
1,4,4,10	0.66667	0.25926
1,4,5,10	0.55556	0.26392
1,4,6,10	0.44444	0.26392
1,4,7,10	0.33333	0.25926
1,4,8,10	0.22222	0.24968

n = 10, K = 3

r-values	f	σ f
2,2,9,10	0.25000	0.24296
2,2,10,10	0.12500	0.22140
2,3,3,10	0.87500	0.27043
2,3,4,10	0.75000	0.28106
2,3,5,10	0.62500	0.28603
2,3,6,10	0.50000	0.28565
2,3,7,10	0.37500	0.27990
2,3,8,10	0.25000	0.28260
2,3,9,10	0.12500	0.27081
2,3,10,10	0.00000	0.25345
2,4,4,10	0.62500	0.29204
2,4,5,10	0.50000	0.29756
2,4,6,10	0.37500	0.29789
2,4,7,10	0.25000	0.30619
2,4,8,10	0.12500	0.30117
2,4,9,10	0.00000	0.29166
2,5,5,10	0.37500	0.30369
2,5,6,10	0.25000	0.31732
2,5,7,10	0.12500	0.31799
2,5,8,10	0.00000	0.31458

n = 10, K = 3

r-values	f	σ f
2,6,6,10	0.12500	0.32340
2,6,7,10	0.00000	0.32543
3,0,5,10	1.00000	0.23809
3,0,6,10	0.85714	0.23536
3,0,7,10	0.71429	0.22695

(Continued)

TABLE I
(Continued)

r-values	n = 10, K = 3 f	σ f	r-values	n = 10, K = 3 f	σ f	r-values	n = 10, K = 3 f	σ f
0,3,9,10	0.3	0.18257	1,4,9,10	0.11111	0.23457	3,0,8,10	0.57143	0.21695
0,3,10,10	0.2	0.15275	1,4,10,10	0.00000	0.21276	3,0,9,10	0.42857	0.18962
0,4,4,10	0.7	0.23094	1,5,5,10	0.44444	0.26907	3,0,10,10	0.28571	0.15587
0,4,5,10	0.6	0.23336	1,5,6,10	0.33333	0.26963	3,1,4,10	1.00000	0.27355
0,4,6,10	0.5	0.23094	1,5,7,10	0.22222	0.26565	3,1,5,10	0.85714	0.27941
0,4,7,10	0.4	0.22361	1,5,8,10	0.11111	0.25690	3,1,6,10	0.71429	0.28057
0,4,8,10	0.3	0.21082	1,5,9,10	0.00000	0.24287	3,1,7,10	0.57143	0.28074
0,4,9,10	0.2	0.19149	1,6,6,10	0.22222	0.27076	3,1,8,10	0.42857	0.26877
0,4,10,10	0.1	0.16330	1,6,7,10	0.11111	0.26736	3,1,9,10	0.28571	0.25517
0,5,5,10	0.5	0.23570	1,6,8,10	0.00000	0.25926	3,1,10,10	0.14286	0.23536
0,5,6,10	0.4	0.23336	1,7,7,10	0.00000	0.26450	3,2,3,10	1.00000	0.28965
0,5,7,10	0.3	0.22608	2,0,5,10	1.00000	0.20833	3,2,4,10	0.85714	0.30278
0,5,8,10	0.2	0.21344	2,0,6,10	0.87500	0.20465	3,2,5,10	0.71429	0.31122
0,5,9,10	0.1	0.19436	2,0,7,10	0.75000	0.19320	3,2,6,10	0.57143	0.31857
0,5,10,10	0.0	0.16667	2,0,8,10	0.62500	0.17237	3,2,7,10	0.42857	0.31536
0,6,6,10	0.3	0.23094	2,0,9,10	0.50000	0.10534	3,2,8,10	0.28571	0.31122
0,6,7,10	0.2	0.22361	2,0,10,10	0.37500	0.07365	3,2,9,10	0.14286	0.30278
0,6,8,10	0.1	0.21082	2,1,4,10	1.00000	0.23936	3,2,10,10	0.00000	0.28965

r-values	n = 10, K = 3 f	σ f	r-values	n = 10, K = 3 f	σ f	r-values	n = 10, K = 3 f	σ f
0,6,9,10	0.0	0.19149	2,1,5,10	0.8750	0.22902	3,3,3,10	0.85714	0.31018
0,7,7,10	0.1	0.21602	2,1,6,10	0.75000	0.24116	3,3,4,10	0.71429	0.32546
0,7,8,10	0.0	0.20276	2,1,7,10	0.62500	0.23246	3,3,5,10	0.57143	0.33926
1,0,5,10	1.0	0.18518	2,1,8,10	0.50000	0.21651	3,3,6,10	0.42857	0.34291
1,0,6,10	0.88889	0.18186	2,1,9,10	0.37500	0.19151	3,3,7,10	0.28571	0.34574
1,0,7,10	0.77778	0.17151	2,1,10,10	0.25000	0.17678	3,3,8,10	0.14286	0.34480

1,0,8,10	0.66667	2,2,3,10	0.15270	1.00000	0.25345	3,3,9,10	0.00000	0.34007
1,0,9,10	0.55556	2,2,4,10	0.12159	0.87500	0.26393	3,4,4,10	0.57143	0.34588
1,0,10,10	0.44444	2,2,5,10	0.06172	0.75000	0.26842	3,4,5,10	0.42857	0.35589
1,1,4,10	1.00000	2,2,6,10	0.21276	0.62500	0.26717	3,4,6,10	0.28571	0.36488
3,4,7,10	0.14286	5,3,4,10	0.37017	0.60000	0.46667	9,0,6,10	0.00000	1.77951
3,4,8,10	0.00000	5,3,5,10	0.37192	0.40000	0.48990	9,1,4,10	1.00000	1.77951
3,5,5,10	0.28571	5,3,6,10	0.37104	0.20000	0.52068	9,1,5,10	0.00000	2.18581
3,5,6,10	0.14286	5,3,7,10	0.38223	0.00000	0.54569	9,2,3,10	1.00000	2.02759
3,5,7,10	0.00000	5,4,4,10	0.38978	0.40000	0.49889	9,2,4,10	0.00000	2.33333
3,6,6,10	0.00000	5,4,5,10	0.39555	0.20000	0.53748	9,3,3,10	0.00000	2.38048
4,0,5,10	1.00000	5,4,6,10	0.27778	0.00000	0.56960	0,0,6,9	1.00000	0.21276
4,0,6,10	0.83333	5,5,5,10	0.27592	1.00000	0.57735	0,0,7,9	0.88889	0.19637
4,0,7,10	0.66667	6,0,5,10	0.27027	0.75000	0.41667	0,0,8,9	0.77778	0.17151
4,0,8,10	0.50000	6,0,6,10	0.26058	0.50000	0.45644	0,0,9,9	0.66667	0.13354
4,0,9,10	0.33333	6,0,7,10	0.24637	0.25000	0.43301	0,0,10,9	0.55556	0.06172
4,0,10,10	0.16667	6,0,8,10	0.22680	0.00000	0.45262	0,1,5,9	1.00000	0.24287
4,1,4,10	1.00000	6,0,9,10	0.31914	1.00000	0.47871	0,1,6,9	0.88889	0.23457
4,1,5,10	0.83333	6,1,4,10	0.32710	0.75000	0.47871	0,1,7,9	0.77778	0.22050
4,1,6,10	0.66667	6,1,5,10	0.33178	0.50000	0.52705	0,1,8,9	0.66667	0.19945
4,1,7,10	0.50000	6,1,6,10	0.33333	0.25000	0.52042	0,1,9,9	0.55555	0.16882
4,1,8,10	0.33333	6,1,7,10	0.33178	0.00000	0.54962	0,1,10,9	0.44444	0.12159
4,1,9,10	0.16667	6,1,8,10	0.32710	1.00000	0.58333	0,2,4,9	1.00000	0.25926
4,1,10,10	0.00000	6,2,3,10	0.31914	0.75000	0.50690	0,2,5,9	0.88889	0.25690
4,2,3,10	1.00000	6,2,4,10	0.33793	0.50000	0.56519	0,2,6,9	0.77778	0.24968
4,2,4,10	0.83333	6,2,5,10	0.35428	0.25000	0.57130	0,2,7,9	0.66667	0.23715
4,2,5,10	0.66667	6,2,6,10	0.36711	0.00000	0.60953	0,2,8,9	0.55556	0.21842
4,2,6,10	0.50000	6,2,7,10	0.37679	1.00000	0.65085	0,2,9,9	0.44444	0.19166
4,2,7,10	0.33333	6,3,3,10	0.38356	0.75000	0.57735	0,2,10,9	0.33333	0.15270
4,2,8,10	0.16667	6,3,4,10	0.38756	0.50000	0.59512	0,3,3,9	1.00000	0.26450
4,2,9,10	0.00000	6,3,5,10	0.38889	0.25000	0.64280	0,3,4,9	0.88889	0.26736
4,3,3,10	0.83333	6,3,6,10	0.36289	0.00000	0.69222	0,3,5,9	0.77778	0.26565
4,3,4,10	0.66667	6,4,4,10	0.38356	0.25000	0.65352	0,3,6,9	0.66667	0.25926

(Continued)

TABLE I
(Continued)

n = 10, K = 3			n = 10, K = 3			n = 10, K = 3		
r-values	f	σf	r-values	f	σf	r-values	f	σf
4,3,4,10	0.50000	0.40062	6,4,5,10	0.00000	0.71200	0,3,7,9	0.55556	0.24784
4,3,6,10	0.33333	0.41450	7,0,5,10	1.00000	0.55556	0,3,8,9	0.44444	0.23064
4,3,7,10	0.16667	0.42552	7,0,6,10	0.66667	0.57013	0,3,9,9	0.33333	0.20621
4,3,8,10	0.00000	0.43390	7,0,7,10	0.33333	0.61195	0,3,10,9	0.22222	0.17151
4,4,4,10	0.50000	0.48025	7,0,8,10	0.00000	0.67586	0,4,4,9	0.77778	0.27076

n = 10, K = 3			n = 10, K = 3			n = 10, K = 3		
r-values	f	σf	r-values	f	σf	r-values	f	σf
4,4,5,10	0.33333	0.42913	7,1,4,10	1.00000	0.63828	0,4,5,9	0.66667	0.26963
4,4,6,10	0.16667	0.44675	7,1,5,10	0.66667	0.66975	0,4,6,9	0.55556	0.26392
4,4,7,10	0.00000	0.46148	7,1,6,10	0.33333	0.72293	0,4,7,9	0.44444	0.25331
4,5,5,10	0.16667	0.45361	7,1,7,10	0.00000	0.79349	0,4,8,9	0.33333	0.23715
4,5,6,10	0.00000	0.47466	7,2,3,10	1.00000	0.67586	0,4,9,9	0.22222	0.21419
5,0,5,10	1.00000	0.33333	7,2,4,10	0.66667	0.72293	0,4,10,9	0.11111	0.18186
5,0,6,10	0.80000	0.33333	7,2,5,10	0.33333	0.78829	0,5,5,9	0.55556	0.26907
5,0,7,10	0.60000	0.33333	7,2,6,10	0.00000	0.86780	0,5,6,9	0.44444	0.26392
5,0,8,10	0.40000	0.33333	7,3,3,10	0.66667	0.73981	0,5,7,9	0.33333	0.25391
5,0,9,10	0.20000	0.33333	7,3,4,10	0.33333	0.81901	0,5,8,9	0.22222	0.23843
5,0,10,10	0.00000	0.33333	7,3,5,10	0.00000	0.90948	0,5,9,9	0.11111	0.21631
5,1,4,10	1.00000	0.38297	7,4,4,10	0.00000	0.92296	0,5,10,9	0.00000	0.18518
5,1,5,10	0.80000	0.39440	8,0,5,10	1.00000	0.83333	0,6,6,9	0.33333	0.25926
5,1,6,10	0.60000	0.40552	8,0,6,10	0.50000	0.88192	0,6,7,9	0.22222	0.24968
5,1,7,10	0.40000	0.41633	8,0,7,10	0.00000	1.01379	0,6,8,9	0.11111	0.23457
5,1,8,10	0.20000	0.42687	8,1,4,10	1.00000	0.95743	0,6,9,9	0.00000	0.21276
5,1,9,10	0.00000	0.43716	8,1,5,10	0.50000	1.02740	0,7,7,9	0.11111	0.24034
5,2,3,10	1.00000	0.40552	8,1,6,10	0.00000	1.16667	0,7,8,9	0.00000	0.22529
5,2,4,10	0.80000	0.42688	8,2,3,10	1.00000	1.01379	1,0,6,9	1.00000	0.23936

n = 10, K = 3

r-values	f	σf
5,2,5,10	0.60000	0.44721
5,2,6,10	0.40000	0.46667
5,2,7,10	0.20000	0.48534
5,2,8,10	0.00000	0.50332
5,3,3,10	0.80000	0.43716
1,1,6,9	0.87500	0.26363
1,1,7,9	0.75000	0.24869
1,1,8,9	0.62500	0.22738
1,1,9,9	0.50000	0.19766
1,1,10,9	0.37500	0.15468
1,2,4,9	1.00000	0.29167
1,2,5,9	0.87500	0.28877
1,2,6,9	0.75000	0.28144
1,2,7,9	0.62500	0.26933
1,2,8,9	0.50000	0.25173
1,2,9,9	0.37500	0.22738
1,2,10,9	0.25000	0.193766
1,3,3,9	1.00000	0.29756
1,3,4,9	0.87500	0.30055
1,3,5,9	0.75000	0.29938
1,3,6,9	0.62500	0.29398
1,3,7,9	0.50000	0.28413
1,3,8,9	0.37500	0.26933

n = 10, K = 3

r-values	f	σf
1,3,9,9	0.25000	0.24869
1,3,10,9	0.12500	0.22060

n = 10, K = 3

r-values	f	σf
8,2,4,10	0.50000	1.10554
8,2,5,10	0.00000	1.25830
8,3,3,10	0.50000	1.13039
8,3,4,10	0.00000	1.30171
9,0,5,10	1.00000	1.66667
2,3,6,9	0.57143	0.34194
2,3,7,9	0.42857	0.33423
2,3,8,9	0.28571	0.32261
2,3,9,9	0.14286	0.30838
2,3,10,9	0.00000	0.28966
2,4,4,9	0.71429	0.34960
2,4,5,9	0.57143	0.35496
2,4,6,9	0.42857	0.35400
2,4,7,9	0.28571	0.34960
2,4,8,9	0.14286	0.34318
2,4,9,9	0.00000	0.33333
2,5,5,9	0.42857	0.36034
2,5,6,9	0.28571	0.36234
2,5,7,9	0.14286	0.36246
2,5,8,9	0.00000	0.35952
2,6,6,9	0.14286	0.36867
2,6,7,9	0.00000	0.37192
3,0,6,9	1.00000	0.31914

n = 10, K = 3

r-values	f	σf
3,0,7,9	0.83333	0.29310
3,0,8,9	0.66667	0.26254

n = 10, K = 3

r-values	f	σf
1,0,7,9	0.87500	0.22060
1,0,8,9	0.75000	0.19376
1,0,9,9	0.62500	0.15468
1,0,10,9	0.50000	0.08838
1,1,5,9	1.00000	0.27323
4,1,6,9	0.80	0.42016
4,1,7,9	0.60	0.40596
4,1,8,9	0.40	0.39486
4,1,9,9	0.20	0.38713
4,1,10,9	0.00	0.38297
4,2,4,9	1.00	0.46667
4,2,5,9	0.80	0.46053
4,2,6,9	0.60	0.45743
4,2,7,9	0.40	0.45743
4,2,8,9	0.20	0.46053
4,2,9,9	0.00	0.46667
4,3,3,9	1.00	0.47610
4,3,4,9	0.80	0.47944
4,3,5,9	.060	0.48571
4,3,6,9	0.40	0.49477
4,3,7,9	0.20	0.50649
4,3,8,9	0.00	0.52068
4,4,4,9	0.60	0.49477

n = 10, K = 3

r-values	f	σf
4,4,5,9	0.40	0.51242
4,4,6,9	0.20	0.53216

(Continued)

TABLE I
(*Continued*)

r-values	n = 10, K = 3 f	σf	r-values	n = 10, K = 3 f	σf	r-values	n = 10, K = 3 f	σf
1,4,4,9	0.75000	0.30512	3,0,9,9	0.50000	0.22567	4,4,7,9	0.00	0.55377
1,4,5,9	0.62500	0.30557	3,0,10,9	0.33333	0.18703	4,5,5,9	0.20	0.54045
1,4,6,9	0.50000	0.30190	3,1,5,9	1.00000	0.36340	4,5,6,9	0.00	0.56960
1,4,7,9	0.37500	0.29398	3,1,6,9	0.83333	0.35000	5,0,6,9	1.00	0.47871
1,4,8,9	0.25000	0.28144	3,1,7,9	0.66667	0.33487	5,0,7,9	0.75	0.43800
1,4,9,9	0.12500	0.26363	3,1,8,9	0.50000	0.31672	5,0,8,9	0.50	0.41248
1,4,10,9	0.00000	0.23936	3,1,9,9	0.33333	0.30089	5,0,9,9	0.25	0.40505
1,5,5,9	0.50000	0.30760	3,1,10,9	0.16667	0.26692	5,0,10,9	0.00	0.41667
1,5,6,9	0.37500	0.30557	3,2,4,9	1.00000	0.38889	5,1,5,9	1.00	0.54645
1,5,7,9	0.25000	0.29938	3,2,5,9	0.83333	0.38423	5,1,6,9	0.75	0.52457
1,5,8,9	0.12500	0.28887	3,2,6,9	0.66667	0.37816	5,1,7,9	0.50	0.51707
1,5,9,9	0.00000	0.27323	3,2,7,9	0.50000	0.37060	5,1,8,9	0.25000	0.52457
1,6,6,9	0.25000	0.30512	3,2,8,9	0.33333	0.36571	5,1,9,9	0.00000	0.54657
1,6,7,9	0.12500	0.30055	3,2,9,9	0.16667	0.34731	5,2,4,9	1.00000	0.58333
1,6,8,9	0.00000	0.29167	3,2,10,9	0.00000	0.33793	5,2,5,9	0.75000	0.57509
1,7,7,9	0.00000	0.29756	3,3,3,9	1.00000	0.39674	5,2,6,9	.050000	0.58035
2,0,6,9	1.00000	0.27355	3,3,4,9	0.83333	0.39997	5,2,7,9	0.25000	0.59875

r-values	n = 10, K = 3 f	σf	r-values	n = 10, K = 3 f	σf	r-values	n = 10, K = 3 f	σf
2,0,7,9	0.85714	0.25170	3,3,5,9	0.66667	0.40190	5,2,8,9	0.00000	0.62915
2,0,8,9	0.71429	0.22283	3,3,6,9	0.50000	0.40254	5,3,3,9	1.00000	0.59512
2,0,9,9	0.57143	0.18786	3,3,7,9	0.33333	0.40572	5,3,4,9	0.75000	0.59875
2,0,10,9	0.42857	0.12834	3,3,8,9	0.16667	0.39707	5,3,5,9	0.50000	0.61520
2,1,5,9	1.00000	0.31226	3,3,9,9	0.00000	0.35573	5,3,6,9	0.25000	0.64348
2,1,6,9	0.85714	0.30094	3,4,4,9	0.66667	0.40965	5,3,7,9	0.00000	0.68211
2,1,7,9	0.71429	0.28531	3,4,5,9	0.50000	0.41759	5,4,4,9	0.50000	0.62639

2,1,8,9	0.57143	0.26753	3,4,6,9	0.33333	0.42793	5,4,5,9	0.25000	0.66471
2,1,9,9	0.42857	0.23934	3,4,7,9	0.16667	0.42703	5,4,6,9	0.00000	0.71200
2,1,10,9	0.28571	0.20146	3,4,8,9	0.00000	0.39674	5,5,5,9	0.00000	0.72169
2,2,4,9	1.00000	0.33333	3,5,5,9	0.33333	0.43509	6,0,6,9	1.00000	0.63828
2,2,5,9	0.85714	0.32970	3,5,6,9	0.16667	0.44125	6,0,7,9	0.66667	0.58443
2,2,6,9	0.71429	0.32261	3,5,7,9	0.00000	0.41944	6,0,8,9	0.33333	0.58443
2,2,7,9	0.57143	0.31430	3,6,6,9	0.00000	0.42673	6,0,9,9	0.00000	0.63828
2,2,8,9	0.42857	0.29838	4,0,6,9	1.00	0.38297	6,1,5,9	1.00000	0.72860
2,2,9,9	0.28571	0.27725	4,0,7,9	0.80	0.35100	6,1,6,9	0.66667	0.69979
2,2,10,9	0.14286	0.25170	4,0,8,9	0.60	0.32028	6,1,7,9	0.33333	0.71722
2,3,3,9	1.00000	0.34007	4,0,9,9	0.40	0.29120	6,1,8,9	0.00000	0.77778
2,3,4,9	0.85714	0.34318	4,0,10,9	0.20	0.26432	6,2,4,9	1.00000	0.77778
2,3,5,9	0.71429	0.34305	4,1,5,9	1.00	0.43716	6,2,5,9	0.66667	0.76712
6,2,6,9	0.33333	0.79866	0,4,10,8	0.125	0.20465	2,1,8,8	0.66667	0.32341
06,2,7,9	0.00000	0.79349	0,5,5,8	0.625	0.30369	2,1,9,9	0.50000	0.28328
6,3,3,9	1.00000	0.79349	0,5,6,8	0.500	0.29756	2,1,10,8	0.33333	0.23497
6,3,4,9	0.66667	0.79866	0,5,7,8	0.375	0.28603	2,2,5,8	1.00000	0.41944
6,3,5,9	0.33333	0.84376	0,5,8,8	0.250	0.26842	2,2,6,8	0.83333	0.39890
6,3,6,9	0.00000	0.85346	0,5,9,8	0.125	0.22902	2,2,7,8	0.66667	0.37634
6,4,4,9	0.33333	0.85827	0,5,10,8	0.000	0.20833	2,2,8,8	0.50000	0.35136
6,4,5,9	0.00000	0.88192	0,6,6,8	0.375	0.29204	2,2,9,8	0.33333	0.32341
7,0,6,9	1.00000	0.95743	0,6,7,8	0.250	0.28106	2,2,10,8	0.16667	0.29163
7,0,6,9	1.00000	0.95743	0,6,7,8	0.250	0.28106	2,2,10,8	0.16667	0.29163
7,0,7,9	0.50000	0.88976	0,6,8,8	0.125	0.26393	2,3,4,8	1.00000	0.43390
7,0,8,9	0.00000	1.01379	0,6,9,8	0.000	0.23936	2,3,5,8	0.83333	0.42174
7,1,5,9	1.00	1.09291	0,7,7,8	0.125	0.27043	2,3,6,8	0.66667	0.40783
7,1,6,9	0.50	1.06066	0,7,8,8	0.000	0.25345	2,3,7,8	0.50000	0.40062
7,1,7,9	0.00	1.10924	1,0,7,8	1.00000	0.28966	2,3,8,8	0.33333	0.37634
7,2,4,9	1.00	1.16667	1,0,8,8	0.85714	0.25170	2,3,9,8	0.16667	0.35813
7,2,5,9	0.50	1.16070	1,0,9,8	0.71429	0.20146	2,3,10,8	0.00000	0.33793
7,2,6,9	0.00	1.30171	1,0,10,8	0.57143	0.12834	2,4,4,8	0.83333	0.42783
7,3,3,9	1.00	1.19024	1,1,6,8	1.00000	0.33333	2,4,5,8	0.66667	0.42269

(Continued)

TABLE I
(Continued)

	n = 10, K = 3			n = 10, K = 3			n = 10, K = 3	
r-values	f	σ f	r-values	f	σ f	r-values	f	σ f
7,3,4,9	0.50	1.20761	1,1,7,8	0.58714	0.30838	2,4,6,8	0.50000	0.43033
7,3,5,9	0.00	1.36422	1,1,8,8	0.71429	0.27725	2,4,7,8	0.33333	0.40783
7,4,4,9	0.00	1.38444	1,1,9,8	0.57143	0.23934	2,4,8,8	0.1667	0.39890

	n = 10, K = 3			n = 10, K = 3			n = 10, K = 3	
r-values	f	σ f	r-values	f	σ f	r-values	f	σ f
8,0,6,9	1.00	1.77951	1,1,10,8	0.42857	0.18786	2,4,9,8	0.00000	0.38888
8,0,7,9	0.00	2.02759	1,2,5,8	1.00000	0.35952	2,5,5,8	0.50000	0.44445
8,1,5,9	1.00	2.18581	1,2,6,8	0.58714	0.34318	2,5,6,8	0.33333	0.42269
8,1,6,9	0.00	2.33333	1,2,7,8	0.71429	0.32261	2,5,7,8	0.16667	0.42147
8,2,4,9	1.00	2.33333	1,2,8,8	0.57143	0.29839	2,5,8,8	0.00000	0.41944
8,2,5,9	0.00	2.51661	1,2,9,8	0.42857	0.26753	2,6,6,8	0.16667	0.42873
8,3,3,9	1.00	2.38048	1,2,10,8	0.28571	0.22283	2,6,7,8	0.00000	0.43390
8,3,4,9	0.00	2.60342	1,3,4,8	1.00000	0.37192	3,0,7,8	1.00000	0.40552
0,0,7,8	1.000	0.25345	1,3,5,8	0.58	0.36246	3,0,8,8	0.80000	0.34692
0,0,8,8	0.875	0.22140	1,3,6,8	0.71429	0.34960	3,0,9,8	0.60000	0.28378
0,0,9,0	0.750	0.17678	1,3,7,8	0.57143	0.33423	3,0,10,8	0.40000	0.21208
0,0,10,8	0.625	0.07365	1,3,8,8	0.42857	0.31430	3,1,6,8	1.00000	0.46667
0,1,6,8	1.000	0.29166	1,3,9,8	0.28571	0.28531	3,1,7,8	0.80000	0.42729
0,1,7,8	0.875	0.27081	1,3,10,8	0.14286	0.25170	3,1,8,8	0.60000	0.38941
0,1,8,8	0.750	0.24296	1,4,4,8	0.85714	0.36867	3,1,9,8	0.40000	0.35352
0,1,9,8	0.625	0.19151	1,4,5,8	0.71429	0.36234	3,1,10,8	0.20000	0.32028
0,1,10,8	0.500	0.10534	1,4,6,8	0.57143	0.35400	3,2,5,8	1.00000	0.50332
0,2,5,8	1.000	0.31458	1,4,7,8	0.42857	0.34194	3,2,6,8	0.80000	0.47647
0,2,6,8	1.000	0.30117	1,4,8,8	0.28571	032261	3,2,7,8	0.60000	0.45274
0,2,7,8	0.750	0.28260	1,4,9,8	0.14286	0.30094	3,2,8,8	0.40000	0.43267
0,2,8,8	0.625	0.24694	1,4,10,8	0.00000	0.27355	3,2,10,8	0.20000	0.41676
0,2,9,8	0.500	0.21651	1,5,5,8	0.57143	0.26034	3,2,10,8	0.00000	0.40552

n = 10, K = 3

r-values	f	σf
0,2,10,8	0.375	0.17237
0,3,4,8	1.000	0.32543
0,3,5,8	0.875	0.31799
0,3,6,8	0.750	0.30619
0,3,7,8	0.625	0.27990
0,3,8,8	0.500	0.26021
0,3,9,8	0.375	0.23246
0,3,10,8	0.250	0.19320
0,4,4,8	0.875	0.32340
0,4,5,8	0.750	0.31732
0,4,6,8	0.625	0.29789
0,4,7,8	0.500	0.28565
0,4,8,8	0.375	0.26717
0,4,9,8	0.250	0.24116
3,6,6,8	0.00000	0.55377
4,0,7,8	1.00000	0.50690
4,0,8,8	0.75000	0.42898
4,0,9,8	0.50000	0.36324
4,0,10,8	0.25000	0.31732
4,1,6,8	1.00000	0.58333
4,1,7,8	0.75000	0.53033
4,1,8,8	0.50000	0.49301
4,1,9,8	0.25000	0.47507
4,1,10,8	0.00000	0.47871
4,2,5,8	1.00000	0.62915

n = 10, K = 3

r-values	f	σf
1,5,6,8	0.42857	0.35496
1,5,7,8	0.28571	0.34305
1,5,8,8	0.14286	0.32970
1,5,9,8	0.00000	0.31226
1,6,6,8	0.28571	0.34960
1,6,7,8	0.14286	0.34318
1,6,8,8	0.00000	0.33333
1,7,7,8	1.00000	0.34007
1,7,8,8	0.83333	0.33793
2,0,7,8	0.66667	0.29163
2,0,8,8	0.50000	0.23497
2,0,9,8	1.00000	0.15713
2,0,10,8	0.83333	0.38888
2,1,6,8	0.00000	0.35813
2,1,7,8	0.50000	1.38444
6,3,6,8	0.00000	1.26930
6,4,4,8	0.00000	1.42400
6,4,5,8	1.00000	2.02759
7,0,7,8	0.00000	2.02759
7,0,8,8	1.00000	2.33333
7,1,6,8	0.00000	2.38048
7,1,7,8	1.00000	2.51661
7,2,5,8	0.00000	2.60342
7,2,6,8	1.00000	2.60342
7,3,4,8	0.00000	2.72845
7,3,5,8	1.00000	

n = 10, K = 3

r-values	f	σf
3,3,4,8	1.00000	0.52068
3,3,5,8	0.80000	0.50368
3,3,6,8	0.60000	0.49044
3,3,7,8	0.40000	0.48129
3,3,8,8	0.20000	0.47647
3,3,9,8	0.00000	0.47610
3,4,4,8	0.80000	0.51242
3,4,5,8	0.60000	0.50824
3,4,6,8	0.40000	0.50824
3,4,7,8	0.20000	0.51242
3,4,8,8	0.00000	0.52068
3,5,5,8	0.40000	0.51691
3,5,6,8	0.20000	0.52949
3,5,7,8	0.00000	0.54569
1,2,7,7	0.83333	0.39707
1,2,8,7	0.66667	0.36571
1,2,9,7	0.50000	0.31672
1,2,10,7	0.33333	0.26254
1,3,5,7	1.00000	0.41944
1,3,6,7	0.83333	0.42703
1,3,7,7	0.66667	0.40572
1,3,8,7	0.50000	0.37060
1,3,9,7	0.33333	0.33487
1,3,10,7	0.16667	0.29310
1,4,4,7	1.00000	0.42673

(Continued)

TABLE I
(Continued)

	n = 10, K = 3			n = 10, K = 3			n = 10, K = 3	
r-values	f	σ f	r-values	f	σ f	r-values	f	σ f
4,2,6,8	0.75000	0.59219	7,4,4,8	0.00000	2.76887	1,4,5,7	0.83333	0.44125
4,2,7,8	0.50000	0.57130	0,0,8,7	1.00000	0.28965	1,4,6,7	0.66667	0.42793
4,2,8,8	0.25000	0.56826	0,0,9,7	0.85714	0.23536	1,4,7,7	0.50000	0.40254
4,2,9,8	0.00000	0.58333	0,0,10,7	0.71429	0.15587	1,4,8,7	0.33333	0.37816
4,3,4,8	1.00000	0.65085	0,1,7,7	1.00000	0.34007	1,4,9,7	0.16667	0.35000
4,3,5,8	0.75000	0.62639	0,1,8,7	0.85714	0.30278	1,4,10,7	0.00000	0.31914
4,3,6,8	0.50000	0.61802	0,1,9,7	0.71429	0.25517	1,5,5,7	0.66667	0.43509
4,3,7,8	0.25000	0.62639	0,1,10,7	0.57143	0.18962	1,5,6,7	0.50000	0.41759
4,3,8,8	0.00000	0.65085	0,2,6,7	1.00000	0.37192	1,5,7,7	0.33333	0.40190
4,4,4,8	0.75000	0.63191	0,2,7,7	0.85714	0.34480	1,5,8,7	0.16667	0.38423
4,4,5,8	0.50000	0.64010	0,2,8,7	0.71429	0.31122	1,5,9,7	0.00000	0.36430
4,4,6,8	0.25000	0.66926	0,2,9,7	0.57143	0.26877	1,6,6,7	0.33333	0.40965
4,4,7,8	0.00000	0.69222	0,2,10,7	0.42857	0.21695	1,66,7,7	0.16667	0.39997
4,5,5,8	0.25000	0.68971	0,3,5,7	1.00000	0.38978	1,6,8,7	0.00000	0.38889
4,5,6,8	0.00000	0.71200	0,3,6,7	0.85714	0.37017	1,7,7,7	0.00000	0.39674
5,0,7,8	1.00000	0.67586	0,3,7,7	0.71429	0.34574	2,0,8,7	1.00000	0.40552
5,0,8,8	0.66667	0.56534	0,3,8,7	0.57143	0.31536	2,0,9,7	0.80000	0.32028

	n = 10, K = 3			n = 10, K = 3			n = 10, K = 3	
r-values	f	σ f	r-values	f	σ f	r-values	f	σ f
5,0,9,8	0.33333	0.51984	0,3,9,7	0.42857	0.28074	2,0,10,7	0.60000	0.21208
5,0,10,8	0.00000	0.55556	0,3,10,7	0.28571	0.22695	2,1,7,7	1.00000	0.47610
5,1,6,8	1.00000	0.77778	0,4,4,7	1.00000	0.39555	2,1,8,7	0.80000	0.41676
5,1,7,8	0.66667	0.70175	0,4,5,7	0.85741	0.38223	2,1,9,7	0.60000	0.35352
5,1,8,8	0.33333	0.68393	0,4,6,7	0.71429	0.36488	2,1,10,7	0.40000	0.28378
5,1,9,8	0.00000	0.72860	0,4,7,7	0.57143	0.34291	2,2,6,7	1.00000	0.52068
5,2,5,8	1.00000	0.83887	0,4,8,7	0.42857	0.31857	2,2,7,7	0.80000	0.47647

n = 10, K = 3

r-values	f	σf
5,2,6,8	0.66667	0.78480
5,2,7,8	0.33333	0.78480
5,2,8,8	0.00000	0.83887
5,3,4,8	1.00000	0.86780
5,3,5,8	0.66667	0.83065
5,3,6,8	0.33333	0.84539
5,3,7,8	0.00000	0.90948
5,4,4,8	0.66667	0.84539
5,4,5,8	0.33333	0.87410
5,4,6,8	0.00000	0.94933
5,5,5,8	0.00000	0.96225
6,0,7,8	1.00000	01.01379

n = 10, K = 3

r-values	f	σf
6,0,8,8	0.50000	0.84984
6,0,9,8	0.00000	0.95743
6,0,9,8	1.00000	1.16667
6,1,7,8	0.50000	1.05409
6,1,8,8	0.00000	1.16667
6,2,5,8	1.00000	1.25830
6,2,6,8	0.50000	1.17851
6,3,7,8	0.00000	1.30171
6,3,4,8	1.00000	1.30171
6,3,5,8	0.50000	1.24722
3,0,9,7	0.75000	0.39198
3,0,10,7	0.50000	0.27003
3,1,7,7	1.00000	0.59512
3,1,8,7	0.75000	0.51454
3,1,9,7	0.50000	0.44488

r-values	f	σf
0,4,9,7	0.28571	0.28057
0,4,10,7	0.14286	0.23536
0,5,5,7	0.71429	0.37104
0,5,6,7	0.57143	0.35589
0,5,7,7	0.42857	0.33927
0,5,8,7	0.28571	0.31122
0,5,9,7	0.14285	0.37941
0,5,10,7	0.00000	0.23809
0,6,6,7	0.42857	0.34588
0,6,7,7	0.28571	0.32546
0,6,8,7	0.14286	0.30278
0,6,9,7	0.00000	0.27355

n = 10, K = 3

r-values	f	σf
0,7,7,7	0.14286	0.31018
0,7,8,7	0.00000	0.28965
1,0,8,7	1.00000	0.33793
1,0,9,7	0.83333	0.26692
1,0,10,7	0.66667	0.18793
1,1,7,7	1.00000	0.35573
1,1,8,7	0.83333	0.34731
1,1,9,7	0.66667	0.30089
1,1,10,7	0.50000	0.22567
1,2,6,7	1.00000	0.39674
5,3,7,7	0.00000	1.36422
5,4,4,7	1.00000	1.38444
5,4,5,7	0.50000	1.29636
5,4,6,7	0.00000	1.42400
5,5,5,7	0.00000	1.44338

r-values	f	σf
2,2,8,7	0.60000	0.43267
2,2,9,7	0.40000	0.48941
2,2,10,7	0.20000	0.34692
2,3,5,7	1.00000	0.54569
2,3,6,7	0.80000	0.51242
2,3,7,7	0.60000	0.48129
2,3,8,7	0.40000	0.45274
2,3,9,7	0.20000	0.42729
2,3,10,7	0.00000	0.40552
2,4,4,7	1.00000	0.55377
2,4,5,7	0.80000	0.52949
2,4,6,7	0.60000	0.50824

n = 10, K = 3

r-values	f	σf
2,4,7,7	0.40000	4.49044
2,4,8,7	0.20000	0.47647
2,4,9,7	0.00000	0.46667
2,5,5,7,	0.60000	0.51691
2,5,6,7	0.40000	0.50824
2,5,7,7	0.20000	0.50368
2,5,8,7	0.00000	0.50332
2,6,6,7	0.20000	0.51242
2,6,7,7	0.00000	0.52068
3,0,8,7	1.00000	0.50690
1,3,7,6	0.80000	0.50649
1,3,8,6	0.60000	0.45743
1,3,9,6	0.40000	0.40596
1,3,10,6	0.20000	0.35100
1,4,5,6	1.00000	0.56960

(Continued)

TABLE I

(Continued)

r-values	n = 10, K = 3 f	σf	r-values	n = 10, K = 3 f	σf	r-values	n = 10, K = 3 f	σf
3,1,10,7	0.25000	0.39198	6,0,8,7	1.00000	2.10818	1,4,6,6	0.80000	0.53216
3,2,6,7	1.00000	0.65085	6,0,9,7	0.00000	1.77951	1,4,7,6	0.60000	0.49477
3,2,7,7	0.75000	0.58999	6,1,7,7	1.00000	2.44949	1,4,8,6	0.40000	0.45743
3,2,8,7	0.50000	0.54327	6,1,8,7	0.00000	2.33333	1,4,9,6	0.20000	0.42016
3,2,9,7	0.25000	0.51454	6,2,6,7	1.00000	2.66667	1,4,10,6	0.00000	0.38297
3,2,10,7	0.00000	0.50690	6,2,7,7	0.00000	2.60342	1,5,5,6	0.80000	0.54045
3,3,5,7	1.00000	0.68211	6,3,5,7	1.00000	2.78887	1,5,6,6	0.60000	0.51242
3,3,6,7	0.75000	0.63533	6,3,6,7	0.00000	2.76887	1,5,7,6	0.40000	0.48571

r-values	n = 10, K = 3 f	σf	r-values	n = 10, K = 3 f	σf	r-values	n = 10, K = 3 f	σf
3,3,7,7	0.50000	0.60403	6,4,4,7	1.00000	2.82427	1,5,8,6	0.20000	0.46053
3,3,8,7	0.25000	0.58999	6,4,5,7	0.00000	2.84800	1,5,9,6	0.00000	0.43716
3,3,9,7	0.00000	0.59512	0,0,9,6	1.00000	0.31914	1,6,6,6	0.40000	0.49477
3,4,4,7	1.00000	0.69222	0,0,10,6	0.83333	0.22680	1,6,7,6	0.20000	0.47944
3,4,5,7	0.75000	0.65683	0,1,8,6	1.00000	0.38889	1,6,8,6	0.00000	0.46667
3,4,6,7	0.50000	0.63191	0,1,9,6	0.83333	0.32710	1,7,7,6	0.00000	0.47610
3,4,7,7	0.25000	0.63533	0,1,10,6	0.66667	0.24637	2,0,9,6	1.00000	0.47871
3,4,8,7	0.00000	0.65085	0,2,7,6	1.00000	0.43390	3,0,10,6	0.75000	0.31732
3,5,5,7	0.50000	0.64818	0,2,8,6	0.83333	0.38756	2,1,8,6	1.00000	0.58333
3,5,6,7	0.25000	0.65683	0,2,9,6	0.66667	0.33178	2,1,9,6	0.75000	0.47507
3,5,7,7	0.00000	0.68211	0,2,10,6	0.50000	0.26058	2,1,10,6	0.50000	0.36324
3,6,6,7	0.00000	0.69222	0,3,6,6	1.00000	0.46148	2,2,7,6	1.00000	0.65085
4,0,8,7	1.00000	0.67586	0,3,7,6	0.83333	0.42552	2,2,8,6	0.75000	0.56826
4,0,9,7	0.66667	0.50917	0,3,8,6	0.66667	0.38356	2,2,9,6	0.50000	0.49301
4,0,10,7	0.33333	0.40062	0,3,9,6	0.50000	0.33333	2,2,10,6	0.25000	0.42998
4,1,7,7	1.00000	0.79349	0,3,10,6	0.33333	0.27027	2,3,6,6	1.00000	0.69222

r-values	f	σf	r-values	f	σf	r-values	f	σf
4,1,8,7	0.66667	0.67586	0,4,5,6	1.00000	0.47466	2,3,7,6	0.75000	0.62639
4,1,9,7	0.33333	0.61864	0,4,6,6	0.83333	0.44675	2,3,8,6	0.50000	0.57130
4,1,10,7	0.00000	0.63828	0,4,7,6	0.66667	0.41450	2,3,9,6	0.25000	0.53033

	n = 10, K = 3			n = 10, K = 3			n = 10, K = 3	
r-values	f	σf	r-values	f	σf	r-values	f	σf
4,2,6,7	1.00000	0.86780	0,4,8,6	0.50000	0.37679	2,3,10,6	0.00000	0.50690
4,2,7,7	0.66667	0.77778	0,4,9,6	0.33333	0.33178	2,4,5,6	1.00000	0.71200
4,2,8,7	0.33333	0.74536	0,4,10,6	0.16667	0.27592	2,4,6,6	0.75000	0.66926
4,2,9,7	0.00000	0.77778	0,5,5,6	0.83333	0.45361	2,4,7,6	0.50000	0.61802
4,3,5,7	1.00000	0.90948	0,5,6,6	0.66667	0.42913	2,4,8,6	0.25000	0.59219
4,3,6,7	0.66667	0.83887	0,5,7,6	0.50000	0.40062	2,4,9,6	0.00000	0.58333
4,3,7,7	0.33333	0.82402	0,5,8,6	0.33333	0.36711	2,5,5,6	0.75000	0.68971
4,3,8,7	0.00000	0.86780	0,5,9,6	0.16667	0.32710	2,5,6,6	0.50000	0.64010
4,4,4,7	1.00000	0.92296	0,5,10,6	0.00000	0.27778	2,5,7,6	0.25000	0.62639
4,4,5,7	0.77778	0.86780	0,6,6,6	0.50000	0.40825	2,5,8,6	0.00000	0.62915
4,4,6,7	0.33333	0.86780	0,6,7,6	0.33333	0.38356	2,6,6,6	0.25000	0.63191
4,4,7,7	0.00000	0.92296	0,6,8,6	0.16667	0.35428	2,7,7,6	0.00000	0.65085
4,5,5,7	0.33333	0.88192	0,6,9,2	0.00000	0.31914	3,0,9,6	1.00000	0.63828
4,5,6,7	0.00000	0.94933	0,7,7,6	0.16667	0.36289	3,0,10,6	0.66667	0.40062
5,0,8,7	1.00000	1.01379	0,7,8,6	0.00000	0.33793	3,1,8,6	1.00000	0.77778
5,0,9,7	0.50000	0.75462	1,0,9,6	1.00000	0.38297	3,1,9,6	0.66667	0.61864
5,0,10,7	0.00000	0.83333	1,0,10,6	0.80000	0.26432	3,1,10,6	0.33333	0.50917
5,1,7,7	1.00000	1.19024	1,1,8,6	1.00000	0.46667	3,2,7,6	1.00000	0.86780
5,1,8,7	0.50000	1.00692	1,1,9,6	0.80000	0.38713	3,2,8,6	0.66667	0.74536
5,1,9,7	0.00000	1.09291	1,1,10,6	0.60000	0.29120	3,2,9,6	0.33333	0.67586
5,2,6,7	1.00000	1.30171	1,2,7,6	1.00000	0.52068	3,2,10,6	0.00000	0.67586
5,2,7,7	0.50000	1.16070	1,2,8,6	0.80000	0.46053	3,3,6,6	1.00000	0.92296
5,2,8,7	0.00000	1.25830	1,2,9,6	0.60000	0.39486	3,3,7,6	0.66667	0.82402

(Continued)

TABLE I
(Continued)

r-values	n = 10, K = 3 f	σf
5,3,5,7	1.00000	1.36422
5,3,6,7	0.50000	1.25277
3,4,5,6	1.00000	0.94933
3,4,6,6	0.66667	0.86780
3,4,7,6	0.33333	0.83887
3,4,8,6	0.00000	0.86780
3,5,5,6	0.66667	0.88192
3,5,6,6	0.33333	0.86780
3,5,7,6	0.00000	0.90948
3,6,6,6	0.00000	0.92296
4,0,9,6	1.00000	0.95743
4,0,10,6	0.50000	0.57735
4,1,8,6	1.00000	0.16667
4,1,9,6	0.50000	0.91287
4,2,7,6	1.00000	1.19024
4,2,8,6	0.50000	1.10554
4,2,9,6	0.00000	1.16667
4,3,6,6	1.0	1.28019
4,3,7,6	0.5	1.22474
4,3,8,6	0.0	1.19024

r-values	n = 10, K = 3 f	σf
4,4,5,6	1.0	1.32288
4,4,6,6	0.5	1.29099
4,4,7,6	0.0	1.28019
4,5,5,6	0.5	1.31233

r-values	n = 10, K = 3 f	σf
1,2,10,6	0.40000	0.32028
1,3,6,6	1.00000	0.55377
0,6,7,5	0.40000	0.46667
0,6,8,5	0.20000	0.42688
0,6,9,5	0.00000	0.38297
0,7,7,5	0.20000	0.43716
0,7,8,5	0.00000	0.40552
1,0,10,5	1.00000	0.41667
1,1,9,5	1.00000	0.54645
1,1,10,5	0.75000	0.40505
1,2,8,5	1.00000	0.62915
1,2,9,5	0.75000	0.52457
1,2,10,5	0.50000	0.41248
1,3,7,5	1.00000	0.68211
1,3,9,5	0.50000	0.51707
1,3,10,5	0.25000	0.43800
1,4,6,5	1.00000	0.71200
1,4,7,5	0.75000	0.64348
1,4,8,5	0.50000	0.58035
1,4,9,5	0.25000	0.52457

r-values	n = 10, K = 3 f	σf
1,4,10,5	0.00000	0.47871
1,5,5,5	1.00000	0.72169
1,5,6,5	0.75000	0.66471
1,5,7,5	0.50000	0.61520

r-values	n = 10, K = 3 f	σf
3,3,8,6	0.33333	0.77778
3,3,9,6	0.00000	0.79349
3,3,9,5	0.00000	1.19024
3,4,6,5	1.00000	1.42400
3,4,7,5	0.50000	1.25277
3,4,8,5	0.00000	1.30171
3,5,5,5	1.00000	1.44388
3,5,6,5	0.50000	1.29636
3,5,7,5	0.00000	1.36422
3,6,6,5	0.00000	1.38444
4,0,10,5	1.00000	1.66667
4,1,9,5	1.00000	2.18581
4,1,10,5	0.00000	2.23607
4,2,8,5	1.00000	2.51661
4,3,7,5	1.00000	2.72845
4,3,8,5	0.00000	2.84800
4,4,6,5	1.00000	2.84800
4,4,7,5	0.00000	3.00000
4,5,5,5	1.00000	2.88675
4,5,6,5	0.00000	3.07318

r-values	n = 10, K = 3 f	σf
0,1,10,4	1.00000	0.47871
0,2,9,4	1.00000	0.58333
0,2,10,4	0.75000	0.45262
0,3,8,4	1.00000	0.65085

n = 10, K = 3

r-values	f	σ f
4,5,6,6	0.0	1.32288
5,0,9,6	1.0	2.23607
5,0,10,6	0.0	1.66667
5,1,8,6	1.0	2.60342
5,1,9,6	0.0	2.18581
5,2,7,6	1.0	2.84800
5,2,8,6	0.0	2.51661
5,3,7,6	0.0	2.72845
5,4,5,6	1.0	3.07318
5,4,6,6	0.0	2.84800
5,5,5,6	0.0	2.88675
0,0,10,5	1.0	0.33333
0,1,9,5	1.0	0.43716
0,1,10,5	0.8	0.33333
0,2,8,5	1.0	0.50332
0,2,9,5	0.8	0.42687
0,2,10,5	0.6	0.33333
0,3,7,5	1.0	0.54569
0,3,8,5	0.8	0.48534

n = 10, K = 3

r-values	f	σ f
1,5,8,5	0.25000	0.57509
1,5,9,5	0.00000	0.54645
1,6,6,5	0.50000	0.62639
1,6,7,5	0.25000	0.59875
1,6,8,5	0.00000	0.58333
1,7,7,5	0.00000	0.59512
2,1,9,5	1.00000	0.72860
2,1,10,5	0.66667	0.51985
2,2,8,5	1.00000	0.83887
2,2,9,5	0.66667	0.68393
2,2,10,5	0.33333	0.56534
2,3,7,5	1.00000	0.90948
2,3,8,5	0.66667	0.78480
2,3,9,5	0.33333	0.70175
2,3,10,5	0.00000	0.67586
2,4,6,5	1.00000	0.94933
2,4,7,5	0.66667	0.84539
2,4,8,5	0.33333	0.78480
2,4,9,5	0.00000	0.77778

n = 10, K = 3

r-values	f	σ f
0,3,9,4	0.75000	0.54962
0,3,10,4	0.50000	0.53301
0,4,7,4	1.00000	0.69222
0,4,8,4	0.75000	0.60953
0,4,9,4	0.50000	0.52042
0,4,10,4	0.25000	0.45644
0,5,7,4	0.75000	0.64280
0,5,8,4	0.50000	0.57130
0,5,9,4	0.25000	0.52705
0,5,10,4	0.00000	0.41667
0,6,6,4	0.75000	0.65352
0,6,7,4	0.50000	0.59512
0,6,8,4	0.25000	0.56519
0,6,9,4	0.00000	0.47871
0,7,7,4	0.25000	0.57735
0,7,8,4	0.00000	0.50690
1,1,10,4	1.00000	0.63828
1,2,9,4	1.00000	0.77778
1,2,10,4	0.66667	0.58443

n = 10, K = 3

r-values	f	σ f
0,3,9,5	0.6	0.41633
0,3,10,5	0.4	0.33333
0,4,6,5	1.0	0.56960
0,4,7,5	0.8	0.52068
0,4,8,5	0.6	0.46667
0,4,9,5	0.4	0.40552
0,4,10,5	0.2	0.33333
0,5,5,5	1.0	0.57735
0,5,6,5	0.8	0.53748

n = 10, K = 3

r-values	f	σ f
2,5,5,5	1.00000	0.96225
2,5,6,5	0.66667	0.87410
2,5,7,5	0.33333	0.83065
2,5,8,5	0.00000	0.83887
2,6,6,5	0.33333	0.84539
2,6,7,5	0.00000	0.86780
3,0,10,5	1.00000	0.83333
3,1,9,5	1.00000	1.09291
3,1,10,5	0.50000	0.75462

n = 10, K = 3

r-values	f	σ f
1,3,8,4	1.00000	0.79349
1,3,9,4	0.66667	0.71722
1,3,10,4	0.33333	0.58443
1,4,7,4	1.00000	0.85346
1,4,8,4	0.66667	0.79866
1,4,9,4	0.33333	0.69979
1,4,10,4	0.00000	0.63828
1,5,6,4	1.00000	0.88192
1,5,7,4	0.66667	0.84376

(Continued)

TABLE I

(Continued)

n = 10, K = 3			n = 10, K = 3			n = 10, K = 3		
r-values	f	σf	r-values	f	σf	r-values	f	σf
0,5,7,5	0.6	0.48990	3,2,8,5	1.00000	1.25830	1,5,8,4	0.33333	0.76712
0,5,8,5	0.40000	0.44721	3,2,9,5	0.50000	1.00692	1,5,9,4	0.00000	0.72860
0,5,9,5	0.20000	0.39440	3,2,10,5	0.00000	1.01379	1,6,6,4	0.66667	0.85827
0,5,10,5	0.00000	0.33333	3,3,7,5	1.00000	1.36422	1,6,7,4	0.33333	0.79866
0,6,6,5	0.60000	0.49889	3,3,8,5	0.50000	1.16070	1,6,8,4	0.00000	0.77778
1,7,7,4	0.00000	0.79349	0,4,10,3	0.33333	0.57013	2,5,8,3	0.0	2.51661
2,1,10,4	1.00000	0.95743	0,5,7,3	1.00000	0.90948	2,6,6,3	1.0	2.76887
2,2,9,4	1.00000	1.16667	0,5,8,3	0.66667	0.78829	2,6,7,3	0.0	2.60342
2,2,10,4	0.50000	0.84984	0,5,9,3	0.33333	0.66975	0,3,10,2	1.0	1.01379
2,3,8,4	1.00000	1.30171	0,5,10,3	0.00000	0.55556	0,4,9,2	1.0	1.16667
2,3,9,4	0.50000	1.05409	0,6,6,3	1.00000	0.92296	0,4,10,2	0.5	0.88192
2,3,10,4	0.00000	1.01319	0,6,7,3	0.66667	0.81901	0,5,8,2	1.0	1.25830

n = 10, K = 3			n = 10, K = 3			n = 10, K = 3		
r-values	f	σf	r-values	f	σf	r-values	f	σf
2,4,7,4	1.00000	1.38444	0,6,8,3	0.33333	0.72293	0,5,9,2	0.5	1.02740
2,4,8,4	0.50000	1.17851	0,6,9,3	0.00000	0.63828	0,5,10,2	0.0	0.83333
2,4,9,4	0.00000	1.16667	0,7,7,3	0.33333	0.73981	0,6,7,2	1.0	1.30171
2,5,6,4	1.00000	1.42400	0,7,8,3	0.00000	0.67586	0,6,8,2	0.5	1.10554
2,5,7,4	0.50000	1.24722	1,2,10,3	1.0	1.01379	0,6,9,2	0.0	0.95743
2,5,8,4	0.00000	1.25830	1,3,9,3	1.0	1.19024	0,7,7,2	0.5	1.13039
2,6,6,4	0.50000	1.26930	1,3,10,3	0.5	0.88976	0,7,8,2	0.0	1.01379
2,6,7,4	0.00000	1.30171	1,4,8,3	1.0	1.30171	1,3,10,2	1.0	2.02759
3,1,10,4	1.00000	1.79951	1,4,9,3	0.5	1.06066	1,4,9,2	1.0	2.33333
3,2,9,4	1.00000	2.33333	1,4,10,3	0.0	0.95743	1,4,10,2	0.0	1.77951
3,2,10,4	0.00000	2.10818	1,5,7,3	1.0	1.36422	1,5,8,2	1.0	2.51661

r-values	f	σf	r-values			r-values		
3,3,8,4	1.00000	2.60342	1,5,8,3	0.5	1.16070	1,5,9,2	0.0	2.18581
3,3,9,4	0.00000	2.44949	1,5,9,3	0.0	1.09291	1,6,7,2	1.0	2.60342
3,4,7,4	1.00000	2.76887	1,6,6,3	1.0	1.38444	1,6,8,2	0.0	2.33333
3,4,8,4	0.00000	2.66667	1,6,7,3	0.5	1.20761	1,7,7,2	0.0	2.38048
3,5,6,4	1.00000	2.84800	1,6,8,3	0.0	1.16667	0,4,10,1	1.0	0.77951
3,5,7,4	0.00000	2.78887	1,7,7,3	0.0	1.19024	0,5,9,1	1.0	2.18581
3,6,6,4	0.00000	2.82427	2,2,10,3	1.0	2.02759	0,5,10,1	0.0	1.66667
0,2,10,3	1.00000	0.67586	2,3,9,3	1.0	2.38048	0,6,8,1	1.0	2.33333
0,3,9,3	1.00000	0.79349	2,3,10,3	0.0	2.02759	0,6,9,1	0.0	1.77951
0,3,10,3	0.66667	0.61195	2,4,8,3	1.0	2.60342	0,7,7,1	1.0	2.38048
0,4,8,3	1.00000	0.86780	2,4,9,3	0.0	2.33333	0,7,8,1	0.0	2.02759
0,4,9,3	0.66667	0.72293	2,5,7,3	1.0	2.72845			

Table for Calculation of Median-Effective Dose*

r-values	f	σ f	d = 0.30103	
			d(f + 0.5)	2.306 d$_{of}$
0,0,5	1.0	(0)	0.45154	(0)
0,1,5	0.8	0.20000	0.39134	0.13884
0,2,5	0.6	0.24495	0.33113	0.17004
0,3,5	0.4	0.24495	0.27093	0.17004
0,4,5	0.2	0.20000	0.21072	0.13884
0,5,5	0.0	(0)	0.15052	(0)
1,0,5	1.0	(0)	0.45154	(0)
1,1,5	0.75	0.25769	0.37629	0.17888
1,2,5	0.5	0.33072	0.30103	0.22958
1,3,5	0.25	0.35904	0.22577	0.24924
1,4,5	0.0	0.35355	0.15052	0.24543
2,0,5	1.0	(0)	0.45154	(0)

TABLE I
(Continued)

r-values	f	σ f	d = 0.30103	
			d(f + 0.5)	2.306 d_of
2,1,5	0.66667	0.36004	0.35120	0.24993
2,2,5	0.33333	0.49065	0.25086	0.34060
2,3,5	0.0	0.57735	0.15052	0.40078
3,0,5	1.0	(0)	0.45154	(0)
3,1,5	0.5	0.58630	0.30103	0.40700
3,2,5	0.0	0.86603	0.15052	0.60118
4,0,5	1.0	(0)	0.45154	(0)
4,1,5	0.0	1.41421	0.15052	0.98171

* Calculation by moving average interpolation for N = 5, K = 2, d = 0.30103 from the formula: $\log m = \log D_a + d(k - 1)/2 + df = \log D_a + d(f + 0.5)$ in this case. $2.306\ \sigma \log m = 2.306\ d_{of}$.

r-values	f	σ f	d = 0.30103	
			d(f + 0.5)	2.306 d_of
0,1,4	1.0	0.35355	0.45154	0.24543
0,2,4	0.75	0.35904	0.37629	0.24924
0,3,4	0.5	0.33072	0.30103	0.22958
0,4,4	0.25	0.25769	0.22577	0.17888
0,5,4	0.0	(0)	0.15052	(0)
1,1,4	1.0	0.47140	0.45154	0.32723
1,2,4	0.6667	0.47791	0.35120	0.33175
1,3,4	0.3333	0.47791	0.25086	0.33175
1,4,4	0.0	0.47140	0.15052	0.32723
2,1,4	1.0	0.70711	0.45154	0.49086
2,2,4	0.5	0.72887	0.30103	0.50596
2,3,4	0.0	0.86603	0.15052	0.60118
3,1,4	1.0	1.41421	0.45154	0.98171

r-values	f	σf	d(f+1)	2.179d_σf
3,2,4	0.0	1.73205	0.15052	1.20235
0,2,3	1.0	0.57735	0.45154	0.40078
0,3,3,	0.66667	0.49065	0.35120	0.35060
0,4,3	0.33333	0.36004	0.25086	0.24993
0,5,3	0.0	(0)	0.15052	(0)
1,2,3	1.0	0.86603	0.45154	1.20235
1,3,3	0.5	0.72887	0.30103	0.50596
1,4,3	0.0	0.70711	0.15052	0.49806
2,2,3	1.0	1.73205	0.45154	1.20235
2,3,3	0.0	1.73205	0.15052	1.20235
0,3,2	1.0	0.86603	0.45154	0.60113
0,4,2	0.5	0.58630	0.30103	0.40700
0,5,2	0.0	(0)	0.15052	(0)
1,3,2	1.0	1.73205	0.45154	1.20235
1,4,2	0.0	1.41421	0.15052	0.98171
0,4,1	1.0	1.41421	0.45154	0.98171
0,5,1	0.0	(0)	0.15052	(0)

* Calculation by moving average interpolation for $N = 5$, $K = 2$, $d = 0.30103$ from the formula: $\log m = \log D_a + d(k - 1)/2 + df = \log D_a + d(f + 0.5)$ in this case. $2.306\ s\ \log m = 2.306\ d_{\sigma f}$.

r-values	f	σf	d(f + 1)	2.179d_σf	r-values	f	σf	d(f + 1)	2.179d_σf
			d = 0.30103					d = 0.30103	
0,0,3,5	0.9	0.24495	0.57196	0.16067	0,2,4,4	0.375	0.40625	0.41392	0.26648
0,0,4,5	0.7	0.20000	0.51175	0.13119	0,2,5,4	0.125	0.30778	0.33866	0.20189
0,0,5,5	0.5	0.0	0.45154	0.0	0,3,3,4	0.375	0.44304	0.41392	0.29061
0,1,2,5	0.9	0.31623	0.57196	0.20743	0,3,4,4	0.125	0.39652	0.33866	0.26010
0,1,3,5	0.7	0.31623	0.51175	0.20743	1,0,4,4	0.83333	0.43744	0.55189	0.28694
0,1,4,5	0.5	0.28284	0.45154	0.18553	1,0,5,4	0.50	0.23570	0.45154	0.15461
0,1,5,5	0.3	0.20000	0.39134	0.13119	1,1,3,4	0.83333	0.59835	0.55189	0.39248
0,2,2,5	0.7	0.34641	0.51175	0.22723	1,1,4,4	0.50	0.52705	0.45154	0.34572
0,2,3,5	0.5	0.34641	0.45154	0.22723	1,1,5,4	0.16667	0.43744	0.35120	0.28694

(Continued)

TABLE I
(*Continued*)

r-values	f	σ f	d(f + 1)	2.179d$_{of}$
			d = 0.30103	
0,2,4,5	0.3	0.31623	0.39134	0.20743
0,2,5,5	0.1	0.24495	0.33113	0.16067
0,3,3,5	0.3	0.34641	0.39134	0.22723
0,3,4,5	0.1	0.31623	0.33113	0.20743
1,0,3,5	0.875	0.30778	0.56443	0.20189
1,0,4,5	0.625	0.26700	0.48917	0.17514
1,0,5,5	0.375	0.15625	0.41392	0.10249
1,1,2,5	0.875	0.39652	0.56443	0.26010
1,1,3,5	0.625	0.40625	0.48917	0.26648
1,1,4,5	0.375	0.38654	0.41392	0.25355
1,1,5,5	0.125	0.33219	0.33866	0.21790
1,2,2,5	0.625	0.44304	0.48917	0.29061
1,2,3,5	0.375	0.46034	0.41392	0.30196
1,2,4,5	0.125	0.45178	0.33866	0.29634
1,3,3,5	0.125	0.48513	0.33866	0.31822
2,0,3,5	0.83333	0.41388	0.55189	0.27148
2,0,4,5	0.50000	0.39087	0.45154	0.25639
2,0,5,5	0.16667	0.34021	0.35120	0.22316
2,1,2,5	0.83333	0.53142	0.55189	0.34858
2,1,3,5	0.50	0.56519	0.45154	0.37073
2,1,4,5	0.16667	0.58134	0.35120	0.38133
2,2,2,5	0.50	0.61237	0.45154	0.40168
2,2,3,5	0.16667	0.67013	0.35120	0.43957
3,0,3,5	0.75	0.63122	0.52680	0.41404
3,0,4,5	0.25	0.67892	0.37629	0.44533
3,1,2,5	0.75	0.80526	0.52680	0.52820
3,1,3,5	0.25	0.91430	0.37629	0.59973

r-values	f	σ f	d(f + 1)	2.179d$_{of}$
			d = 0.30103	
1,2,2,4	0.83333	0.64310	0.55189	0.42184
1,2,3,4	0.50	0.62361	0.45154	0.40905
1,2,4,4	1.16667	0.59835	0.35120	0.39248
1,3,3,4	0.16667	0.64310	0.35120	0.42184
2,0,4,4	0.75	0.64348	0.52680	0.42209
2,0,5,4	0.25	0.47598	0.37629	0.31222
2,1,3,4	0.75	0.88829	0.52680	0.58267
2,1,4,4	0.25	0.85239	0.37629	0.55912
2,2,2,4	0.75	0.95607	0.52680	0.62713
2,2,3,4	0.25	0.98821	0.37629	0.64821
3,0,4,4	0.5	1.27475	0.45154	0.83616
3,1,3,4	0.5	1.76777	0.45154	1.15956
3,2,2,4	0.5	1.90394	0.45154	1.24888
0,0,5,3	0.83333	0.34021	0.55189	0.22316
0,1,4,3	0.83333	0.58134	0.55134	0.38133
0,1,5,3	0.50000	0.39087	0.45154	0.25639
0,2,3,3	0.83333	0.67013	0.55189	0.43957
0,2,4,3	0.50000	0.56519	0.45154	0.37073
0,2,5,3	0.16667	0.41388	0.35120	0.27148
0,3,3,3	0.50000	0.61237	0.45154	0.40168
0,3,4,3	0.16667	0.53142	0.35120	0.34858
1,0,5,3	0.75	0.47598	0.52680	0.31222
1,1,4,3	0.75	0.85239	0.52680	0.55912
1,1,5,3	0.25	0.64348	0.37629	0.42209
1,2,3,3	0.75	0.98821	0.52680	0.64821
1,2,4,3	0.25	0.88829	0.37629	0.58267
1,3,3,3	0.25	0.95609	0.37629	0.62713

3,2,2,5	0.25	0.98028	0.37629	0.64301	2,0,5,3	0.5	0.86602	0.45154	0.56806
4,0,3,5	0.5	1.32288	0.45154	0.86774	0,1,5,2	0.75	0.67892	0.52680	0.44533
4,1,2,5	0.5	1.64831	0.45154	1.08776	0,2,4,2	0.25	0.91430	0.37629	0.59973
0,0,4,4	0.875	0.33219	0.56443	0.21790	0,2,5,2	0.25	0.63122	0.37629	0.41404
0,0,5,4	0.625	0.15625	0.48917	0.10249	0,3,3,2	0.75	0.98028	0.52680	0.64301
0,1,3,4	0.875	0.45178	0.56443	0.29634	0,3,4,2	0.25	0.80526	0.37625	0.52820
0,1,4,4	0.625	0.38654	0.48917	0.25355	1,1,5,2	0.5	1.27475	0.45154	0.83616
0,1,5,4	0.375	0.26700	0.41392	0.17514	1,2,4,2	0.5	1.76777	0.45154	1.15956
0,2,2,4	0.875	0.48513	0.56443	0.31822	1,3,3,2	0.5	1.90394	0.45154	1.24888
0,2,3,4	0.625	0.46034	0.48917	0.30196	0,2,5,1	0.5	1.32288	0.45154	0.86774
0,3,4,1			0.5	1.65831	0,2,5,1	0.45154			0.45154
						1.08776			

* Calculation by moving average interpolation for $N = 5$, $K = 3$, $d = 0.30103$ from the formula: $\log m = \log D_a + d(k - 1)/2 + df = \log D_a + d(f + 1)$ in this case. $2.179\,\sigma \log m = 2.179\,\sigma f(d)$.

			d = 0.30103	
r-values	f	σ f	d(f + 0.5)	2.447$d_{σf}$
0,0,4	1.0	(0)	0.45154	(0)
0,1,4	0.75	0.25000	0.37629	0.18416
0,2,4	0.5	0.28868	0.30103	0.21255
0,3,4	0.25	0.25000	0.22577	0.18416
0,4,4	0.0	(0)	0.15052	(0)
1,0,4	1.0	(0)	0.45154	(0)
1,1,4	0.66667	0.35138	0.35120	0.25833
1,2,4	0.33333	0.44444	0.25086	0.32738
1,3,4	0.0	0.47140	0.15052	0.34724
2,0,4	1.0	(0)	0.45154	(0)
2,1,4	0.5	0.57735	0.30103	0.42529
2,2,4	0.0	0.81650	0.15052	0.60145
3,0,4	1.0	(0)	0.45154	(0)
3,1,4	0.0	1.41421	0.15052	0.04174

(Continued)

TABLE I
(Continued)

r-values	f	σ f	d = 0.30103	
			d(f + 0.5)	2.447 d$_{\sigma f}$
0,1,3	1.0	0.47140	0.45154	0.34724
0,2,3	0.66667	0.44444	0.35120	0.32738
0,3,3	0.33333	0.35138	0.25086	0.25883
0,4,3	0.0	(0)	0.15052	(0)
1,1,3	1.0	0.70711	0.45154	0.52087
1,2,3	0.5	0.6770	0.30103	0.49869
1,3,3	0.0	0.70711	0.15052	0.52087
2,1,3	1.0	1.41421	0.45154	1.04174
2,2,3	0.0	1.63299	0.15052	1.20289
0,2,2	1.0	0.81650	0.45154	0.60145
0,3,2	0.5	0.57735	0.30103	0.42529
0,4,2	0.0	(0)	0.15052	(0)
1,2,2	1.0	1.63299	0.45154	1.20289
1,3,2	0.0	1.41421	0.15052	1.04174
0,3,1	1.0	1.41421	0.45154	1.04174
0,4,1	0.0	(0)	0.15052	(0)

* Calculation by moving average interpolation for N = 4, K = 2, d = 0.30103 from the formula: log m = log D$_a$ + d(k − 1)/2 + df = log D$_a$ + d(f + 0.5). 2.447 σ = 2.447 σf(d).

r-values	f	σ f	d = 0.30103	
			df	3.182 d$_{\sigma f}$
0,4	0.5	(0)	0.15052	(0)
1,4	0.33333	0.22222	0.10034	0.21286
2,4	0.0	0.57735	0.0	0.55303
0,3	0.66667	0.22222	0.20069	0.21286
1,3	0.5	0.34355	0.15052	0.33866
2,3	0.0	1.15470	0.0	1.10606

r-values	f	σ f	df	
0,2	1.0	0.57735	0.30103	0.55303
1,2	1.0	1.15470	0.30103	1.10606

r-values	f	σ f	df	$2.776d_{\sigma f}$
		(0)		(0)
0,5	0.5	0.15625	0.15052	0.13057
1,5	0.375	0.34021	0.11289	0.28430
2,5	0.16667	0.15625	0.05017	0.13057
0,4	0.625	0.23570	0.18814	0.19696
1,4	0.5	0.47599	0.15052	0.39776
2,4	0.25	0.34021	0.07526	0.28430
0,3	0.83333	0.47599	0.25086	0.39776
1,3	0.75	0.34021	0.22577	0.28430
2,3	0.5	0.86603	0.15052	0.72371

* Calculation by moving average interpolation for N = 4 or 5, K = 1, d = 0.30103 from the formula: log m = log D_a + d(k − 1)/2 + df = log D_a + df in this case. N = 4; upper section; N = 5; lower section.

d = 0.30103

r-values	f	σ f	d(f + 1)	$2.262d_{\sigma f}$
0,0,2,4	1.00000	0.28868	0.60206	0.19657
0,0,3,4	0.75000	0.2500	0.52680	0.17023
0,0,4,4	0.50000	0.0	0.45154	0.0
0,1,1,4	1.00000	0.35355	0.60206	0.24074
0,1,2,4	0.75000	0.38188	0.52680	0.26003
0,1,3,4	0.50000	0.35355	0.45154	0.24074
0,1,4,4	0.25000	0.25000	0.37629	0.17023
0,2,2,4	0.50000	0.40825	0.45154	0.27799
0,2,3,4	0.25000	0.28188	0.37629	0.26003
0,2,4,4	0.00000	0.28868	0.30103	0.19657
0,3,3,4	0.00000	0.35355	0.30103	0.24074
1,0,2,4	1.00000	0.38490	0.60206	0.26209
1,0,3,4	0.66667	0.35136	0.50172	0.23925
1,0,4,4	0.33333	0.22222	0.40137	0.15132

d = 0.30103

r-values	f	σ f	d(f + 1)	$2.262d_{\sigma f}$
0,1,4,3	0.33333	0.35136	0.40137	0.23925
0,2,2,3	0.66667	0.58794	0.50172	0.40035
0,2,3,3	0.33333	0.52116	0.40137	0.35487
0,2,4,3	0.00000	0.38490	0.30103	0.26209
0,3,3,3	0.00000	0.47140	0.30103	0.26209
1,0,3,3	1.00000	0.70711	0.60206	0.48149
1,0,4,3	0.50000	0.35355	0.45154	0.24074
1,1,2,3	1.00000	0.91287	0.60206	0.62160
1,1,3,3	0.50000	0.79057	0.45154	0.53832
1,1,4,3	0.00000	0.70711	0.30103	0.48149
1,2,2,3	0.50000	0.88976	0.45154	0.60586
1,2,3,3	0.00000	0.91287	0.30103	0.62160
2,0,3,3	1.00000	1.41421	0.60206	0.96298
2,0,4,3	0.00000	1.15470	0.30103	0.78627

(Continued)

TABLE I
(Continued)

r-values	f	σ f	d = 0.30103 d(f + 1)	d = 0.30103 2.262d$_{\sigma f}$	r-values	f	σ f	d = 0.30103 d(f + 1)	d = 0.30103 2.262d$_{\sigma f}$
1,1,1,4	1.00000	0.47140	0.60206	0.32099	2,1,2,3	1.00000	1.82574	0.60206	1.24320
1,1,2,4	0.66667	0.52116	0.50172	0.35487	2,1,3,3	0.00000	1.82574	0.30103	1.24320
1,1,3,4	0.33333	0.52116	0.40137	0.35487	2,2,2,3	0.00000	2.00000	0.30103	1.36186
1,1,4,4	0.00000	0.47140	0.30103	0.32099	0,0,4,2	1.00000	0.57735	0.60206	0.39313
1,2,2,4	0.33333	0.58794	0.40137	0.40035	0,1,3,2	1.00000	0.91287	0.60206	0.62160

* Calculation by moving average interpolation for N = 4, K = 3, d = 0.30103 from the formula: log m = log D_a = d(k − 1)/2 + δ

r-values	f	σ f	d = 0.30103 d(f + 1)	d = 0.30103 2.179d$_{\sigma f}$	r-values	f	σ f	d = 0.30103 d(f + 1)	d = 0.30103 2.179d$_{\sigma f}$
1,2,3,4	0.00000	0.60858	0.30103	0.41440	0,1,4,2	0.50000	0.57735	0.45154	0.39313
2,0,2,4	1.00000	0.57735	0.60206	0.39313	0,2,2,2	1.00000	1.00000	0.60206	0.68093
2,0,3,4	0.50000	0.57735	0.45154	0.39313	0,2,3,2	0.50000	0.81650	0.45154	0.55598
2,0,4,4	0.00000	0.57735	0.30103	0.39313	0,2,4,2	0.00000	0.57735	0.30103	0.39313
2,1,1,4	1.00000	0.70711	0.60206	0.48149	0,3,3,2	0.00000	0.70711	0.60206	0.48149
2,1,2,4	0.50000	0.81650	0.45154	0.55598	1,0,4,2	1.00000	1.15470	0.60206	0.78627
2,1,3,4	0.00000	0.91287	0.60206	0.62160	1,1,3,2	1.00000	1.82574	0.30103	1.24320
2,2,2,4	0.00000	1.00000	0.30103	0.68093	1,1,4,2	0.00000	1.41421	0.60206	0.96298
3,0,2,4	1.00000	1.15470	0.60206	0.78627	1,2,2,2	1.00000	2.00000	0.30103	1.36186
3,0,3,4	0.00000	1.41421	0.30103	0.96298	1,2,3,2	1.00000	1.82574	0.60206	1.24320
3,1,1,4	1.00000	1.41421	0.60206	0.96298	0,2,3,1	1.00000	1.82574	0.60206	1.24320
3,1,2,4	0.00000	1.82574	0.30103	1.24320	0,2,4,1	0.00000	1.15470	0.80103	0.78627
0,0,3,3	1.00000	0.47140	0.60206	0.32099	0,3,3,1	0.00000	1.41421	0.30103	0.96298
0,0,4,3	0.66667	0.22222	0.50172	0.15132	0,1,4,1	1.00000	1.41421	0.60206	0.96298
0,1,2,3	1.00000	0.60858	0.60206	0.41440					
0,1,3,3	0.66667	0.52116	0.50172	0.35487					

* Calculation by moving average interpolation for N = 4, K = 3, d = 0.30103 from the formula: log m = log D_a = d(k − 1)/2 + δ

TABLE J

Critical Values for the Wilcoxon Rank Sum Test (N_1 = Larger Group)

P = .05 one-sided, P = .10 two-sided

N_1	$N_2 = 3$	$N_2 = 4$	$N_2 = 5$	$N_2 = 6$	$N_2 = 7$	$N_2 = 8$	$N_2 = 9$	$N_1 = 10$	$N_2 = 11$	$N_2 = 12$	$N_2 = 13$	$N_2 = 14$
$N_1 = N_2$	6,15	12,24	19,36	28,50	39,66	52,84	66,105	83,127	101,152	121,179	143,208	167,239
$N_1 = N_2 + 1$	7,17	13,27	20,40	30,54	41,71	54,90	69,111	86,134	105,159	125,187	148,216	172,248
$N_1 = N_2 + 2$	7,20	14,30	22,43	32,58	43,76	57,95	72,117	89,141	109,166	129,195	152,225	177,257
$N_1 = N_2 + 3$	8,22	15,33	24,46	33,63	46,80	60,100	75,123	93,147	112,174	134,202	157,233	182,266
$N_1 = N_2 + 4$	9,24	16,36	25,50	35,67	48,85	62,106	78,129	96,154	116,181	138,210	162,241	187,275
$N_1 = N_2 + 5$	9,27	17,39	26,54	37,71	50,90	65,111	81,135	100,160	120,188	142,218	166,250	192,284
$N_1 = N_2 + 6$	10,29	18,42	27,58	39,75	52,95	67,117	84,141	103,167	124,195	147,225	171,258	197,293
$N_1 = N_2 + 7$	11,31	19,45	29,61	41,79	54,100	70,122	87,147	107,173	128,202	151,233	176,266	203,301
$N_1 = N_2 + 8$	11,34	20,48	30,65	42,84	57,104	73,127	90,153	110,180	132,209	155,241	181,274	208,310
$N_1 = N_2 + 9$	12,36	21,51	32,68	44,88	59,109	75,133	93,159	114,186	136,216	159,249	185,283	213,319
$N_1 = N_2 + 10$	13,38	22,54	33,72	46,92	61,114	78,138	96,165	117,193	139,224	164,256	190,291	218,328
$N_1 = N_2 + 11$	13,41	23,57	34,76	48,96	63,119	80,144	100,170	120,200	143,231	168,264	195,299	223,337
$N_1 = N_2 + 12$	14,43	24,60	36,79	50,100	65,124	83,149	103,176	124,206	147,238	172,272	199,308	228,346
$N_1 = N_2 + 13$	15,45	25,63	37,83	52,104	68,128	86,154	106,182	127,213	151,245	177,279	204,316	234,354
$N_1 = N_2 + 14$	15,48	26,66	39,86	53,109	70,133	88,160	109,188	131,219	155,252	181,287	209,324	239,363
$N_1 = N_2 + 15$	16,50	27,69	40,90	55,113	72,138	91,165	112,194	134,226	159,259	185,295	214,332	244,372
$N_1 = N_2 + 16$	17,52	28,72	42,93	57,117	74,143	94,170	115,200	138,232	163,266	190,302	218,341	249,381
$N_1 = N_2 + 17$	17,55	29,75	43,97	59,121	77,147	96,176	118,206	141,239	167,273	194,310	223,349	254,390
$N_1 = N_2 + 18$	18,57	30,78	44,101	61,125	79,152	99,181	121,212	145,245	171,280	198,218	228,357	260,398
$N_1 = N_2 + 19$	19,59	31,81	46,104	62,130	81,157	102,186	124,218	148,252	175,287	203,325	233,365	265,407
$N_1 = N_2 + 20$	19,62	32,84	47,108	64,134	83,162	104,192	127,224	152,258	178,295	207,333	237,374	270,416
$N_1 = N_2 + 21$	20,64	33,87	49,111	66,138	86,166	107,197	130,230	155,265	182,302	211,341	242,382	275,425
$N_1 = N_2 + 22$	21,66	34,90	50,115	68,142	88,171	109,203	133,236	159,271	186,309	216,348	247,390	280,434
$N_1 = N_2 + 23$	21,69	35,93	52,118	70,146	90,176	112,208	136,242	162,278	190,316	220,357	252,398	285,443
$N_1 = N_2 + 24$	22,71	37,95	53,122	72,150	92,181	115,213	139,248	166,284	194,323	224,364	257,406	291,451
$N_1 = N_2 + 25$	23,73	38,98	54,126	73,155	94,186	117,219	142,254	169,291	198,330	229,371	261,415	296,460

(Continued)

TABLE J

(Continued)

P = .05 one-sided, P = .10 two-sided

N_1	$N_2 = 15$	$N_2 = 16$	$N_2 = 17$	$N_2 = 18$	$N_2 = 19$	$N_2 = 20$	$N_2 = 21$	$N_1 = 22$	$N_2 = 23$	$N_2 = 24$	$N_2 = 25$
$N_1 = N_2$	192,273	220,308	249,346	280,386	314,427	349,471	386,517	424,566	465,618	508,668	552,723
$N_1 = N_2 + 1$	198,282	226,318	256,356	287,397	321,439	356,484	394,530	433,579	474,630	517,683	562,738
$N_1 = N_2 + 2$	203,292	232,328	262,367	294,408	328,451	364,496	402,543	442,592	483,644	527,697	572,753
$N_1 = N_2 + 3$	209,301	238,338	268,378	301,419	336,462	372,508	410,556	450,606	492,658	536,712	582,768
$N_1 = N_2 + 4$	215,310	244,348	275,388	308,430	343,474	380,520	418,569	459,619	501,672	546,726	592,783
$N_1 = N_2 + 5$	220,320	250,358	281,399	315,441	350,486	387,533	427,581	468,632	511,685	555,741	602,798
$N_1 = N_2 + 6$	226,329	256,368	288,409	322,452	358,497	395,545	435,594	476,646	520,699	565,755	612,813
$N_1 = N_2 + 7$	231,339	262,378	294,420	329,463	365,509	403,557	443,607	485,659	529,713	574,770	622,828
$N_1 = N_2 + 8$	237,348	268,388	301,430	336,474	372,521	411,569	451,620	494,672	538,727	584,784	632,843
$N_1 = N_2 + 9$	242,358	274,398	307,441	342,486	480,532	419,581	459,633	502,686	547,741	594,798	642,858
$N_1 = N_2 + 10$	248,367	280,408	314,451	349,497	387,544	426,594	468,645	511,699	556,744	603,813	652,873
$N_1 = N_2 + 11$	254,376	286,418	320,462	356,508	394,556	434,606	476,658	520,712	565,769	613,827	662,888
$N_1 = N_2 + 12$	259,386	292,428	327,472	363,519	402,567	442,618	484,671	528,726	574,783	622,842	672,903
$N_1 = N_2 + 13$	265,395	298,438	333,483	370,530	409,579	450,630	492,684	537,739	584,796	632,856	682,918
$N_1 = N_2 + 14$	270,405	304,448	340,493	377,541	416,591	458,642	501,696	546,752	593,810	642,870	692,933
$N_1 = N_2 + 15$	276,414	310,458	346,504	384,552	424,602	465,655	509,709	554,766	602,824	651,885	702,948
$N_1 = N_2 + 16$	282,423	316,468	353,514	391,563	431,614	473,667	517,722	563,779	611,838	661,899	712,963
$N_1 = N_2 + 17$	287,433	322,478	359,525	398,574	438,626	481,679	526,734	572,792	620,852	670,914	723,977
$N_1 = N_2 + 18$	293,442	328,488	366,535	405,585	446,637	489,691	534,747	581,805	629,866	680,928	733,992
$N_1 = N_2 + 19$	299,451	334,498	372,546	412,596	453,649	497,703	542,760	589,819	639,879	690,942	743,1007
$N_1 = N_2 + 20$	304,461	340,508	379,556	419,607	461,660	505,715	550,773	598,832	648,893	699,957	753,1022
$N_1 = N_2 + 21$	310,470	347,517	385,568	426,618	468,672	512,728	559,785	607,845	657,907	709,971	763,1037
$N_1 = N_2 + 22$	315,480	353,527	392,577	433,629	475,684	520,740	567,798	615,859	666,921	718,986	773,1052
$N_1 = N_2 + 23$	321,489	359,537	398,588	439,641	483,695	528,752	575,811	624,872	675,935	728,1000	783,1067
$N_1 = N_2 + 24$	327,498	365,547	405,598	446,652	490,707	536,764	583,824	633,885	684,949	738,1014	793,1082
$N_1 = N_2 + 25$	332,508	371,557	411,609	453,663	498,718	544,776	592,836	642,898	694,962	747,1027	803,1097

P = .025 one-sided, P = .05 two-sided

N_1	$N_2 = 3$	$N_2 = 4$	$N_2 = 5$	$N_2 = 6$	$N_2 = 7$	$N_2 = 8$	$N_2 = 9$	$N_1 = 10$	$N_2 = 11$	$N_2 = 12$	$N_2 = 13$	$N_2 = 14$
$N_1 = N_2$	5,16	11,25	18,37	26,52	37,68	49,87	63,108	79,131	96,157	116,184	137,214	160,246
$N_1 = N_2 + 1$	6,18	12,28	19,41	28,56	39,73	51,93	66,114	82,138	100,164	120,192	141,223	165,255
$N_1 = N_2 + 2$	6,21	12,32	20,45	29,61	41,78	54,98	69,121	85,145	103,172	124,200	146,231	170,264
$N_1 = N_2 + 3$	7,23	13,35	21,49	31,65	43,83	56,104	71,127	88,152	107,179	128,208	150,240	174,274
$N_1 = N_2 + 4$	7,26	14,38	22,53	32,70	45,88	58,110	74,133	91,159	110,187	131,217	154,249	179,283
$N_1 = N_2 + 5$	8,28	15,41	24,56	34,74	46,94	61,115	77,139	94,166	114,194	135,225	159,257	184,292
$N_1 = N_2 + 6$	8,31	16,44	25,60	36,78	48,99	63,121	79,146	97,173	118,201	139,233	163,266	189,301
$N_1 = N_2 + 7$	9,33	17,47	26,64	37,83	50,104	65,127	82,152	101,179	121,209	143,241	168,274	194,310
$N_1 = N_2 + 8$	10,35	17,51	27,68	39,87	52,109	68,132	85,158	104,186	125,216	147,249	172,283	198,320
$N_1 = N_2 + 9$	10,38	18,54	29,71	41,91	54,114	70,138	88,164	107,193	128,224	151,257	176,292	203,329
$N_1 = N_2 + 10$	11,40	19,57	30,75	42,96	56,119	72,144	90,171	110,200	132,231	155,265	181,300	208,338
$N_1 = N_2 + 11$	11,43	20,60	31,79	44,100	58,124	75,149	93,177	113,207	135,239	159,273	185,309	213,347
$N_1 = N_2 + 12$	12,45	21,63	32,83	45,105	60,129	77,155	96,183	117,213	139,246	163,281	190,317	218,356
$N_1 = N_2 + 13$	12,48	22,66	33,87	47,109	62,134	80,160	99,189	120,220	143,253	167,289	194,326	222,366
$N_1 = N_2 + 14$	13,50	23,69	35,90	49,113	64,139	82,166	101,196	123,227	146,261	171,297	198,335	227,375
$N_1 = N_2 + 15$	13,53	24,72	36,94	50,118	66,144	84,172	104,202	126,234	150,268	175,305	203,343	232,384
$N_1 = N_2 + 16$	14,55	24,76	37,98	52,122	68,149	87,177	107,208	129,241	153,276	179,313	207,352	237,393
$N_1 = N_2 + 17$	14,58	25,79	38,102	53,127	70,154	89,183	110,214	132,248	157,283	183,321	212,360	242,402
$N_1 = N_2 + 18$	15,60	26,82	40,105	55,131	72,159	92,188	113,220	136,254	161,290	187,329	216,369	247,411
$N_1 = N_2 + 19$	15,63	27,85	41,109	57,135	74,164	94,194	115,227	139,261	164,298	191,337	221,377	252,420
$N_1 = N_2 + 20$	16,65	28,88	42,113	58,140	76,169	96,200	118,233	142,268	168,305	195,345	225,386	256,430
$N_1 = N_2 + 21$	16,68	29,91	43,117	60,144	78,174	99,205	121,239	145,275	171,313	199,353	229,395	261,439
$N_1 = N_2 + 22$	17,70	30,94	45,120	61,149	80,179	101,211	124,245	148,282	175,320	203,361	234,403	266,448
$N_1 = N_2 + 23$	17,73	31,97	46,124	63,153	82,184	103,217	127,251	152,288	179,327	207,369	238,412	271,457
$N_1 = N_2 + 24$	18,75	31,101	47,128	65,157	84,189	106,222	129,258	155,295	182,335	211,377	243,420	276,466
$N_1 = N_2 + 25$	18,78	32,104	48,132	66,162	86,194	108,228	132,264	158,302	186,342	216,384	247,429	281,475

(Continued)

TABLE J

(Continued)

$P = .025$ one-sided, $P = .05$ two-sided

N_1	$N_2 = 15$	$N_2 = 16$	$N_2 = 17$	$N_2 = 18$	$N_2 = 19$	$N_2 = 20$	$N_2 = 21$	$N_1 = 22$	$N_2 = 23$	$N_2 = 24$	$N_2 = 25$
$N_1 = N_2$	185,280	212,316	240,355	271,395	303,438	337,483	373,530	411,579	451,630	493,683	536,739
$N_1 = N_2 + 1$	190,290	217,327	246,366	277,407	310,450	345,495	381,543	419,593	460,644	502,698	546,754
$N_1 = N_2 + 2$	195,300	223,337	252,377	284,418	317,462	352,508	389,556	428,606	468,659	511,713	555,770
$N_1 = N_2 + 3$	201,309	229,347	258,388	290,430	324,474	359,521	397,569	436,620	477,673	520,728	565,785
$N_1 = N_2 + 4$	206,319	234,358	264,399	297,441	331,486	367,533	404,583	444,634	486,687	529,743	574,801
$N_1 = N_2 + 5$	211,329	240,368	271,409	303,453	338,498	374,546	412,596	452,648	494,702	538,758	584,816
$N_1 = N_2 + 6$	216,339	245,379	277,420	310,464	345,510	381,559	420,609	460,662	503,716	547,773	593,832
$N_1 = N_2 + 7$	221,349	251,389	283,431	316,476	351,523	389,571	428,622	469,675	512,730	556,788	603,847
$N_1 = N_2 + 8$	227,358	257,399	289,442	323,487	358,535	396,584	436,635	477,689	520,745	565,803	612,863
$N_1 = N_2 + 9$	232,368	262,410	295,453	329,499	365,547	403,597	443,649	485,703	529,759	575,817	622,878
$N_1 = N_2 + 10$	237,378	268,420	301,464	336,510	372,559	411,609	451,662	493,717	538,773	584,832	632,893
$N_1 = N_2 2 + 11$	242,388	274,430	307,475	342,522	379,571	418,622	459,675	502,730	546,788	593,847	641,909
$N_1 = N_2 + 12$	248,397	279,441	313,486	349,533	386,583	426,634	467,688	510,744	555,802	602,862	651,924
$N_1 = N_2 + 13$	253,407	285,451	319,496	355,545	393,595	433,647	475,701	518,758	564,816	611,877	660,940
$N_1 = N_2 + 14$	258,417	291,461	325,508	362,556	400,607	440,660	482,715	526,772	572,831	620,892	670,955
$N_1 = N_2 + 15$	263,427	296,472	331,519	368,568	407,619	448,672	490,728	535,785	581,845	629,907	679,971
$N_1 = N_2 + 16$	269,436	302,482	338,529	375,579	414,631	455,685	498,741	543,799	590,859	638,922	689,986
$N_1 = N_2 + 17$	274,446	308,492	344,540	381,591	421,643	463,697	506,754	551,813	699,873	648,936	699,1001
$N_1 = N_2 + 18$	279,456	314,502	350,551	388,602	428,655	470,710	514,767	560,826	607,888	657,951	708,1017
$N_1 = N_2 + 19$	284,466	319,513	356,652	395,613	435,667	477,723	522,780	568,840	616,902	666,966	718,1032
$N_1 = N_2 + 20$	290,475	325,523	362,573	401,625	442,679	485,735	530,793	576,854	625,916	675,981	727,1048
$N_1 = N_2 + 21$	295,485	331,533	368,584	408,636	449,691	492,748	537,807	584,868	633,931	684,996	737,1063
$N_1 = N_2 + 22$	300,495	336,544	374,595	414,648	456,703	500,760	545,820	593,881	642,945	693,1011	747,1078
$N_1 = N_2 + 23$	306,504	342,554	374,595	414,648	456,703	500,760	545,820	593,881	642,945	693,1011	747,1078
$N_1 = N_2 + 24$	311,514	348,564	387,616	427,671	470,727	515,785	561,846	609,909	660,973	712,1040	766,1109
$N_1 = N_2 + 25$	316,524	353,575	393,629	434,682	477,739	522,798	569,859	618,922	668,988	721,1055	775,1125

P = .005 one-sided, P = .01 two-sided

N_1	$N_2 = 3$	$N_2 = 4$	$N_2 = 5$	$N_2 = 6$	$N_2 = 7$	$N_2 = 8$	$N_2 = 9$	$N_1 = 10$	$N_2 = 11$	$N_2 = 12$	$N_2 = 13$	$N_2 = 14$
$N_1 = N_2$	5,16	9,27	15,40	23,55	33,72	44,92	57,114	71,139	88,165	106,194	126,225	148,258
$N_1 = N_2 + 1$	5,19	10,30	16,44	24,60	34,78	46,98	59,121	74,146	91,173	109,203	130,234	152,268
$N_1 = N_2 + 2$	5,22	10,34	17,48	25,65	36,83	47,105	61,128	76,154	94,181	113,211	133,244	156,278
$N_1 = N_2 + 3$	5,25	11,37	18,52	27,69	37,89	49,111	63,135	79,161	97,189	116,220	137,253	160,288
$N_1 = N_2 + 4$	6,27	11,41	19,56	28,74	39,94	51,117	65,142	82,168	100,197	119,229	141,262	164,298
$N_1 = N_2 + 5$	6,30	12,44	19,61	29,79	40,100	53,123	68,148	84,176	102,206	123,273	144,272	168,308
$N_1 = N_2 + 6$	6,33	12,48	20,65	30,84	42,105	55,129	70,155	87,183	105,214	126,246	148,281	172,318
$N_1 = N_2 + 7$	6,36	13,51	21,69	31,89	43,111	57,135	72,162	89,191	108,222	129,255	152,290	176,328
$N_1 = N_2 + 8$	7,38	13,5	22,73	32,94	45,116	59,141	79,182	97,213	117,246	139,281	163,318	189,357
$N_1 = N_2 + 9$	7,41	14,58	23,77	34,98	46,122	61,147	77,175	95,205	114,238	136,272	159,309	185,347
$N_1 = N_2 + 10$	7,44	15,61	24,81	35,108	48,127	62,154	79,182	97,213	117,246	139,281	163,318	189,357
$N_1 = N_2 + 11$	8,46	15,65	25,85	36,108	49,133	64,160	81,189	100,220	120,254	143,289	167,327	193,367
$N_1 = N_2 + 12$	8,49	16,68	26,89	37,113	51,138	66,166	83,196	103,227	123,262	146,298	171,336	197,377
$N_1 = N_2 + 13$	8,52	16,72	26,94	38,118	52,144	68,172	86,202	105,235	126,270	150,306	175,345	201,387
$N_1 = N_2 + 14$	9,54	17,75	27,98	40,122	54,149	70,178	88,209	108,242	129,278	153,315	178,355	205,397
$N_1 = N_2 + 15$	9,57	17,79	28,102	41,127	55,155	72,184	90,216	110,250	132,286	156,324	182,364	210,406
$N_1 = N_2 + 16$	9,60	18,82	92,106	42,132	57,160	74,190	93,222	113,257	136,293	160,332	186,373	214,416
$N_1 = N_2 + 17$	9,63	19,85	30,110	43,137	59,165	76,196	95,229	116,264	139,301	163,341	190,382	218,426
$N_1 = N_2 + 18$	10,65	19,89	31,114	45,141	60,171	78,202	97,236	118,272	142,309	167,349	194,391	222,436
$N_1 = N_2 + 19$	10,68	20,92	32,118	46,146	62,176	80,208	99,243	121,279	145,317	170,358	197,401	226,446
$N_1 = N_2 + 20$	10,71	20,96	33,122	47,151	63,182	82,214	102,249	124,286	148,325	173,367	201,410	231,455
$N_1 = N_2 + 21$	11,73	21,99	33,127	48,156	65,187	83,221	104,256	126,294	151,333	177,375	205,419	235,465
$N_1 = N_2 + 22$	11,76	21,103	34,131	49,161	66,193	85,227	106,263	129,301	154,341	180,384	209,428	239,475
$N_1 = N_2 + 23$	11,79	22,106	35,135	51,165	68,198	87,233	109,269	132,308	157,349	184,392	213,437	243,485
$N_1 = N_2 + 24$	12,81	23,109	36,139	52,170	70,203	89,239	111,276	134,316	160,357	187,401	216,447	247,495
$N_1 = N_2 + 25$	12,84	23,113	37,143	53,175	71,209	91,245	113,283	173,323	163,365	191,409	220,456	252,504

(Continued)

TABLE J
(Continued)

P = .005 one-sided, P = .01 two-sided

N₁	N₂ = 15	N₂ = 16	N₂ = 17	N₂ = 18	N₂ = 19	N₂ = 20	N₂ = 21	N₁ = 22	N₂ = 23	N₂ = 24	N₂ = 25
N₁ = N₂	171,294	196,332	223,372	252,414	283,458	316,504	250,553	386,604	424,657	464,712	506,679
N₁ = N₂ + 1	176,304	201,343	229,383	258,426	289,471	322,518	357,567	393,619	432,672	472,728	514,786
N₁ = N₂ + 2	180,315	206,354	234,395	264,438	295,484	329,531	364,581	401,633	440,687	480,744	532,802
N₁ = N₂ + 3	184,326	211,365	239,407	269,451	301,497	335,545	371,595	408,648	447,703	489,759	531,819
N₁ = N₂ + 4	189,336	216,376	245,418	275,463	307,510	342,558	378,609	415,663	455,718	497,775	540,835
N₁ = N₂ + 5	194,346	221,387	250,430	281,475	314,522	348,572	385,623	423,677	463,733	505,791	549,851
N₁ = N₂ + 6	198,357	226,398	255,442	287,487	320,535	355,585	392,637	430,692	471,748	513,807	557,868
N₁ = N₂ + 7	203,367	231,409	261,453	292,500	326,548	361,599	399,651	438,706	479,763	521,823	566,884
N₁ = N₂ + 8	207,378	236,420	266,465	298,512	332,561	368,612	405,666	445,721	486,779	530,838	575,900
N₁ = N₂ + 9	212,388	241,431	271,477	304,524	338,574	374,626	412,680	452,736	494,794	538,854	583,917
N₁ = N₂ + 10	216,399	245,443	277,488	310,536	34,587	381,639	419,694	460,750	502,890	546,870	592,933
N₁ = N₂ + 11	221,409	250,454	282,500	315,549	351,599	388,652	426,708	467,765	510,824	554,886	601,949
N₁ = N₂ + 12	225,420	255,465	287,512	321,561	357,612	394,666	433,722	475,779	518,839	563,901	609,966
N₁ = N₂ + 13	230,430	260,476	293,623	327,573	363,625	401,679	440,736	482,794	526,854	571,917	618,982
N₁ = N₂ + 14	235,440	265,487	298,535	333,585	369,638	407,693	447,750	490,808	533,870	579,933	627,998
N₁ = N₂ + 15	239,451	270,498	303,547	338,598	375,651	414,706	454,764	497,823	541,885	587,949	635,1015
N₁ = N₂ + 16	244,461	275,509	309,558	344,610	381,664	421,719	462,777	504,838	549,900	596,964	644,1031
N₁ = N₂ + 17	248,472	280,520	314,570	350,622	388,676	427,733	469,791	512,852	557,915	604,980	653,1047
N₁ = N₂ + 18	253,482	285,531	320,581	356,634	394,689	434,746	476,805	519,867	585,930	612,996	661,1064
N₁ = N₂ + 19	257,493	290,542	325,593	362,646	400,702	440,760	483,819	527,881	573,945	620,1012	670,1080
N₁ = N₂ + 20	262,503	295,553	330,605	367,659	406,715	447,773	490,833	534,896	480,961	629,1027	679,1096
N₁ = N₂ + 21	267,513	300,564	336,616	373,671	413,727	454,786	497,847	542,910	588,976	637,1043	687,1113
N₁ = N₂ + 22	271,524	305,575	341,628	379,683	419,740	460,800	504,861	549,925	596,991	645,1059	696,1129
N₁ = N₂ + 23	276,534	310,586	347,639	385,695	425,753	467,813	511,875	556,940	604,1006	654,1074	705,1145
N₁ = N₂ + 24	280,545	315,597	352,651	391,707	431,766	474,826	518,889	564,954	612,1021	662,1090	714,1161
N₁ = N₂ + 25	285,555	320,608	357,663	397,719	438,778	480,840	525,903	571,969	620,1106	670,1106	722,1178

P = .01 one-sided, P = .02 two-sided

N_1	$N_2 = 3$	$N_2 = 4$	$N_2 = 5$	$N_2 = 6$	$N_2 = 7$	$N_2 = 8$	$N_2 = 9$	$N_1 = 10$	$N_2 = 11$	$N_2 = 12$	$N_2 = 13$	$N_2 = 14$
$N_1 = N_2$	5,16	10,26	16,39	24,54	34,71	46,90	59,112	74,136	91,162	110,190	130,221	153,253
$N_1 = N_2 + 1$	5,19	10,30	17,43	26,58	36,76	48,96	62,118	77,143	94,170	113,199	134,230	157,263
$N_1 = N_2 + 2$	6,21	11,33	18,47	27,63	38,81	50,102	64,125	80,150	97,178	117,207	138,239	161,273
$N_1 = N_2 + 3$	6,24	12,36	19,51	28,68	39,87	52,108	66,132	83,157	101,185	120,216	142,248	166,282
$N_1 = N_2 + 4$	6,27	12,40	20,55	30,72	41,92	54,114	69,138	85,165	104,193	124,224	146,257	170,292
$N_1 = N_2 + 5$	7,29	13,43	21,59	31,77	43,97	56,120	71,145	88,172	107,201	128,232	150,266	174,302
$N_1 = N_2 + 6$	7,32	14,46	22,63	32,82	44,103	58,126	74,151	91,179	110,209	131,241	154,275	179,311
$N_1 = N_2 + 7$	7,35	14,50	23,67	34,86	46,108	60,132	76,158	94,186	113,217	135,249	158,284	183,321
$N_1 = N_2 + 8$	8,37	15,53	24,71	35,91	48,113	62,138	79,164	97,193	117,224	138,258	162,293	188,330
$N_1 = N_2 + 9$	8,40	16,56	25,75	36,96	49,119	64,144	81,171	100,200	120,232	142,266	166,302	192,340
$N_1 = N_2 + 10$	9,42	16,60	26,79	38,100	51,124	66,150	83,178	102,208	123,240	146,274	170,311	196,350
$N_1 = N_2 + 11$	9,45	17,63	27,83	39,105	53,129	68,156	86,184	105,215	126,248	149,283	174,320	201,359
$N_1 = N_2 + 12$	9,48	18,66	28,87	40,110	55,134	71,161	88,191	108,222	130,255	153,291	178,329	205,369
$N_1 = N_2 + 13$	10,50	18,70	29,91	42,114	56,140	73,167	91,197	111,229	133,263	157,299	182,338	210,378
$N_1 = N_2 + 14$	10,53	19,73	30,95	43,119	58,145	75,173	93,204	114,236	136,271	160,308	186,347	214,388
$N_1 = N_2 + 15$	10,56	20,76	31,99	45,123	60,150	77,179	96,210	117,243	139,279	164,316	190,356	219,397
$N_1 = N_2 + 16$	11,58	20,80	32,103	56,128	61,156	79,185	98,217	120,250	143,286	168,324	194,365	223,407
$N_1 = N_2 + 17$	11,61	21,83	33,107	47,133	63,161	81,191	101,223	122,258	146,294	171,333	198,374	228,416
$N_1 = N_2 + 18$	12,63	22,86	34,111	49,137	65,166	83,197	103,230	125,265	149,302	175,341	203,382	232,426
$N_1 = N_2 + 19$	12,66	23,89	35,115	50,142	67,171	85,203	106,236	128,272	152,310	179,349	207,391	236,436
$N_1 = N_2 + 20$	12,69	23,93	36,119	1,147	68,177	87,177	108,243	131,279	156,317	182,358	211,400	241,445
$N_1 = N_2 + 21$	13,71	24,96	37,123	53,151	70,182	90,214	111,249	134,286	159,325	186,366	215,409	245,455
$N_1 = N_2 + 22$	13,74	25,99	39,127	54,156	72,187	92,220	113,256	137,293	162,333	190,374	219,418	250,464
$N_1 = N_2 + 23$	14,76	25,103	39,131	56,160	74,192	94,226	116,262	140,300	165,341	193,383	223,427	254,474
$N_1 = N_2 + 24$	14,79	26,106	40,135	57,165	75,198	96,232	118,269	143,307	169,348	197,391	227,436	259,483
$N_1 = N_2 + 25$	14,82	27,109	41,139	58,170	77,203	98,238	121,275	145,315	172,356	201,399	231,445	263,493

(Continued)

TABLE J
(Continued)

P = .01 one-sided, P = .02 two-sided

N_1	$N_2 = 15$	$N_2 = 16$	$N_2 = 17$	$N_2 = 18$	$N_2 = 19$	$N_2 = 20$	$N_2 = 21$	$N_1 = 22$	$N_2 = 23$	$N_2 = 24$	$N_2 = 25$
$N_1 = N_2$	177,288	202,326	230,365	260,406	291,450	324,496	359,544	396,594	435,646	476,700	518,757
$N_1 = N_2 + 1$	181,299	208,336	235,376	266,418	297,463	331,509	367,557	404,608	443,661	484,716	527,773
$N_1 = N_2 + 2$	186,309	213,347	241,388	272,430	304,475	338,522	374,571	412,622	451,676	493,731	536,789
$N_1 = N_2 + 3$	191,319	218,358	247,399	278,442	310,488	345,535	381,585	419,637	459,691	501,747	545,805
$N_1 = N_2 + 4$	196,329	223,369	253,410	284,454	317,500	352,548	388,599	427,651	467,706	510,762	554,821
$N_1 = N_2 + 5$	200,340	228,380	258,422	290,466	323,513	359,561	396,612	435,665	476,720	518,778	563,837
$N_1 = N_2 + 6$	205,350	234,390	264,433	296,478	330,525	365,575	403,626	442,680	484,735	527,793	572,853
$N_1 = N_2 + 7$	210,360	239,401	269,445	302,490	336,538	372,588	410,640	450,694	492,750	535,809	581,869
$N_1 = N_2 + 8$	215,370	244,412	275,456	308,502	343,550	379,601	418,653	458,708	500,765	544,824	590,885
$N_1 = N_2 + 9$	220,380	249,423	281,457	314,514	349,563	386,614	425,667	466,722	508,780	553,839	599,901
$N_1 = N_2 + 10$	225,390	255,433	286,479	320,526	356,575	393,627	432,681	473,737	516,795	561,855	608,917
$N_1 = N_2 + 11$	229,401	260,444	292,490	326,538	362,588	400,640	440,694	481,751	524,810	570,870	617,933
$N_1 = N_2 + 12$	234,411	265,455	298,501	332,550	369,600	407,653	447,708	489,765	533,824	578,886	626,949
$N_1 = N_2 + 13$	239,421	270,466	303,513	338,562	375,613	414,666	454,722	497,779	541,839	587,901	635,965
$N_1 = N_2 + 14$	244,431	276,476	309,524	344,574	382,625	421,679	462,735	504,794	549,854	596,916	644,981
$N_1 = N_2 + 15$	249,441	281,487	315,535	350,586	388,638	428,692	469,749	512,808	557,869	604,932	653,997
$N_1 = N_2 + 16$	254,451	286,498	320,547	357,597	395,650	434,706	476,763	520,822	565,884	613,947	662,1013
$N_1 = N_2 + 17$	259,461	291,509	326,558	363,609	401,663	441,719	484,776	528,836	574,898	621,963	671,1029
$N_1 = N_2 + 18$	263,472	297,519	332,569	369,621	408,675	448,732	491,790	535,851	582,913	630,978	680,1045
$N_1 = N_2 + 19$	268,482	302,530	337,581	375,633	414,688	455,745	498,804	543,865	590,928	639,993	689,1061
$N_1 = N_2 + 20$	273,492	307,541	343,592	381,645	421,700	462,758	506,817	551,879	598,943	647,1009	698,1077
$N_1 = N_2 + 21$	278,502	312,552	349,603	387,657	427,713	469,771	513,831	559,893	606,958	656,1024	707,1093
$N_1 = N_2 + 22$	283,512	318,562	355,614	393,669	434,725	476,784	520,845	567,907	615,972	665,1039	716,1109
$N_1 = N_2 + 23$	288,522	323,573	361,626	399,681	440,738	483,797	528,858	574,922	623,987	673,1055	726,1124
$N_1 = N_2 + 24$	293,532	328,584	366,637	405,693	447,750	490,810	535,872	582,936	631,1002	682,1070	735,1140
$N_1 = N_2 + 25$	298,542	334,594	372,648	412,704	453,763	497,823	543,885	590,950	639,1017	691,1085	744,1156

Appendix 2

Definition of Terms

Accuracy—The degree to which a measurement reflects the true value of a variable.

Assumptions—Accepted properties of populations, samples, or of the variables that make them up. Every test or technique has a set of underlying assumptions without which it is not completely valid. Robust tests suffer less from violations of assumptions.

Continuous—Data that can be considered as points on an unbroken line. Meters, minutes, and kilograms are generally measured so that they are examples of continuous data.

Degree of freedom (df)—Choices or freedom to vary. If testing one group, the dfs are $N - 1$. If two groups, it is $N - 2$. Each additional group "costs" another df.

Dependent variable—Response or predicted value (such as the number of animals dying given a certain dose).

Discontinuous—Data based on measurements that can only be expressed as whole numbers or units, such as numbers of animals. Also called discrete data.

False negative—When there is an actual effect (difference between groups) but it is statistically judged not to be significant. See type II error.

Frequency distribution—The manner in which data points are distributed in accordance with their values. A histogram is a plot of a frequency distribution. The Gaussian curve is a plot of the frequency distribution of a normal distribution.

Homoscedasticity—The condition when the variability of the data of two or more groups, expressed as standard deviations, is equivalent.

Hypothesis—A statement of the proposed relationship between two sets of variables. Usually presented as the null hypothesis. H0, which states that there is no relationship.

Independence—An assumption basic to many statistical techniques and to experimental design: that the measurement of any one datum (or collection of any sample) does not affect or is not affected by that of another sample.

Independent variable—Treatment or predictor variables.

Kurtosis—Peakedness. A leptokurtic curve is very thin and pointed. A normal curve is mesokurtic. A platykurtic curve is flattened.

Mean—Usually refers to the arithmetic mean. The average computed as the sum of values divided by the total number of values.

Measurement scales—Quantal/nominal: classified as one sort or another such as dead or alive. Ordinal: numbers reflect the rank order of the values but not the distance betweeen values. Numerical/ interval: each value is by definition a precise degree greater or lesser than every other value.

Median—The middle value in a sample or population. Exactly half the values are greater and half are less.

Mode—The most common value in a sample or population; the value that occurs the most often.

Multivariate—The study of one or more independent and one or more dependent variables and the relationships between them.

N—The number of cases in a sample, such as the number of animals in a treatment group.

Normal—The most common frequency distribution, which describes such variables as heights, body weights, and red blood cell counts. The distribution is such that 68% of the population is within ±1.0 standard deviation of the mean and 95% of the population is within ±1.96 standard deviation of the mean.

Outlier—An extreme value that seemingly is not part of the data set (in comparison to the other data).

Parameters—Values that describe characteristics of a population: mean, standard deviation, variance, proportion, number of cases.

Population—All the possible values in a set, such as all second graders in the United States.

Power—The probability of rejecting the null hypothesis when it is actually false. This is equal to 1.0 minus the probability of making a type II error, or $1 - a$.

Precision—The reproducibility of measurement of a variable.

Quartile—A portion of a sample or population that contains one quarter of all values.

Range—The entire expanse of values in a sample or population; the interval that includes all possible values.

Rank—To arrange all the values in a data set from lowest to highest (or from highest to lowest). Once a data set is so arranged, then the order in which values occur in the set (first, second, third, etc.) is the value's assigned rank (1, 2, 3, etc.).

Residual—data – fit.

Residuals—What remains after a summary or fitted model has been subtracted out of the data according to the equation.

Resistance—Insensitivity to localized "misbehavior" in data. Resistant methods pay much attention to the main body of the data and little to outliers. The median is a resistant statistic, whereas the sample mean is not.

Robustness—Generally implies insensitivity to departures from the assumptions underlying the use of a method.

Rounding—Reducing the number of digits in a number in such a manner that the information present in those digits that are dropped is reflected in a modification of the last digit that is retained.

Sample—A subset of a population on which data actually are collected; usually designed to be randomly collected and therefore unbiased.

Semi-quartile distance—That portion of the range that contains the middle two quarters of all values.

Sensitivity—Ability to detect relationships between variables. Must strike a balance between the levels of different types of errors it will accept.

Significance—A measure of the probability (p) that an occurrence is due to random chance. An acceptable level of significance is usually $p \geq 0.05$; that is, by random chance such a set of events would occur no more than 5% of the time.

Skewness—Tendency of a frequency distribution to be to one side or other of the mean, to the left or the right.

Standard deviation—A measure of the dispersion of data in a normal population or sample. It is calculated as the square root of $x - 2/ N - 1$.

Type I error—Rejecting the null hypothesis when it is, in fact, true. Probability of this is called the alpha level. Also called a false positive.

Type II error—Not rejecting the null hypothesis when it is false. The probability of this is called the beta level. Also called a false negative.

Univariate—The study of one independent and one dependent variable and the relationship between these.

Appendix 3

Greek Alphabet, Abbreviations, and Symbols

Greek Alphabet

Greek Character		Greek Name	Use in Statistics
α	Aa	Alpha	Type I error; false-negative rate
β	Bb	Beta	Type II error; false-positive rate
γ	Gg	Gamma	Number of "hits" in risk assessment needed
δ	Dd	Delta	Rate of change
ε	Ee	Epsilon	
ζ	Zz	Zeta	
η	Hh	Eta	
θ	qt	Theta	
ι	Ii	Iota	
κ	Kk	Kappa	
λ	Ll	Lambda	
μ	Mm	Mu	Lowercase letter: arithmetic mean of population; also micro or 1×10^{-6}
ν	Nn	Nu	
Ξ	Xx	Xi	
	Oo	Omicron	
π	Pp	Pi	Constant $= 3.14159...$
ρ	Rr	Rho	
Σ	Ss	Sigma	Capital letter: summation; lowercase: standard deviation
τ	Tt	Tau	
υ	Uu	Upsilon	
φ	Vv	Phi	
χ	Cc	Chi	
ψ	Yy	Psi	
ω	Ww	Omega	

Mathematical Symbols

Symbol	Meaning
∞	Infinity
lim	Limiting value
$a \sim b$	a approximately equal to b
$a > b$	a greater than b
$a < b$	a smaller than b
$a \neq b$	a not equal to b
$\lvert a \rvert$	Absolute value of a; this is always positive
$\sum_{i\,xi}^{k}$	Sum of all values $x_1, x_2, x_3, \ldots$, of all values of x_1 from $i = 1$ to $i = k$ inclusive, or $\sum_{i\,xi}^{k} = x_1 + x_2 + x_3 + \cdots + x_k$
$\int$	Indefinite integral
$\int_{a}^{b}$	Definite integral, or integral between $x = a$ and $x = b$
$a^b = c$	ab, read as "a to the power b," known as involution, a is the base, b the exponent; ab, or c, is the bth power of a
$^b a = c$	$^b a$ is the bth root of a, b being known as the root exponent; in the special case of $^2 a = c$, a or c is known as the square root of a, and the root exponent is omitted
log	Base ten logarithm
ln	Natural logarithm
$x!$	x factorial, equal to $x\,(x-1)(x-2)\ldots(3)\,(2)\,(1)$
x	Arithmetic mean of sample
xg	Geometric mean of sample

Abbreviations

ANCOVA	Analysis of covariance
ANOVA	Analysis of variance
CV	Coefficient of variation
df	Degree of freedom
EDA	Exploratory data analysis
N	Number of data in group
Q	Quartile
QD	Semi-quartile distance (also called quartile deviation)
SAR	Structure–activity relationship
SD	Standard deviation
SEM	Standard error of measurement

Appendix 4

Practice Problems and Solutions

Problem 1: Mean and Standard Deviation

Body weights were collected from five groups of male rats in an acute oral toxicity study. These weights were:

Group A	Group B	Group C	Group D	Group E
204.7	196.7	208.9	206.5	205.9
198.6	207.9	204.7	202.3	187.6
210.3	204.5	204.1	208.7	182.4
207.4	210.2	192.3	191.1	192.5
202.9	203.9	206.7	189.6	203.7

Calculate the mean and standard deviation for each of these groups and for the entire set of male rats.

Solution:

	Group A	Group B	Group C	Group D	Group E	Combined
Mean	204.78	204.64	203.34	199.64	194.42	201.36
SD	4.446	5.127	6.452	8.803	10.156	7.806

Problem 2: Mean and Standard Deviation

During the course of a chronic study, a portion of the female animals from four dose groups were necropsied and their spleens were weighed. These weights were:

Group A	Group B	Group C	Group D
2.14	1.97	1.92	1.87
2.12	2.14	2.14	1.74

2.21	2.16	1.85	1.98
2.11	1.87	1.92	1.89
1.99	2.17	2.17	2.01
2.05	2.04	1.97	2.03
2.04	2.01	1.89	1.95
2.00	2.06	2.05	1.86
1.97	2.19		
2.08	2.03		

Calculate the mean and standard deviation for each of these groups and for the entire set of spleens.

Solution:

	Group A	Group B	Group C	Group D	Combined
Mean	2.071	2.064	1.989	1.916	2.016
SD	0.07549	0.10178	0.11850	0.09561	0.11284

Problem 3: Mean and Standard Deviation

In a cytotoxicity assay, the number of viable cells in samples from each of three treatment groups was counted. These cell counts were:

Control	Treatment A	Treatment B
362	318	312
258	247	287
347	219	317
319	308	321
297	251	374
278	293	285
321	286	326

Calculate the mean and standard deviation for each of these groups and for the combined set of 21 samples.

Solution:

	Control	Treatment A	Treatment B	Combined
Mean	311.7	274.6	317.4	301.2
SD	36.86	36.23	29.70	38.02

Problem 4: Mean and Standard Deviation

As part of a metabolism study, urine samples were collected from a group of animals at different times and the concentration of metabolite present in each sample was determined. These concentrations were:

Animal	Hour 1	Hour 2	Hour 3	Hour 4
1	18	47	104	59
2	60	59	98	48
3	19	46	105	64
4	34	55	118	84
5	12	31	107	51
6	26	56	121	69
7	21	38	111	67
8	29	61	83	63
9	28	63	116	56
10	23	64	112	72

Calculate the mean and standard deviation for each of the intervals and for each animal across the sampling period.

Solution:

Animal	Mean	SD
1	57	35.749
2	66.25	21.854
3	58.5	36.097
4	72.75	36.473
5	50.25	41.048
6	68	39.657
7	59.25	39.382
8	59	22.331
9	65.75	36.755
10	67.75	36.482

	Hour 1	Hour 2	Hour 3	Hour 4
Mean	27	52	107.5	63.3
SD	13.19	11.15	11.09	10.58

Problem 5: Mean and Standard Deviation

Body weights of dogs were collected predose and for each of five days during an acute toxicity test. These weights were:

Animal	Predose	Day 1	Day 2	Day 3	Day 4	Day 5
1	10.4	10.1	10.3	10.0	9.8	10.4
2	9.1	9.4	9.3	9.5	8.9	8.7
3	9.9	9.9	10.3	10.7	9.8	9.6
4	10.8	10.5	10.8	10.3	10.0	10.1
5	9.4	9.3	9.5	9.4	9.3	9.6
6	8.6	8.5	8.3	8.0	8.0	7.7
7	9.7	10.2	10.0	10.0	9.7	9.9
8	11.3	11.2	11.0	10.8	11.2	11.3

Calculate the mean and standard deviation for each day and also for each animal across the course of the study.

Solution:

Animal	Mean	SD
1	10.17	0.242
2	9.15	0.308
3	10.03	0.398
4	10.42	0.343
5	9.42	0.117
6	8.18	0.343
7	9.92	0.194
8	11.13	0.197

	Predose	Day 1	Day 2	Day 3	Day 4	Day 5
Mean	9.90	9.89	9.94	9.84	9.59	9.66
SD	0.90	0.82	0.88	0.90	0.92	1.09

Problem 6: Mean and Standard Deviation

In a reproductive study, the numbers of live births per litter were recorded for three sets of rats: one control and one each with two different compounds. These birth counts were:

Control	Compound A	Compound B
7	7	9
12	7	10
8	6	13
11	4	12
10	8	16
8	8	6
9	9	11
9	5	12

Calculate the mean and standard deviation of the litter size for the control and each compound.

Solution:

	Control	Compound A	Compound B
Mean	9.25	6.75	11.13
SD	1.669	1.669	2.949

Problem 7: Mean and Standard Deviation

For each of the following days, calculate the mean and standard deviation of the food eaten.

Animal	Predose (Day 0)	Dose Day (Day 1)	Washout Day (Day 2)	Washout Day (Day 3)
1	3.8	1.9	1.7	4.2
2	4.6	0.8	2.3	4.5
3	4.5	3.4	3.6	4.3
4	2.6	1.6	3.3	5.2
5	2.1	1.4	2.4	3.6
6	3.4	2.3	3.2	4.3
7	5.6	0.7	4.1	5.8
8	4.3	1.0	2.4	4.7
9	4.0	0.7	3.0	3.2
10	3.4	1.8	2.9	3.9

Solution:

	Day 0	Day 1	Day 2	Day 3
Mean	3.83	1.56	2.89	4.37
SD	1.02	0.85	0.71	0.75

Problem 8: Mean and Standard Deviation

Calculate the mean and standard deviation of the following dose per body weight values:

Group A	Group B	Group C	Group D	Group E
2.73	2.62	2.79	2.75	2.75
2.65	2.77	2.73	2.70	2.50
2.80	2.73	2.72	2.78	2.43
2.77	2.80	2.56	2.55	2.57
2.71	2.72	2.76	2.53	2.72

Solution:

	Group A	Group B	Group C	Group D	Group E
Mean	2.730	2.729	2.711	2.662	2.592
SD	0.059	0.068	0.086	0.117	0.135

Problem 9: Standard Error of the Mean

Using the data presented below on spleen weights, calculate the standard error of the mean for each group and for the entire data set.

Group A	Group B	Group C	Group D
2.14	1.97	1.92	1.87
2.12	2.14	2.14	1.74
2.21	2.16	1.85	1.98
2.11	1.87	1.92	1.89
1.99	2.17	2.17	2.01
2.05	2.04	1.97	2.03
2.04	2.01	1.89	1.95
2.00	2.06	2.05	1.86
1.97	2.19		
2.08	2.03		

Solution:

	Group A	Group B	Group C	Group D	Combined
SEM	0.02387	0.03219	0.04189	0.03380	0.01881

Problem 10: Standard Error of the Mean

Using the data presented below, calculate the standard error of the mean for each interval and for each animal. Compare these SEMs with the SDs for the same sets of data.

Animal	Hour 1	Hour 2	Hour 3	Hour 4
1	18	47	104	59
2	60	59	98	48
3	19	46	105	64
4	34	55	118	84
5	12	31	107	51
6	26	56	121	69
7	21	38	111	67
8	29	61	83	63
9	28	63	116	56
10	23	64	112	72

Solution:

Animal	SEM	SD
1	17.87	35.749
2	10.93	21.854
3	18.05	36.097
4	18.24	36.473
5	20.52	41.048
6	19.83	39.657
7	19.69	39.382
8	11.17	22.331
9	18.38	36.755
10	18.24	36.482

	Hour 1	Hour 2	Hour 3	Hour 4
SEM	4.17	3.52	3.51	3.35
SD	13.19	11.15	11.09	10.58

Problem 11: Standard Error of the Mean

Using the data presented below, calculate the standard error of the mean for each group and for the entire data set.

Control	Treatment A	Treatment B
362	318	312
258	247	287
347	219	317
319	308	321
297	251	374
278	293	285
321	286	326

Solution:

	Control	Treatment A	Treatment B	Combined
SEM	13.93	13.70	11.23	8.30

Problem 12: Standard Error of the Mean

Using the data presented in Problem 5, calculate the standard error of the mean for each day and for each animal. Compare these SEMs with the SDs for the same sets of data.

Solution:

Animal	SEM	SD
1	0.10	0.242
2	0.13	0.308
3	0.16	0.398
4	0.14	0.343
5	0.05	0.117
6	0.14	0.343
7	0.08	0.194
8	0.08	0.197

	Predose	Day 1	Day 2	Day 3	Day 4	Day 5
SEM	0.32	0.29	0.31	0.32	0.33	0.38
SD	0.90	0.82	0.88	0.90	0.92	1.09

Problem 13: Median and Semi-Quartile Distance

For each of the following sets of data, determine the median and the semi-quartile distance.

Set A: 1, 13, 47, 26, 63, 34, 3, 18, 29, 9, 31, and 39.
Set B: 53, 9, 4, 18, 25, 30, 29, 15, 41, 23, 33, 15, and 31.
Set C: 16, 43, 51, 86, 24, 26, 31, 91, 59, 68, 53, 47, 29, 46, 5, and 55.
Set D: 4, 11, 17, 19, 13, 27, 15, 6, 22, 7, 18, 23, and 15.

Solution:

Set	Lower	Median	Upper	Semi-Quartile Distance
A	10	27.5	37.75	13.875
B	15	25	32	8.5
C	26.75	46.5	58	15.625
D	9	15	20.5	5.75

Problem 14: Median and Semi-Quartile Distance

In performing an evaluation of the neurobehavioral effects of a series of solvents, righting reflex scores were obtained and recorded for three separate groups of animals. These values are presented below:

Control	Low Dose	High Dose
0	3	3
0	1	7
1	0	4
0	3	5
0	2	8
2	1	3
1	3	4
0	2	5
0	0	8
0	1	4

Determine the median righting reflex scores and semi-quartile distances for the test groups.

Solution:

Set	Lower	Median	Upper	Semi-Quartile Distance
Control	0	0	1	0.5
Low dose	0.75	1.5	3	1.125
High dose	3.75	4.5	7.25	1.75

Problem 15: Median and Semi-Quartile Distance

Using the data presented below on rat litter sizes, calculate median litter sizes and semi-quartile distances for the test groups.

Control	Compound A	Compound B
7	7	9
12	7	10
8	6	13
11	4	12
10	8	16
8	8	6
9	9	11
9	5	12

Solution:

Set	Lower	Median	Upper	Semi-Quartile Distance
Control	8	9	10.75	1.375
Compound A	5.25	7	8	1.375
Compound B	9.25	11.5	12.75	1.75

Problem 16: Median and Semi-Quartile Distance

Using the following mortality data, calculate the median time to death and the semi-quartile distances for both groups.

Hours to Death (Group A)		Hours to Death (Group B)	
4	3	7	4
6	6	3	3
7	1	6	1
7	7	7	2
2	7	2	5
5	4	5	4
7	6		
7	3		

Solution:

Group	Lower	Median	Upper	Semi-Quartile Distance
A	3.25	6	7	1.875
B	2.25	4	9.75	3.75

Problem 17: Geometric Mean

Using the data presented below, calculate both the arithmetic and geometric means for each of the four data sets.

Set A: 1, 13, 47, 26, 63, 34, 3, 18, 29, 9, 31, and 39.
Set B: 53, 9, 4, 18, 25, 30, 29, 15, 41, 23, 33, 15, and 31.
Set C: 16, 43, 51, 86, 24, 26, 31, 91, 59, 68, 53, 47, 29, 46, 5, and 55.
Set D: 4, 11, 17, 19, 13, 27, 15, 6, 22, 7, 18, 23, and 15.

Solution:

	Set A	Set B	Set C	Set D
Mean	26.08	25.08	45.63	15.15
Geometric mean	16.99	21.08	38.07	13.35

Problem 18: Geometric Mean

As part of an inhalation study, samples of dust particles were collected from three different generation systems. The aerodynamic diameters of portions of each sample were determined and are presented below:

Generator A	Generator B	Generator C
18	4	1
250	1	7
164	29	6
142	19	9
187	16	9
129	23	1
178	26	5
225	11	8
453	18	12
46	37	39

Calculate the arithmetic and geometric means of particle diameters for each of three generation systems.

Solution:

	Generator A	Generator B	Generator C
Mean	179.20	18.40	9.70
G. Mean	135.65	13.15	6.03

Problem 19: Geometric Mean

Using the data presented below, calculate the geometric mean for each animal and for each dosing day.

Animal	Predose (Day 0)	Dose Day (Day 1)	Washout Day (Day 2)	Washout Day (Day 3)
1	3.8	1.9	1.7	4.2
2	4.6	0.8	2.3	4.5
3	4.5	3.4	3.6	4.3
4	2.6	1.6	3.3	5.2
5	2.1	1.4	2.4	3.6
6	3.4	2.3	3.2	4.3
7	5.6	0.7	4.1	5.8
8	4.3	1.0	2.4	4.7
9	4.0	0.7	3.0	3.2
10	3.4	1.8	2.9	3.9

Solution:

Animal	Geometric Mean
1	2.68
2	2.48
3	3.92
4	2.91
5	2.24
6	3.22
7	3.11
8	2.64
9	2.28
10	2.88

	Predose (Day 0)	Dose Day (Day 1)	Washout Day (Day 2)	Washout Day (Day 3)
Geometric Mean	3.70	1.37	2.81	4.31

Problem 20: Geometric Mean

Particle sizes were collected from dust taken in three different areas of a manufacturing plant. For each area, calculate the mean and geometric mean sizes.

Area A	Area B	Area C
3.4	3.4	2.3
0.9	249.5	13.9
19.8	79.0	11.4
19.0	131.7	18.6
9.3	154.3	17.4
14.6	105.9	1.5
37.2	151.5	12.1
12.1	20.0	15.0
14.9	183.8	24.5
44.7	19.0	59.9
12.8		3.4
20.5		9.9
18.6		

Solution:

	Area A	Area B	Area C
Mean	17.52	109.81	15.83
Geometric Mean	12.68	65.81	10.40

Problem 21: Coefficient of Variation

Using the data presented below, calculate the coefficient of variation for the five groups of animal weights.

Group A	Group B	Group C	Group D	Group E
204.7	196.7	208.9	206.5	205.9
198.6	207.9	204.7	202.3	187.6
210.3	204.5	204.1	208.7	182.4
207.4	210.2	192.3	191.1	192.5
202.9	203.9	206.7	189.6	203.7

Solution:

	Group A	Group B	Group C	Group D	Group E
CV	0.02	0.03	0.03	0.04	0.05

Problem 22: Coefficient of Variation

Using the data presented below, calculate the coefficient of variation for the particle size measurements from each of the three generators.

Generator A	Generator B	Generator C
18	4	1
250	1	7
164	29	6
142	19	9
187	16	9
129	23	1
178	26	5
225	11	8
453	18	12
46	37	39

Solution:

	Generator A	Generator B	Generator C
CV	0.67	0.60	1.12

Problem 23: Coefficient of Variation

Using the data presented below, calculate the coefficient of variation for the particle size measurements from each of the three generators.

Area A	Area B	Area C
3.4	3.4	2.3
0.9	249.5	13.9
19.8	79.0	11.4
19.0	131.7	18.6
9.3	154.3	17.4
14.6	105.9	1.5
37.2	151.5	12.1
12.1	20.0	15.0
14.9	183.8	24.5
44.7	19.0	59.9
12.8		3.4
20.5		9.9
18.6		

Solution:

	Area A	Area B	Area C
CV	0.69	0.73	0.98

Problem 24: Coefficient of Variation

Calculate the coefficients of variation for the following hematology values, both by animal and by day:

Animal	Predose (Day 0)	Dose (Day 1)	Washout (Day 2)	Washout (Day 3)
1	2.92	2.26	1.87	4.07
2	3.87	0.78	2.21	4.82
3	5.13	4.04	3.75	4.15
4	2.57	1.98	3.17	5.16
5	1.90	1.73	2.82	3.05
6	3.12	1.86	3.42	5.06
7	6.23	0.66	5.05	6.43
8	5.07	1.13	2.12	5.82
9	3.08	0.82	2.77	2.42
10	3.44	2.07	2.66	3.41

Solution:

Animal	CV
1	0.35
2	0.61
3	0.14
4	0.43
5	0.28
6	0.39
7	0.59
8	0.64
9	0.44
10	0.23

	Day 0	Day 1	Day 2	Day 3
CV	0.36	0.58	0.31	0.28

Problem 25: Chauvenet's Criterion

During an acute oral study, several technicians weighed members of groups of animals. Use Chauvenet's criteria to identify outliers from each of the data sets.

Group A	Group B	Group C
164.08	177.58	209.84
163.32	168.28	223.46
162.98	172.78	214.46
167.80	180.94	213.14
165.16	110.44	204.46

Solution:

	Group A	Group B	Group C
Reject(s)	None	110.44	227.46

Problem 26: Chauvenet's Criterion

During the course of a dermal study, blood samples were collected and analyzed for methemoglobin levels. The cooximeter used for these measurements was jammed with blood clots several times during the measurements and it is suspected that this may have interfered with the proper evaluation of some samples. Using Chauvenet's criterion, identify outlier values from the following three sets of data.

Control	Low Dose	High Dose
0.16	0.94	5.64
0.14	0.86	4.97
0.21	0.65	4.85
0.23	0.42	0.01
0.19	0.47	6.12
0.11	0.54	5.89
0.12	0.02	5.94
0.26	0.46	3.87
0.20	0.75	5.61
0.17	0.01	1.04

Solution:

	Control	Low Dose	High Dose
Reject(s)	0.26	0.26	0.01

Note that in the high dose group, if 0.01 is rejected as an outlier, then the value 1.04 becomes an outlier.

Problem 27: Chauvenet's Criterion

Five compounds were tested by inoculating growth media with the compounds and bacteria and counting the number of colonies after incubation. Use Chauvenet's criterion to identify possible outliers.

Compound 1	Compound 2	Compound 3	Compound 4	Compound 5
7	7	7	7	9
12	7	12	7	10
8	7	7	9	13
11	12	7	10	12
10	8	6	17	16
8	4	4	5	6
9	11	8	13	11
9	10	8	12	12
	8	9	16	
	9	5	6	
	9		11	
			12	

Solution:

	Compound 1	Compound 2	Compound 3	Compound 4	Compound 5
Reject(s)	None	12, 4	12	17	None

Note that for Compound 2, rejecting 12 and 4 leaves 11 as a reject of the revised data; for Compound 3, rejecting 12 leaves 4 as an outlier; and for Compound 4, rejecting 17 means 16 becomes an outlier. Chauvenet's criterion is most likely not appropriate for Compound 2, since it has more than one extreme value.

Problem 28: Chauvenet's Criterion

Visually examine the data below. Would it be reasonable to use Chauvenet's criterion on it or not? Include reasons.

Generator A	Generator B	Generator C
18	4	1
250	1	7
164	29	6
142	19	9
187	16	9
129	23	1
178	26	5
225	11	8
453	18	12
46	37	39

Answer the same questions for the following sets of data:

W		X	Y	Z
110	101	560	3	9
87	106	590	10	10
78	110	597	10	9
110	108	592	11	10
115	126	633	11	26
70	103	752	12	10
118	87	840	12	11
60	88	1013	13	22
152	115	972		10
104	84	1098		8
92	103	1034		9
124	102			10
	117			11

Solution:

First set: All three sets do not appear to be normally distributed and there is more than one obvious outlier in each.

    W: Chauvenet's criterion is not usable — more than 20 values.

    X: Chauvenet's criterion is not usable — not normally distributed.

    Y: Chauvenet's criterion is usable.

    Z: Chauvenet's criterion is not usable — more than one outlier.

Problem 29: Rounding and Truncating Numbers

Using the rules and procedures presented in the text, present the following ten values as both rounded and truncated to three digits.

A	163.32	F	165.49
B	14.51	G	0.0953
C	10.50	H	1.0953
D	165.42	I	14.18
E	0.0987	J	11.56

Solution:

	Rounded	Truncated
A	163	163
B	14.5	14.15
C	10.5	10.5
D	165	165
E	0.0987	0.0987
F	165	165
G	0.0953	0.0953
H	1.10	1.09
I	14.2	14.1
J	11.6	11.5

Problem 30: Rounding and Truncating Numbers

Using the rules and procedures presented in the text, present the following six values as both rounded and truncated to four digits.

A	1053.26	D	19.9850
B	0.5326	E	162.49
C	19.9950	F	162.51

Solution:

	Rounded	Truncated
A	1053	1053
B	0.5326	0.5326
C	20.00	19.99
D	19.99	19.98
E	162.5	162.4
F	162.5	162.5

Problem 31: Rounding and Truncating Numbers

Using the common rounding rule presented in the text, present the following three sets of values as rounded to whole numbers.

Set A	Set B	Set C
10.2	9.4	6.5
10.6	10.7	12.9
9.2	11.0	5.5
10.3	7.3	13.7
10.9	7.4	10.4
10.0	11.8	9.7
8.5	11.8	11.9
9.4		

Solution:

Set A	Set B	Set C
10	9	6
11	11	13
9	11	6
10	7	14
11	7	10
10	12	10
8	12	12
9		

Remember that values of "$n.5$" are rounded up if n is odd and down if n is even.

Problem 32: Rounding and Truncating Numbers

Using the common rounding rule presented in the text, present the following set of values as rounded to 2 digits: 205, 221, 163, 27, 52, 116, 2, 264, 82, 242, 215, 94, 256, and 262.

Solution:

200, 220, 160, 30, 50, 120, 0, 260, 80, 240, 220, 90, 260, and 260

Note that 27 rounds to 30 because the set of data has to be rounded uniformly. The same applies to the values 52 (rounds to 50) and 2 (rounds to 0).

Problem 33: Sufficient Sample Size

Using the data presented below to estimate the sample standard deviation, calculate the group sample size necessary to have a p level of 0.05 (two-sided) for each group and also for the combined data. The acceptable range of variations in weights is 5 grams and $\beta = 0.10$.

Group A	Group B	Group C	Group D	Group E
204.7	196.7	208.9	206.5	205.9
198.6	207.9	204.7	202.3	187.6
210.3	204.5	204.1	208.7	182.4
207.4	210.2	192.3	191.1	192.5
202.9	203.9	206.7	189.6	203.7

Solution:

	Group A	Group B	Group C	Group D	Group E	Combined
Calculated	8.3080	11.0480	17.4962	32.5700	43.3512	25.6102
Size	9	12	18	33	44	26

Problem 34: Sufficient Sample Size

Using the data presented below to estimate the sample standard deviation, calculate the group sample size necessary to have a p level of 0.01 (two-sided) for each group and for the combined data. The acceptable range of variations in spleen weights is 0.10 grams and $\beta = 0.10$.

Group A	Group B	Group C	Group D
2.14	1.97	1.92	1.87
2.12	2.14	2.14	1.74
2.21	2.16	1.85	1.98
2.11	1.87	1.92	1.89
1.99	2.17	2.17	2.01
2.05	2.04	1.97	2.03
2.04	2.01	1.89	1.95
2.00	2.06	2.05	1.86
1.97	2.19		
2.08	2.03		

Solution:

	Group A	Group B	Group C	Group D	Combined
Calculated	8.4794	15.4138	20.8940	13.6017	18.9457
Size	9	16	21	14	19

Problem 35: Sufficient Sample Size

Using the same data and procedures as detailed in Problem 33 and Problem 34, calculate the group sample size necessary to have a p level of 0.01 (one-sided) for both sets of controls and experimental groups. Set $\beta = 0.05$.

Solution:

	Group A	Group B	Group C	Group D	Group E	Combined
Calculated	12.46931	16.58175	26.25986	48.88381	65.06523	38.43799
Size	13	17	27	49	66	39

	Group A	Group B	Group C	Group D	Combined
Calculated	8.98716	16.33686	22.14524	14.41618	20.08028
Size	9	17	23	15	21

Problem 36: Sufficient Sample Size

Using the following data, calculate the sample size necessary to find a significant difference. Assume a two-tailed test at $p = 0.05$ and $\beta = 0.10$.

Group	Δ	Standard Deviation
A	1	2.0
B	2	2.0
C	5	2.0
D	1	4.0
E	2	4.0
F	5	4.0

Solution:

Group	Calculated	N
A	42.0297	43
B	10.5074	11
C	1.6812	2
D	168.1187	169
E	42.0297	43
F	6.7247	7

Problem 37: Sufficient Sample Size

Your company has a drug to treat an extremely rare, fatal disease. The known compound, S001, extends the life expectancy by an average of 3.4 months. The standard deviation of the life expectancy from the first stages to death is 6.3 months for untreated patients and 7.9 months for S001. It is estimated that only 12 patients can be recruited for the initial trial. For the combinations of α and β shown in the table below and for both given standard deviations, use the sample size formula to back-calculate the minimum differences in survival times that the new compound would have to cause in order to be detectable.

β	χ
0.10	0.20
0.05	0.20
0.10	0.05
0.05	0.05

Solution:

		SD	
β	χ	6.3	7.9
0.10	0.20	4.5	5.7
0.05	0.20	5.1	6.4
0.10	0.05	6.0	7.5
0.05	0.05	6.6	8.2

Problem 38: Sufficient Sample Size

Referring to the previous problem (Problem 37): the company executives want to be able to show a two-month difference in survival times. If all other factors stay the same, what would α and β have to be in order to show this difference. Hint: first find the equation for the total $t_{1-\alpha/2} + t_{1-\beta}$ allowed and then assume $\alpha = \beta$ and calculate the maximum of those parameters allowed.

Solution:

For SD = 6.3: $t_{1-\alpha/2} + t_{1-\beta} = 1.100$, so $\alpha = \beta = 0.40$.

For SD = 7.9: $t_{1-\alpha/2} + t_{1-\beta} = 0.879$, so $\alpha = \beta = 0.45$.

Problem 39: Scattergram

Prepare scattergrams of the following data sets such that the nature of the distribution and occurrence of outliers can be evaluated.

Group A: 1.6, 2.4, 3.3, 3.5, 4.1, 4.36, 4.4, 4.7, 5.2, 5.3, 5.5, 5.6, 5.7, 6.3, 6.4, 6.6, 6.8, 7.2, 7.5, 8.3, 9.2.

Group B: 1.0, 2.1, 3.2, 4.1, 5.2, 6.3, 7.0, 7.6, 7.7, 7.8, 8.2, 8.3, 8.4, 8.5, 8.6, 8.7, 8.9, 9.1, 9.2, 9.5, 10.0.

Solution:

Group A:

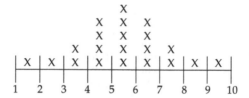

Group B:

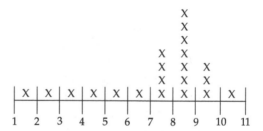

Problem 40: Scattergram

Prepare scattergrams for the data from the following data:

Compound 1	Compound 2	Compound 3	Compound 4	Compound 5
7	7	7	7	9
12	7	12	7	10
8	7	7	9	13
11	12	7	10	12
10	8	6	17	16
8	4	4	5	6
9	11	8	13	11
9	10	8	12	12
	8	9	16	
	9	5	6	
	9		11	
			12	

Solution:

Compound 1:

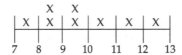

Compound 2:

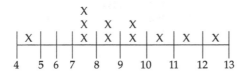

Compound 3:

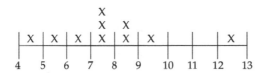

Compound 4:

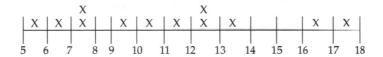

Compound 5:

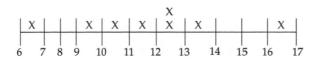

Problem 41: Scattergram

Prepare scattergrams using the following data sets. For set W the interval size should be 10; for X the interval should be 100; for Y the interval should be 1; and for Z the interval should be 2.

W		X	Y	Z
110	101	560	3	9
87	106	590	10	10
78	110	597	10	9
110	108	592	11	10
115	126	633	11	26
70	103	752	11	10
118	87	840	12	11
60	88	1013	12	22
152	115	972	13	10
104	84	1098		8
92	103	1034		9
124	102			10
	117			11

Solution:

W:

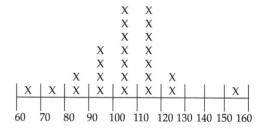

X:

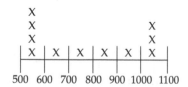

Y:

Z:

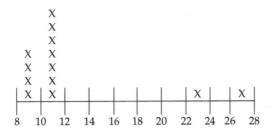

Problem 42: Bartlett's Homogeneity of Variance

Calculate Bartlett's homogeneity of variance for the body weights of the male and female rats presented below and state whether (at a $p \leq 0.05$ level) the variances are such that analysis of variance could be employed in further analysis.

Males			
Group A	Group B	Group C	Group D
194.4	201.8	216.0	235.4
189.3	230.8	225.5	207.8
195.1	190.4	219.2	221.5
212.5	201.7	218.0	210.1
223.3	218.8	208.5	204.2
195.5	216.7	214.4	213.5
203.2	178.9	222.2	191.4
187.5	225.3	213.5	214.1
186.1	231.7	195.6	204.5
220.4	223.8	243.2	225.9

Females			
Group A	Group B	Group C	Group D
152.4	142.3	129.7	141.6
154.8	145.8	134.8	124.4
140.9	140.9	146.8	134.7
137.7	138.6	152.9	148.8
135.6	139.2	168.5	135.9
136.6	138.0	154.0	132.8
148.2	140.8	144.1	136.9
141.5	142.2	147.3	145.2
143.1	153.9	149.8	144.5
141.4	141.8	146.2	140.6

Solution:

Males:	Yes	$\chi^2 = 1.779$
		$p = 0.620$
Females:	Yes	$\chi^2 = 5.912$
		$p = 0.116$

Problem 43: Bartlett's Homogeneity of Variance

Calculate Bartlett's homogeneity of variance for the male and female brain weights presented below and state whether (at a $p \leq 0.01$ level) the variances are such that analysis of variance could be employed in further analysis.

Males			
Group A	Group A	Group A	Group A
1.70	1.66	1.82	1.79
1.70	1.78	1.81	1.72
1.70	1.69	1.80	1.81
1.77	1.74	1.89	1.72
1.81	1.73	1.60	1.73
1.71	1.81	1.73	1.71
1.70	1.72	1.80	1.74
1.70	1.81	1.77	1.87
1.71	1.80	1.80	1.82
1.72	1.86	1.79	1.88

Females			
Group A	Group A	Group A	Group A
1.63	1.65	1.60	1.63
1.63	1.50	1.62	1.66
1.66	1.55	1.66	1.65
1.58	1.58	1.71	1.64
1.57	1.56	1.70	1.62
1.66	1.64	1.68	1.58
1.65	1.72	1.69	1.63
1.67	1.62	1.63	1.69
1.65	0.58	1.68	1.60
1.61	1.70	1.67	1.68

Solution:

Males:	Yes	$\chi^2 = 3.892$
		$p = 0.273$
Females:	No	$\chi^2 = 69.991$
		$p \leq 0.001$

Problem 44: Bartlett's Homogeneity of Variance

Calculate Bartlett's homogeneity of variance and the probability for the plasma bilirubin values presented below:

Control	10 mg/kg	20 mg/kg	40 mg/kg	80 mg/kg
6.74	5.49	6.64	6.57	5.49
5.33	5.30	5.85	5.73	5.83
5.45	6.45	6.00	6.89	5.74
6.67	5.70	6.35	6.29	5.76
6.17	5.93	5.37	5.41	5.72
5.40	4.76	5.46	4.41	4.35
5.66	5.63	5.81	5.48	
6.93	7.38	7.09		
	6.11	6.53		
	6.71			

Solution:

$$\chi^2 = 1.451$$
$$p = 0.835$$

Problem 45: Bartlett's Homogeneity of Variance

Given the following male and female urine albumin measurements, use Bartlett's homogeneity of variance to decide whether ANOVA can be used on the values at the $p \leq 0.10$ level:

Males Group A	Group B	Group C	Group D
2.88	3.06	2.77	2.05
3.37	3.55	3.30	3.08
3.76	4.15	3.30	2.63
2.71	2.41	2.16	2.96
2.36	3.93	2.91	3.25
2.81	2.14	2.37	3.23
2.35	3.73	3.33	3.23
2.85	2.49	2.80	2.97
3.93	4.37		2.88
3.34			

| Females | | | |
Group A	Group B	Group C	Group D
3.44	2.94	2.65	3.43
3.53	3.51	3.82	2.30
2.76	3.56	2.35	1.92
3.50	3.14	2.36	3.20
3.95	3.37	2.48	3.35
3.17	1.91	2.80	2.60
3.40	1.80	3.65	1.78
3.36	1.96	3.83	2.36
4.13	3.72		3.38
4.25			

Solution:

Males:	No	$\chi^2 = 5.222$
		$p = 0.156$
Females:	No	$\chi^2 = 2.362$
		$p \leq 0.501$

Problem 46: Sign Test

In reviewing a study conducted in another laboratory, you want to ensure that animals were assigned randomly to each of the five test groups. Using a sign test, evaluate the validity of randomization of the five groups of animals presented below:

Group I	Group II	Group III	Group IV	Group V
17	19	20	22	36
21	27	26	23	38
24	37	34	25	44
30	50	45	29	49
31	52	54	32	51
40	53	57	33	69
41	68	60	35	71
43	73	67	39	79
55	81	75	42	83
59	82	85	47	88

Solution:

Group I:	8–, 2+
Group II:	3–, 7+
Group III:	3–, 7+
Group IV:	9–, 1+
Group V:	2–, 8+
Conclusion:	Assignment was not random.

Problem 47: Sign Test

While reviewing acute studies on a series of related compounds, you notice that the incidence of clinical observations by some technicians seems higher than others. Given the number of times in each study that each of four technicians have recorded seeing effects during clinical tests, and using a sign test as a quick and easy assessment, evaluate the possibility that the different technicians are not equivalent in their performance of clinical observations.

Technician A	Technician B	Technician C	Technician D
12	4	31	7
15	25	27	9
10	32	13	6
26	6	29	5
15	16	24	14
16	26	28	21
10	15	30	11
1	36	9	19
8	21	8	0
18	7	10	15

Solution:

Technician A:	5–, 3+, 2 midpoint
Technician B:	3–, 6+, 1 midpoint
Technician C:	4–, 6+
Technician D:	7–, 2+, 1 midpoint
Conclusion:	There is no clear pattern associated with any technician.

Problem 48: Arcsine Transformation

Determine the arcsine-transformed values (in degrees) of the following ten incidence rates.

A	13/50
B	3/10
C	4/25
D	7/49
E	3/21
F	5/50
G	10/18
H	12/100
I	12/50
J	7/50

Solution:

A	30.66
B	33.21
C	23.58
D	22.21
E	22.21
F	18.43
G	48.19
H	20.27
I	29.33
J	21.97

Problem 49: Arcsine Transformation

Determine the arcsine-transformed values (in radians) for the following lethality rates. Do as both proportion alive and proportion dead.

Test	Alive	Dead
I	18	14
II	75	2
III	14	56
IV	51	25
V	96	7
VI	88	44
VII	94	10
VIII	55	85
IX	83	55
X	67	74

Solution:

Test	Alive	Dead
I	0.858	0.713
II	1.413	0.158
III	0.466	1.105
IV	0.964	0.606
V	1.314	0.257
VI	0.953	0.618
VII	1.256	0.315
VIII	0.678	0.893
IX	0.889	0.682
X	0.759	0.812

Problem 50: Probit Transformation

Determine the probit transformation values associated with the ten incidences presented below:

A	7/10
B	2/100
C	3/9
D	8/30
E	49/50
F	7/29
G	5/12
H	7/28
I	34/46
J	99/100

Solution:

A	5.5244
B	2.9463
C	4.5684
D	4.3781
E	7.0537
F	4.2969
G	4.7904
H	4.3255
I	5.6403
J	7.3263

Problem 51: Probit Transformation

Calculate the probit transformation for the incidence values from Problem 49. Do the calculations both for portion alive and portion dead.

Solution:

Test	Alive	Dead
I	5.1575	4.8425
II	6.9600	3.0400
III	4.1584	5.8416
IV	5.4445	4.5555
V	6.5052	3.4948
VI	5.4307	4.5693
VII	6.3041	3.6959
VIII	4.7269	5.2731
IX	5.2575	4.7425
X	4.9385	5.0615

Problem 52: Fisher's Exact Test

Using Fisher's exact test, determine if any of the tumor incidences presented below are statistically higher in test group animals.

Present the exact one-tailed p-values for each of the nine test group incidences.

Site	Control Males	Control Females	Test Group Males	Test Group Females
Liver	2/50	4/50	17/50	8/50
Pituitary	4/50	6/49	7/50	5/50
Lung	2/50	1/50	7/50	6/50
Kidney	4/50	1/50	15/50	14/50
Mammary	NA	12/50	NA	19/50

Solution:

Site	Test Group Males	Test Group Females
Liver	0.0001	0.1783
Pituitary	0.2623	0.4856
Lung	0.0798	0.0559
Kidney	0.0047	0.0002
Mammary	NA	0.0971

Problem 53: Fisher's Exact Test

Calculate and present the Fisher's exact values for the following incidence rates.

A.	0/10 vs. 3/10	F.	6/12 vs. 12/12
B.	0/10 vs. 4/10	G.	5/15 vs. 9/15
C	3/10 vs. 7/10	H.	5/15 vs. 10/15
D.	3/10 vs. 8/10	I.	5/15 vs. 11/15
E.	2/10 vs. 6/8	J.	5/20 vs. 10/20

Solution:

A:	0.1052632
B:	0.0433436
C:	0.0894477
D:	0.0348893
E:	0.0306458
F:	0.0068650
G:	0.1361517
H:	0.0715555
I:	0.0327977
J:	0.0953963

Problem 54: Fisher's Exact Test

Using Fisher's exact test, determine which of the five following incidence rates is significantly different from a control rate of $1/25$ at $p \leq 0.05$.

A	3/10
B	4/10
C	7/25
D	10/25
E	20/100

What do these results suggest about the influence of differences in group sizes on test results?

Solution:

A:	Not significant
B:	Significant
C:	Significant
D:	Significant
E:	Significant

As the group numbers rise, the range of proportions considered significant gets smaller, i.e., the estimate gets more accurate (discriminating).

Problem 55: Fisher's Exact Test

Using Fisher's exact test, determine which of the five following incidence rates is significantly different from a control rate of $9/100$ at $p \leq 0.05$.

A	9/50
B	13/100
C	18/100
D	8/50
E	13/96

Solution:

A:	No
B:	No
C:	Yes; $p \leq 0.048250$
D:	No
E:	No

Problem 56: Fisher's Exact Test

Using Fisher's exact test, determine the one-tailed and two-tailed p-values for the following compounds:

	Successful Pregnancies after Mating			
Control	Compound A	Compound B	Compound C	Compound D
14/48	7/36	3/60	4/44	12/21

Solution:

	Compound A	Compound B	Compound C	Compound D
One-tailed	0.2236	0.0007	0.0140	0.0271
Two-tailed	0.4456	0.0009	0.0186	0.0340

Problem 57: Fisher's Exact Test

The number of misreads out of 40 tests each on four analytical machines was monitored. Using Fisher's exact test at a 20% probability, determine if any of the technicians have a significantly different record than the most senior (Tech 1). Give the probabilities.

			Misreads				
Machine	Tech 1	Tech 2	Tech 3	Tech 4	Tech 5	Tech 6	Tech 7
A	5	3	11	12	2	16	6
B	4	3	12	7	5	10	5
C	6	5	6	8	1	4	2
D	6	10	5	15	17	1	2

Solution:

			p-Values			
Machine	Tech 2	Tech 3	Tech 4	Tech 5	Tech 6	Tech 7
A	0.7119	*0.1610*	*0.0993*	0.4315	*0.0100*	1.0000
B	1.0000	*0.0482*	0.3553	1.0000	*0.1395*	1.0000
C	0.7589	1.0000	0.7695	*0.1084*	0.7370	0.2633
D	0.2817	0.7589	*0.0406*	*0.0126*	*0.1084*	0.2633

Note: Italics indicate significance at $p \leq 0.20$.

Problem 58: Fisher's Exact Test

In a clinical study with three groups, the ethnicities (self-rated) of the participants, as well as the census information for the surrounding area, are as follows:

Ethnicity	Control	Low Dose	High Dose	Population (per 100)
Anglo-Saxon	8	5	3	40
African-American	4	6	1	19
Asian	0	3	4	13
Hispanic	7	6	10	16
Other	1	0	2	10

Group the participants into two groups by ethnicity, e.g., Anglo-Saxon vs. nonAnglo-Saxon, and use Fisher's exact test to determine if the participants of each group vary from the population at a 95% confidence level. Do so for each listed ethnicity.

Solution:

Ethnicity	Control	Low Dose	High Dose
Anglo-Saxon vs. non	No: 1.000	No: 0.312	Yes: 0.041
African-American vs. non	No: 1.000	No: 0.364	No: 0.190
Asian vs. non	No: 0.122	No: 0.730	No: 0.481
Hispanic vs. non	No: 0.063	No: 0.201	Yes: 0.002
Other vs. non	No: 0.689	No: 0.210	No: 1.000

Problem 59: Fisher's Exact Test

A chemotherapy drug is evaluated against a control and against a standard treatment in transgenic mice. The results are divided into two groups: significant improvement (tumor size reduction of 50% or greater) or no significant improvement. Using Fisher's exact test, determine if the two drug treatments improve survival compared to controls (at the $p \leq 0.05$ level). Also determine if the two treatments differ from each other at the $p \leq 0.05$ level.

Result	Drug N	Drug S	Control
Significant improvement	8	16	38
No significant improvement	48	35	41

Solution:

Note: "Improve survival" = one-tailed test; "treatments differ" = two-tailed test.

	Drug N	Drug S
Drug S	0.03950	—
Control	0.00003	0.04320

Both treatments represent an improvement over controls. Drug N represents an improvement over Drug S. Note that the two-tailed test of Drug S over the control is not significant at the 0.05 level: $p = 0.06965$.

Problem 60: 2 × 2 Chi-Square

Using a 2 × 2 chi-square test, determine which of the following incidence rates are significantly different ($p \leq 0.05$) from a control rate of 6/15.

A	9/15
B	10/15
C	15/20
D	21/30
E	18/30

Solution:

A:	NS ($p = 0.273$)
B:	NS ($p = 0.143$)
C:	Significant ($p = 0.036$)
D:	NS ($p = 0.053$)
E:	NS ($p = 0.205$)

Problem 61: 2 × 2 Chi-Square

Using a 2 × 2 chi-square test, evaluate the incidence rates presented below for significance at $p \le 0.10$, $p \le 0.05$, and $p \le 0.01$.

A	0/10 vs. 3/10	F	6/12 vs. 12/12
B	0/10 vs. 4/10	G	5/15 vs. 9/15
C	3/10 vs. 7/10	H	5/15 vs. 10/15
D	3/10 vs. 8/10	I	5/15 vs. 11/15
E	2/10 vs. 6/8	J	5/20 vs. 10/20

Solution:

A:	$0.10 > p$ ($= 0.060$) > 0.05
B:	$0.05 > p$ ($= 0.025$) > 0.01
C:	$0.10 > p$ ($= 0.074$) > 0.05
D:	$0.05 > p$ ($= 0.025$) > 0.01
E:	$0.05 > p$ ($= 0.020$) > 0.01
F:	$0.01 > p$ ($= 0.005$)
G:	p ($= 0.143$) > 0.10
H:	$0.10 > p$ ($= 0.068$) > 0.05
I:	$0.05 > p$ ($= 0.028$) > 0.01
J:	p ($= 0.102$) > 0.10

Problem 62: 2 × 2 Chi-Square

Using a 2 × 2 chi-square test, evaluate the incidence rates presented below for significance at $p \leq 0.05$ and $p \leq 0.01$.

A	9/50
B	13/100
C	18/100
D	8/50
E	13/96

Solution:

A:	$p (= 0.110) > 0.05$
B:	$p (= 0.366) > 0.05$
C:	$p (= 0.063) > 0.05$
D:	$p (= 0.202) > 0.05$
E:	$p (= 0.314) > 0.05$

Problem 63: 2 × 2 Chi-Square

Using a 2 × 2 chi-square test, evaluate the significance of differences for each of the four following sets of data.

A	Control	8/13	Test	13/18
B	Control	7/13	Test	12/16
C	Control	13/49	Test	10/49
D	Control	7/92	Test	10/80

Solution:

A:	$p (= 0.530) > 0.05$
B:	$p (= 0.350) > 0.05$
C:	$p (= 0.475) > 0.05$
D:	$p (= 0.284) > 0.05$

Problem 64: 2 × 2 Chi-Square

Using a 2 × 2 chi-square test, give the chi-square calculated value and the probability of difference from 17/95 for each of the following sets of data.

A	3/10
B	15/50
C	60/200
D	7/10
E	10/20
F	34/95

Solution:

Group	Chi-Square Value	p
A	0.860	0.354
B	2.791	0.095
C	4.893	0.027
D	13.931	<0.001
E	9.479	0.002
F	7.746	0.005

Problem 65: 2 × 2 Chi Square

A reverse mutation assay with a control substance showed 7 mutants out of 33 colonies. Assuming the next assay also evaluates 33 colonies, what are the ranges of the number of mutants that would be significantly different in a 2 × 2 chi-square test at the $p \le 0.10$, $p \le 0.05$, and $p \le 0.01$ levels?

Solution:

p	Number Needed
0.100	≥ 14 or ≤ 2
0.050	≥ 15 or ≤ 1
0.001	≥ 21

Problem 66: 2 × 2 Chi-Square

A previous study showed that an antibiotic caused rashes in 8 of 40 people. It is suspected that this was due to contamination and that the normal rate of reaction is 4%. What is the minimum number of people needed in a second study to show a difference in a 2 × 2 chi-square at $p \le 05$? Assume that the normal reaction rate is exactly what will be found in the second study and ignore (truncate) fractions of a person in the calculations.

Solution:

Since we are ignoring fractions of a person, it's 50. Otherwise, 30 would be enough.

Problem 67: 2 × 2 Chi-Square

Using a 2 × 2 chi-square test, give the chi-square calculated value and the probability of difference for each of the following sets of data.

A	21/37	8/23
B	26/59	14/45
C	22/46	32/98
D	28/68	20/59
E	17/26	23/73
F	9/20	27/77

Solution:

Group	Chi-Square Value	p
A	2.742	0.098
B	1.811	0.178
C	3.075	0.080
D	0.712	0.399
E	9.138	0.003
F	0.671	0.413

Problem 68: R × C Chi-Square

In developing a new test method, each of the three technicians used four different techniques to measure one of the endpoints. For each technician and technique we determine the number of times the accuracy of the measurements is unacceptable. These are presented in the table below.

Are there significant differences between techniques based on R × C chi-square calculations?

	Technique				
	A	B	C	D	Total per Technician
Technician 1	10	5	15	12	42
Technician 2	10	12	21	19	62
Technician 3	13	10	17	20	60
Total per technique	33	27	53	51	164

Solution:

Six degrees of freedom; chi-square = 2.3704.
Not significant (test statistic at $p \leq 0.05 = 12.592$).

Problem 69: R × C Chi-Square

In evaluating a Drosophila mutagenesis test, the following frequencies of eye color were recorded.

Eye Color	Treatment				Row Total
	A	B	C	D	
Black	11	10	9	16	46
Red	9	38	52	54	153
Albino	12	24	56	57	149
Barred	16	18	34	74	142
Column total	48	90	151	201	490

Were there any significant relationships between treatment and eye color at a level of $p \leq 0.05$?

Solution:

Nine degrees of freedom; chi-square = 32.1917.
$p \leq 0.001$ (test statistic at this level is 27.877).

Problem 70: R × C Chi-Square

In evaluating the aquatic environmental effects of production plant effluents, we must determine if there is a relationship between the populations of fish in proximity to effluent streams in local rivers. A species capture study was performed and the data from it are presented below.

Fish Species	Site			Row Total
	Upstream	Plant 1	Plant 2	
A	54	15	17	86
B	27	25	22	74
C	14	27	36	77
D	32	16	28	76
E	73	47	53	173
F	9	12	37	58
Column Total	209	142	193	544

Is there a statistical relationship at the level of $p < 0.05$ in an R × C chi-square test?

Solution:

Ten degrees of freedom; chi-square = 62.1093.
$p \leq 0.001$ (test statistic at this level is 29.588).

Problem 71: R × C Chi-Square

Using an R×C chi-square test, examine the tumor incidence presented below
to determine if there is a relationship between treatment and the pattern of
tumors in the female at a level of $p \leq 0.05$.

Site	Control Males	Control Females	Test Group Males	Test Group Females
Liver	2/50	4/50	17/50	8/50
Pituitary	4/50	6/49	7/50	5/50
Lung	2/50	1/50	7/50	6/50
Kidney	4/50	1/50	15/50	14/50
Mammary	NA	12/50	NA	19/50

Solution:

Four degrees of freedom; chi-square = 8.7093.
Not significant (test statistic at $p \leq 0.05$ is 9.488).

Problem 72: R × C Chi-Square

Soil samples from four locations were plated and, after incubation, colonies
of six different types of bacteria were counted. Use an R × C chi-square test
to determine the probability of the distribution.

Location	Type A	Type B	Type C	Type D	Type E	Type F
001	7	20	8	22	18	17
002	19	22	19	13	12	18
003	18	18	14	19	13	19
004	12	9	10	15	22	19

Solution:

Chi-square value: 23.455.
Probability: 0.075.

Problem 73: R × C Chi-Square

Adverse events are monitored for a drug trial in which there is a test substance group, a comparator group, and a control group. Use an R × C chi-square to determine if there are significant differences in the adverse events between the groups at the 0.05 level.

AE	Active	Treatment Comparator	Control
Nausea	5	17	7
Headache	0	7	8
Dizziness	5	5	6
Lassitude	6	3	4
Blurred vision	1	3	2
Fever	1	4	1

Solution:

No: chi-square value = 15.113 at 10 df, $p = 0.128$.

Problem 74: R × C Chi-Square

Several different strains of a single genus of bacteria were grown with three different variations on a substance and a control. Colonies were then counted by morphology. Use an R × C chi-square to calculate the probability of there being a significant difference. Then use the same technique comparing the active substances to the control.

Colony Morphology	Control	Substance K123	K234	K345
Small, homogeneous, round, white	9	9	13	19
Small, wrinkled, round, white	3	10	6	5
Small, homogeneous, uneven, white	7	13	10	21
Large, homogeneous, round, yellow	24	17	22	23
Small, bullseye, uneven, white	17	9	16	14
Small, homogeneous, round, clear	3	15	13	17
Large, wrinkled, uneven, yellow	25	8	16	22
Large, homogeneous, round, white	10	24	9	15

Solution:

| | | Test | | |
Parameter	All	C vs. K123	C vs. K234	C vs. K345
Chi-square value	45.477	31.544	10.423	16.642
Df	21	7	7	7
p	0.001	<0.001	0.166	0.020

Problem 75: R × C Chi-Square

Patients were tested for the effects of a diet designed to control kidney stones in combination with a drug for the same endpoint. After a month on the various plans, urine was sampled for three days and the total number of microcrystals was counted. Use an R × C chi-square test to determine if there was any difference in treatments.

| | Treatment | | | |
| | Normal Diet | | Restricted Diet | |
Microcrystal Total	Control	Active	Control	Active
0	3	0	3	5
1	4	1	4	3
2	5	1	4	2
3	5	2	7	1
4	1	5	2	1
5	3	5	1	0
6	1	0	1	0
7	0	1	1	3

Solution:

Yes: chi-square value = 34.335 at 21 df, $p = 0.033$.

Problem 76: Wilcoxon Rank Sum

Segmented RBC counts were determined for blood samples collected from rats. These are presented in the table below:

Group A	Group B
6	6
11	4
8	11
8	14
10	13
9	10
13	15
5	12
11	22
15	15

Using the Wilcoxon rank sum, determine if these groups are different at $p \leq 0.05$.

Solution:

Group A = 86.5.
Group B = 123.5.
Test statistics = 79, 131 for $p \leq 0.05$.
The results are not significant.

Problem 77: Wilcoxon Rank Sum

Using the Wilcoxon rank sum test, compare the treatment group presented below with the control group.

Control	Treatment
12	4
11	9
9	1
6	13
12	10
9	18
10	17
9	14
6	12
8	16

Are these groups different at $p \leq 0.05$?

Solution:

Control = 85.
Treatment = 125.
Test statistics = 79, 131 for $p \leq 0.05$.
These results are not significant.

Problem 78: Wilcoxon Rank Sum

Using the Wilcoxon rank sum test, evaluate the difference between the following two sets of lymphocyte counts at a level of $p \leq 0.01$.

Group A	Group B
92	93
84	87
84	95
80	94
89	88
83	84
85	89
86	91
82	97
85	87
81	92

Solution:

Group A = 81.
Group B = 172.
Test statistics = 96, 157 for $p \leq 0.05$.
Test statistics = 88, 165 for $p \leq 0.01$.

Problem 79: Wilcoxon Rank Sum

Compare the following two sets of implantation site counts at levels of $p \leq 0.05$ and $p \leq 0.01$ using the Wilcoxon rank sum test.
    Control: 13, 15, 12, 9, 12, 13, 14, 14, 10, 14, 13, 14, 12, 11, 8, 16, 12, 13, 13, 14, and 12.
    Test: 11, 13, 12, 12, 11, 14, 7, 10, 11, 13, 12, 13, 9, 12, 12, 11, 10, 8, 10, 13, and 11.

Solution:

Group A = 550.
Group B = 353.
Test statistics = 373, 530 for $p \leq 0.05$.
Groups are significantly different at $p \leq 0.05$.

Problem 80: Wilcoxon Rank Sum

Cytotoxicity of three test substances, along with a control and a comparator, was tested by determining how many bacterial colonies would grow after treatment. Using Wilcoxon rank sum, determine if any compound is cytotoxic; i.e., different from the control at the $p \leq 0.05$ level (one-sided test).

	Colony Counts			
Control	Comparator	C111	C222	C333
22	9	9	13	19
33	3	10	6	5
12	7	13	10	21
7	8	17	12	23
27	2	9	16	14
16	3	15	13	17
14	25	8	16	22
19	10	24	9	15

Solution:

	Control/C111	Control/C222	Control/C333
Measured rank sum	81/55	85.5/50.5	69.5/66.5
Limits (from table)	52/84	52/84	52/84
Significant ($p \leq 0.05$)	No	Yes	No

Problem 81: Wilcoxon Rank Sum

Using the data from Problem 80, determine if the substances differ in cytotoxicity from the comparator at the $p \leq 0.10$ level (two-sided test).

Solution:

	Comparator/C111	Comparator/C222	Comparator/C333
Measured rank sum	55/81	48/88	48/88
Limits (from table)	52/84	52/84	52/84
Significant ($p \leq 0.10$)	No	Yes	Yes

Problem 82: Wilcoxon Rank Sum

The time (in minutes) that rats took to go through a maze was noted for three groups of rats. Using Wilcoxon rank sum, determine whether there were differences between groups at the $p \leq 0.05$ level (two-sided test). Compare two groups at a time.

Untreated	Stimulant	Hallucinogen
1	8	19
8	16	18
23	9	15
20	2	17
13	7	16
11	7	23
6	16	4
19	15	26
14		8
11		

Solution:

	Unt/Stim	Unt/Hall	Stim/Hall
Measured rank sum	105.5/65.5	87.5/98	51/102
Limits (from table)	54/98	66/114	51/93
Significant ($p \leq 0.05$)	No	No	No

Problem 83: Wilcoxon Rank Sum

Drivers were given either a liter of water or a liter of beer to drink. Twenty minutes later they drove around a test track lined with cones and the number of cones that were hit were noted. Are the beer drinkers significantly different from the water drinkers at the $p \leq 0.05$, $p \leq 0.025$, and $p \leq 0.01$ levels (one-sided) using the Wilcoxon rank sum test?

Water: 19, 7, 13, 5, 14, 17, 24, 10, 20, 18, 4, 8, 6, 19, 21.

Beer: 20, 28, 18, 39, 14, 29, 18, 9, 27, 22, 17, 26, 3, 19, 14.

After the test, it was found out that the beer drinker who only knocked down three cones was actually given water. Move that score to the water category and recalculate.

Solution:

$N_1 = N_2 = 15$	$p \le 0.05$	$p \le 0.025$	$p \le 0.01$
Measured rank sum	185/280	185/280	185/280
Limits (from table)	192/273	185/280	177/288
Significant	Yes	No	No

$N_1 = 16; N_2 = 14$	$p \le 0.05$	$p \le 0.025$	$p \le 0.01$
Measured rank sum	186/279	186/279	186/279
Limits (from table)	177/257	170/264	161/273
Significant	Yes	Yes	Yes

Problem 84: Mann–Whitney U Test

Use the Mann–Whitney U test to compare the two groups in Problem 78.

Solution:

UA = 106.
UB = 15.
The test statistics are 35 ($p \le 0.05$) and 26 ($p \le 0.01$). The groups are different at $p \le 0.05$ but not at $p \le 0.01$.

Problem 85: Mann–Whitney U Test

Use the Mann–Whitney U test to compare the two groups in Problem 79.

Solution:

UC = 122.
UT = 316.
The test statistics are 146 ($p \le 0.05$) and 122 ($p \le 0.01$). The groups are different at $p \le 0.05$ but not at $p \le 0.01$.

Problem 86: Mann–Whitney U Test

Use the Mann–Whitney U test to compare the two groups in Problem 80, exactly as stated in the problem.

Solution:

	Control/C111	**Control/C222**	**Control/C333**
Calculated values	19/45	14.5/49.5	30.5/33.5
Limits (from table)	16	16	16
Significant ($p \leq 0.05$)?	Yes	Yes	Yes

Problem 87: Mann–Whitney U Test

Use the Mann–Whitney U test to compare the two groups in Problem 83, exactly as stated in the problem.

Solution:

$N_1 = N_2 = 15$	$p \leq 0.05$	$p \leq 0.025$	$p \leq 0.01$
Calculated values	160/65	160/65	160/65
Limits (from table)	73	65	57
Significant?	Yes	Yes	Yes

$N_1 = 16; N_2 = 14$	$p \leq 0.05$	$p \leq 0.025$	$p \leq 0.01$
Calculated values	174/50	174/50	174/50
Limits (from table)	72	65	57
Significant?	Yes	Yes	Yes

Problem 88: Kruskal–Wallis Nonparametric ANOVA

Righting reflex scores were determined for each animal study in a study dosing. Are there any significant differences between the scores from the four groups of animals presented below? Use the Kruskal–Wallis nonparametric ANOVA to evaluate.

Control	1 mg/kg	5 mg/kg	15 mg/kg
0	1	1	2
0	1	2	2
0	1	2	3
0	2	2	3
0	2	2	3
0	2	2	4
1	3	3	8
1	3	3	8
1	3	3	8
1	3	4	8

Solution:

With 3 degrees of freedom, H = 25.5764, $p \leq 0.01$.

Problem 89: Kruskal-Wallis Nonparametric ANOVA

As in Problem 88, use the Kruskal–Wallis test to determine if there are any significant differences between groups in the following data.

Control	Low Dose	Mid Dose	High Dose
0	1	2	4
0	1	3	5
0	2	3	5
0	2	3	6
0	2	4	6
0	2	4	6
0	2	4	6
1	2	4	7
1	3	5	7
1	3	5	7

How do the arithmetic means of these groups vary from those in Problem 88? What is the major difference between the two sets of data?

Solution:

With 3 degrees of freedom, H = 34.7914, $p \leq 0.01$.

Problem 90: Kruskal–Wallis Nonparametric ANOVA

Use the Kruskal–Wallis test to compare the three sets of implantation site counts below. Are any of the groups significantly different from another group at $p \leq 0.05$?

Control	Low Dose	High Dose
8	7	3
9	8	3
10	9	4
11	9	4
11	10	5
12	10	6
12	10	6
12	11	6
12	11	7
13	11	7
13	11	7
13	11	7
13	11	7
13	11	8
14	12	8
14	12	10
14	12	10
15	12	10
15	13	11
16	14	13

Solution:

With 2 degrees of freedom, H = 31.0235, $p \leq 0.01$.

Problem 91: Kruskal–Wallis Nonparametric ANOVA

The number of live fetuses produced by female rats administered a control and two different substances were determined. Using the Kruskal–Wallis nonparametric ANOVA, determine if there is a significant group difference at the $p \leq 0.01$ level.

Control	Compound X	Compound Y
7	7	9
12	7	10
8	6	13
11	4	12
10	8	16
8	8	6
9	9	11
9	5	12

Solution:

With 2 degrees of freedom, H = 10.365, $p = 0.0052$; significant.

Problem 92: Kruskal–Wallis Nonparametric ANOVA

Two compounds and a comparator were tested for their ability to promote seizures in seizure-prone mice. Using the Kruskal–Wallis nonparametric ANOVA, determine if there is a significant group difference at the $p \leq 0.05$ level.

Pro1	Pro2	Comparator
9	13	19
10	6	21
13	10	14
17	12	17
9	16	22
15	13	15
24	16	
	9	

Solution:

With 2 degrees of freedom, H = 6.266, p = 0.0425; significant.

Problem 93: Kruskal–Wallis Nonparametric ANOVA

Eight compounds are being tested for their ability to cause place preference in rats. The number of minutes spent in the dosing chamber was recorded. Using the Kruskal–Wallis nonparametric ANOVA, determine if there is a significant group difference at the $p \leq 0.05$ level.

S001	S002	S003	S004	S005	S006	S007	S008
9	3	7	24	17	3	13	10
9	10	13	17	9	15	8	24
13	6	10	22	16	13	16	9
19	5	21	23	14	17	22	15
6	18			10		17	12
	18			13			

Solution:

With 7 degrees of freedom, H = 11.4343, p = 0.1208; not significant.

Problem 94: Student's *t*-Test

Compare the following two sets of body weights using the Student's *t*-test. Are they significantly different? At what level?

Control	Treatment
148.6	154.3
154.7	149.4
151.7	153.6
144.7	151.8
155.8	154.0
160.8	155.1
151.4	151.1
152.1	152.3
148.3	147.3
150.1	156.7

Solution:

$t = 0.4425$ with 18 degrees of freedom; not significant at $p \leq 0.05$.

Problem 95: Student's *t*-Test

Compare the following two sets of body weights using the Student's *t*-test at a level of $p \leq 0.05$.

Control	Treatment
224.7	217.3
213.3	217.8
226.3	220.8
229.6	228.3
213.8	220.8
209.2	223.5
208.4	209.6
214.0	212.8
207.2	222.2
210.9	218.8

Solution:

$t = 1.1267$ with 18 degrees of freedom; not significant at $p \leq 0.05$.

Problem 96: Student's *t*-Test

Using Student's *t*-test, compare the following two sets of creatinine values presented below at the level of $p \leq 0.05$.

Control	Treatment
201.3	113.3
171.6	128.7
149.6	141.9
206.8	152.9
192.5	150.7

Solution:

$t = 3.636$ with 8 degrees of freedom; not significant at $p \leq 0.05$.

Problem 97: Student's *t*-Test

Using Student's *t*-test, compare the following two sets of alkaline phosphatase data.

A	B
531	504
500	530
498	470
538	528
519	510
582	506
538	542
519	496
612	526
540	491

Solution:

$t = 2.092$ with 18 degrees of freedom; not significant at $p < 0.05$.

$$(t18df = 2.101)$$

Problem 98: Student's *t*-Test

Using the Student's *t*-test, compare the following two sets of alkaline phosphatase data. How does this data set differ from that in Problem 97?

A	B
521	494
500	540
498	470
548	538
509	500
582	496
548	542
509	486
612	546
550	491
567	
508	

Solution:

$t = 1.9447$ with 20 degrees of freedom; not significant at $p \leq 0.05$. Although means and standard deviations are nearly equal to those of the groups in Problem 97, the group sizes here are unequal.

Problem 99: Student's *t*-Test

Using the Student's *t*-test, compare the following sets of body weight data against the control at the $p \leq 0.05$ level.

Control	Tmt 1	Tmt 2	Tmt 3	Tmt 4	Tmt 5	Tmt 6
2.8	3	3	3.2	3.1	3.4	2.8
2.8	2.9	3	3	3.1	3.2	2.5
2.5	2.5	2.7	2.9	3.1	3.2	2.7
2.3	2.5	2.5	2.7	2.9	3.1	2.3
3	3.1	3.1	3.3	3.2	3.6	3.1
3.6	3.7	3.6	3.7	3.9	4.1	3.6
3.1	3.1	3.2	3.3	3.4	3.6	2.8
2.7	2.8	2.9	3	3.1	3.5	2.5
2.5	2.6	2.6	2.7	2.9	3.1	2.8
2.8	2.9	3	3	3.1	3.2	3

Solution:

	Tmt 1	Tmt 2	Tmt 3	Tmt 4	Tmt 5	Tmt 6
t-Value	0.4023	0.6735	1.2525	1.7863	2.6794	0
Significant?	No	No	No	Yes	Yes	No

Problem 100: Cochran *t*-Test

Compare the following unequal groups of data using the Cochran *t*-test. Are they significantly different? At what level?

A	B
521	504
510	527
498	470
538	524
519	510
572	502
538	525
519	490
602	538
540	
557	
518	
544	
543	

Solution:

$t_{obs} = 2.6817$.

$t' = 2.2315$ (at $p < 0.05$), 3.180 (at $p \leq 0.0$).

Groups are significantly different at $0.05 > p > 0.01$.

Problem 101: Cochran *t*-Test

Using the Cochran *t*-test, compare the two groups of serum glucose values presented below. Are they significantly different at a $p \leq 0.05$ level?

Group A	Group B
166	179
157	180
168	169
174	182
178	188
174	177
168	186
170	166
169	
174	
168	

Solution:

$t_{obs} = 2.746$.
$t' = 3.401$ (at $p \leq 0.01$).
Groups are significantly different at $0.05 > p > 0.01$.

Problem 102: Cochran *t*-Test

Using the Cochran *t*-test, compare the potassium levels in the dosed groups with the control groups (separately) for the following data. Determine the *t* values and the level of significance in the following ranges: $p \leq 0.001$, $p \leq 0.01$, $p \leq 0.05$, $p \leq 0.10$, or $p > 0.10$.

No Treatment	Vehicle Treatment	Dose 0.001 mg/kg	Dose 0.010 mg/kg	Dose 0.100 mg/kg
4.9	4.9	4.4	5.7	4.3
3.8	5.9	4.1	4.7	4.2
5.2	4.8	4.0	5.4	4.9
4.8	5.6	5.0	6.1	4.7
4.7	5.2	4.6	5.2	4.6
4.4	5.9		5.8	4.5
5.8	5.5		4.9	4.2
4.6	4.8		5.7	4.3
4.9	5.9			4.5
4.5				4.4
5.3				

Solution:

	Versus No Treatment			Versus Vehicle		
	Dose 0.001 mg/kg	Dose 0.010 mg/kg	Dose 0.100 mg/kg	Dose 0.001 mg/kg	Dose 0.010 mg/kg	Dose 0.100 mg/kg
t value	1.6267	2.7204	2.0166	4.0397	0.2098	5.3382
p	> 0.10	0.05	0.10	0.05	> 0.10	0.001

Problem 103: Cochran *t*-Test

Use the data from Problem 97 to calculate a Cochran *t*-value and *t'*. Are these the same as the Student's *t*-value and *t* distribution values? Explain why or why not.

Solution:

$t_{obs} = 2.092.$
$t' = 2.2622$ (at $p \leq 0.05$).
t_{obs} is the same because the formula for the Cochran t is the same for that of Student's t when $N_1 = N_2$. t' is not the same but if the variances were the same between groups, it would be.

Problem 104: *F*-Test

Using the *F*-test, evaluate the two data sets in Problems 100 and 101 for homogeneity of variance. Use the first nine pairs of values in 100 and the first eight pairs in 101.

Solution:

A: For the Problem 100 data set, $F = 2.38669$. The test statistic for 8 and 8 df at $p \leq 0.05$ is 4.43, so the groups are suitably homogeneous.

B: For the Problem 101 data set, $F = 1.43376$. The test statistic for 7 and 7 df at $p \leq 0.05$ is 4.99, so the groups are suitably homogeneous.

Problem 105: *F*-Test

Using the *F*-test, evaluate the two data sets in Problem 97 and Problem 98 for homogeneity of variance.

Solution:

A: For the Problem 97 data set, $F = 2.666$. The test statistic for 9 and 9 df at $p \leq 0.05$ is 4.03, so the groups are suitably homogeneous.

B: For the Problem 98 data set, $F = 1.6740$. The test statistic for 9 and 9 df at $p \leq 0.05$ is 4.03, so the groups are suitably homogeneous.

Problem 106: *F*-Test

Using the data from Problem 99, compare Tmt 4 and Tmt 5 against the control values. Are these significant at the $p \leq 0.05$ level?

Solution:

Tmt 4 vs. control: $F = 1.5992$ (df = 9 for both), so not significant (critical value 3.18).

Tmt 5 vs. control: $F = 1.3739$ (df = 9 for both), so not significant (critical value 3.18).

Problem 107: *F*-Test

Using the data from Problem 102, compare each treated group to each control. Use $p \leq 0.01$ to test for significance.

Solution:

	Versus No Treatment			Versus Vehicle		
	Dose 0.001 mg/kg	Dose 0.010 mg/kg	Dose 0.100 mg/kg	Dose 0.001 mg/kg	Dose 0.010 mg/kg	Dose 0.100 mg/kg
F-value	1.6846	1.1949	5.2935	1.3957	1.0101	4.3858
$p \leq 0.01$?	No	No	Yes	No	No	No

Problem 108: ANOVA

Using an analysis of variance, analyze the body weights from the following four groups of rats to determine if there are any statistically significant differences between groups.

Group A	Group B	Group C	Group D
148.8	146.8	148.7	148.0
153.2	157.3	154.5	144.8
148.9	143.5	152.3	148.7
141.8	156.6	139.4	150.7
149.8	148.0	146.9	146.7
152.2	151.6	151.7	146.1
147.7	150.4	149.8	146.7
145.4	155.1	144.0	151.9
144.2	144.1	150.5	142.4
146.2	145.8	151.3	152.3
145.7	149.2	144.0	150.1
152.1	151.1	148.9	146.8

Solution:

$F = 0.71025$, making consultation of a table unnecessary. There is no significant difference between groups.

Problem 109: ANOVA

Using an analysis of variance, compare the body weights from five groups of rats to determine if there are any statistically significant differences between groups. Data are presented below.

Group A	Group B	Group C	Group D	Group E
326.2	314.5	324.9	306.2	274.0
296.1	312.5	316.2	314.6	279.2
316.9	312.2	286.4	302.5	292.2
328.2	333.7	334.7	317.8	293.9
327.8	298.3	303.88	319.3	274.2
311.1	320.7	325.5	300.1	285.5
340.9	324.4	314.4	316.6	279.5
327.2	324.4	348.4	301.3	290.9
314.8	331.1	284.1	333.4	276.7
325.5	335.1	317.8	303.5	297.7
326.6	331.6	311.5	305.3	252.8
305.1	303.4	306.2	292.9	271.0
302.2	333.9	325.1	283.8	265.7
343.6	301.1	322.4	323.9	277.9
335.5	335.7	270.2	310.4	273.2

Solution:

$F = 21.2951$, making the groups significantly different at a level of $p \leq 0.001$.

Problem 110: ANOVA

Using an analysis of variance, compare the following four sets of alkaline phosphatase data and determine if there are any significant differences between the groups.

Group A	Group B	Group C	Group D
418	437	380	272
367	413	501	334
328	348	356	368
276	307	386	357
421	416	341	448
312	388	413	390
279	426	332	376
306	401	373	314
342	384	370	517
379	359	400	340

Solution:

$F = 1.4936$, making the groups not significantly different at a 0.05 level (that is, $p \geq 0.05$).

Problem 111: ANOVA

Using an analysis of variance, compare the SGPT values presented below for five groups of rats.

Group A	Group B	Group C	Group D	Group E
65	53	65	79	80
66	65	83	73	70
59	58	66	56	87
46	88	57	58	86
77	133	82	56	76
52	58	69	85	71
48	69	86	72	84
52	66	77	66	102
63	58	68	59	93
76	66		74	

Solution:

$F = 3.1928$, making the groups significantly different at a level of $p \leq 0.05$.

Problem 112: ANOVA

Using the data from Problem 99, run an ANOVA. Would *post hoc* tests be reasonable (at the $p \leq 0.05$ level)?

Solution:

$F = 4.1965$, critical value 2.2464, $p = 0.0013$, so *post hoc* tests are appropriate.

Problem 113: ANOVA

Using the data from Problem 102, run an ANOVA. Would *post hoc* tests be reasonable (at the $p \leq 0.01$ level)? If the previous answer is yes, what test would be appropriate (from Figure 2.2)?

Solution:

$F = 9.9898$, critical value 2.6190, $p < 0.001$, so *post hoc* tests are appropriate. Given that group counts are not close, Duncan's is appropriate.

Problem 114: Duncan's Multiple Range Test

Using Duncan's multiple range test, compare the groups in Problem 108 to determine which, if any, are significantly different at the $p \leq 0.05$ level.

Solution:

None of the groups are different from the others at $p \leq 0.05$.

Problem 115: Duncan's Multiple Range Test

Using Duncan's multiple range test, compare the groups in Problem 109 to determine which, if any, are significantly different.

Solution:

Using Duncan's multiple range, group E is significantly different from all other groups at $p \leq 0.001$. Group D is significantly different from groups A and B at $p \leq 0.05$.

Problem 116: Duncan's Multiple Range Test

Using Duncan's multiple range test, compare the groups in Problem 111 to determine which, if any, are significantly different.

Solution:

Using Duncan's multiple range, group E is significantly different from group A at $p \leq 0.01$.

Problem 117: Duncan's Multiple Range Test

Using Duncan's multiple range test, compare the groups in Problem 99 to determine which, if any, are significantly different.

Solution:

```
Duncan's Multiple Range Test for Value
     Alpha                              0.05
     Error Degrees of Freedom             63
     Error Mean Square              0.110206

Duncan Grouping       Mean     N      Group
             A       3.4000    10      tmt5
       B,    A       3.1800    10      tmt4
       B,    C       3.0800    10      tmt3
       B,    C       2.9600    10      tmt2
       B,    C       2.9100    10      tmt1
             C       2.8100    10    control
             C       2.8100    10      tmt6
```

Problem 118: Duncan's Multiple Range Test

Using Duncan's multiple range test, compare the groups in Problem 102 to determine which, if any, are significantly different.

Solution:

```
Duncan's Multiple Range Test for Value
     Alpha                                  0.05
     Error Degrees of Freedom                 38
     Error Mean Square                  0.190756
     Harmonic Mean of Cell Sizes        7.974225
          NOTE: Cell sizes are not equal.
Duncan Grouping       Mean        N      Group
          A          5.4375       8       d010
          A          5.3889       9     vehicle
          B          4.8091      11    no_treat
          B          4.4600      10       d100
          B          4.4200       5       d001
```

Problem 119: Scheffé's Method

Using Scheffe's method, compare the groups in Problem 110 to determine which, if any, are significantly different.

Solution:

Using Scheffé's procedure, no groups are significantly different from any others at $p \leq 0.05$.

Problem 120: Scheffé's Method

Using Scheffé's method, compare the groups in Problem 111 to determine which, if any, are significantly different.

Solution:

Using Scheffé's procedure, group E is significantly different from group A at $p \leq 0.05$.

Problem 121: Scheffé's Method

Using Scheffé's method, compare the groups in Problem 99 to determine which, if any, are significantly different.

Solution:

```
                    Scheffe's Test for Value
        Alpha                                          0.05
        Error Degrees of Freedom                         63
        Error Mean Square                          0.110206
        Critical Value of F                         2.24641
        Minimum Significant Difference               0.5451
Scheffe Grouping              Mean        N           Group
            A               3.4000       10            tmt5
       B,   A               3.1800       10            tmt4
       B,   A               3.0800       10            tmt3
       B,   A               2.9600       10            tmt2
       B,   A               2.9100       10            tmt1
       B                    2.8100       10         control
       B                    2.8100       10            tmt6
```

Problem 122: Scheffé's Method

Using Scheffé's method, compare the groups in Problem 102 to determine which, if any, are significantly different.

Statistics and Experimental Design

Solution:

```
                    Scheffe's Test for Value
        Alpha                                        0.05
        Error Degrees of Freedom                       38
        Error Mean Square                        0.190756
        Critical Value of F                       2.61899
        Minimum Significant Difference              0.708
        Harmonic Mean of Cell Sizes              7.974225
                NOTE: Cell sizes are not equal.
Scheffe Grouping           Mean         N             Group
            A            5.4375         8             d010
            A            5.3889         9          vehicle
      B,    A            4.8091        11         no_treat
      B                  4.4600        10             d100
      B                  4.4200         5             d001
```

Problem 123: Dunnet's *t*-Test

Using Dunnet's *t*-test, compare the groups in Problem 109 to determine which, if any, are significantly different. For the purposes of this problem, assume group A is the control group while groups B–E are successively higher-dosage groups.

Solution:

Using Dunnet's *t*-test, the only comparison is vs. group A, which is designated the control. Group E is significantly different from group A at $p \leq 0.01$.

Problem 124: Dunnet's *t*-Test

Using Dunnet's *t*-test, compare the groups in Problem 111 to determine which, if any, are significantly different. For the purposes of this problem, assume group A is the control group while groups B–E are groups with successively higher inhalation exposure levels.

Solution:

Using Dunnet's *t*-test, the only comparison is vs. group A, which is designated the control. Group E is significantly different from group A at $p \leq 0.01$.

Problem 125: Dunnet's *t*-Test

Using Dunnet's *t*-test, compare the groups in Problem 99 to determine which, if any, are significantly different from the control at the $p \leq 0.05$ level.

Solution:

```
               Dunnett's t-Tests for Value
          Alpha                                          0.05
          Error Degrees of Freedom                         63
          Error Mean Square                           0.110206
          Critical Value of Dunnett's t                2.63801
          Minimum Significant Difference                0.3916
Comparisons significant at the 0.05 level are indicated by ***.
```

Group Comparison		Difference Between Means	Simultaneous 95% Confidence Limits		
tmt5	- control	0.5900	0.1984	0.9816	***
tmt4	- control	0.3700	-0.0216	0.7616	
tmt3	- control	0.2700	-0.1216	0.6616	
tmt2	- control	0.1500	-0.2416	0.5416	
tmt1	- control	0.1000	-0.2916	0.4916	
tmt6	- control	0.0000	-0.3916	0.3916	

Problem 126: Dunnet's *t*-Test

Using Dunnet's *t*-test, compare the groups in Problem 102 to determine which, if any, are significantly different from the two controls (test separately) at the $p \leq 0.05$ level.

Solution:

```
               Dunnett's t-Tests for Value
Alpha                                                    0.05
Error Degrees of Freedom                                   38
Error Mean Square                                    0.190756
Critical Value of Dunnett's t                         2.56870
```

Group Comparison		Difference Between Means	Simultaneous 95% Confidence Limits		
d010	- no_treat	0.6284	0.1071	1.1497	***
vehicle	- no_treat	0.5798	0.0755	1.0841	***
d100	- no_treat	-0.3491	-0.8393	0.1411	
d001	- no_treat	-0.3891	-0.9942	0.2160	

```
               Dunnett's t-Tests for Value
Alpha                                                    0.05
Error Degrees of Freedom                                   38
Error Mean Square                                    0.190756
Critical Value of Dunnett's t                         2.55473
```

Group Comparison		Difference Between Means	Simultaneous 95% Confidence Limits		
d010	- vehicle	0.0486	-0.4936	0.5908	
no_treat	- vehicle	-0.5798	-1.0813	-0.0783	***
d100	- vehicle	-0.9289	-1.4416	-0.4162	***
d001	- vehicle	-0.9689	-1.5912	-0.3465	***

Problem 127: ANCOVA

Use analysis of covariance to adjust for body weight (BW) effects and determine if there is any difference between combined kidney weights of the 4 groups of animals below at a $p \leq 0.05$ level.

Control		Low Dose		Mid Dose		High Dose	
BW	Kidney Wt	BW	Kidney Wt	BW	Kidney Wt	BW	Kidney Wt
184.2	1.74	174.3	1.63	182.0	1.64	176.4	1.55
183.5	1.73	178.6	1.67	179.8	1.62	167.4	1.47
186.9	1.76	184.2	1.72	174.2	1.56	172.4	1.53
188.8	1.78	183.9	1.72	179.3	1.61	171.5	1.51
188.6	1.77	182.4	1.70	178.6	1.63	177.3	1.56
189.7	1.79	192.7	1.80	180.7	1.62	173.6	1.53
194.4	1.83	186.3	1.73	176.3	1.59	172.9	1.52
192.6	1.82	194.8	1.83	186.5	1.69	168.1	1.48

Solution:

Using analysis of covariance, there was a significant difference between groups of kidney weights (adjusted for body weight) at $p \leq 0.001$ ($F = 163.2$).

Problem 128: ANCOVA

Using analysis of covariance to adjust for body weight (BW) effects, determine if there is any difference between spleen weights of the 4 groups of animals presented below at a $p \leq 0.05$ level.

Control		Low Dose		Mid Dose		High Dose	
BW	Spleen Wt	BW	Spleen Wt	BW	Spleen Wt	BW	Spleen Wt
142.2	0.40	147.5	0.45	138.6	0.45	132.2	0.49
146.9	0.42	143.6	0.43	142.9	0.46	130.2	0.47
138.7	0.39	140.4	0.43	130.3	0.42	134.6	0.50
144.3	0.41	138.2	0.41	134.6	0.43	125.4	0.46
140.5	0.40	133.9	0.40	146.7	0.47	127.8	0.47

Solution:

Using analysis of covariance, there was a significant difference between groups of spleen weights (adjusted for body weight) at $p \leq 0.001$ ($F = 296.6$). How would you determine which groups are significantly different?

Problem 129: ANCOVA

We would like to determine if a pollutant had an effect on the average height of pine trees over a period of time. We measure both the ages (in years, by counting rings) and height of two random samples of pines (one from the contaminated area and the other from a noncontaminated area). These data are presented below.

Control		Contaminated	
Age (Years)	Height (Feet)	Age (Years)	Height (Feet)
71	79.0	59	77.6
71	80.9	63	87.8
78	74.9	65	77.6
79	89.5	66	80.7
81	88.7	70	72.7
82	71.0	71	85.4
83	82.3	74	75.0
84	79.9	76	73.0
88	82.9	79	88.5
88	87.7	80	87.5
88	87.7	83	88.8
90	74.6	84	73.2
90	75.8	85	87.7
90	88.5	87	92.0

Using analysis of covariance to adjust for age, determine if there is a statistically significant effect.

Solution:

Using analysis of covariance to adjust for age, there is no significant difference between tree heights in the two samples ($F = 0.617$).

Problem 130: ANCOVA

Use analysis of covariance to calculate the significance of sex and dose on the predose body weight for the following data at the $p \leq 0.05$ level:

Dose	Sex	Predose	Week 1	Week 2	Week 3
0	Female	2.6	2.7	3.0	3.1
0	Female	2.6	2.7	2.8	2.9
0	Female	2.6	2.7	3.0	3.1
0	Male	2.7	2.7	3.0	3.3
0	Male	2.8	2.9	3.3	3.2
0	Male	2.8	2.8	2.9	3.0
3	Female	2.5	2.7	2.9	3.1
3	Female	2.8	2.7	2.8	3.1
3	Female	2.5	2.6	2.8	2.9

3	Male	2.8	2.8	3.0	3.1
3	Male	2.8	2.8	2.9	3.0
3	Male	2.9	2.9	3.0	3.2
9	Female	2.5	2.6	2.8	3.0
9	Female	2.8	2.9	3.1	3.1
9	Female	2.6	2.8	2.9	2.8
9	Male	2.6	2.6	2.8	2.8
9	Male	3.0	3.1	3.2	3.3
9	Male	2.5	2.6	2.7	2.8

Solution:

As shown by the SAS results below, the groups' differences are sex based, but not dose or dose and sex based.

```
                    The ANOVA Procedure
Dependent Variable: Value
                                Sum of
Source                 DF       Squares     Mean Square   F-Value   Pr > F
Model                   5     0.13777778    0.02755556      1.27    0.3375
Error                  12     0.26000000    0.02166667
Corrected Total        17     0.39777778

         R-Square        Coeff Var    Root MSE    Value Mean
         0.346369        5.474232     0.147196     2.688889

Source                 DF     Anova SS     Mean Square   F-Value   Pr > F
Dose                    2    0.00777778    0.00388889      0.18    0.8379
Sex                     1    0.10888889    0.10888889      5.03    0.0447
Dose and sex            2    0.02111111    0.01055556      0.49    0.6260
```

Problem 131: ANCOVA

Further analyze the data from Problem 130 to determine the significance of time in the analysis.

Solution:

As shown by the SAS results below, the groups' differences are sex and time based. No other factor or combination of factors is significant.

```
                    The ANOVA Procedure
Dependent Variable: Value
                                Sum of
Source                 DF       Squares     Mean Square   F-Value   Pr > F
Model                  23     1.82319444    0.07926932      3.30    0.0002
Error                  48     1.15333333    0.02402778
Corrected Total        71     2.97652778

         R-Square        Coeff Var    Root MSE    Value Mean
         0.612524        5.425690     0.155009     2.856944

Source                 DF     Anova SS     Mean Square   F-Value   Pr > F
Dose                    2    0.03527778    0.01763889      0.73    0.4852
```

Sex	1	0.17013889	0.17013889	7.08	0.0106
Dose and sex	2	0.07194444	0.03597222	1.50	0.2340
Week	3	1.44708333	0.48236111	20.08	<.0001
Dose and week	6	0.06583333	0.01097222	0.46	0.8367
Sex and week	3	0.02152778	0.00717593	0.30	0.8262
Dose and sex and week	6	0.01138889	0.00189815	0.08	0.9979

Problem 132: Linear Regression

Using the data below, perform a linear regression of tree age vs. height. What is the predicted height of a 75-year-old tree?

Age (Years)	Height (Feet)
71	79.0
71	80.9
78	74.9
79	89.5
81	88.7
82	71.0
83	82.3
84	79.9
88	82.9
88	87.7
88	87.7
90	74.6
90	75.8
90	88.5

Solution:

A 75-year-old tree would be predicted to have a height of 80.77 feet. The correlation, however, is not particularly good ($p = 0.1197$).

Problem 133: Linear Regression

Use linear regression to prepare a dose response curve for the following study data.

Dose (mg/kg)	RBC Survival (Days, Mean)
0	59
10	57
20	55
40	44
60	38
100	22
120	17

What dose would be predicted to produce a mean survival of 50 days?

Solution:

The regression should predict that a dose of 27.55 will produce a survival time of 50 days.

Problem 134: Linear Regression

The following dose–response mortality data were obtained in an oral study.

Dose (mg/kg)	Mortality (%)
10	0
20	10
40	20
60	30
80	30
140	80
160	100

Using a probit transformation for response and log-transformation for dose, perform a linear regression to estimate the LD_{50} for this compound.

Solution:

The LD_{50} is predicted to be 66.67 mg/kg.

Problem 135: Linear Regression

The following dose–response mortality data were obtained in an inhalation study.

Concentration (ppm)	Mortality (%)
10	0
30	10
60	20
120	30
240	60
360	80
480	80
600	100

Using a log-transformation for dose and a probit transformation for response, perform a linear regression to estimate the LC_{50} for this compound.

Solution:

The LC_{50} is predicted to be 153.60 ppm.

Problem 136: Linear Regression

For the following data, estimate the equation for the model and the week 6 weights for each dose level using a linear regression. Count males and females together.

Dose	Sex	Predose	Week 1	Week 2	Week 3
0	Female	2.6	2.7	3.0	3.1
0	Female	2.6	2.7	2.8	2.9
0	Female	2.6	2.7	3.0	3.1
0	Male	2.7	2.7	3.0	3.3
0	Male	2.8	2.9	3.3	3.2
0	Male	2.8	2.8	2.9	3.0
3	Female	2.5	2.7	2.9	3.1
3	Female	2.8	2.7	2.8	3.1
3	Female	2.5	2.6	2.8	2.9
3	Male	2.8	2.8	3.0	3.1
3	Male	2.8	2.8	2.9	3.0
3	Male	2.9	2.9	3.0	3.2
9	Female	2.5	2.6	2.8	3.0
9	Female	2.8	2.9	3.1	3.1
9	Female	2.6	2.8	2.9	2.8
9	Male	2.6	2.6	2.8	2.8
9	Male	3.0	3.1	3.2	3.3
9	Male	2.5	2.6	2.7	2.8

Solution:

Dose	Model	Week 6 Estimated
0	Weight = 2.65833 + 0.15000 * Week	3.55833
3	Weight = 2.67833 + 0.12000 * Week	3.39833
9	Weight = 2.67167 + 0.10500 * Week	3.30167

Problem 137: Linear Regression

Repeat the regression from Problem 136, but separate out males and females. Comment on the results.

528 Statistics and Experimental Design

Solution:

Dose	Sex	Model	Week 6 Estimated
0	F	Weight = 2.58667 + 0.15333 * Week	3.50665
0	M	Weight = 2.73000 + 0.14667 * Week	3.61002
3	F	Weight = 2.56333 + 0.14667 * Week	3.44335
3	M	Weight = 2.79333 + 0.09333 * Week	3.35331
9	F	Weight = 2.65000 + 0.11667 * Week	3.35002
9	M	Weight = 2.69333 + 0.09333 * Week	3.25331

Males started out with a higher mean weight than females, but weight gain was slower for males than females for every dose level, leading to similar weights at week 6 (projected). Higher doses had lower weight gain when adjusting for sex.

Problem 138: Linear Regression

Using the data from Problem 136, first calculate the weight gain at week 3 for each animal (week 3 – predose). Then, plot the gain vs. the dose and fit a linear regression to the model (weight gain) = (dose 0 value) + dose * slope. What dose (if any) is projected to stop weight gain?

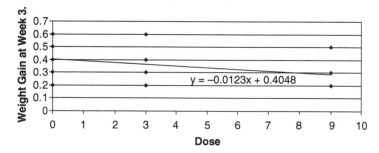

Solution:

The dose predicted to stop weight gain is 32.91.

Problem 139: Moving Average

Using the moving average method, predict the LD_{50} from the data in Problem 134. Assume 10 animals per group.

Solution:

LD_{50} = 80.0 mg/kg with a 95% confidence interval of 58.1 to 110.3 mg/kg. Note that you must use only the 20, 40, 80, and 160 mg/kg dose groups in the computation.

Problem 140: Moving Average

Using the moving average method, predict the LC_{50} from the data in Problem 135. Assume ten animals per group.

Solution:

$LC_{50} = 190.5$ ppm with a 95% confidence interval of 106.6 to 340.2 ppm. Note that you may use either the 60, 120, 240, or 480 exposure groups (which were used for these values) or the 30, 60, 120, or 240 groups (which are less desirable, as they do not cover the entire range of doses).

Problem 141: Moving Average

Using the moving average method, estimate the ED_{50} for the following oral study data. Assume eight animals per group.

Dose (mg/kg)	Effected
2	0/8
4	1/8
8	2/8
16	3/8
20	3/8
40	4/8
60	6/8
100	8/8

Solution:

Cannot be calculated from the data set by this method.

Problem 142: Moving Average

Using the data from Problem 141 and assuming that the lowest four doses were the only ones, calculate the ED_{50}.

Solution:

The tables do not have $n = 8$, but the relationship is obviously log-linear, so $ED_{50} = 32$.

Problem 143: Moving Average

Using the moving average method, estimate the ED_{50} for the following oral study data. Assume five animals per group.

Dose	Effected
3	0/5
9	1/5
27	2/5
81	5/5

Solution:

D 0.4771; $d = 0.4771$; $f = 0.9000$, so $ED_{50} = 41.89$.

Problem 144: Correlation Coefficient

Using the data presented in Problem 141, calculate the correlation coefficient for a linear regression.

Solution:

$$r = 0.96869$$

Problem 145: Correlation Coefficient

Using the data presented in Problem 135, calculate the linear correlation coefficient.

Solution:

$$r = 0.973$$

Problem 146: Correlation Coefficient

Using the data presented in Problem 132, calculate the linear correlation coefficient.

Solution:

$$r = 0.1197$$

Problem 147: Correlation Coefficient

Given the following data, calculate the correlation coefficient between dose and weight changes for the three weeks shown.

Dose	Sex	Week 1 Δ	Week 2 Δ	Week 3 Δ
0	Female	0.1	0.4	0.5
0	Female	0.1	0.2	0.3
0	Female	0.1	0.4	0.5
0	Male	0	0.3	0.6
0	Male	0.1	0.5	0.4
0	Male	0	0.1	0.2
3	Female	0.2	0.4	0.6
3	Female	−0.1	0	0.3
3	Female	0.1	0.3	0.4
3	Male	0	0.2	0.3
3	Male	0	0.1	0.2
3	Male	0	0.1	0.3
9	Female	0.1	0.3	0.5
9	Female	0.1	0.3	0.3
9	Female	0.2	0.3	0.2
9	Male	0	0.2	0.2
9	Male	0.1	0.2	0.3
9	Male	0.1	0.2	0.3

Solution:

	Week 1 Δ	Week 2 Δ	Week 3 Δ
r	0.2390	−0.4316	−0.3539

Problem 148: Correlation Coefficient

Repeat the calculation from Problem 147, but this time separate the sexes. Did the fit get worse or improve (using absolute value)?

Solution:

	Week 1 Δ	Week 2 Δ	Week 3 Δ
Females	0.218218	−0.04963	−0.35714
	worse	worse	improved
Males	0.377964	−0.24815	−0.39703
	improved	worse	improved

Problem 149: Correlation Coefficient

Make up a set of points and calculate the correlation coefficient for it. Then double it by duplicating each point and calculate the correlation coefficient again. Compare the two values.

Solution:

The values will be the same.

Problem 150: Kendall's Rank Correlation

Using Kendall's rank correlation, determine if the height of the trees in the following data is rank correlated to the age.

Age (Years)	Height (Feet)
59	77.6
63	87.8
65	77.6
66	80.7
70	72.7
71	85.4
74	75.0
76	73.0
79	88.5
80	87.5
83	88.8
84	73.2
85	87.7
87	92.0

Solution:

Kendall's rank correlation $\tau = 0.2857$.

Problem 151: Kendall's Rank Correlation

Use Kendall's rank correlation to determine if crown-rump length is inversely correlated to test error rate in the data set below.

Crown-Rump Length	Number of Errors
5	24
9	18
6	24
8	19
10	16
5	26
11	17
13	14
6	23
8	22
12	15

Solution:

Kendall's rank correlation $\tau = 0.8909$.

Problem 152: Kendall's Rank Correlation

Using the data from Problem 147, calculate Kendall's rank correlation for each of the three weeks to the dose. Do not differentiate by sex.

Solution:

	Week 1 Δ	Week 2 Δ	Week 3 Δ
τ	0.15320	−0.15369	−0.31492

Problem 153: Kendall's Rank Correlation

Repeat the calculation from Problem 152, but this time separate the sexes. Did the fit get worse or improve (using absolute value)?

Solution:

	Week 1 Δ	Week 2 Δ	Week 3 Δ
Females	0.21517	−0.22222	−0.28109
	improved	improved	worse
Males	0.27217	−0.07407	−0.29630
	improved	worse	worse

Problem 154: Cox–Stuart Trend Analysis

Perform a Cox–Stuart trend analysis to determine if there is dose responsive increase in tumor incidence over time associated with the compound.

	Total Animals with Tumors			
Month of Study	Control	Low Dose	Mid Dose	High Dose
10	0	0	0	0
11	1	1	2	2
12	1	3	6	6
13	1	4	10	10
14	2	6	11	13
15	3	7	13	16
16	4	9	15	18
17	5	10	16	20
18	5	10	20	22
19	8	12	22	24
20	11	13	24	26
21	15	14	27	28
22	18	18	29	30
23	24	23	31	33
24	32	29	33	35

Solution:

Cumulative scores and their probabilities under the sign test are
  Low dose   6+, 9–: Sum = 3– ($p = 0.30$).
  Mid dose   16+, 14–: Sum = 2+ ($p = 0.40$).
  High dose   18+, 14–: Sum = 4+ ($p = 0.25$).
There is a nonsignificant trend across the doses.

Problem 155: Cox–Stuart Trend Analysis

Perform a Cox–Stuart trend analysis to determine if there is a dose responsive increase in the incidence of malformed rat pups associated with being fed either of two synthetic diets over ten generations of animals.

	Total Malformed Pups		
Generation	Natural Diet	Synthetic A	Synthetic B
1	2	0	1
2	4	2	3
3	6	4	5
4	7	7	8

5	9	10	9
6	11	12	11
7	14	15	13
8	17	18	16
9	19	20	18
10	21	23	22

Solution:

Cumulative scores and their probabilities under the sign test are (compared to the natural diet):

Synthetic diet A: 4+ ($p = 0.09$).
Synthetic diet B: 4 +, 2– ($p = 0.38$).

Problem 156: Cochran–Armitage Trend Test

Given the doses and number of patients having adverse events shown in the following table, calculate the Cochran–Armitage trend values.

Dosage	No Events	Events
0	25	5
0.2	23	7
0.4	22	8
0.8	16	14
1.6	9	21

Solution:

```
Cochran-Armitage Trend Test
Statistic (Z)               -4.9121.
Asymptotic Test
One-sided Pr  <    Z         <0.0001.
Two-sided Pr  >   |Z|        <0.0001.
Exact Test
One-sided Pr  <=   Z        5.711E-07.
Two-sided Pr  >=  |Z|       6.555E-07.
         Sample Size = 150.
```

Problem 157: Cochran–Armitage Trend Test

Given the number of months since treatment and the number of mice with tumors remaining shown in the following table, calculate the Cochran–Armitage trend values.

Time	Tumors	No Tumors
0	20	80
1	21	79
2	20	80
3	17	83
4	16	84
5	14	86
6	13	87

Solution:

```
Cochran-Armitage Trend Test
Statistic (Z)            1.9492.
Asymptotic Test
One-sided Pr  >  Z        0.0256.
Two-sided Pr  > |Z|       0.0513.
Exact Test
One-sided Pr  >= Z            .
Two-sided Pr  >=|Z|           .
          Sample Size = 700.
```

Problem 158: Life Table Analysis

Using the data presented below, perform a life table analysis to determine if there is a time-adjusted, compound-related effect on either survival or tumor incidences at $p \leq 0.10$.

Interval (Months)	Control Animals Alive at Beginning of Interval	Withdrawn	Dead	Animals with Tumors	Test Animals Alive at Beginning of Interval	Withdrawn	Dead	Animals with Tumors
8–9	120	0	0	0	120	0	2	0
9–10	120	0	1	0	118	0	3	1
10–11	119	0	4	1	115	0	2	0
11–12	116	10	0	2	113	10	4	2
12–13	106	0	1	0	99	0	3	1
13–14	105	0	2	2	96	1	4	2
14–15	103	0	4	2	91	0	6	3
15–16	99	2	4	2	85	0	4	2
16–17	93	0	4	2	81	2	6	2
17–18	89	0	6	3	73	1	9	3
18–19	83	20	4	4	63	20	10	3
19–20	59	0	8	4	33	2	12	3
Total animals found with tumors				**22**				**22**

Solution:

	Control							
Interval (Months)	Alive	Withdrawn	Dead	At Risk	Proportion Dead	Probability of Survival	Cumulative Proportion Surviving	Standard Error
8–9	120	0	0	120	0.0000	1.0000	1.0000	0.000000
9–10	120	0	1	120	0.0083	0.9917	1.0000	0.008299
10–11	119	0	4	119	0.0336	0.9664	0.9917	0.018395
11–12	116	10	0	111	0.0000	1.0000	0.9583	0.019035
12–13	106	0	1	106	0.0094	0.9906	0.9583	0.021064
13–14	105	0	2	105	0.0190	0.9810	0.9493	0.024760
14–15	103	0	4	103	0.0388	0.9612	0.9312	0.030838
15–16	99	2	4	98	0.0408	0.9592	0.8950	0.036695
16–17	93	0	4	93	0.0430	0.9570	0.8585	0.042225
17–18	89	0	6	89	0.0674	0.9326	0.8216	0.048986
18–19	83	20	4	73	0.0548	0.9452	0.7662	0.056343
19–20	59	0	8	59	0.1356	0.8644	0.7242	0.068129

Interval (Months)	At Risk	Proportion Dead	Probability of Survival	Cumulative Proportion Surviving	Standard Error	Difference	SD	T	p
8–9	120	0.0167	0.9833	1.0000	0.011686	0.0000	0.1081	0.0000	0.0000
9–10	118	0.0254	0.9746	0.9833	0.018551	−0.0167	0.1639	−0.1017	0.0810
10–11	115	0.0174	0.9826	0.9583	0.022325	−0.0333	0.2018	−0.1652	0.1312
11–12	108	0.0370	0.9630	0.9417	0.028442	−0.0167	0.2179	−0.0765	0.0610
12–13	99	0.0303	0.9697	0.9068	0.033423	−0.0515	0.2334	−0.2208	0.1748
13–14	95.5	0.0419	0.9581	0.8793	0.038869	−0.0700	0.2522	−0.2774	0.2185
14–15	91	0.0659	0.9341	0.8425	0.045964	−0.0887	0.2771	−0.3202	0.2512
15–16	85	0.0471	0.9529	0.7869	0.052216	−0.1081	0.2982	−0.3626	0.2831
16–17	80	0.0750	0.9250	0.7499	0.058619	−0.1086	0.3176	−0.3420	0.2677
17–18	72.5	0.1241	0.8759	0.6937	0.067679	−0.1279	0.3416	−0.3745	0.2920
18–19	53	0.1887	0.8113	0.6075	0.082575	−0.1587	0.3727	−0.4257	0.3296
19–20	32	0.3750	0.6250	0.4929	0.106633	−0.2313	0.4180	−0.5533	0.4199

So there is no significant difference in survival between treatments. Calculations for tumor incidence are similar and also nonsignificant.

Problem 159: Life Table Analysis

Using the data presented below, perform a life table analysis to determine if there is a time-adjusted, compound-related effect on either survival or tumor incidences.

Interval (Months)	Control			Treatment		
	Alive at Start	Withdrawn	Dead	Alive at Start	Withdrawn	Dead
0–1	100	0	1	100	0	1
1–2	99	1	1	99	0	1
2–3	98	0	3	98	0	1

3–4	95	10	2	97	10	1
4–5	83	0	3	86	0	2
5–6	80	0	5	84	0	2
6–7	75	10	4	82	0	3
7–8	61	0	6	69	0	5
8–9	55	0	5	64	0	4
9–10	50	0	8	60	0	9
10–11	42	0	7	51	0	10
11–12	35	10	9	41	10	12
Total dead			**54**			**51**

Solution:

				Control			
Alive	Withdrawn	Dead	At Risk	Proportion Dead	Probability of Survival	Cumulative Proportion Surviving	Standard Error
100	0	1	100	0.0100	0.9900	1.0000	0.009950
99	1	1	98.5	0.0102	0.9898	0.9900	0.014177
98	0	3	98	0.0306	0.9694	0.9799	0.022262
95	10	2	90	0.0222	0.9778	0.9500	0.027306
83	0	3	83	0.0361	0.9639	0.9288	0.033827
80	0	5	80	0.0625	0.9375	0.8953	0.042603
75	10	4	70	0.0571	0.9429	0.8393	0.051044
61	0	6	61	0.0984	0.9016	0.7914	0.061939
55	0	5	55	0.0909	0.9091	0.7135	0.073504
50	0	8	50	0.1600	0.8400	0.6487	0.085445
42	0	7	42	0.1667	0.8333	0.5449	0.102432
35	10	9	30	0.3000	0.7000	0.4541	0.120014

						Treatment					
Alive	With-drawn	Dead	At Risk	Proportion Dead	Probability of Survival	Cumulative Proportion Surviving	Standard Error	Difference	Sd	t′	p
100	0	1	100	0.0100	0.9900	1.0000	0.009950	0.0000	0.1411	0.0000	0.0000
99	0	1	99	0.0101	0.9899	0.9900	0.014141	0.0000	0.1683	0.0000	0.0000
98	0	1	98	0.0102	0.9898	0.9800	0.017407	0.0001	0.1992	0.0003	0.0002
97	10	1	92	0.0109	0.9891	0.9700	0.020481	0.0200	0.2186	0.0917	0.0731
86	0	2	86	0.0233	0.9767	0.9595	0.025945	0.0306	0.2445	0.1252	0.0997
84	0	2	84	0.0238	0.9762	0.9371	0.030807	0.0419	0.2709	0.1546	0.1228
82	0	3	82	0.0366	0.9634	0.9148	0.036800	0.0755	0.2964	0.2548	0.2011
69	0	5	69	0.0725	0.9275	0.8814	0.047216	0.0900	0.3304	0.2724	0.2147
64	0	4	64	0.0625	0.9375	0.8175	0.056507	0.1040	0.3606	0.2884	0.2269
60	0	9	60	0.1500	0.8500	0.7664	0.068919	0.1178	0.3929	0.2997	0.2356
51	0	10	51	0.1961	0.8039	0.6514	0.085672	0.1066	0.4337	0.2457	0.1941
41	10	12	36	0.3333	0.6667	0.5237	0.105925	0.0697	0.4753	0.1465	0.1165

No significant treatment-related, time-based differences.

Problem 160: Life Table Analysis

For the following survival table for angina patients, find the year in which the cumulative probability of survival goes below 50%.

Interval (Years)	Alive	Lost to Followup	Died
0–1	577	9	20
1–2	548	7	69
2–3	472	38	36
3–4	398	22	36
4–5	340	17	35
5–6	288	11	30
6–7	247	19	8
7–8	220	33	0
8–9	187	23	15
9–10	149	8	7

Solution:

The last year (as shown).

Interval (Years)	Alive	Lost to Followup	Died	At Risk	Proportion Dead	Probability of Survival	Cumulative Proportion Surviving
0–1	577	9	20	572.5	0.0349	0.9651	1.0000
1–2	548	7	69	544.5	0.1267	0.8733	0.9651
2–3	472	38	36	453	0.0795	0.9205	0.8428
3–4	398	22	36	387	0.0930	0.9070	0.7758
4–5	340	17	35	331.5	0.1056	0.8944	0.7036
5–6	288	11	30	282.5	0.1062	0.8938	0.6293
6–7	247	19	8	237.5	0.0337	0.9663	0.5625
7–8	220	33	0	203.5	0.0000	1.0000	0.5436
8–9	187	23	15	175.5	0.0855	0.9145	0.5436
9–10	149	8	7	145	0.0483	0.9517	0.4971

Problem 161: Risk Ratios

Using the data below, calculate risk ratios (treated to untreated) and 95% two-sided confidence intervals for each trial of seizure medication. Do not use a correction factor.

Study	Treated		Untreated	
	Seizure	No Seizure	Seizure	No Seizure
A	94	63	85	115
B	11	17	10	3
C	35	47	21	28
D	14	36	15	7
E	17	25	11	13
F	9	12	19	20
G	42	111	57	23
H	20	21	20	10
I	10	27	10	7
J	28	39	73	20
K	7	19	12	9
L	5	9	23	6
M	291	1084	434	193
N	2	8	16	5
O	11	27	16	13

Solution:

Study	RR	Lower	Upper
A	1.41	1.15	1.73
B	0.51	0.30	0.88
C	1.00	0.66	1.50
D	0.41	0.24	0.70
E	0.88	0.50	1.56
F	0.88	0.49	1.59
G	0.39	0.29	0.52
H	0.73	0.49	1.09
I	0.46	0.24	0.89
J	0.53	0.39	0.72
K	0.47	0.23	0.98
L	0.45	0.22	0.93
M	0.31	0.27	0.34
N	0.26	0.07	0.93
O	0.52	0.29	0.95

Problem 162: Risk Ratios

Using the data below, calculate risk ratios (exposed to control) and 95% confidence intervals for each trial of potential carcinogen. Use a 0.5 correction factor.

Study	Exposed Colonies		Control Colonies	
	Mutant	Normal	Mutant	Normal
1	34	30	10	41
2	9	5	2	8

3	19	7	11	28
4	15	2	3	18
5	10	5	2	8
6	36	19	13	40
7	12	12	4	14
8	6	6	1	8
9	30	23	15	41
10	19	11	5	35
11	163	81	62	253

Solution:

Study	RR	Lower	Upper
1	2.63	1.46	4.73
2	2.79	0.88	8.85
3	2.51	1.46	4.32
4	5.41	2.03	14.40
5	2.89	0.92	9.08
6	2.61	1.58	4.30
7	2.11	0.86	5.18
8	3.33	0.69	16.06
9	2.08	1.28	3.37
10	4.69	2.06	10.68
11	3.37	2.66	4.29

Problem 163: Risk Reduction

Using the data from Problem 161, calculate the risk reduction for each group, as well as a 90% two-sided confidence interval. Do not use a correction factor.

Solution:

Study	RD	s^2_{RD}	Lower	Upper
A	0.17	0.0028	0.09	0.26
B	-0.38	0.0222	-0.62	-0.13
C	0.00	0.0080	-0.15	0.15
D	-0.40	0.0139	-0.60	-0.21
E	-0.05	0.0161	-0.26	0.16
F	-0.06	0.0181	-0.28	0.16
G	-0.44	0.0039	-0.54	-0.34
H	-0.18	0.0135	-0.37	0.01
I	-0.32	0.0196	-0.55	-0.09
J	-0.37	0.0054	-0.49	-0.25
K	-0.30	0.0192	-0.53	-0.07
L	-0.44	0.0221	-0.68	-0.19
M	-0.48	0.0005	-0.52	-0.45
N	-0.56	0.0246	-0.82	-0.30
O	-0.26	0.0139	-0.46	-0.07

Problem 164: Risk Reduction

Using the data from Problem 162, calculate the risk reduction for each group, as well as a 90% two-sided confidence interval. Use a 0.5 correction factor.

Solution:

Study	RD	s_{RD}^2	Lower	Upper
1	0.33	0.0069	0.19	0.47
2	0.41	0.0314	0.11	0.70
3	0.43	0.0126	0.25	0.62
4	0.70	0.0127	0.52	0.89
5	0.43	0.0301	0.14	0.71
6	0.40	0.0075	0.26	0.54
7	0.26	0.0195	0.03	0.49
8	0.35	0.0320	0.06	0.64
9	0.29	0.0080	0.15	0.44
10	0.49	0.0104	0.33	0.66
11	0.47	0.0014	0.41	0.53

Problem 165: Incidence Rates

Further research on the data from Problem 161 has revealed the study durations and true number of events for each study. Calculate the incidence rates and 99% confidence intervals for them.

	Treated		Untreated		
Study	Patients	Seizures	Patients	Seizures	Duration (Months)
A	157	1032	200	1090	6
B	28	333	13	454	24
C	82	903	49	605	12
D	50	71	22	64	3
E	42	202	24	80	6
F	21	18	39	30	1
G	153	1001	80	1420	12
H	41	198	30	249	12
I	37	568	17	220	24
J	67	43	93	119	1
K	26	196	21	256	12
L	14	18	29	130	3
M	1375	3256	627	4694	12
N	10	100	21	859	24
O	38	228	29	177	12

Solution:

	Treated Rate (Seizures/Person/Month)			Untreated Rate (Seizures/Person/Month)		
Study	Rate	Low	High	Rate	Low	High
A	1.10	1.01	1.19	0.91	0.84	0.98
B	0.50	0.43	0.57	1.45	1.29	1.64
C	0.92	0.84	1.00	1.03	0.93	1.14
D	0.47	0.35	0.64	0.97	0.71	1.34
E	0.80	0.67	0.96	0.55	0.42	0.74
F	0.86	0.47	1.57	0.77	0.48	1.23
G	0.55	0.50	0.59	1.48	1.38	1.58
H	0.40	0.33	0.48	0.69	0.59	0.81
I	0.64	0.57	0.71	0.54	0.45	0.64
J	0.64	0.43	0.95	1.28	1.01	1.62
K	0.63	0.52	0.75	1.02	0.87	1.19
L	0.44	0.24	0.80	1.50	1.20	1.88
M	0.20	0.19	0.21	0.62	0.60	0.65
N	0.42	0.32	0.54	1.71	1.56	1.86
O	0.50	0.42	0.59	0.51	0.42	0.62

Problem 166: Incidence Rates

Using the following tumor incidence counts in mice, calculate the rate of occurrence and a 90% two-sided confidence interval.

Study	Tumors	Time (Years)	N
001	216	2	55
002	45	1	24
003	500	2	125
004	740	1	421
005	71	2	21
006	280	1	227
007	123	1	108
008	79	2	23
009	126	1	71
010	41	2	15
011	181	2	60
012	63	1	36

Solution:

Study	Rate	Low	High
001	1.96	1.76	2.20
002	1.88	1.47	2.40
003	2.00	1.86	2.15
004	1.76	1.65	1.87

005	1.69	1.39	2.05
006	1.23	1.12	1.36
007	1.14	0.98	1.32
008	1.72	1.43	2.07
009	1.77	1.53	2.05
010	1.37	1.06	1.77
011	1.51	1.33	1.70
012	1.75	1.42	2.15

Problem 167: Unweighted Mean Difference

For the following human sera chemistry values calculate the mean difference (postdose − predose), estimated population variance, and variance of the difference. Assume normality and homogeneity of variance and calculate pooled variance with an N of 15 in each case.

Parameter	Predose Mean	Predose Variance	Postdose Mean	Postdose Variance
Sodium	144	6	134	7
K+	5.45	0.11	5.04	0.13
Cl–	109	8	97	9
Calcium	14.4	6.0	12.6	6.6
Phosphorus	4.90	0.25	4.38	0.29

Solution:

Parameter	Mean Difference	Estimated Population Variance	Variance of the Difference
Sodium	−10	6.5	0.9
K+	−0.41	0.120	0.016
Cl–	−12	8.5	1.1
Calcium	−1.8	6.30	0.84
Phosphorus	−0.52	0.270	0.036

Problem 168: Unweighted Mean Difference

The effects of an altered diet, increased exercise, and a cholesterol-lowering drug were tested in dogs. Calculate the mean difference from the control group for the data given below, then use those calculations to estimate a 90% confidence interval.

Treatment	N	Cholesterol Mean	Cholesterol Standard Deviation
Control	20	43	17.98
Exercise	14	59	26.98
Diet	28	42	11.57
Drug	11	55	26.01
Exercise + Diet	21	34	26.16
Exercise + Drug	15	37	7.50
Diet + Drug	29	50	10.87
All treatments	17	43	15.27

Solution:

Treatment	Mean Difference	Estimated Population Variance	Variation of the Difference	Lower CL	Upper CL
Control	NA	NA	NA	NA	NA
Exercise	16	487.67	59.22	3.34	28.66
Diet	−1	212.10	18.18	−8.01	6.01
Drug	12	445.09	62.72	−1.03	25.03
Exercise + diet	−9	508.44	49.63	−20.59	2.59
Exercise + drug	−6	209.99	24.50	−14.14	2.14
Diet + drug	7	201.08	16.99	0.22	13.78
All treatments	0	282.09	30.70	−9.11	9.11

Problem 169: Weighted Mean Differences

For the data from Problem 167, calculate the weighted mean differences and the variance of that measure, using the first equation given.

Solution:

Parameter	Mean Difference	Variance of the Difference
Sodium	−10.7	2.06
K+	−3.2	0.31
Cl−	−11.3	2.25
Calcium	−2.0	0.20
Phosphorus	−2.7	0.26

Problem 170: Weighted Mean Differences

From the data for Problem 168, calculate the weighted mean differences. Use the simplified equation for the variance and then calculate the 99% confidence intervals.

Solution:

Treatment	Weighted Mean Difference	Variance of the Difference.	Lower CL	Upper CL
Control	NA	NA	NA	NA
Exercise	2.08	0.12	1.18	2.98
Diet	−0.23	0.09	−0.99	0.52
Drug	1.52	0.14	0.55	2.48
Exercise + diet	−1.28	0.10	−2.08	−0.47
Exercise + drug	−1.21	0.12	−2.09	−0.33
Diet + drug	1.70	0.08	0.95	2.45
All treatments	0.00	0.11	−0.85	0.85

Problem 171: χ^2 Homogeneity Test

For the data presented in Problem 168, first calculate the weighted mean treatment effect (including the control) and then perform a χ^2 homogeneity test.

Solution:

The weighted mean treatment effect is 42.376 and the χ^2 test value is 1.727 at $8 - 1 = 7$ degrees of freedom. The critical value is 2.833, so the test indicates homogeneity in the treatment effects.

Problem 172: χ^2 Homogeneity Test

For the data presented in Problem 166, calculate the weighted mean treatment effect and then perform an χ^2 homogeneity test. Use the natural logarithm of the rates of occurrence for ease of calculation.

Solution:

The weighted mean of the natural logarithm of the treatment effect is 0.5220 and the χ^2 test value is 72.171 at $12 - 1 = 11$ degrees of freedom. The critical value is 5.578, so the test is highly significant for heterogeneity.

Problem 173: Mantel–Haenszel Method

For the following data, calculate the mean, variance (of the logarithm), and 95% confidence interval using the Mantel–Haenszel method. Use a 0.5 correction factor.

Study	Treated	Untreated	Affected	Not Affected
501	20	20	2	1
502	40	20	4	1
503	20	40	2	1
504	100	100	13	5
505	10	50	1	2
506	50	10	7	0

Solution:

M–H ratio: 2.5411. Variance: 0.1326. CI: 1.2448 to 5.1872.

Problem 174: Mantel–Haenszel Method

Using the data from Problem 161, calculate the mean, variance (of the logarithm), and 90% confidence interval using the Mantel–Haenszel method. Do not use a correction factor.

Solution:

M–H ratio: 0.2564. Variance: 0.0057. CI: 0.2266 to 0.2902.

Problem 175: Mantel–Haenszel Method

Using the data from Problem 162, calculate the mean, variance (of the logarithm), and 99% confidence interval using the Mantel–Haenszel method. Use a 0.5 correction factor.

Solution:

M–H ratio: 6.5802. Variance: 0.0175. CI: 4.6794 to 9.2530.

Problem 176: Peto's Method

Using the data from Problem 173, calculate the mean, variance (of the logarithm), and confidence interval using Peto's method. Use the specified α; do not use a correction factor.

Solution:

Peto's ratio: 2.4197. Variance: 0.1120. CI: 1.2556 to 4.6631.

Problem 177: Peto's Method

Using the data from Problem 161, calculate the mean, variance (of the logarithm), and confidence interval using Peto's method. Use the specified α; do not use a correction factor.

Solution:

Peto's ratio: 0.2367. Variance: 0.0055. CI: 0.2095 to 0.2673.

Problem 178: Peto's Method

Using the data from Problem 174, calculate the mean, variance (of the logarithm), and confidence interval using Peto's method. Use the specified α; do not use a correction factor.

Solution:

Peto's ratio: 5.8155. Variance: 0.0139. CI: 4.7892 to 7.0617.

Problem 179: Fisher's Sum of Logarithms

For the following study results, calculate whether the combination is significant at the $\alpha = 0.05$ level.

Study	p-Value
X001	0.94
X002	0.83
X003	0.95
X004	0.66
X005	0.68
X006	0.98
X007	0.71
X008	0.68

Solution:

Test value: 3.6981; df = 16; significant ($p > 0.999$).

Problem 180: Fisher's Sum of Logarithms

For the following study results, calculate whether the combination is significant at the $\alpha = 0.10$ level.

Study	Result
A	0.76
B	0.65
C	0.77
D	0.66
E	0.64
F	0.40
G	0.46
H	0.80
I	0.53
J	0.40

Solution:

Test value: 10.5910; df = 20; significant ($p = 0.9561$).

Problem 181: Effects of Exposure

For the following list of adverse events, determine the whole population's proportion affected and then the treated population's proportion affected.

Adverse Event	Treated (33)		Control (26)	
	Effect	No Effect	Effect	No Effect
Injection site irritation	12	21	5	21
Sniffing behavior	18	15	0	26
Staggering	2	31	1	25
Emesis	6	27	14	12
Excessive salivation	20	13	9	17

Solution:

Adverse Event	Population	Treated
Injection site irritation	0.2881	0.3636
Sniffing behavior	0.3051	0.5455
Staggering	0.0508	0.0606
Emesis	0.3390	0.1818
Excessive salivation	0.4915	0.6061

Problem 182: Assays

Given the sensitivity and specificity of the following assays and a population prevalence of the condition of 1.2%, determine what percentage of those diagnosed positive will actually have the condition. Based on the results, which one of sensitivity or specificity is more critical for conditions with a low prevalence rate?

Assay	Sensitivity (%)	Specificity (%)
1	90	94
2	98	85
3	80	99.5
4	95	95

Solution:

Assay	Condition (%)
1	15.41
2	7.35
3	66.02
4	18.75

Specificity is more important to isolate a population with the condition.

Problem 183: Assays

The treatment for the condition in Problem 182 requires surgery, but the ethics committee will only recommend it if the chance of having the condition is greater than 95%. What is the minimum number of tests needed, and in which order would they be needed to isolate a population with ≥95% portion of the condition? Give the percentages also. Assume that repeats of the same test do not give additional information and that results are independent; i.e., the tests' sensitivity and specificity are the same no matter what subset of the original population is tested.

Solution:

Assays	Condition (%)
1 and 3	96.68
3 and 4	97.36

Two tests as shown above. The order the tests are given does not matter.

Problem 184: Assays

Given the assays listed in Problem 182, calculate what percentage of the population who tested negative for each assay actually does have the condition. Given the results, which one of sensitivity or specificity is more important to finding the condition? Hint: this requires calculating the proportion with the condition who are misdiagnosed and the proportion without the condition who were properly diagnosed.

Solution:

Assay	Condition (%)
1	0.13
2	0.03
3	0.24
4	0.06

Sensitivity is more important to proper diagnosis.

Problem 185: Discrete Results

A specific transgenic mouse strain reliably lives to age 6 months but also reliably develops a variety of tumors. Historical data on the proportion of tumor counts for this strain at 6 months are listed below, along with data for a recent test batch. Use the historical data as a prior and calculate posterior distribution of the tumors. Would the results be different if the current batch were used as the prior and the historical proportions were used as the data set?

Tumors (Count)	Historical (Portion)	Test Animals (#)
0	0.10	11
1	0.15	17
2	0.20	25
3	0.15	14
4	0.15	15
5	0.10	8
6	0.10	12
7+	0.05	6

Solution:

Tumors (Count)	Historical (Portion)	Test Animals (#)	Individual	Posterior
0	0.10	11	1.10	0.0719
1	0.15	17	2.55	0.1667
2	0.20	25	5.00	0.3268
3	0.15	14	2.10	0.1373
4	0.15	15	2.25	0.1471
5	0.10	8	0.80	0.0523
6	0.10	12	1.20	0.0784
7+	0.05	6	0.30	0.0196

The posterior does not depend on the order of the data and the prior in calculations, so the results would be identical.

Problem 186: Discrete Results

The experiment in Problem 185 was repeated with larger numbers of mice. Use the posterior of 13.5 as the prior and the data below to generate a new posterior distribution.

Tumors (Count)	Test Animals (#)
0	119
1	155
2	216
3	207
4	172
5	136
6	103
7+	62

Solution:

Tumors (Count)	Historical (Portion)	Test Animals (#)	Individual	Posterior
0	0.0719	119	8.56	0.0489
1	0.1667	155	25.84	0.1476
2	0.3268	216	70.59	0.4031
3	0.1373	207	28.42	0.1623
4	0.1471	172	25.30	0.1445
5	0.0523	136	7.11	0.0406
6	0.0784	103	8.08	0.0461
7+	0.0196	62	1.22	0.0069

Problem 187: Discrete Results

Using the data from Problems 185 and 186, add the results from the first and second tests and use the original historical proportions as a prior to generate another posterior distribution. Are the results the same as those of Problem 186 or not and why are they that way?

Solution:

Additive results

Tumors (Count)	Historical (Portion)	Test Animals (#)	Individual	Posterior
0	0.10	130	13.00	0.0732
1	0.15	172	25.80	0.1454
2	0.20	241	48.20	0.2715
3	0.15	221	33.15	0.1868
4	0.15	187	28.05	0.1580
5	0.10	144	14.40	0.0811
6	0.10	115	11.50	0.0648
7+	0.05	68	3.40	0.0192

The results are not the same because (generally):

Prior * Data Set 1 * Data Set 2 Prior * (Data Set 1 + Data Set 2)

In other words, in the additive case, the historical proportions were used equally for both populations (historical weight of about 0.5), while in the sequential case the historical proportions were used first for one set, then the results on the other (historical weight of about 0.33).

Problem 188: Flat Priors

Assume that a flat prior is used to calculate the posterior for data set with the distributions listed below. What are the distributions of the respective posteriors?

Distributions: Flat, binomial, normal, beta, chi-square, and *F*.

Solution:

Flat priors do not change the distribution of the data; therefore, the posteriors are distributed the same as the data's distribution.

Distributions: Flat → flat, binomial → binomial, normal → normal, beta → beta, chi-square → chi-square, and F → F.

Problem 189: Normal Distributions: Known Population and Known Prior

Assume a prior with mean weight of 7.4 and a variance of 3.35. For each of the known population means and variances listed below, calculate the posterior mean and variance, using the formula for the case where means and variances are fully known.

Population	Mean	Variance
Brown	5.8	3.28
Gray	2.6	2
Blue	12.8	4.88
Pink	7.0	3.84
Green	16.4	12.38

Solution:

Population	Posterior Mean	Posterior Variance
Brown	6.59	1.66
Gray	4.39	1.25
Blue	9.60	1.99
Pink	7.21	1.79
Green	9.32	2.64

Problem 190: Normal Distributions: Sample Population and Known Prior

Using the data given in Problem 189, assume that the population variances are accurate but that the means are based on random samples drawn from the population. Given the number of samples drawn listed below, calculate the posterior means, variances of the means, and sample variances.

Population	N
Brown	33
Gray	25
Blue	28
Pink	44
Green	6

Solution:

Population	Posterior Mean	Posterior Variance of the Mean	Posterior Variance
Brown	5.85	0.10	0.55
Gray	2.71	0.08	0.39
Blue	12.53	0.17	0.88
Pink	7.01	0.09	0.56
Green	12.97	1.28	3.13

Problem 191: Normal Distributions: Sample Population and Known Prior

Using the posteriors from Problem 190, calculate the 95% credible intervals of the means for the posteriors. Do two sets of calculations, first under the assumption of known population variance, then under the assumption that the variances given are sample variances. Which intervals are larger and why?

Solution:

Population	Population Variances		Sample Variances	
	Lower CL	Upper CL	Lower CL	Upper CL
Brown	5.66	6.04	5.65	6.11
Gray	2.56	2.87	2.55	3.70
Blue	12.21	12.86	12.19	12.90
Pink	6.84	7.18	6.84	7.22
Green	10.47	15.47	9.69	15.78

The estimates based on sample variances are larger than those based on population variances. The sample variance estimates are based on the t distribution while the population variance estimates are based on the z distribution. Since t-values are larger than z-values at the same level, the sample variances are larger.

Problem 192: Normal Distributions: Difference between Means

The effect of compound CmpX on body weights was investigated in rats. Rats were fed various levels of CmpX or a placebo for 10 weeks and body weights were taken at the end of the study. The supplier estimates that the rat body weights are normally distributed and gave an estimate of 1.66 for the mean and 0.0441 for the variance for rat weights, for the appropriate age rats. Using the supplier's estimate as the prior, first calculate the posterior means and variances of the means, and then calculate the 90% credible interval for the difference between the actively dosed groups and the placebo. Are any of the differences significant at this level?

Treatment	N	Mean	Standard Deviation
Placebo	21	1.53	0.22
0.01 mg/kg CmpX	10	1.59	0.16
0.03 mg/kg CmpX	27	1.56	0.28
0.10 mg/kg CmpX	31	1.46	0.14

Solution:

Treatment	Posterior Mean	Posterior Variance of the Mean	Lower CL	Upper CL
Placebo	1.54	0.0022	NA	NA
0.01 mg/kg CmpX	1.59	0.0024	−0.0580	0.1727
0.03 mg/kg CmpX	1.57	0.0027	−0.0880	0.1474
0.10 mg/kg CmpX	1.46	0.0006	−0.1625	0.0153

There are no significant differences at this level, although the difference between the placebo and the highest dosed group approaches statistical significance.

Problem 193: Satterthwaite's Degrees of Freedom Estimation

Use the means and standard deviations from Problem 192 to Satterthwaite's estimates of N for the difference between the three dosed groups and the placebo. For this exercise, use the given standard deviations for the groups, not the calculated posteriors. Compare the results with the degrees of freedom that would be used in assumptions of equal variance.

Solution:

Treatment	Satterthwaite's Estimation	Usual df
Placebo	NA	NA
0.01 mg/kg CmpX	23	29
0.03 mg/kg CmpX	45	46
0.10 mg/kg CmpX	30	50

Problem 194: Statistics of Beta Distributions

Assuming beta distributions defined by the a and b values in the table below, calculate the mean, mode, N_{eq}, and variance of the distribution.

Identifier	a	b
A	5	11
B	8	17
C	13	10
D	10	22
E	20	9
F	12	7
G	7	22

Solution:

Identifier	Mean	Mode	N_{eq}	Variance
A	0.313	0.286	17	0.013
B	0.320	0.304	26	0.008
C	0.565	0.571	24	0.010
D	0.313	0.300	33	0.007
E	0.690	0.704	30	0.007
F	0.632	0.647	20	0.012
G	0.241	0.222	30	0.006

Problem 195: Statistics of Beta Distributions

Given the already calculated statistics below, calculate the a and b values needed to make a beta distribution that fits.

Identifier	Mean	Mode	N_{eq}	Variance
H	0.4	0.35	—	—
I	0.4	—	12	—
J	0.4	—	—	0.010
K	—	0.35	12	—
L	—	0.35	—	0.010
M	—	—	12	0.010

Solution:

Identifier	a	b
H	2.4	3.6
I	4.4	6.6
J	9	13.5
K	4.4	6.6
L	8	14
M	1.5	9.5

Problem 196: Normal Approximation of the Beta Distribution

For the beta distribution in Problem 194 and Problem 195, which one(s) can use the normal approximation for confidence intervals?

Solution:

C and D only.

Problem 197: Updating Beta Distributions

Using the beta distributions in Problem 194 and Problem 195 as priors, calculate the posteriors for the following trials:

Prior to Use	N	Successes
A	10	30
B	13	35
C	8	13
D	11	17
E	15	24
F	15	30
G	14	28
H	2	8

I	13	18
J	7	19
K	4	20
L	10	18
M	18	25

Solution:

Prior Used	Posterior *a*	Posterior *b*
A	15	31
B	21	39
C	21	15
D	21	28
E	35	18
F	27	22
G	21	36
H	4.4	9.6
I	17.4	11.6
J	16	25.5
K	8.4	22.6
L	18	22
M	19.5	16.5

Problem 198: Confidence Intervals of Beta Distributions

For the following defined beta distributions, calculate the 90% (two-sided) confidence interval using the normal approximation.

Identifier	a	b
N	10	27
O	17	14
P	17	18
Q	10	11
R	26	14
S	22	18

Solution:

Identifier	Mean	Variance	Lower	Upper
N	0.27027	0.00519	0.1518	0.3888
O	0.54839	0.00774	0.4037	0.6931
P	0.48571	0.00694	0.3487	0.6227
Q	0.47619	0.01134	0.3010	0.6513
R	0.65	0.00555	0.5275	0.7725
S	0.55	0.00604	0.4222	0.6778

Index to Problem Set

Technique	Problem Nos.
Fisher's Sum of Logarithms	179–180
Effects of Exposure	181
Assay Evaluation	182–184
Discrete Results	185–187
Flat Priors	188
Normal Distributions: Known Population and Known Prior	189
Normal Distributions: Sample Population and Known Prior	190–191
Normal Distributions: Difference between Means	192
Satterthwaite's Degrees of Freedom Estimation	193
Statistics of Beta Distributions	194–195
Normal Approximation of the Beta Distribution	196
Updating Beta Distributions	197
Confidence Intervals of Beta Distributions	198

Index